Springer Tracts in Modern Physics

Volume 260

AF180856

Honorary Editor

G. Höhler, Karlsruhe, Germany

Series editors

Atsushi Fujimori, Tokyo, Japan
Johann H. Kühn, Karlsruhe, Germany
Thomas Müller, Karlsruhe, Germany
Frank Steiner, Ulm, Germany
William C. Stwalley, Storrs, CT, USA
Joachim E. Trümper, Garching, Germany
Peter Wölfle, Karlsruhe, Germany
Ulrike Woggon, Berlin, Germany

For further volumes:
http://www.springer.com/series/426

Springer Tracts in Modern Physics

Springer Tracts in Modern Physics provides comprehensive and critical reviews of topics of current interest in physics. The following fields are emphasized: Elementary Particle Physics, Condensed Matter Physics, Light Matter Interaction, Atomic and Molecular Physics, Complex Systems, Fundamental Astrophysics.

Suitable reviews of other fields can also be accepted. The Editors encourage prospective authors to correspond with them in advance of submitting a manuscript. For reviews of topics belonging to the above mentioned fields, they should address the responsible Editor as listed below.

Special offer: For all clients with a print standing order we offer free access to the electronic volumes of the Series published in the current year.

Elementary Particle Physics, Editors

Johann H. Kühn

Institut für Theoretische Teilchenphysik
Karlsruhe Institut für Technologie KIT
Postfach 69 80
76049 Karlsruhe, Germany
Phone: +49 (7 21) 6 08 33 72
Fax: +49 (7 21) 37 07 26
Email: johann.kuehn@KIT.edu
www-ttp.physik.uni-karlsruhe.de/~jk

Thomas Müller

Institut für Experimentelle Kernphysik
Karlsruhe Institut für Technologie KIT
Postfach 69 80
76049 Karlsruhe, Germany
Phone: +49 (7 21) 6 08 35 24
Fax: +49 (7 21) 6 07 26 21
Email: thomas.muller@KIT.edu
www-ekp.physik.uni-karlsruhe.de

Fundamental Astrophysics, Editor

JoachimE. Trümper

Max-Planck-Institut für Extraterrestrische Physik
Postfach 13 12
85741 Garching, Germany
Phone: +49 (89) 30 00 35 59
Fax: +49 (89) 30 00 33 15
Email: jtrumper@mpe.mpg.de
www.mpe-garching.mpg.de/index.html

Solid State and Optical Physics

Ulrike Woggon

Institut für Optik und Atomare Physik
Technische Universität Berlin
Straße des 17. Juni 135
10623 Berlin, Germany
Phone: +49 (30) 314 78921
Fax: +49 (30) 314 21079
Email: ulrike.woggon@tu-berlin.de
www.ioap.tu-berlin.de

Condensed Matter Physics, Editors

Atsushi Fujimori

Editor for The Pacific Rim

Department of Physics
University of Tokyo
7-3-1 Hongo, Bunkyo-ku
Tokyo 113-0033, Japan
Email: fujimori@phys.s.u-tokyo.ac.jp
http://wyvern.phys.s.u-tokyo.ac.jp/welcome_en.html

Peter Wölfle

Institut für Theorie der Kondensierten Materie
Karlsruhe Institut für Technologie KIT
Postfach 69 80
76049 Karlsruhe, Germany
Phone: +49 (7 21) 6 08 35 90
Phone: +49 (7 21) 6 08 77 79
Email: peter.woelfle@KIT.edu
www-tkm.physik.uni-karlsruhe.de

Complex Systems, Editor

Frank Steiner

Institut für Theoretische Physik
Universität Ulm
Albert-Einstein-Allee 11
89069 Ulm, Germany
Phone: +49 (7 31) 5 02 29 10
Fax: +49 (7 31) 5 02 29 24
Email: frank.steiner@uni-ulm.de
www.physik.uni-ulm.de/theo/qc/group.html

Atomic, Molecular and Optical Physics

William C. Stwalley

University of Connecticut
Department of Physics
2152 Hillside Road, U-3046
Storrs, CT 06269-3046, USA
Phone: +1 (860) 486 4924
Fax: +1 (860) 486 3346
Email: w.stwalley@uconn.edu
www-phys.uconn.edu/faculty/stwalley.html

Kamakhya P. Ghatak · Sitangshu Bhattacharya

Heavily-Doped 2D-Quantized Structures and the Einstein Relation

Springer

Kamakhya P. Ghatak
Department of Electronics and Communication Engineering
National Institute of Technology
Agartala, Jirania
India

Sitangshu Bhattacharya
Department of Electrical and Communication Engineering
Indian Institute of Information Technology
Allahabad
India

ISSN 0081-3869 ISSN 1615-0430 (electronic)
ISBN 978-3-319-38127-5 ISBN 978-3-319-08380-3 (eBook)
DOI 10.1007/978-3-319-08380-3
Springer Cham Heidelberg New York Dordrecht London

© Springer International Publishing Switzerland 2015
Softcover reprint of the hardcover 1st edition 2015
This work is subject to copyright. All rights are reserved by the Publisher, whether the whole or part of the material is concerned, specifically the rights of translation, reprinting, reuse of illustrations, recitation, broadcasting, reproduction on microfilms or in any other physical way, and transmission or information storage and retrieval, electronic adaptation, computer software, or by similar or dissimilar methodology now known or hereafter developed. Exempted from this legal reservation are brief excerpts in connection with reviews or scholarly analysis or material supplied specifically for the purpose of being entered and executed on a computer system, for exclusive use by the purchaser of the work. Duplication of this publication or parts thereof is permitted only under the provisions of the Copyright Law of the Publisher's location, in its current version, and permission for use must always be obtained from Springer. Permissions for use may be obtained through RightsLink at the Copyright Clearance Center. Violations are liable to prosecution under the respective Copyright Law.
The use of general descriptive names, registered names, trademarks, service marks, etc. in this publication does not imply, even in the absence of a specific statement, that such names are exempt from the relevant protective laws and regulations and therefore free for general use.
While the advice and information in this book are believed to be true and accurate at the date of publication, neither the authors nor the editors nor the publisher can accept any legal responsibility for any errors or omissions that may be made. The publisher makes no warranty, express or implied, with respect to the material contained herein.

Printed on acid-free paper

Springer is part of Springer Science+Business Media (www.springer.com)

This book is dedicated to my mother
Late Mira Ghatak, a research
mathematician whose only wish
is to enjoy the books of his son
published from Springer-Verlag
and for which no tears are enough
to wash her lotus feet

Kamakhya P. Ghatak

I dedicate this monograph
to my beautiful and wonderful
wife Rekha Verma, the source
of inspiration for playing
in this creative play

Sitangshu Bhattacharya

Preface

The combination of the concept of asymmetry of the wave-vector space of charge carriers in semiconductors with modern techniques of fabricating nanostructured materials such as MBE, MOCVD and FLL in one, two and three dimensions (such as quantum wells (QWs), doping superlattices, accumulation and inversion layers, quantum well superlattices, carbon nanotubes, quantum wires, quantum wire superlattices, magnetic quantization, magneto size quantization, quantum dots, magneto accumulation and inversion layers, magneto NIPIs, magneto quantum well superlattices, quantum dot superlattices and other field aided low-dimensional systems) spawns not only useful quantum effect devices but also unearths new concepts in the realm of low-dimensional solid-state science and related disciplines. These semiconductor nanostructures occupy a central position in the entire arena of condensed matter science in general, by their own right and find extensive applications in quantum registers, quantum switches, quantum sensors, quantum logic gates, quantum well and quantum wire transistors, quantum cascade lasers, heterojunction field-effect transistors, high-speed digital networks, high-frequency microwave circuits, high-resolution terahertz spectroscopy, superlattice photo-oscillator, advanced integrated circuits, superlattice photocathodes, resonant tunneling diodes and transistors, thermoelectric devices, superlattice coolers, thin film transistors, intermediate-band solar cells, micro-optical systems, high performance infrared imaging systems, band-pass filters, thermal sensors, optical modulators, optical switching systems, single electron electronics, molecular electronics, nanotube-based diodes and other nanoelectronic devices. Knowledge regarding these quantized structures may be gained from original research contributions in scientific journals, various patents, personal communications, proceedings of the conferences/seminars, review articles and different research monographs [1] respectively. In this context, it may be noted that the available reports on the said areas cannot afford to cover even an entire chapter regarding the Einstein Relation (ER) for the diffusivity-mobility ratio of carriers in heavily doped (HD) two-dimensional (2D) quantized structures and the single first book on ER [2] does not contain even a paragraph regarding this important specialized topic of research and, after 30 years of continuous effort, we see that the complete investigations of the ER comprising the whole set of materials and allied sciences is really a sea and is a permanent member of the domain of impossibility theorems.

It is well known that the ER occupies a central position in the whole field of solid-state device electronics and the related sciences since the diffusion constant (a quantity very useful for device analysis where exact experimental determination is rather difficult) can be obtained from this ratio by knowing the experimental values of the mobility. The classical value of the ER is equal to $(k_B T/|e|)$, (k_B, T and $|e|$ are Boltzmann's constant, temperature and the magnitude of the carrier charge respectively). This relation in this form was first introduced by Einstein to studythe diffusion of gas particles and is known as the Einstein relation [2, 3]. It appears that the ER increases linearly with increasing T and is independent of electron concentration. This relation is applicable for both types of charge carriers only under nondegenerate carrier concentration although its validity has been suggested erroneously for degenerate materials [4]. Landsberg first pointed out that the ER for degenerate semiconductors is essentially determined by their energy band structures [5, 6]. This relation is useful for semiconductor homostructures [7, 8], semiconductor–semiconductor heterostructures [9, 10], metals–semiconductor heterostructures [11–19] and insulator–semiconductor heterostructures [20–23]. The nature of the variations of the ER under different physical conditions has been studied in the literature [1–3, 5, 6, 24–49]. Incidentally, A. N. Chakravarti (a recognized leading expert of ER in general) and his research group are still contributing significantly under his able leadership regarding this pinpointed research topic on ER from 1972 [2, 24, 25–28, 34, 39–49] and some of the significant features, which have emerged from these studies, are:

(a) The ER increases monotonically with increasing carrier concentration in bulk semiconductors and the nature of these variations is significantly influenced by the band structures of different materials.
(b) The ER increases with the increasing quantizing electric field as in inversion layers.
(c) The ER oscillates with the inverse quantizing magnetic field under magnetic quantization due to the Shubnikov-de Haas effect.
(d) The ER shows composite oscillations with the various controlled quantities of semiconductor superlattices.
(e) In ultrathin films, quantum wires and other field assisted low-dimensional systems, the value of the ER changes appreciably with the external variables depending on the nature of quantum confinements of different materials.

The ER depends on the density-of-states (DOS) function, which, in turn, is significantly affected by the different carrier energy spectra of different semiconductors having various band structures. In recent years, various energy wave-vector dispersion relations of carriers of different materials have been proposed [50], which have created interest in studying the ER in HD 2D-quantized structures. It is well known that heavy doping and carrier degeneracy are the keys to unlock the important properties of semiconductors and they are especially instrumental in dictating the characteristics of Ohmic and Schottky contacts respectively [11–19, 51]. It is an amazing fact that although heavily doped semiconductors (HDS) have been investigated in the literature the study of carrier transport in such materials through proper formulation of the Boltzmann transport equation which needs, in turn, *the corresponding HD carrier energy spectra is still one of the open research problems.*

It is well known that band tails are being formed in the forbidden zone of the HDS and can be explained by the overlapping of the impurity band with the conduction and valence bands [52]. Kane [53] and Bonch Bruevich [54] have independently derived the theory of band tailing for semiconductors having unperturbed parabolic energy bands. Kane's model [53] was used to explain the experimental results on tunneling [55] and the optical absorption edges [56, 57] in this context. Halperin and Lax [58] developed a model for band tailing applicable only to the deep tailing states. Although Kane's concept is often used in the literature for the investigation of band tailing [59, 60], it may be noted that this model [53, 61] suffers from serious assumptions in the sense that the local impurity potential is assumed to be small and slowly varying in space coordinates [60]. In this respect, the local impurity potential may be assumed to be a constant. In order to avoid these approximations, we have developed in this book, the electron energy spectra for HDS for studying the ER based on the concept of the variation of the kinetic energy [52, 60] of the electron with the local point in space coordinates. This kinetic energy is then averaged over the entire region of variation using a Gaussian-type potential energy. On the basis of the *E-k* dispersion relation, we have obtained the electron statistics for different HDS for the purpose of numerical computation of the respective ERs. It may be noted that a more general treatment of many-body theory for the DOS of HDS merges with one-electron theory under macroscopic conditions [52]. Also, the experimental results for the Fermi energy and others are the average effect of this macroscopic case. So, the present treatment of the one-electron system is more applicable to the experimental point of view and it is also easy to understand the overall effect in such a case [62]. In a HDS, each impurity atom is surrounded by electrons, assuming a regular distribution of atoms and it is screened independently [59, 61, 63]. The interaction energy between electrons and impurities is known as the impurity screening potential. This energy is determined by the inter-impurity distance and the screening radius (popularly known as the Debye screening length). The screening length changes with the band structure. Furthermore, these entities are important for HDS in characterizing the semiconductor properties [64, 65] and the modern electronic devices [59, 66]. The works on Fermi energy and the screening length in an n-type GaAs have already been initiated in the literature [67], based on Kane's model. Incidentally, the limitations of Kane's model [53, 60], as mentioned above, are also present in their studies.

At this point, it may be noted that many band tail models are proposed using Gaussian distribution of the impurity potential variation [53, 60]. From the very start, we have used Gaussian band tails to obtain the *exact E-k dispersion relations* for HD nonlinear optical, III–V, II–VI, IV–VI, stressed Kane-type semiconductors, Te, GaP, $PtSb_2$, Bi_2Te_3, Ge and GaSb respectively. Our method is not related with the DOS technique as used in the aforementioned works. From the electron energy spectrum, one can obtain the DOS but the DOS technique, as used in the literature cannot provide the *E-k* dispersion relation. *Therefore, our study is more fundamental than those in the existing literature, because the Boltzmann transport equation, which controls the study of the charge transport properties of the semiconductor devices, can be solved if and only if the E-k dispersion relation is known*. We wish to note that many authors have used the Gaussian function for the impurity potential distribution. It has been widely used since 1963 when Kane first proposed it and we will use the Gaussian distribution for the present study.

This book contains ten chapters where the Appendices A to E are placed chronologically in chapters five to ten respectively, is partially based on our ongoing researches on the ER of HDS from 1990, and an attempt has been made to present a cross section of the ER for wide range of HDS and their quantized-structures with varying carrier energy spectra under various physical conditions. The first chapter deals with the influence of quantum confinement on the ER in non-parabolic HDS. First, we study the ER in QWs of HD nonlinear optical materials on the basis of a generalized electron dispersion law introducing the anisotropies of the effective masses and the spin orbit splitting constants, respectively, together with the inclusion of the crystal field splitting within the framework of the *k.p* formalism. *We observe that the appearance of the complex electron dispersion law in HDS instead of real one occurs from the existence of the poles in the finite complex plane of the corresponding electron energy spectrum in the absence of band tails.* It may be noted that the complex band structures have already been studied for bulk semiconductors and superlattices without heavy doping [69] and bears no relationship to the complex electron dispersion law as formulated in this book. The physical picture behind the existence of the complex energy spectrum in HD nonlinear optical semiconductors is the interaction of the impurity atoms in the tails with the splitting constants of the valance bands. The more the interaction, the more the prominence of the complex part than the other case. In the absence of band tails, there is no interaction of impurity atoms in the tails with the spin orbit constants and, consequently, the complex part vanishes. Besides, the complex spectra are not related to same evanescent modes in the band tails and the conduction bands. In this context it is worth remarking that the concept of effective electron mass (EEM) is one of the basic pillars in the whole set of materials science in general [68]. One important consequence of the HDS forming band tails is that *the EEM exists in the forbidden zone, which is impossible without the effect of band tailing. In the absence of band tails, the effective mass in the band gap of semiconductors is infinity. Besides, depending on the type of the unperturbed carrier energy spectrum, the new forbidden zone will appear within the normal energy band gap for HDS.* The results of HD III–V (e.g., InAs, InSb, GaAs etc.), ternary (e.g., $Hg_{1-x}Cd_xTe$, etc.), quaternary (e.g., $In_{1-x}Ga_xAs_{1-y}P_y$ lattice matched to InP, etc.) compounds form a special case of our generalized analysis under certain limiting conditions. The ER in HD QWs of II–VI, IV–VI, stressed Kane-type semiconductors, Te, GaP, $PtSb_2$, Bi_2Te_3, Ge and GaSb has been investigated by formulating the respective appropriate HD energy band structure. The importance of the aforementioned semiconductors has been described in the same chapter. As a collateral study we shall observe that the EEM in such QWs becomes a function of size quantum number, the Fermi energy, the scattering potential and other constants of the system which is the intrinsic property of such 2D electrons.

With the advent of modern experimental techniques of fabricating nanomaterials, it is possible to grow semiconductor superlattices (SLs) composed of alternative layers of two different degenerate layers with controlled thickness [70]. These structures have found wide applications in many new devices such as photodiodes [71], photo-resistors [72], transistors [73], light emitters [74], tunneling devices [75], etc. [76–87]. The investigations of the physical properties of narrow gap SLs have increased extensively since they are important for opto-electronic devices and because of the quality of heterostructures involving narrow gap materials have been improved. It may be noted in this context that the doping superlattices are crystals with a periodic sequence of ultrathin film layers [88, 89] of the same semiconductor with the intrinsic layer in-between together with the opposite sign of doping. All the donors are positively charged and all the acceptors negatively. This periodic space charge causes a periodic space charge potential which quantizes the motions of the carriers in the z-direction together with the formation of the subband energies. The electronic structures of the doping superlattices differ radically from the corresponding bulk semiconductors as stated below:

(a) Each band is split into mini-bands;
(b) The magnitude and the spacing of these mini-bands may be designed by the choice of the superlattices parameters; and
(c) The electron energy spectrum of the nippy crystal becomes two-dimensional leading to the step functional dependence of the DOS function.

In the second chapter, the ER in doping superlattices of HD nonlinear optical, III–V, II–VI, IV–VI and stressed Kane-type semiconductors has been investigated. In this case we note that the EEM in such doping supper-lattices becomes a function of nipi subband index, surface electron concentration, Fermi energy, the scattering potential and other constants of the system which is the intrinsic property of such 2D-quantized systems.

In recent years, there has been considerable interest in the study of the inversion layers, which are formed at the surfaces of semiconductors in metal-oxide-semiconductor field-effect transistors (MOSFET) under the influence of a sufficiently strong electric field applied perpendicular to the surface by means of a large gate bias. In such layers, the carriers form a two-dimensional gas and are free to move parallel to the surface while their motion is quantized in the direction perpendicular to it leading to the formation of electric subbands [90]. Although considerable work has already been done regarding the various physical properties of different types of inversion layers having various band structures, nevertheless, it appears from the literature that there lies scopes in the investigations made while the interest for studying different other features of accumulation layers is becoming increasingly important. In the third chapter, the ER in accumulation layers of HD nonlinear optical, III–V, II–VI, IV–VI, stressed Kane-type semiconductors and Ge, have been investigated. For the purpose of relative comparisons, we have also studied the ER in inversion layers of the aforementioned materials. *It is interesting to note that the EEM in such layers is a function of electric subband index, surface electric field, Fermi energy, the scattering potential and other constants of the system which is the intrinsic property of such 2D electrons.*

Chapter four suggests the experimental determinations of 2D and 3D ERs for HDS and contains six related applications of the content of this book. Our suggestion for the experimental determination of the ERs and the theoretical formula for degenerate tetragonal compunds (e.g., Cd_3As_2*) based on our generalized analysis incorporating all types of anisotropies of the energy band structure agree well with each other and are discussed in this chapter.* Chapter five contains the conclusion and the scope for future research.

It may be noted that the effects of quantizing magnetic field (B) on the band structures of compound semiconductors are more striking than the parabolic one and are easily observed in experiments. A number of interesting physical features originate from the significant changes in the basic energy wave-vector relation of the carriers caused by the magnetic field. The valuable information could be obtained from experiments under magnetic quantization regarding the important physical properties such as Fermi energy and effective masses of the carriers, which affect almost all the transport properties of the electron devices [91] of various materials having different carrier dispersion relations [92]. The ER in the presence of magnetic quantization is a tensor quantity and we take that particular element of the ER which is in the direction of magnetic field only $(D/\mu)_{zz}$. *Appendix A studies the ER in HD nonlinear optical, III–V, IV–VI, stressed compounds, n-Te, n-GaP,* $PtSb_2$*, n-Ge, II–V semiconductors and Lead Germanium Telluride under magnetic quantization respectively.* In this *appendix weobserve that the EEM depends on Landau quantum number in addition to Fermi energy and the other system constants due to the specific band structures of the HD materials together with the fact that EEM exists in the band gap due to the presence of finite scattering potential as noted already.*

It is well known that Keldysh [93] first suggested the fundamental concept of a superlattice (SL), although it was successfully experimental realized by Esaki and Tsu [94]. The importance of SLs in the field of nanoelectronics has already been described in [95–97]. The most extensively studied III–V SL is that consisting of alternate layers of GaAs and $Ga_{1-x}Al_xAs$ owing to the relative ease of fabrication. The GaAs layers form quantum wells and $Ga_{1-x}Al_xAs$ form potential barriers. The III–V SLs are attractive for the realization of high speed electronic and optoelectronic devices [98]. In addition to SLs with usual structure, SLs with more complex structures such as II–VI [99], IV–VI [100] and HgTe/CdTe [101] SLs have also been proposed. The IV–VI SLs exhibit quite different properties compared to the III–V SL due to the peculiar band structure of the constituent materials [102]. The epitaxial growth of II–VI SL is a relatively recent development and the primary motivation for studying the mentioned SLs made of materials with large band gap is in their potential for optoelectronic operation in the blue [102]. HgTe/CdTe SLs have aroused a great deal of attention since 1979 as promising new materials for long wavelength infrared detectors and other electro-optical applications [103]. Interest in Hg-based SLs has been further increased as new properties with potential device applications were revealed [103, 104]. These features arise from the unique zero band gap material HgTe [105] and the direct band gap semiconductor CdTe that can be described by the three-band mode of Kane [106]. The combination of the aforementioned materials with specified dispersion relation makes HgTe/CdTe SL very attractive, especially because of the possibility to tailor the material properties for various applications by varying the energy band constants of the SLs. In addition, for effective mass SLs, the electronic subbands appear continually in real space [107].

We note that all the aforementioned SLs have been proposed with the assumption that the interfaces between the layers are sharply defined, of zero thickness, i.e., devoid of any interface effects. The SL potential distribution may be then considered as a one-dimensional array of rectangular potential wells. The aforementioned advanced experimental techniques may produce SLs with physical interfaces between the two materials crystallographically abrupt; adjoining their interface will change at least on an atomic scale. As the potential form changes from a well (barrier) to a barrier (well), an intermediate potential region exists for the electrons. The influence of finite thickness of the interfaces on the electron dispersion law is very important, since the electron energy spectrum governs the electron transport in SLs.

In Appendix B, we study the ER under magnetic quantization in III–V, II–VI, IV–VI, HgTe/CdTe and strained layer HD SLs with graded interfaces. We also investigate the ER in III–V, II–VI, IV–VI, HgTe/CdTe and strained layer effective mass HD SLs in the presence of quantizing magnetic field respectively. *This appendix explores the fact that the EEM becomes a function of the Fermi energy, Landau quantum number, scattering potential and the magnetic field in all the cases which are the characteristic features of such superlattices. We present a simplified analysis of the ER in superlattices of HD nonparabolic semiconductors under magnetic quantization, which is a huge topic of research by its own right.*

It is worth remarking that the influence of crossed electric and quantizing magnetic fields on the transport properties of semiconductors having various band structures are relatively less investigated compared with the corresponding magnetic quantization, although the crossfields are fundamental with respect to the addition of new physics and the related experimental findings. It is well known that in the presence of electric field (E_o) along x-axis and the quantizing magnetic field (B) along z-axis, the dispersion relations of the conduction electrons in semiconductors become modified and for which the electron moves in both z and y directions. The motion along y-direction is purely due to the presence of E_0 along x-axis and in the absence of electric field, the effective electron mass along y-axis tends to infinity which indicates the fact that the electron motion along y-axis is forbidden. The effective electron mass of the isotropic, bulk semiconductors having parabolic energy bands exhibits mass anisotropy in the presence of crossfields and this anisotropy depends on the electron energy, the magnetic quantum number, the electric and the magnetic fields, respectively, although the effective electron mass along z- axis is a constant quantity. In 1966, Zawadzki and Lax [108] formulated the electron dispersion law for III–V semiconductors in accordance with the two-band model of Kane under crossfields configuration which generates the interest to study this particular topic of solid-state science in general [109]. *Appendix C investigates the ER under crossfield configuration in HD nonlinear optical, III–V, II–VI, IV–VI, stressed Kane-type semiconductors and their ultrathin films counterparts. This appendix tells us that the EEM in all the cases is a function of the size quantum number, the finite scattering potential, the magnetic quantum number and the Fermi energy even for HD semiconductors whose bulk electrons in the absence of band tails are defined by the parabolic energy bands.*

With the advent of nano-devices, the build-in electric field becomes so large that the electron energy spectrum changes fundamentally instead of being invariant and Appendix D investigates the ER under intense electric field in bulk specimens of HD III–V, ternary and quaternary semiconductors. This appendix also explores the influence of electric field on the ER based on HD new dispersion law under magnetic quantization, size quantization, accumulation layers, HD doping superlattices and effective mass HD superlattices under magnetic quantization. *It is interesting to note that the EEM depends on the strong electric field* (which is not observed elsewhere) together with the fact that the EEM in accumulation layers, HD doping superlattices and effective mass HD superlattices depend on the respective quantum numbers in addition to the Fermi energy, the scattering potential and other system constants which are the characteristics features of such heterostructures.

With the advent of nano-photonics, there has been considerable interest in studying the optical processes in semiconductors and their nanostructures in the presence of intense light waves [110]. *It appears from the literature that investigations in the presence of external intense photo-excitation have been carried out on the assumption that the carrier energy spectra are invariant quantities under strong external light waves, which is not fundamentally true.* The physical properties of semiconductors in the presence of strong light waves which alter the basic dispersion relations have relatively been much less investigated in [111, 112] as compared with the cases of other external fields needed for the characterization of low-dimensional semiconductors. Appendix E of this book studies the influence of light waves on the ER in HD opto-electronic semiconductors by formulating new electron dispersion relation within the framework of *k.p* formalism. The same appendix explores the opto ER for HD opto-electronic materials under magnetic quantization, crossfields configuration, size quantization, doping superlattices and effective mass superlattices respectively. It is interesting to note that the EEM is a function of incident light intensity and wave length (not observed elsewhere) together with the fact that the EEM in superlattices and crossfields configuration depend on quantum numbers, Fermi energy, scattering potential and other system constants which are the characteristic features in this case. In these appendices, no graphs together with results and discussions are presented since we strongly feel that the readers should not lose a chance to enjoy the complex computer algorithm to investigate the ER in the respective cases generating new physics and thereby transforming each Appendix into a monograph by considering various materials having different dispersion relations.

It is needless to say that this monograph is based on the *'iceberg principle'* [113] and the rest of which will be explored by researchers from different appropriate fields. Since there is no existing report devoted solely to the study of ER for HD 2D-quantized structures to the best of our knowledge, we hope that this book will be a useful reference source for the present and the next generation of readers and researchers of solid-state and allied sciences in general. Since the production of an error-free first edition of any book from every point of view is a permanent member in the domain of impossibility theorems, therefore in spite of our joint concentrated efforts for a couple of years together with the seasoned team of Springer, the same stands very true for this monograph also. Various expressions and a few chapters of this book appear for the first time in printed form.Suggestions from readers for the development of the book will be highly appreciated for the purpose of inclusion in future editions, if any. In this book, from chapter one till the end, we have presented *200 open research problems* for graduate students, Ph.D. aspirants, researchers and engineers in this pinpointed research topic. We strongly hope that alert readers of this monograph will not only solve the said problems by removing all the mathematical approximations and establishing the appropriate uniqueness conditions, but will also generate new research problems both theoretical and experimental and, thereby, transform this bried monograph into a solid book. Incidentally, our readers after reading this book will easily understand how little is presented and how much more is yet to be investigated in this exciting topic which is the signature of coexistence of new physics, advanced mathematics combined with the inner fire for performing creative research in this context from young scientists, since like Kikoin [114] we feel that *A young scientist is no good if his teacher learns nothing from him and gives his teacher nothing to be proud of.* We emphatically stress that the problems presented here form an integral part of this book and will be useful for readers to initiate their own contributions on the ER in HDS and their quantized counterparts, since like Sakurai [115] we firmly believe *The reader who has read the book but cannot do the exercise has learned nothing.* It is nice to note that if we assign the alphabets A to Z, the positive integers from 1 to 26, chronologically, then the word *ATTITUDE* receives the perfect score 100 and is the vital quality needed from the readers since *attitude* is the ladder on which all the other virtues mount.

In this monograph, we have investigated various dispersion relations of different HD quantized structures and the corresponding carrier statistics to study the concentration dependence of the ER in HD quantum confined materials. *Besides, the expressions of effective electron mass and the subband energy have been formulated throughout this monograph as a collateral study, for the purpose of in-depth investigations of the said important pinpointed research topics.* Thus, in this book, readers will get much information regarding the influence of quantization in HD low-dimensional materials having different band structures. For the enhancement of the materials aspect, we have considered various materials having the same dispersion relation to study the influence of energy band constants of the different HDS on ER. Although the name of the book is extremely specific, from the content one can easily infer that it should be useful in graduate courses on condensed matter physics, materials science, modern physics of materials, solid-state electronics, nano-science and technology and solid-state sciences and devices in many universities and institutions in addition to both Ph.D. students and researchers in the aforementioned fields. *Last but not the least, we do hope that our humble effort will kindle the desire to delve deeper into this fascinating and deep topic by anyone engaged in materials research and device development either in academics or in industry.*

References

1. K.P. Ghatak, S. Bhattacharya, *Debye Screening Length: Effects of Nanostructured Materials*, Springer Tracts in Modern Physics, vol. 255 (Springer, Heidelberg, 2014); S. Bhattacharya, K.P. Ghatak, *Effective Electron Mass in Low Dimensional Semiconductors*, Springer Series in Materials Sciences,vol. 167 (Springer, Heidelberg, 2013); S. Bhattacharya, K.P Ghatak, *Fowler-Nordheim Field Emission: Effects in Semiconductor Nanostructures*, Springer Series in Solid State Sciences, vol. 170 (Springer, Heidelberg, 2012); K.P. Ghatak, S. Bhattacharya, *Thermo Electric Power in Nano structured Materials Strong Magnetic Fields*, Springer Series in Materials Science, vol. 137 (Springer, Heidelberg, 2010); K.P. Ghatak, D. De, S. Bhattacharya, *Photoemission from Optoelectronic Materials and Their Nanostructures*, Springer Series in Nanostructure Science and Technology (Springer, Heidelberg, 2009)
2. K.P. Ghatak, S. Bhattacharya, D. De, *Einstein Relation in Compound Semiconductors and Their Nanostructures*, Springer Series in Materials Science, vol. 116 (Springer, Heidelberg, 2009)
3. A. Einstein, Ann. der Physik. **17**, 549 (1905); H. Kroemer, IEEE Trans. Electron. Dev. **25**, 850 (1978); W. Nernst, Z. Phys. Chem. **2**, 613 (1888); J.S. Townsend, Trans. R. Soc. A **193**, 129 (1900); C. Wagner, Z. Physik, Chem. B **21**, 24 (1933); C. Herring, M.H. Nichols, Rev. Mod. Phys. **21**, 185 (1949); P.T. Landsberg, *Thermodynamics and Statistical Mechanics* (Oxford University Press, Oxford, 1978); in *Recombination in Semiconductors* (Cambridge University Press, UK, 1991);
4. R.W. Lade, Proc. IEEE **52**, 743 (1965)
5. P.T. Landsberg, Proc. R. Soc. A **213**, 226 (1952); Proc. Phys. Soc. A **62**, 806 (1949)
6. P.T. Landsberg, Eur. J. Phys. **2**, 213 (1981)
7. C.H. Wang, A. Neugroschel, IEEE Electron Device Lett. ED **11**, 576 (1990)
8. I.Y. Leu, A. Neugroschel, IEEE Trans. Electron Dev. ED **40**, 1872 (1993)
9. F. Stengel, S.N. Mohammad, H. Morkoç, J. Appl. Phys. **80**, 3031 (1996)
10. H.J. Pan et al., Semicond. Sci. Technol. **15**, 1101 (2000)
11. S.N. Mohammad, J. Appl. Phys. **95**, 4856 (2004)
12. V.K. Arora, Appl. Phys. Lett. **80**, 3763 (2002)
13. S.N. Mohammad, J. Appl. Phys. **95**, 7940 (2004)
14. S.N. Mohammad, Philos. Mag. **84**, 2559 (2004)
15. S.N. Mohammad, J. Appl. Phys. **97**, 063703 (2005)
16. K. Suzue et al., J. Appl. Phys. **80**, 4467 (1996)
17. S.N. Mohammad et al., Electron. Lett. **32**, 598 (1996)

18. Z. Fan et al., J. Electron. Mater. **25**, 1703 (1996)
19. C. Lu et al., J. Appl. Phys. **91**, 9216 (2002)
20. S.G. Dmitriev, Yu V. Markin, Semiconductors **34**, 931 (2000)
21. M. Tao et al., Philos. Mag. B **73**, 723 (1996)
22. D.G. Park et al., J. Vac. Sci. Technol. B **14**, 2674 (1996)
23. Z. Chen et al., Appl. Phys. Lett. **69**, 230 (1996)
24. A.N. Chakravarti, D.P. Parui, Phys. Lett. A **40**, 113 (1972)
25. A.N. Chakravarti, D.P. Parui, Phys. Lett. A **43**, 60 (1973); B.R. Nag, A.N. Chakravarti, Solid State Electron. **18**, 109 (1975), Phys. Status Solidi. A **22**, K153 (1974)
26. B.R. Nag, A.N. Chakravarti, P.K. Basu, Phys. Status Solidi A **68**, K75 (1981)
27. B.R. Nag, A.N. Chakravarti, Phys. Status Solidi A **67**, K113 (1981)
28. A.N. Chakravarti, B.R. Nag, Phys. Status Solidi A **14**, K55 (1972); Int. J. Electron **37**, 281 (1974); Phys.StatusSolidi A **14**, K23 (1972); A.N. Chakravarti, D.P. Parui, Can. J. Phys. **51**, 451 (1973). D. Mukherjee, A.N. Chakravarti, B.R. Nag, Phys. Stat. Sol A **26**, K27 (1974); A.N. Chakravarti et al., Appl. Phys. A **26**, 165 (1981); S. Ghosh, A.N. Chakravarti, Phys. Status Solidi B **147**, 355 (1988); K.P. Ghatak et al., Appl. Phys. **23**, 241 (1980); A.N. Chakravarti et al., Phys. Stat. Sol B **118**, 843 (1983);A.N. Chakravarti et al., Phys. Stat. Sol B **111**, K61 (1982);A.N. Chakravarti et al., Phys. Stat. Sol B **108**, 609 (1981);A.N. Chakravarti et al., Phys. Stat. Sol B **105**, K55 (1981);A.N. Chakravarti et al., Phys. Stat. Sol B **59**, K211 (1980); A.N. Chakravarti et al., Acta Phys. Pol. A **58**, 251 (1980); K.P. Ghatak et al., Phys. Stat. Sol B **99**, K55 (1980); A.N. Chakravarti et al., Czech J. Phys. B **30**, 1161 (1980); A.N. Chakravarti et al., Acta Phys. Pol. A **60**, 151 (1981); A.N. Chakravarti et al., Czech J. Phys. B **31**, 905 (1981); A.N. Chakravarti et al., Czech J. Phys. B **31**, 1138 (1981)
29. P.T. Landsberg, J. Appl. Phys. **56**, 1696 (1984); P.T. Landsberg, A.G. Guy, Phys. Rev. B **28**, 1187 (1983);P.T. Landsberg, Phys. Rev. B **33**, 8321 (1986); Y. Roichman, N. Tessler, Appl. Phys. Lett. **80**, 1948 (2002); J.M.H. Peters, Eur. J. Phys. **3**, 19 (1982); H. Van Cong, S. Brunet, S. Charar, Phys. Status Solidi B **109**, K1 (1982); H. Van Cong, Phys. Status Solidi A **56**, 395 (1979); H. Van Cong, Solid State Electron. **24**, 495 (1981)
30. S.N. Mohammad, S.T.H. Abidi, J. Appl. Phys. **61**, 4909 (1987); Solid State Electron. **27**, 1153 (1985); J. Appl. Phys. **56**, 3341 (1984); M.A. Sobhan, S.N. Mohammad, J. Appl. Phys. **58**, 2634 (1985); S.N. Mohammad, A.V. Bemis, IEEE Trans. Electron. Dev. ED **39**, 282 (1992); S.N. Mohammad, R.L. Carter, Philos. Mag. B **72**, 13 (1995); S.N. Mohammad, Solid State Electron. **46**, 203 (2002); S.N. Mohammad et al., Appl. Phys. Lett. **56**, 937 (1990)
31. P.T. Landsberg, S.A. Hope, Solid State Electron. **20**, 421 (1977); S.A. Hope, G. Feat, P.T. Landsberg, J. Phys. A: Math. Gen. **14**, 2377 (1981)
32. W. Elsäber, E.O. Göbel, Electron. Lett. **19**, 335 (1983); R. Hilfer, A. Blumen, Phys. Rev. A **37**, 578 (1988); T.G. Castner, Phys. Rev. B **55**, 4003 (1997); E. Barkai, V.N. Fleurov, Phys. Rev. E. **58**, 1296 (1998); T.H. Nguyen, S.K. O'Leary, Appl. Phys. Lett. **83**, 1998 (2003); T.H. Nguyen, S.K. O'Leary, J. Appl. Phys. **98**, 076102 (2005); C.G. Rodrigues, Á.R. Vasconcellos, R. Luzzi, J. Appl.Phys. **99**, 073701 (2006)
33. R.K. Jain, Phys. Status Solidi A **42**, K221 (1977); B.A. Aronzon, E.Z. Meilikhov, Phys. Status Solidi A **19**, 313 (1973)
34. S. Choudhury et al., J. Comput. Theor. Nanosci. **5**, 375 (2008); S.M. Adhikari, K.P. Ghatak, Quantum Matter **2**, 296 (2013); S. Pahari et al., Physica B **405**, 4064 (2010); K.P. Ghatak et al., SuperlatticesMicrostruct. **46**, 387 (2009)
35. J.P. Bouchaud, A. Georges, Phys. Rep. **195**, 127 (1996); V. Blickle et al., Phys. Rev. Lett. **98**, 210601 (2007); Y. Kang, E. Jean, C. M. Fortmonn, Appl. Phys. Lett. **88**, 112110 (2006); F. Neumann, Y.A. Genenko, H.V. Seggern, J. Appl. Phys. **99**, 013704 (2006); J. van de Lagemaat, Phys. Rev. B. **73**, 235319 (2005); Q. Gu et al., Phys. Rev. Lett. **76**, 3196 (1996); M.Y. Azbel, Phys. Rev. B **46**, 15004 (1992)
36. A.H. Marshak, Solid State Electron. **30**, 1089 (1987); A.H. Marshak, C.M.V. Vliet, Proc. IEEE **72**, 148 (1984); C.M.V. Vliet, A. van der Zeil, Solid State Electron. **20**, 931 (1977)

37. A. Khan, A. Das, Appl. Phys. A **89**, 695 (2007)
38. O. Madelung, *Semiconductors: Data Handbook*, 3rd ed. (Springer, 2004); M. Krieehbaum et al., IEEE QE **24** 1727 (1988)
39. K.P. Ghatak, S.N. Biswas, J. Appl. Phys. **70**, 4309 (1991); K.P. Ghatak, B. Mitra, M. Mondal, Ann. der Physik **48**, 283 (1991); B. Mitra, K.P. Ghatak, Phys. Scr. **42**, 103 (1990); B. Mitra, K.P. Ghatak, Phys. Letts A **135**, 397 (1989); K.P. Ghatak, N. Chattopadhyay, M. Mondal, J. Appl. Phys. **63**, 4536 (1988); J. Low Temp. Phys. **73**, 321 (1988)
40. K.P. Ghatak, D. Bhattacharyya, Phys. Letts A **184**, 366 (1994)
41. K.P. Ghatak, D. Bhattacharyya, Phys. Scr. **52**, 343 (1995); M. Mondal, K.P. Ghatak, J. Magn. Magn. Mater. **62**, 115 (1986)
42. K.P. Ghatak, B. Nag, D. Bhattacharyya, J. Low Temp. Phys. **14**, 1 (1995)
43. K.P. Ghatak, M. Mondal, Thin Solid Films **148**, 219 (1987)
44. K.P. Ghatak, S.N. Biswas, Nanostruct. Mater. **2**, 91 (1993)
45. K.P. Ghatak, *Influence of Band Structure on Some Quantum Processes in Nonlinear optical Semiconductors*, D. Eng. Thesis (Jadavpur University, Kolkata, India, 1991)
46. K.P. Ghatak, N. Chattropadhyay, M. Mondal, Appl. Phys. A**44**, 305 (1987)
47. S.N. Biswas, K.P. Ghatak, in *Proceedings of the Society of Photo-optical and Instrumentation Engineers (SPIE), Quantum Well and Superlattice Physics*, vol. 792 (1987) p. 239; K.P. Ghatak, M. Mondal,S. Bhattacharyya, SPIE **1284**, 113 (1990); K.P. Ghatak, S. Bhattacharyya, M. Mondal, SPIE **1307**, 588 (1990); K.P. Ghatak,B. De, in*Defect Engineering in Semiconductor Growth, Processing and Device Technology Materials Research Society Proceedings (MRS) Spring Meeting* **262**, 911 (1992); S. Bhattacharya, K.P. Ghatak, S.N. Biswas, Optoelectronic Materials, Devices, Packaging Interconnects, SPIE **836**, 72 (1988)
48. M. Mondal, K.P. Ghatak, J. Phys. C: Solid State. **20**, 1671 (1987); M. Mondal. S.N. Banik, K.P. Ghatak, Can. J. Phys. **67**, 72 (1989); K.P. Ghatak, M. Mondal, J. Appl. Phys. **70**, 1277 (1992)
49. M. Mondal, K.P. Ghatak, J. Phys. C (Sol. State.). **20**, 1671 (1987); M. Mondal. S.N. Banik, K.P. Ghatak, Can. J. Phys. **67**, 72 (1989); K.P. Ghatak, M. Mondal, J. Appl. Phys. **70**, 1277 (1992)
50. S.K. Biswas, A.R. Ghatak, A. Neogi, A. Sharma, S. Bhattacharya, K.P. Ghatak; Physica E: Low-Dimen. Syst. Nanostruct. **36**, 163 (2007); K.P. Ghatak, B. De, M Mondal; Phys. Stat. Solidi. (b) **165**, K53 (1991); K.P. Ghatak, J.Y. Siddiqui, B. Nag, Phys. Lett. Section A: Gen, At. Solid Stat. Phys., **282**, 428 (2001); B. Nag, K.P. Ghatak, J. Phys. Chem. Sol. **58**, 427 (1997); K.P. Ghatak, J.P. Banerjee, B. Nag, J. Appl. Phys. **83**, 1420 (1998); K.P. Ghatak, B. Nag, Nanostruct. Materials **10**, 923 (1998); K.P. Ghatak, J.P. Banerjee, D. Bhattacharyya, B. Nag, Nanotechnol. **7,** 110 (1996); B. Nag, K.P. Ghatak, Phys. Scr. **54**, 657 (1996); A.K. Sreedhar, S.C. Gupta, Phys. Rev. B **5,** 3160 (1972); R.W. Keyes, IBM. J. Res. Dev. **5**, 266 (1961) and Sol. Stat. Phys. **20**, 37 (1967); L.J. Singh, S. Choudhary, A. Mallik, K.P. Ghatak, J. Comput. Theor. Nanosci. **2**, 287 (2005); K.P. Ghatak, D.K. Basu, B. Nag, J. Phys. Chem. Sol. **58**, 133 (1997); K.P. Ghatak, B. Mitra, Phys. Scr. **46**, 182 (1992); K.P. Ghatak, A. Ghosal, Phys. Stat. Sol. (b), **170**, K27 (1992); K.P. Ghatak, S. Biswas, Phys. Stat. Sol. (b), **140**, K107 (1987); S. Debbarma, A.Bhattacharjee, S. Battacharya, A. Mondal, K.P. Ghatak, J. Adv. Phys. **1**, 84 (2012); N. Paitya, S. Bhattacharya, D. De, K.P. Ghatak, Adv. Sci. Eng. Med. **4**, 96 (2012); N. Paitya, S. Bhattacharya, D. De, K.P. Ghatak, Quantum Matter. **1**, 63 (2012); S.M. Adhikari, A. Sakar, K.P. Ghatak, Quantum Matter. **2**, 455 (2013); S. Bhattachrya, N. Paitya, K.P. Ghatak., J. Comput. Theor. Nanosci. **10**, 1999 (2013); S.M. Adhikari, K.P. Ghatak., J. Adv. Phys. **2**, 130 (2013); K.P. Ghatak, P.K. Bose, S. Bhattacharya, A. Bhattacharjee, D. De, S. Ghosh, S. Debbarma, N. Paitya, Quantum Matter. **2**, 83 (2013); S. M. Adhikari, D. De, J.K. Baruah, S. Chowdhury, K.P. Ghatak, **1**, 57 (2013); S.M. Adhikari, K.P. Ghatak, J. Nanoeng. Nanomanufact. **3**, 48 (2013); S. Bhattacharya, D.

De, S. Ghosh, K.P. Ghatak, J. Comput. Theor. Nanosci. **10**, 664 (2013); N. Paitya, K P. Ghatak, J. Nanoeng. Nanomanufact. **2**, 347 (2012); S. Bhattacharya, D. De, N. Paitya, S.M. Adhikari, K.P. Ghatak, Adv. Sci. Lett. **16**, 348 (2012)

51. S.N. Mohammad, J. Appl. Phys. **63**, 1614 (1988); S.N. Mohammad, J. Appl. Phys. **68**, 1710 (1990)
52. R.K. Willardson, A.C. Beer, *Semiconductors and Semimetals* (eds.) (New York, Academic, 1966), vol. 1, p. 102
53. E.O. Kane, Phys. Rev. **131**, 79 (1963); Phys. Rev. B., **139**, 343 (1965)
54. V.L. Bonch Bruevich, Sov. Phys. Sol. Stat. **4**, 53 (1963)
55. R A. Logan, A.G. Chynoweth, Phys. Rev. **131**, 89 (1963)
56. C.J. Hwang, J. Appl. Phys. **40**, 3731 (1969)
57. J.I. Pankove, Phys. Rev. A **130**, 2059 (1965)
58. B.I. Halperin, M. Lax, Phys. Rev. **148**, 722 (1966)
59. R.A. Abram, G.J. Rees, B.L.H. Wilson, Adv. Phys. **27**, 799 (1978)
60. B.I. Shklovskii, A.L. Efros, in *Electronic Properties of Doped Semiconductors*, Springer series in Solid State Sciences, vol. **45** (Springer, Berlin, 1984)
61. E.O. Kane, Solid State Electron. **28,** 3 (1985)
62. P.K. Chakraborty, J.C. Biswas, J. Appl. Phys. **82**, 3328 (1997)
63. B.R. Nag, in, *Electron Transport in Compound Semiconductors*, Springer Series in Solid state Sciences, vol. 11, (Springer, Heidelberg, 1980)
64. P.E. Schmid, Phys. Rev. B **23**, 5531 (1981)
65. Jr. G.E. Jellison, F.A. Modine, C.W. White, R.F. Wood, R.T. Young, Phys. Rev. Lett. **46**, 1414 (1981)
66. V.I. Fistul, *HD Semiconductors* (New York, Plenum, 1969), Chap. 7
67. C.J. Hwang, J. Appl. Phys. **41**, 2668 (1970); W. Sritrakool, H.R. Glyde, V. Sa Yakanit Can. J. Phys. **60**, 373 (1982); H. Ikoma, J. Phys. Soc. Japan **27**, 514 (1969)
68. A.N. Chakravarti, A.K. Chowdhury, K.P. Ghatak, S. Ghosh, A. Dhar, Appl. Phys. **25**, 105 (1981); A.N. Chakravarti, K.P. Ghatak, K.K. Ghosh, S. Ghosh, A. Dhar, Zeitschrift für Physik B Conden. Matter. **47**, 149 (1982); A.N. Chakravarti, K.P. Ghatak, A. Dhar, K.K. Ghosh, Czecho. J. Phys. **33**, 65 (1983); A.N. Chakravarti, K.P. Ghatak, K.K. Ghosh, H.M. Mukherjee, Phys. Stat. Solidi. (b) **116**, 17 (1983); M. Mondal, K.P. Ghatak, Phys. Status Sol. (b) **123**, K143 (1984); M. Mondal, K.P. Ghatak, Phys. Stat. Solidi. (b) **128**, K133 (1985); M. Mondal, K.P. Ghatak, Phys. Stat. Solidi. (b) **129**, 745 (1985); M. Mondal, K.P. Ghatak, Phys. Stat. Solidi. (b) **135**, K21 (1986); M. Mondal, K.P. Ghatak, Czech. J. Phys. B **36**, 1389 (1986); M. Mondal, N. Chattopadhyay, K.P. Ghatak, J. Low. Temp. Phys. **66**, 131 (1987); M. Mondal, K.P. Ghatak, Phys. Lett. A **131**, 529 (1988); B. Mitra, K.P. Ghatak, Sol. Stat. Electron **32**, 515 (1989); K.P. Ghatak, A. Ghoshal, B. Mitra, Il Nuovo Cimento., D **13**, 867 (1991); K.P. Ghatak, A. Ghoshal, Phys. Status Solidi (a) **126**, K53 (1991); K.P. Ghatak, B. Mitra Int. J. Electron. **72**, 541 (1992); K.P. Ghatak, A. Ghoshal, B. Mitra, Il Nuovo Cimento. D **14**, 903 (1992); E.A. Aruhanov, A.F. Knyazev, A.N. Nateprov, S.I. Radautsan, Sov. Phys. Semicond. **15**, 828 (1981); P.N. Hai, W.M. Chen, I.A. Buyanova, H.P. Xin, C.W. Tu, Appl. Phys. Letts. **77**, 1843 (2000); D.P. DiVincenzo, E.J. Mele, Phys. Rev. B, **29**, 1685 (1984); P. Perlin, E. Litwin-Staszewska, B. Suchanek, W. Knap, J. Camassel, T. Suski, R. Piotrzkowski, I. Grzegory, S. Porowski, E. Kaminska, J.C. Chervin, Appl. Phys. Letts **68,** 1114 (1996); F. Masia, G. Pettinari, A. Polimeni, M. Felici, A. Miriametro, M. Capizzi, A. Lindsay, S.B. Healy, E.P. O'Reilly, A. Cristofoli, G. Bais, M. Piccin, S. Rubini, F. Martelli, A. Franciosi, P.J. Klar, K. Volz, W. Stolz, Phys. Rev. B **73**, 073201 (2006); S.L. Sewall, R.R. Cooney, P. Kambhampati, Appl. Phys. Letts **94,** 243116 (2009); G.E. Smith, Phys. Rev. Letts **9**,) (1962); D. Schneider, D. Rurup, A. Plichta, H.U. Grubert, A. Schlachetzki, K. Hansen, Z. Phys. B. **95** 281 (1994); L.G. Caron, J.P. Jay Gerin, M.J. Aubin, Phys. Rev. B, **15**, 3879 (1977); D. Bhattacharyya, R. Pal, S. Chaudhuri, A.K. Pal, Vaccum **44,** 803 (1993); B. Slomski, F. Meier, J. Osterwalder, J.H. Dil, Phys. Rev., B **83**, 035409 (2011); L.M. Wei, K.H. Gao, X.Z. Liu, W.Z. Zhou, L.J. Cui, Y.P. Zeng, G. Yu, R. Yang, T. Lin,

L.Y. Shang, S.L. Guo, N. Dai, J.H. Chu, D.G. Austing, J. Appl. Phys **110**, 063707 (2011); K. Karra, M. Stopa, X. Ying, H.D. Drew, S. Das Sarma, Phys. Rev. B **42**, 9732 (1990); B. Rößner, G. Isella, H. V. Känel; Appl. Phys. Letts **82**, 754 (2003); T.M. Kim, M. Jung, K.H. Yoo, J. Phys. Chem. Sol. **61**, 1769 (2000); N. Kotera, H. Arimoto, N. Miura, K. Shibata, Y. Ueki, K. Tanaka, H. Nakamura, T. Mishima, K. Aiki, M. Washima, Phys. E **11**, 219 (2001); K. Tanaka, N. Koteraa, H. Nakamura; Microelectron. Eng. **47**, 309 (1999); G. Wang, T. Cagin, Appl. Phys. Letts **89**, 152101 (2006); Z.S. Gribnikov, R.R. Bashirov, H. Eisele, V.V. Mitin, G.I. Haddad; Phys. E **12**, 276 (2002); R. Chen, K.K. Bajaj, Phys. Rev. B **50**, 1949 (1994); E. Borovitskaya, M.S. Shur, Sol. Stat. Electro. **44**, 1293 (2000); L.C.O. Dacal, E.A. de Andrada e Silva, Brazilian J. Phys. **36**, 910 (2006); A. Sasaki, Surface Science, **174** 624 (1986); J.C. Maan, M. Altarelli, H. Sigg, P. Wyder, L.L. Chang, L. Esaki, Surf. Sci. **113**, 347 (1982); K. El Shabrawy, K. Maharatna, D. Bagnall, B.M. Al-Hashimi, IEEE Trans. Nanotechnol. **9**, 184 (2010); N. Neophytou, A. Paul, M.S. Lundstrom, G. Klimeck, IEEE Trans. Electron. Dev. **55**, 866 (2008); E. Zheng, C. Rivas, R. Lake, K. Alam, T.B. Boykin, G. Klimeck, IEEE Trans. Electron. Dev. **52**, 1097 (2005); J. Wang, A. Rahman, A. Ghosh, G. Klimeck, M.S. Lundstrom, IEEE Trans. Electron. Dev. **52**,) (2005); J.A. Yan, L. Yang, M.Y. Chou, Phys. Rev. B **76**, 115319 (2007); M.O. Baykan, S.E. Thompson, T. Nishida, J. Appl. Phys. **108**, 092716 (2010); R.N. Sajjad, K. Alam, J. Appl. Phys. **105**, 044307 (2009); K.H. Hong, J. Kim, S.H. Lee, J.K. Shin, Nano. Lett. **8**, 1335 (2008); P.W. Leu, B. Shan, K. Cho, Phys. Rev. B **73**, 195320 (2006); V. Sverdlov, *Strain-Induced Effects in Advanced MOSFETs* (Springer-Verlag, Wein, 2011); K. Uchida, A. Kinoshita, M. Saitoh, Int. Electron. Dev. Meet. (2006), **1019**; R.N. Sajjad, K. Alam, Q.D.M. Khosru, Semicond. Sci. Technol. **24**, 045023 (2009); M.S. Lundstrom, J. Guo, *Nanoscale Transistors: Device Physics Modeling and Simulation* (Springer, USA, 2006); M.F. Ng, L. Zhou, S.W. Yang, L.Y. Sim, V.B.C. Tan, P. Wu, Phys. Rev. B **76**, 155435 (2007); D. Kienle, K.H. Bevan, G.C. Liang, L. Siddiqui, J.I. Cerda, A.W. Ghosh, J. Appl. Phys. **100**, 043715 (2006); J. Cerda, F. Soria, Phys. Rev. B **61**, 7965 (2000); P. Pulay, Chem. Phys. Lett. **73** 393 (1980); S. Jin, M.V. Fischetti, T.W. Tang, IEEE Trans. Electron. Dev. **55**, 2886 (2008); R.N. Sajjad, S. Bhowmick, Q. Khosru, IEEE International Conference Electron Devices and Solid-State Circuits (EDSSC). **1** (2008); J.L.P.J. van der Steen, D. Esseni, P. Palestri, L. Selmi, R. J.E. Hueting, IEEE, Trans. Electron. Dev. **54**, 1843 (2007); F. Stern, W.E. Howard, Phys. Rev. **163**, 816 (1967); P. Logan, X. Peng, Phys. Rev. B. **80**, 115322 (2009); C.G.V. de Walle, R.M. Martin, Phys. Rev. B **34**, 5621 (1986); J.C. Hensel, G. Feher, Phys. Rev. **129**, 10411062 (1963); B.K. Ridley, *Quantum Processes in Semiconductors*, 4th edn. (Oxford University Press, Oxford, 1999); J.H. Davis, *Physics of Low Dimensional Semiconductors* (Cambridge University Press, UK, 1998); M. Schaden, K.F. Zhao, Z. Wu, Phys. Rev. A **76**, 062502 (2007); T. Kawarabayashi, T. Ohtsuki, Phys. Rev. B **51** 10897 (1995); B. Laikhtman, Phys. Rev. Lett. **72**, 1060 (1994); A. Houghton, J.R. Senna, S.C. Ying, Phys. Rev. B **25**, 6468 (1982); I.M. Tsidilkovski, *Band Structures of Semiconductors,* (Pergamon Press, London 1982); K.P. Ghatak, M. Mondal, Z.F. Physik B **B69** 471 (1988); B. Mitra, A. Ghoshal, K.P. Ghatak, Nouvo Cimento D **12D**, 891 (1990); K.P. Ghatak, S.N. Biswas, Nonlinear Opt. Quantum Opts. **12** 83 (1995); K.P. Ghatak, S.N. Biswas, Proc. SPIE, **1484**, 149 (1991); M. Mondal, K.P. Ghatak, *Graphite Intercalation Compounds: Science and Applications*, *MRS Proceedings*, ed. by M. Endo, M.S. Dresselhaus, G. Dresselhaus, MRS Fall Meeting, **EA 16** 173 (1988); K.P. Ghatak, M. Mondal, Z. fur, Nature A, **41A**, 881 (1986); C.C. Wu, C.J. Lin, J. Low Temp. Phys. **57**, 469 (1984); M.H. Chen, C.C. Wu, C.J. Lin, J. Low Temp. Phys. **55**, 127 (1984)

69. V. Heine, Proc. Phys. Soc. **81**, 300 (1963); J.N. Schulman , Y.C. Chang, Phys. Rev. B, **24**, 4445 (1981)
70. N.G. Anderson, W.D. Laidig, R.M. Kolbas, Y.C. Lo, J. Appl. Phys. **60** 2361 (1986); N. Paitya, K.P. Ghatak, J. Adv. Phys. **1**, 161 (2012); S. Bhattacharya, D. De, S.M. Adhikari, K.P. Ghatak Superlatt. Microst. **51**, 203 (2012); D. De, S. Bhattacharya, S.M. Adhikari, A. Kumar, P.K. Bose, K.P. Ghatak, Beilstein J. Nanotechnol. **2**, 339 (2012); D. De, A. Kumar,

S.M. Adhikari, S. Pahari, N. Islam, P. Banerjee, S.K. Biswas, S. Bhattacharya, K.P. Ghatak, Superlatt. Microstruct. **47**, 377 (2010); S. Pahari, S. Bhattacharya, S. Roy, A. Saha, D. De, K.P. Ghatak, Superlatt. Microstruct. **46**, 760 (2009); S. Pahari, S. Bhattacharya, K.P. Ghatak, J. Comput. Theor. Nanosci. **6**, 2088 (2009); L.J. Singh, S. Choudhury, D. Baruah, S.K. Biswas, S. Pahari, K.P. Ghatak, Phys. B: Conden. Matter **368** 188 (2005); S. Chowdhary, L.J. Singh, K.P. Ghatak, Phys. B: Conden. Matter **365**, 5 (2005); K.P. Ghatak, J. Mukhopadhyay J.P. Banerjee, SPIE Proc Ser. **4746** 1292 (2002); K.P. Ghatak, S. Dutta, D.K. Basu, B. Nag, Il Nuovo Cimento. D **20**, 227 (1998); K.P. Ghatak, B. De, Mat. Resc. Soc. Proc. **300**, 513 (1993); K.P. Ghatak, B. Mitra, Il Nuovo Cimento. D **15**, 97 (1993); K.P. Ghatak, Int. Soci. Opt. Photon. Proc. Soc. Photo Opt. Instrum. Eng. **1626**, 115 (1992); K.P. Ghatak, A Ghoshal, S.N. Biswas, M. Mondal, Proc. Soc. Photo Opt. Instru. Eng. **1308**, 356 (1990); S.N. Biswas, K.P. Ghatak, Int. J. Electron. Theor. Exp. **70** 125 (1991); B. Mitra, K.P. Ghatak, Phys. Lett. A. **146**, 357 (1990); B. Mitra, K.P. Ghatak, Phys. Lett. A. **142**, 401 (1989); K.P. Ghatak, B. Mitra, A Ghoshal, Phy. Stat. Sol. (b) **154**, K121 (1989); B. Mitra, K.P. Ghatak, Phys. Status Solidi (b). **149**, K117 (1988); S. Bhattacharyya, K.P. Ghatak, S. Biswas, OE/Fibers' 87, Int. Soc. Opt. Photonics 73 (1987); K.P. Ghatak, A.N. Chakravarti, Phys. Stat. Solidi. (b). **117**, 707 (1983)

71. F. Capasso, Semicond. Semimetals **22**, 2 (1985)
72. F. Capasso, K. Mohammed, A.Y. Cho, R. Hull, A.L. Hutchinson, Appl. Phys. Lett. **47**, 420 (1985)
73. F. Capasso, R.A. Kiehl, J. Appl. Phys. **58**, 1366 (1985)
74. K. Ploog, G.H. Doheler, Adv. Phys. **32**, 285 (1983)
75. F. Capasso, K. Mohammed, A.Y. Cho, Appl. Phys. Lett. 478 (1986)
76. R. Grill, C. Metzner, G.H. Döhler, Phys. Rev. B **63**, 235316 (2001); Phys. Rev. B **61,** 15614 (2000)
77. A.R. Kost, M.H. Jupina, T.C. Hasenberg, E.M. Garmire, J. Appl. Phys. **99**, 023501 (2006)
78. A.G. Smirnov, D.V. Ushakov, V.K. Kononenko, Proc. SPIE **4706** 70 (2002)
79. D.V. Ushakov, V.K. Kononenko, I.S. Manak, Proc. SPIE **4358**, 171 (2001)
80. J.Z. Wang, Z.G. Wang, Z.M. Wang, S.L. Feng, Z. Yang, Phys. Rev. B **62**, 6956 (2000)
81. A.R. Kost, L. West, T.C. Hasenberg, J.O. White, M. Matloubian, G.C. Valley, Appl. Phys. Lett. **63**, 3494 (1993)
82. S. Bastola, S.J. Chua, S.J. Xu, J. Appl. Phys. **83**, 1476 (1998)
83. Z.J. Yang, E.M. Garmire, D. Doctor, J. Appl. Phys. **82**, 3874 (1997)
84. G.H. Avetisyan, V.B. Kulikov, I.D. Zalevsky, P.V. Bulaev, Proc. SPIE **2694** 216 (1996)
85. U. Pfeiffer, M. Kneissl, B. Knüpfer, N. Müller, P. Kiesel, G.H. Döhler, J.S. Smith, Appl. Phys. Lett. **68** 1838 (1996)
86. H.L. Vaghjiani, E.A. Johnson, M.J. Kane, R. Grey, C.C. Phillips, J. Appl. Phys. **76**, 4407 (1994)
87. P. Kiesel, K.H. Gulden, A. Hoefler, M. Kneissl, B. Knuepfer, S.U. Dankowski, P. Riel, X.X. Wu, J.S. Smith, G.H. Doehler, Proc. SPIE **85** 278 (1993)
88. G.H. Doheler, Phys. Scr. **24**, 430 (1981)
89. S. Mukherjee, S.N. Mitra, P.K. Bose, A.R. Ghatak, A. Neoigi, J.P. Banerjee, A. Sinha, M. Pal, S. Bhattacharya, K.P. Ghatak, J. Compu. Theor. Nanosci, **4**, 550 (2007)
90. J.J. Quinn, P.J. Styles (ed.) *Electronic Properties of Quasi Two Dimensional Systems* (North Holland, Amsterdam, 1976); G.A. Antcliffe, R.T. Bate, R.A. Reynolds, in*Proceedings of the International Conference,Physics of Semi-metals and Narrow-Gap Semiconductors,* ed. by D.L. Carter, R.T. Bate (Pergamon Press, Oxford, 1971). p. 499; G. Paasch, T. Fiedler, M. Kolar, I. Bartos, Phys. Status Sol. (b) **118**, 641 (1983); Th. Lindner, G. Paasch, J. Appl. Phys. **102** 054514 (2007); S. Lamari, J. Appl. Phys. **91**, 1698 (2002); K.P. Ghatak, M. Mondal, J. Appl. Phys. **70**, 299 (1991); K.P. Ghatak, S.N. Biswas, J. Vac. Sci. Technol. **7B**, 104 (1989); B. Mitra, K.P. Ghatak, Sol. State Electron. **32**, 177 (1989); K.P. Ghatak, M. Mondal, J. Appl. Phys. **62**, 922 (1987); M. Mondal, K.P. Ghatak, Phys. Scr. **31**, 613 (1985); K.P. Ghatak, M. Mondal, Z. Phys. B **64**, 223 (1986); K.P. Ghatak, S.N. Biswas, Sol. State

Electron. **37**, 1437 (1994); D.R. Choudhury, A.K. Chowdhury, K.P. Ghatak, A.N. Chakravarti, Phys. Stat. Solidi. (b) **98** K141 (1980); A.N. Chakravarti, A.K. Chowdhury, K.P. Ghatak, Phys. Stat. Solidi. (a) **63**, K97 (1981); M. Mondal, K.P. Ghatak, Acta Phys. Polon. A **67** 983 (1985); M. Mondal, K.P. Ghatak, Phys. Stat. Solidi. (b) **128**, K21 (1985); M. Mondal, K.P. Ghatak, Phys. Stat. Solidi. (a) **93**, 377 (1986); K.P. Ghatak, M. Mondal, Phys. Stat. Solidi. (b) **135**, 819 (1986); M. Mondal, K.P. Ghatak, Phys. Stat. Solidi. (b) **139**, 185 (1987); K.P. Ghatak, N. Chatterjee, M. Mondal, Phys. Stat. Solidi. (b) **139**, K25 (1987); K.P. Ghatak, M. Mondal, Phys. Stat. Solidi. (b) **148**, 645 (1988); K.P. Ghatak, A. Ghosal Phys. Stat. Solidi. (b) **151**, K135 (1989); K.P. Ghatak, N. Chattopadhyay, M. Mondal, Appl. Phys. A **48**, 365 (1989)

91. S. Debbarma, K.P. Ghatak, Rev. Theor. Sci. (2014) In the press
92. K.P. Ghatak and S.N. Biswas, *Long Wave Length Semiconductor Devices, Materials and Processes MRS Symposium Proceedings*. MRS Spring Meeting. **216**, 465 (1990); K.P. Ghata, B. De, *Modern Persepctive on Thermoelectrics and Related Materials*, MRS Symposium Proceedings, Spring Meeting **234**, 55 (1991); *Modern Persepctive on Thermoelectrics and Related Materials*, MRS Symposium Proceedings, Spring Meeting **234**, 59 (1991); K.P. Ghatak, Proc. SPIE, USA, *Fiber Optic and Laser Sensors IX* **1584**, 435 (1992); K.P. Ghatak, A.K. Chowdhury, S. Ghosh, A.N. Chakravarti, Czecho. J. Phys. B **30**, 925 (1980); K.P. Ghatak, A.K. Chowdhury, S. Ghosh, A.N. Chakravarti, Acta Phys. Polon. A **58**, 389 (1980); A.N. Chakravarti, K.P. Ghatak, K.K. Ghosh, S. Ghosh, A. Dhar, Phys. Stat. Solidi. (b) **103**, K55 (1981); A.N. Chakravarti, K.P. Ghatak, S. Ghosh, A.K. Chowdhury, Phys. Stat. Solidi. (b) **109**, 705 (1982); K.P. Ghatak, K.K. Ghosh, H.M. Mukherjee, A.N. Chakravarti, Phys. Stat. Solidi. (b) **110**, 323 (1982); K. P.Ghatak, S. Bhattacharya, S. Bhowmik, R. Benedictus, S. Chowdhury, J. Appl. Phys. **103**, 094314 (2008); K.P. Ghatak, S. Bhattacharya, S. K.Biswas, A. De, A.K. Dasgupta, Phys. Scr. **75**, 820 (2007); A.N. Chakravarti, K.P. Ghatak, G.B. Rao, K.K. Ghosh, Phys. Stat. Solidi. (b) **112** 75 (1982); M. Mondal, K.P. Ghatak, Phys. Stat. Solidi. (b) **120** K63 (1983); M. Mondal, K.P. Ghatak, Acta Phys. Polon. A **66** 539 (1984); M. Mondal, K.P. Ghatak, Acta Phys. Polon. A **66**, 47 (1984); M. Mondal, K.P. Ghatak, Phys. Stat. Solidi. (b) **126**, K41 (1984); M. Mondal, K.P. Ghatak, Phys. Stat. Solidi. (b) **126**, K47 (1984); M. Mondal, S. Banik, K. P. Ghatak, J. Low. Temp. Phys. **74**, 423 (1989); B. Mitra, K.P. Ghatak, Phys. Scr. **40**, 776 (1989); B. Mitra, K.P. Ghatak, Phys. Lett. A **141**, 81 (1989); K.P. Ghatak, M. Mondal, Phys. Stat. Solidi. (b) **160**, 673 (1990); K.P. Ghatak, M. Mondal, J. Appl. Phys. **69**, 1666 (1991); K.P. Ghatak, Il Nuovo Cimento D **14**, 1187 (1992); B. Nag, K.P. Ghatak, J. Phys. Chem. Sol. **59**, 713 (1998); P.K. Chakraborty, K.P. Ghatak J. Phys. D. Appl. Phys. **32**, 2438 (1999); P.K. Chakraborty, S.K. Biswas, K.P. Ghatak, Physica B: Condens Matter, **352**, 111 (2004)
93. L.V. Keldysh, Soviet Phys. Solid State **4**, 1658 (1962)
94. L. Esaki and R. Tsu, IBM J. Res. Dev. **14**, 61 (1970)
95. R. Tsu, *Superlattices to Nanoelectronics* (Elsevier, Boston, 2005); E.L. Ivchenko, G. Pikus, *Superlattices and Other Heterostructures* (Springer, Berlin, 1995)
96. G. Bastard,*Wave Mechanics Applied to Heterostructures, Editions de Physique* (Les Ulis, France, 1990)
97. K.P. Ghatak, B. De, *MRS Proceedings* (Cambridge University Press, Cambridge, 1994), vol.65, p.299; K.P. Ghatak, M. Mondal, Phys. Stat. Solidi. (b) **175**, 113 (1993); K.P Ghatak, B. De, *MRS Proceedings* (Cambridge University Press, Cambridge, 1992), vol. 242, p. 377; K.P. Ghatak, San Diego,'91, San Diego, CA, Int. Soc. Opt. Photonics,p. 282 (1991); K.P. Ghatak, Il Nuovo Cimento D **13**, 1321 (1991); B. Mitra, K.P. Ghatak, Phys. Stat. Solidi. (b) **164**, K13 (1991); K.P. Ghatak, Properties of II–VI semiconductors: bulk crystals, epitaxial films, quantum well structures, and dilute magnetic systems. Mater. Res. Soc. **161**, 371 (1990); K.P. Ghatak, S.N. Biswas, *MRS Proceedings* (Cambridge University Press, Cambridge, 1989), p. 161; S. Bhattacharyya, K.P. Ghatak, S. Biswas, OE/Fibers' **87**, 73 (1987); M. Mondal, K.P. Ghatak, Phys. Stat. Solidi. (b) **122**, K93 (1984)

98. K.V. Vaidyanathan, R.A. Jullens, C.L. Anderson, H. L. Dunlap, Solid State Electron. **26**, 717 (1983)
99. B.A. Wilson, IEEE, J. Quantum Electron. **24**, 1763 (1988)
100. M. Krichbaum, P. Kocevar, H. Pascher, G. Bauer, IEEE, J. Quantum Electron. **24**, 717 (1988)
101. J.N. Schulman, T.C. McGill, Appl. Phys. Letts **34**, 663 (1979)
102. H. Kinoshita, T. Sakashita, H. Fajiyasu, J. Appl. Phys. **52**, 2869 (1981)
103. L. Ghenin, R.G. Mani, J.R. Anderson, J.T. Cheung, Phys. Rev. B **39**, 1419 (1989)
104. C.A. Hoffman, J.R. Mayer, F.J. Bartoli, J.W. Han, J.W. Cook, J.F. Schetzina, J.M. Schubman, Phys. Rev. B. **39**, 5208 (1989)
105. V.A. Yakovlev, Sov. Phys. Semicond. **13**, 692 (1979)
106. E.O. Kane., J. Phys. Chem. Solids **1**, 249 (1957)
107. H. Sasaki, Phys. Rev. B **30**, 7016 (1984)
108. W. Zawadzki, B. Lax, Phys. Rev. Lett. **16**, 1001 (1966)
109. K.P. Ghatak, J.P. Banerjee, B. Goswami, B. Nag, Nonlinear Opt. Quantum Opt. **16**, 241 (1996); K.P. Ghatak, M. Mitra, B. Goswami, B. Nag, Nonlinear Opt. Quantum Opt. **16**, 167 (1996); B. Mitra, A. Ghoshal , K.P. Ghatak, Phys. Stat. Solidi. (b), **154**, K147 (1989); K.P. Ghatak, B. Goswami, M. Mitra, B. Nag, Nonlinear Opt. Quantum Opt. **16**, 9 (1996); S. Bhattaacharya, S. Choudhary, S. Ghoshal, S.K.Bishwas, D. De, K.P. Ghatak, J. Comp. Theor. Nanosci. **3**, 423 (2006); M. Mondal, K.P. Ghatak, Ann. der Physik. **46**, 502 (1989); K.P. Ghatak, OE Fiber-DL, 435 (1991); K.P. Ghatak, B. Mitra, Int. J. Electron. Theor. Exp. **70**, 343 (1991); K.P. Ghatak, Proc. Soc. Photo-Optical Instrum. Eng. USA **1280**, 53 (1990); B. Mitra, K.P. Ghatak, Phys. Lett. A **137**, 413 (1989); M. Mondal, K.P. Ghatak, Annalen der Physik, **501**, 502 (1989); M. Mondal, K.P. Ghatak, Phys. Stat. Solidi. (b) **147** K179 (1988); S. Biswas, N. Chattopadhyay, K.P. Ghatak, Int. Soc. Opt. Photon.Proc. Soc. Photo-Optical Instrum. Eng. USA **836** 175 (1987)
110. K.P. Ghatak, S. Bhattacharya, S. Pahari, D. De, S. Ghosh, M. Mitra; Annalen der Physik **17**, 5 (2008)
111. P.K. Bose, N. Paitya, S. Bhattacharya, D. De, S. Saha, K.M. Chatterjee, S. Pahari, K.P. Ghatak, Quantum Matter **1**, 89 (2012); S. Bhattarchya, D. De, S. Ghosh, P. Banerjee, S. Saha, M. Mitra, B. Nag, M. Pal, S.K. Biswas, K.P. Ghatak, J. Comput. Theor. Nanosci. **7**, 1066 (2010); K.P. Ghatak, S. Bhattacharya, S. Pahari, S.N. Mitra, P.K. Bose, D. De, J. Phys. Chem. Solids **70**, 122 (2009); S. Bhattacharya, N .C. Paul, D De, K.P. Ghatak, Physica B: Condens. Matter **403**, 4139 (2008)
112. K.P. Ghatak, S. Bhattacharya, D. De, R. Sarkar, S. Pahari, A. Dey, A.K. Dasgupta, J. Comput. Theor. Nanosci. **5**, 1345 (2008); K.P. Ghatak, S. Bhattacharya, S. Pahari, D. De, S. Ghosh, M. Mitra, Annalen der Physik **17**, 195 (2008); K.P. Ghatak, S. Bhattacharya, J. Appl. Phys. **102**, 073704 (2007); S. Mukherjee, D. De, D. J. Mukherjee, S. Bhattacharya, A. Sinha, K.P. Ghatak, Physica B: Condens. Matter **393**, 347 (2007)
113. A. Pais, J. *Robert Oppenheimer* (Oxford University Press, Oxford, 2006), p. xviii
114. I.K. Kikoin, *Science for Everyone: Encounters with Physicists and Physics* (Mir Publishers, Russia, 1989), p. 154
115. J.J. Sakurai, *Advanced Quantum Mechanics* (Addision–Wesley Publishing Company, Boston, 1967)

Acknowledgments

I wish to record my speechless gratitude to A.N Chakravarti, my Ph.D. thesis advisor and an *ardent advocator of the concept of theoretical minimum of Landau,* who convinced a 21 years old Circuit theorist that theoretical solid state science, in general, is the hidden dual dance of quantum mechanics, statistical mechanics together with advanced mathematical techniques, and even to appreciate the incredible beauty, he placed a stiff note for me to derive all the equations in the Monumental Course of Theoretical Physics, the Classics of Landau–Lifshitz together with the two-volume classics of Morse–Feshbach 35 years ago. I offer my thanks to Late B.R Nag of the Institute of Radio-physics and Electronics of the University of Calcutta, to whom I am grateful as research worker, for his three books on semiconductor science in general and more than 200 research papers (many of them are in *honor's class*) which still fire my imagination. I am grateful to Late S.C Dasgupta and M. Mitra (both of them were formidable Applied Mathematicians with deep physical insight) of the Department of Mathematics of the then Bengal Engineering College, Shibpur, Howrah (presently Indian Institute of Engineering Science and Technology) for teaching me deeply the various methods of different branches of Applied Mathematics with special emphasis on *analytic number theory* when I was in my teens and pursuing the bachelor degree in the branch of Electronics and Telecommunication Engineering 40 years ago. I am indebted to Late C.K Majumdar of the Department of Physics of the University of Calcutta to light the fire for Theoretical Physics. I am grateful to P.N. Singh, the Ex-Chairman of NIT Agartala for his encouragement and enthusiastic support and the Director, G. Mugeraya for his keen academic insight and the creation of congenial atmosphere to promote research and higher learning among the faculty members. I offer special thanks to all my friends and colleagues for the last 40 years from my college days till date for forming my inner fire to solve independently the problems from five volumes Berkeley Physics Course, three volumes Sakurai, ten volumes of Addison-Wesley Modular Series on Solid State Devices, two volumes Cohen-Tannoudji et al., three volumes Edwards, three volumes Carslaw, six volumes Guillemin, three volumes Tuttle, two volumes Scott, two volumes Budak, five volumes Chen, five volumes Smirnov……….. It is curious to note that they are insisting me in the real sense of the term to verify all the entries of Gradshteyn–Ryzhik and three volumes Bateman manuscript project for the last 40 years. Also they are relentlessly guarding me intellectually so that my education should not be transformed into hedu-cation where as my nothingness should be no-thingness.

It is well-known that academic output of a good quality scholar is the response of a good quality RC coupled amplifier in the mid-band zone where as the same for a fine researcher is the Dirac's delta function. Incidentally, I can safely say that I belong to neither as a consequence of the academic surroundings, which this life presents to me. I express my warm gratitude to H.L. Hartnagel, D. Bimberg, W.L. Freeman and W. Schommers for various academic interactions spanning over the last three decades. I remember the sweet memories of Late P.T. Landsberg, P.N. Robson, I.M. Stephenson, V.S. Letokhov, J. Bodnar, J.S. Blakemore and K.F. Brennan with true reverence for inspiring me with the ebullient idea that *the publications of the research monographs from internationally reputed Publishers containing innovative concepts is of prime importance to excel in creative research activity.* The well-known scientist Late P.N. Butcher has been a steady hidden force since 1983 before his sad demise with respect to our scripting the series in band structure dependent properties of nano-structured materials. He insisted me repeatedly regarding it and to tune with his high rigorous academic standard, myself with my prominent members of my research group wrote [1] as the first one, [2] as the second one, [3] as the third one, [4] as the fourth one, [5] as the fifth one, [6] as the sixth one and the present monograph as the seventh one.

Incidentally, one young theoretician friend of mine often tells me that many works in theoretical semiconductor science are based on the following seven principles:

1. Principles of interchange of the summation and the integration processes and unconditioned convergences of the series and the integrals.
2. Principles of placing the necessary and sufficient conditions of a proof in the band gap regime.
3. Principles of random applications of one electron theory and super-position theorem in studying the properties of semiconductors, although the many body effects are very important together with the fact that the nature is fundamentally non-linear in nature.
4. Principles of using the invariant band structure concept of semiconductors even in the presence of strong external fields (light, electric, heavy doping etc.) and the random applications of perturbation theory, which is in a sense quantum mechanical Taylor series without considering the related consequences.
5. Principle of random applications of the binomial theorem without considering the important concept of branch cut.
6. Principle of little discussion regarding the whole set of materials science comprising of different compounds having various band structures under different physical conditions as compared with the simplified two band model of Kane for III–V semiconductors.
7. Principle of using the Fermi's golden rule, the band structure, and the related features, which are valid for non-degenerate semiconductors to materials having degenerate carrier concentrations directly.

Although my friend is a purist in his conjecture, there are no doubt certain elements of truth inside his beautiful comments. We jointly hope that our readers will present their intricate and advanced theories after paying due weightage of his aforementioned seven principles.

I offer special thanks to Late N. Guhachoudhury of Jadavpur University for instilling in me the thought that the *academic output* = *((desire* × *determination* × *dedication)* − *(false enhanced self-ego pretending like a true friend although a real unrecognizable foe))*. I must not allow even a thanks to my beloved better half for really forming the backbone of my long unperturbed research career, since in accordance with "Sanatan" Hindu Dharma, the fusion of marriage has transformed us to form a single entity, where the individuality is being lost. I am grateful to all the members of my research group for not only quantum confining me in the infinitely deep quantum wells of Ramanujan and Rabindranath but also inspiring me to teach quantum mechanics and related topics from the eight volumes classics of Greiner et al.

I must express my gratitude to S. Debbarma, of Computer Science and Engineering Department of my present working place, one of the strong members of my research group, for offering important suggestions for the condensed presentation of this monograph. I offer special thanks to P. Bhardwaj, K. Sarkar, M. Chakraborty, S. Chakrabarti, B. Chatterjee and other members of my research team for placing their combined effort towards the development of this book in the DO-LOOP of a computer and critically reading the manuscript in its last phase before sending the same to C. Ascheron, Executive Editor, Physics, Springer Verlag. Last but not the least, I am grateful for ever to our life long time tested friend S. Sanyal, Principal, Lake School for Girls, Kolkata for not only motivating me at rather turbulent times of my academic carrier but also instilling in me the concept that *the ratio of total accumulated knowledge in my chosen field of research to my own contribution in my topic of research tends to infinity at any time.*

Kamakhya P. Ghatak

In-spite of many hurdles, the completion of this monograph owes to my teacher S. Mahapatra, at the Center for Electronics Design and Technology at Indian Institute of Science, Bangalore, with whom I have learned to perform quality research. My sincere gratitude is also to my friend R.C. Mallik of Thermoelectric Materials and Device Laboratory, Department of Physics of the same Institute for his constant inspiration. I offer special thanks to all my friends at my Department and Institute for standing by my side at difficult times of my research life. I am indebted to the Department of Science and Technology, India, for sanctioning the project and the fellowship under INSPIRE Faculty award- "IFA-12ENG-39"-2012 under which this monograph has been completed. I am grateful to the faculty members of the Department of Electrical and communication Engineering, Indian Institute of Information Technology, Allahabad, my present working place. *As always, I am ever grateful in the real sense of the term to the first author, my friend, philosopher, mentor and Ph.D. thesis advisor whose mental wave function is the admixture of that of a mathematical physicist and a poet.*

Sitangshu Bhattacharya

We are grateful to C. Ascheron, Executive Editor, Physics, Springer Verlag, in the real sense of the term for his inspiration and priceless technical assistance from the very start of our first monograph from Springer. We express our gratitude to P. Wölfle, the Editor of Condensed Matter Theory of Springer Tracts of Modern Physics for his valuable guidance. We owe a lot to A. Duhm, Associate Editor Physics, Springer, and E. Suer, assistant to Ascheron. Naturally, the authors are responsible for non-imaginative shortcomings although we hope that our readers are in tune with the fact that *Exposition, criticism, appreciation is the work for second-rate minds* [7]. We firmly believe that our Mother Nature has propelled this joint collaboration in her own unseen way in spite of several insurmountable obstacles.

Tripura, Uttar Pradesh, India, March 2014

Kamakhya P. Ghatak
Sitangshu Bhattacharya

References

1. K.P. Ghatak, S. Bhattacharya, D. De, *Einstein Relation in Compound Semiconductors and Their Nanostructures;* Springer Series in Materials Science **116** (Springer, Heidelberg, 2009)
2. K.P. Ghatak, D. De, S. Bhattacharya, *Photoemission from optoelectronic materials and their nanostructures,* Springer Series in Nanostructure Science and Technology (Springer, Heidelberg, 2009)
3. K.P. Ghatak, S. Bhattacharya, *Thermo Electric Power In Nano structured Materials Strong Magnetic Fields,* Springer Series in Materials Science **137** (Springer, Heidelberg, 2010)
4. S. Bhattacharya and K.P Ghatak, *Fowler-Nordheim Field Emission: Effects in semiconductor nanostructures,* Springer Series in Solid state Sciences **170** (Springer, Heidelberg, 2012)
5. S. Bhattacharya and K.P Ghatak, *Effective electron mass in low dimensional semiconductors,* Springer Series in Materials Sciences 167 (Springer, Heidelberg, 2013)
6. K.P Ghatak, S. Bhattacharya, *Debye Screening Length: Effects of Nanostructured Materials;* Springer Tracts in Modern Physics 255 (Springer, Heidelberg, 2014)
7. G.H. Hardy, *A Mathematician's Apology* (Cambridge University Press, 1990), pp. 61

Contents

About the Authors

Kamakhya Prasad Ghatak is the First Recipient of the Degree of Doctor of Engineering of Jadavpur University in 1991 since the University inception in 1955 and in the same year he received the Indian National Science Academy visiting fellowship to IIT-Kharagpur. He is the principal co-author of more than 300 research papers on Semiconductor Nano-science and Technology in peer-reviewed International Journals and more than 50 research papers in the Proceedings of SPIE, MRS etc. and is the supervisor of more than two dozens of Ph.D. candidates in various aspects of quantized materials and materials science in general. He is the principal co-author of the six research monographs among them three from Springer series in Material Science (Vols. 116, 137, 167), one form Springer Series in Nanostructure Science and Technology, one form Springer Series in Solid-State Sciences (Vol. 170), and one form Springer Tracts in Modern Physics (vol. 255). The All Indian Council for Technical Education has selected the first Research and Development project in his life for the best project award in Electronics and in 2011, the University Grant Commission recommended a research project to him and placed him at the top in the list of awardees. His teaching interests are Non-Linear Circuits, Non-Linear Control Theory, Non-Linear Mechanics, Non- Linear Integral and Differential equations and present research interests are semiconductor nano science and number theory respectively. His brief CV has been enlisted in many biographical references of USA and UK. After serving 18 years as a Professor in Department of Electronic Science of the University of Calcutta, from March 2012 he has joined as a Professor in the Department of Electronics and Communication Engineering of National Institute of Technology, Agartala, India and for more please log in to http://www.amazon.com/Kamakhya-Prasad-Ghatak/e/B003B09OEY.

Sitangshu Bhattacharya obtained his M.Sc. and Ph.D. degrees in the Department of Electronic Science of the University of Calcutta and Jadavpur University in 2003 and 2009 respectively. He is the co-author of more than 50 scientific research papers in electro-thermal transport phenomena in semiconductor nanostructures in international peer reviewed journals and the said six research monographs. His present research interest is in electro-thermal management in quantum effect devices and interconnects. Presently he is the Assistant Professor in the Department of Electrical and Communication Engineering in Indian Institute of Information Technology, Allahabad, Uttar Pradesh, India.

Symbols

α	Band nonparabolicity parameter
a	The lattice constant
a_0, b_0	The widths of the barrier and the well for super-lattice structures
A_0	The amplitude of the light wave
$\vec{A}$	The vector potential
$A(E, n_z)$	The area of the constant energy 2D wave vector space for ultrathin films
B	Quantizing magnetic field
B_2	The momentum matrix element
b	Bandwidth
c	Velocity of light
C_1	Conduction band deformation potential
C_2	A constant which describes the strain interaction between the conduction and valance bands
ΔC_{44}	Second order elastic constant
ΔC_{456}	Third order elastic constant
δ	Crystal field splitting constant
Δ_o	Interface width
$\Delta\left(\frac{1}{B}\right)$	Period of SdH oscillation
d_0	Super-lattice period
D	Diffusion constant
$\frac{D}{\mu}$	Einstein relation/Diffusivity-mobility ratio in semiconductors
$D_0(E)$	Density-of-states (DOS) function
$D_B(E)$	DOS function in magnetic quantization
$D_B(E, \lambda)$	DOS function under the presence of light waves
d_x, d_y, d_z	Nano thickness along the x, y and z-directions
$\Delta_{\parallel}$	Spin-orbit splitting constants parallel
$\Delta_{\perp}$	Spin-orbit splitting constants perpendicular to the C-axis
Δ	Isotropic spin-orbit splitting constant
d^3k	Differential volume of the k space
$\in$	Energy as measured from the center of the band gap
ε	Trace of the strain tensor

ε_0	Permittivity of free space
ε_∞	Semiconductor permittivity in the high frequency limit
ε_{sc}	Semiconductor permittivity
ΔE_g	Increased band gap
$\|e\|$	Magnitude of electron charge
E	Total energy of the carrier
E_0, ζ_0	Electric field
E_g	Band gap
E_i	Energy of the carrier in the ith band
E_{ki}	Kinetic energy of the carrier in the ith band
E_F	Fermi energy
$\bar{E}_{FB}$	Fermi energy in the presence of cross-fields configuration
$\bar{E}_{F0}$	Fermi energy in the electric quantum limit
$\bar{E}_0$	Energy of the electric sub-band in electric quantum limit
E_{FB}	Fermi energy in the presence of magnetic quantization
E_n	Landau sub band energy
E_{Fs}	Fermi energy in the presence of size quantization
E_{Fis}, E_{Fiw}	Fermi energy under the strong and weak electric field limit
$\bar{E}_{Fs}, \bar{E}_{Fw}$	Fermi energy in the Accumulation layer under the strong and weak electric field quantum limit
$\bar{E}_{0s}$ $\bar{E}_{0w}$	Sub band energy under the strong and weak electric field quantum limit
$\bar{E}_{Fn}$	Fermi energy for nipis
E_{FSL}	Fermi energy in super-lattices
$\vec{\varepsilon}_s$	Polarization vector
E_{FQWSL}	Fermi energy in quantum wire super-lattices with graded interfaces
E_{F_L}	Fermi energy in the presence of light waves
$E_{F_{BL}}$	Fermi energy under quantizing magnetic field in the presence of light waves
E_{F2DL}	2D Fermi energy in the presence of light waves
E_{F1DL}	1D Fermi energy in the presence of light waves
E_{g_0}	Un-perturbed band-gap
$Erfc$	Complementary error function
Erf	Error function
E_{F_h}	Fermi energy of HD materials
$\bar{E}_{hd}$	Electron energy within the band gap
F_s	Surface electric field
$F(V)$	Gaussian distribution of the impurity potential
$F_j(\eta)$	One parameter Fermi-Dirac integral of order j
f_0	Equilibrium Fermi-Dirac distribution function of the total carriers
f_{0i}	Equilibrium Fermi-Dirac distribution function of the carriers in the ith band

g_v	Valley degeneracy
G	Thermoelectric power under classically large magnetic field
G_0	Deformation potential constant
g^*	Magnitude of the band edge g-factor
h	Planck's constant
$\hat{H}$	Hamiltonian
$\hat{H}'$	Perturbed Hamiltonian
$H(E - E_n)$	Heaviside step function
$\hat{i}$, $\hat{j}$ and $\hat{k}$	Orthogonal triads
i	Imaginary unit
I	Light intensity
j_{ci}	Conduction current contributed by the carriers of the ith band
k	Magnitude of the wave vector of the carrier
k_B	Boltzmann's constant
λ	Wavelength of the light
$\bar{\lambda}_0$	Splitting of the two spin-states by the spin-orbit coupling and the crystalline field
$\bar{l}, \bar{m}, \bar{n}$	Matrix elements of the strain perturbation operator
L_x, L_z	Sample length along x and z directions
L_0	Super-lattices period length
L_D	Debye screening length
m_1	Effective carrier masses at the band-edge along x direction
m_2	Effective carrier masses at the band-edge along y direction
m_3	The effective carrier masses at the band-edge along z direction
m_2'	Effective- mass tensor component at the top of the valence band (for electrons) or at the bottom of the conduction band (for holes)
m_i^*	Effective mass of the ith charge carrier in the ith band
$m_{\parallel}^*$	Longitudinal effective electron masses at the edge of the conduction band
$m_{\perp}^*$	Transverse effective electron masses at the edge of the conduction band
m_c	Isotropic effective electron masses at the edge of the conduction band
$m_{\perp,1}^*, m_{\parallel,1}^*$	Transverse and longitudinal effective electron masses at the edge of the conduction band for the first material in super-lattice
m_r	Reduced mass
m_v	Effective mass of the heavy hole at the top of the valance band in the absence of any field
n	Landau quantum number
n_x, n_y, n_z	Size quantum numbers along the x, y and z-directions
n_{1D}, n_{2D}	1D and 2D carrier concentration

n_{2Ds}, n_{2Dw}	2D surface electron concentration under strong and weak electric field
$\bar{n}_{2Ds}, \bar{n}_{2Dw}$	Surface electron concentration under the strong and weak electric field quantum limit
n_i	Mini-band index for nipi structures
$N_{nipi}(E)$	DOS function for nipi structures
$N_{2DT}(E)$	2D DOS function
$N_{2D}(E, \lambda)$	2D DOS function in the presence of light waves
$N_{1D}(E, \lambda)$	1D DOS function in the presence of light waves
n_0	Total electron concentration
$\bar{n}_0$	Electron concentration in the electric quantum limit
n_i	Carrier concentration in the ith band
P	Isotropic momentum matrix element
P_n	Available noise power
$P_{\parallel}$	Momentum matrix elements parallel to the direction of crystal axis
$P_{\perp}$	Momentum matrix elements perpendicular to the direction of crystal axis
$\vec{r}$	Position vector
S_i	Zeros of the Airy function
$\vec{s}_0$	Momentum vector of the incident photon
t	Time scale
t_c	Tight binding parameter
T	Absolute temperature
$\tau_i(E)$	Relaxation time of the carriers in the ith band
$u_1(\vec{k}, \vec{r}), u_2(\vec{k}, \vec{r})$	Doubly degenerate wave functions
$V(E)$	Volume of k space
V_0	Potential barrier encountered by the electron
$V(\vec{r})$	Crystal potential
x, y	Alloy compositions
z_t	Classical turning point
μ_i	Mobility of the carriers in the ith band
μ	Average mobility of the carriers
$\zeta(2r)$	Zeta function of order $2r$
$\Gamma(j + 1)$	Complete Gamma function
η	Normalized Fermi energy
η_g	Impurity scattering potential
ω_0	Cyclotron resonance frequency
θ	Angle
μ_0	Bohr magnetron
ω	Angular frequency of light wave
$\uparrow', \downarrow'$	Spin up and down function

Chapter 1
The ER in Quantum Wells of HD Non-parabolic Semiconductors

1.1 Introduction

In recent years, with the advent of fine lithographical methods [1, 2] molecular beam epitaxy [3], organometallic vapor-phase epitaxy [4], and other experimental techniques, the restriction of the motion of the carriers of bulk materials in one (QWs, doping super-lattices, accumulation, and inversion layers), two (nanowires) and three (quantum dots, magneto-size quantized systems, magneto inversion layers, magneto accumulation layers, quantum dot super-lattices, magneto QW super-lattices, and magneto doping superlattices) dimensions have in the last few years, attracted much attention not only for their potential in uncovering new phenomena in nano-science but also for their interesting quantum device applications [5–8]. In QWs, the restriction of the motion of the carriers in the direction normal to the film (say, the z direction) may be viewed as carrier confinement in an infinitely deep 1D rectangular potential well, leading to quantization [known as quantum size effect (QSE)] of the wave vector of the carriers along the direction of the potential well, allowing 2D carrier transport parallel to the surface of the film representing new physical features not exhibited in bulk semiconductors [9–13]. The low-dimensional hetero-structures based on various materials are widely investigated because of the enhancement of carrier mobility [14].These properties make such structures suitable for applications in QWs lasers [15], hetero-junction FETs [16, 17], high-speed digital networks [18–21], high-frequency microwave circuits [22], optical modulators [23], optical switching systems [24], and other devices. The constant energy 3D wave-vector space of bulk semiconductors becomes 2D wave-vector surface in QWs due to dimensional quantization. Thus, the concept of reduction of symmetry of the wave-vector space and its consequence can unlock the physics of low-dimensional structures. In this chapter, we study the ER in QWs of HD non-parabolic semiconductors having different band structures in the presence of Gaussian band tails. At first we shall investigate the ER in QWs of HD nonlinear optical compounds which are being used in nonlinear optics and light emitting diodes [25]. The quasi-cubic model can be used to investigate the symmetric properties of both the bands at the zone center of wave vector space of the same

K. P. Ghatak and S. Bhattacharya, *Heavily-Doped 2D-Quantized Structures and the Einstein Relation*, Springer Tracts in Modern Physics 260,
DOI 10.1007/978-3-319-08380-3_1, © Springer International Publishing Switzerland 2015

compound. Including the anisotropic crystal potential in the Hamiltonian, and special features of the nonlinear optical compounds, Kildal [26] formulated the electron dispersion law under the assumptions of isotropic momentum matrix element and the isotropic spin-orbit splitting constant, respectively, although the anisotropies in the two aforementioned band constants are the significant physical features of the said materials [27–29]. In Sect. 1.2.1, the ER in QWs of HD nonlinear optical semiconductors has been investigated on the basis of newly formulated HD dispersion relation of the said compound by considering the combined influence of the anisotropies of the said energy band constants together with the inclusion of the crystal field splitting respectively within the framework of $\vec{k} \cdot \vec{p}$ formalism. The III-V compounds find applications in infrared detectors [30], quantum dot light emitting diodes [31], quantum cascade lasers [32], QWs wires [33], optoelectronic sensors [34], high electron mobility transistors [35], etc. The electron energy spectrum of III-V semiconductors can be described by the three- and two-band models of Kane [36–38], together with the models of Stillman et al. [39], Newson and Kurobe [40] and, Palik et al. [41] respectively. In this context it may be noted that the ternary and quaternary compounds enjoy the singular position in the entire spectrum of optoelectronic materials. The ternary alloy $Hg_{1\text{-}x}Cd_xTe$ is a classic narrow gap compound. The band gap of this ternary alloy can be varied to cover the spectral range from 0.8 to over 30 μm [42] by adjusting the alloy composition. $Hg_{1\text{-}x}Cd_xTe$ finds extensive applications in infrared detector materials and photovoltaic detector arrays in the 8–12 μm wave bands [43]. The above uses have generated the $Hg_{1\text{-}x}Cd_xTe$ technology for the experimental realization of high mobility single crystal with specially prepared surfaces. The same compound has emerged to be the optimum choice for illuminating the narrow sub-band physics because the relevant material constants can easily be experimentally measured [44]. Besides, the quaternary alloy $In_{1\text{-}x}Ga_xAs_yP_{1\text{-}y}$ lattice matched to InP, also finds wide use in the fabrication of avalanche photo-detectors [45], hetero-junction lasers [46], light emitting diodes [47] and avalanche photodiodes [48], field effect transistors, detectors, switches, modulators, solar cells, filters, and new types of integrated optical devices are made from the quaternary systems [49]. It may be noted that all types of band models as discussed for III-V semiconductors are also applicable for ternary and quaternary compounds. In Sect. 1.2.2, the ER in QWs of HD III-V, ternary and quaternary semiconductors has been studied in accordance with the corresponding HD formulation of the band structure and the simplified results for wide gap materials having parabolic energy bands under certain limiting conditions have further been demonstrated as a special case in the absence of heavy doping and thus confirming the compatibility test. The II-VI semiconductors are being used in nano-ribbons, blue green diode lasers, photosensitive thin films, infrared detectors, ultra-high-speed bipolar transistors, fiber optic communications, microwave devices, solar cells, semiconductor gamma-ray detector arrays, semiconductor detector gamma camera and allow for a greater density of data storage on optically addressed compact discs [50–57]. The carrier energy spectra in II-VI compounds are defined by the Hopfield model [58] where the splitting of the two-spin states by the spin-orbit coupling and the crystalline field has been taken into account. Section 1.2.3 contains the investigation of the ER in QWs of HD II-VI compounds.

Lead Chalcogenides (PbTe, PbSe, and PbS) are IV-VI non-parabolic semiconductors whose studies over several decades have been motivated by their importance in infrared IR detectors, lasers, light-emitting devices, photo-voltaic, and high temperature thermo-electrics [59–63]. PbTe, in particular, is the end compound of several ternary and quaternary high performance high temperature thermoelectric materials [64–68]. It has been used not only as bulk but also as films [69–72], QWs [73] super-lattices [74, 75] nanowires [76] and colloidal and embedded nano-crystals [77–80], and PbTe films doped with various impurities have also been investigated [81–88]. These studies revealed some of the interesting features that had been seen in bulk PbTe, such as Fermi level pinning and, in the case of superconductivity [89]. In Sect. 1.2.4, the 2D ER in QWs of HD IV-VI semiconductors has been studied taking PbTe, PbSe, and PbS as examples. The stressed semiconductors are being investigated for strained silicon transistors, quantum cascade lasers, semiconductor strain gages, thermal detectors, and strained-layer structures [90–93]. The ER in QWs of HD stressed compounds (taking stressed n-InSb as an example) has been investigated in Sect. 1.2.5. The vacuum deposited Tellurium (Te) has been used as the semiconductor layer in thin-film transistors (TFT) [94] which is being used in CO_2 laser detectors [95], electronic imaging, strain sensitive devices [96, 97], and multichannel Bragg cell [98]. Section 1.2.6 contains the investigation of ER in QWs of HD Tellurium. The n-Gallium Phosphide (n-GaP) is being used in quantum dot light emitting diode [99], high efficiency yellow solid state lamps, light sources, high peak current pulse for high gain tubes. The green and yellow light emitting diodes made of nitrogen-doped n-GaP possess a longer device life at high drive currents [100–102]. In Sect. 1.2.7, the ER in QWs of HD n-GaP has been studied. The Platinum Antimonide ($PtSb_2$) finds application in device miniaturization, colloidal nanoparticle synthesis, sensors and detector materials and thermo-photovoltaic devices [103–105]. Section 1.2.8 explores the ER in QWs of HD $PtSb_2$.Bismuth telluride (Bi_2Te_3) was first identified as a material for thermoelectric refrigeration in 1954 [106] and its physical properties were later improved by the addition of bismuth selenide and antimony telluride to form solid solutions. The alloys of Bi_2Te_3 are useful compounds for the thermoelectric industry and have been investigated in the literature [107–111]. In Sect. 1.2.9, the ER in QWs of HD Bi_2Te_3 has been considered. The usefulness of elemental semiconductor Germanium is already well known since the inception of transistor technology and, it is also being used in memory circuits, single photon detectors, single photon avalanche diode, ultrafast optical switch, THz lasers and THz spectrometers [112–115]. In Sect. 1.2.10, the ER has been studied in QWs of HD Ge. Gallium Antimonide (GaSb) finds applications in the fiber optic transmission window, hetero-junctions, and QWs. A complementary hetero-junction field effect transistor in which the channels for the p-FET device and the n-FET device forming the complementary FET are formed from GaSb. The band gap energy of GaSb makes it suitable for low power operation [116–121]. In Sect. 1.2.11, the ER in QWs of HD GaSb has been studied. Section 1.3 contains the result and discussions pertaining to this chapter. The last Sect. 1.4 contains open research problems.

1.2 Theoretical Background

1.2.1 The ER in QWs of HD Non-Linear Optical Semiconductors

The form of **k.p** matrix for nonlinear optical compounds can be expressed extending Bodnar [27] as

$$H = \begin{bmatrix} H_1 & H_2 \\ H_2^+ & H_1 \end{bmatrix} \tag{1.1}$$

where,

$$H_1 \equiv \begin{bmatrix} E_{g_0} & 0 & P_{\|}k_z & 0 \\ 0 & (-2\Delta_{\|}/3) & (\sqrt{2}\Delta_{\perp}/3) & 0 \\ P_{\|}k_z & (\sqrt{2}\Delta_{\perp}/3) & -(\delta+\frac{1}{3}\Delta_{\|}) & 0 \\ 0 & 0 & 0 & 0 \end{bmatrix}, \quad H_2 \equiv \begin{bmatrix} 0 & -f_{,+} & 0 & f_{,-} \\ f_{,+} & 0 & 0 & 0 \\ 0 & 0 & 0 & 0 \\ f_{,+} & 0 & 0 & 0 \end{bmatrix}$$

in which E_{g_0} is the band gap in the absence of any field, P and $P_{\perp}$ are the momentum matrix elements parallel and perpendicular to the direction of crystal axis respectively, δ is the crystal field splitting constant, $\Delta_{\|}$ and $\Delta_{\perp}$ are the spin-orbit splitting constants parallel and perpendicular to the C-axis respectively, $f_{,\pm} \equiv (P_{\perp}/\sqrt{2})(k_x \pm ik_y)$ and $i = \sqrt{-1}$. Thus, neglecting the contribution of the higher bands and the free electron term, the diagonalization of the above matrix leads to the dispersion relation of the conduction electrons in bulk specimens of nonlinear optical semiconductors as

$$\gamma(E) = f_1(E)k_s^2 + f_2(E)k_z^2 \tag{1.2}$$

where

$$\gamma(E) \equiv E(E+E_{g_0})\left[(E+E_{g_0})(E+E_{g_0}+\Delta_{\|}) + \delta(E+E_{g_0}+\frac{2}{3}\Delta_{\|}) + \frac{2}{9}(\Delta_{\|}^2 - \Delta_{\perp}^2)\right]$$

E is the total energy of the electron as measured from the edge of the conduction band in the vertically upward direction in the absence of any quantization, $k_s^2 = k_x^2 + k_y^2$,

$$f_1(E) \equiv \frac{\hbar^2 E_{g0}(E_{g0}+\Delta_\perp)}{[2m_\perp^*(E_{g0}+\frac{2}{3}\Delta_\perp)]}\left[\delta\left(E+E_{g0}+\frac{1}{3}\Delta_\parallel\right)+(E+E_{g0})\left(E+E_{g0}+\frac{2}{3}\Delta_\parallel\right)+\frac{1}{9}\left(\Delta_\parallel^2-\Delta_\parallel^2\right)\right],$$

$$f_2(E) \equiv \frac{\hbar^2 E_{g0}(E_{g0}+\Delta_\parallel)}{\left[2m_\parallel^*(E_{g0}+\frac{2}{3}\Delta_\parallel)\right]}\left[(E+E_{g0})\left(E+E_{g0}+\frac{2}{3}\Delta_\parallel\right)\right],\ \hbar = \mathrm{h}/2\pi,$$

h is Planck's constant and $m_\parallel^*$ and $m_\perp^*$ are the longitudinal and transverse effective electron masses at the edge of the conduction band respectively.

Thus the generalized unperturbed electron energy spectrum for the bulk specimens of the nonlinear optical materials in the absence of band tails can be expressed following (1.2) as

$$\frac{\hbar^2 k_z^2}{2m_\parallel^*}+\left(\frac{b_\parallel\, c_\perp}{b_\perp\, c_\parallel}\right)\frac{\hbar^2 k_s^2}{2m_\perp^*}=\left\{\frac{E(\alpha E+1)(b_\parallel E+1)}{(c_\parallel E+1)}+\frac{\alpha b_\parallel}{c_\parallel}\left[\delta E+\frac{2}{9}\left(\Delta_\parallel^2-\Delta_\perp^2\right)\right]-\left(\frac{2}{9}\right)\frac{\alpha b_\parallel}{c_\parallel}\frac{\left(\Delta_\parallel^2-\Delta_\perp^2\right)}{(c_\parallel E+1)}\right\}$$
$$-\left(\frac{\hbar^2 k_s^2}{2m_\perp^*}\right)\left\{\left(\frac{b_\parallel\, c_\perp}{b_\perp\, c_\parallel}\right)\left[\left(\frac{\delta}{2}+\frac{\Delta_\parallel^2-\Delta_\perp^2}{6\Delta_\parallel}\right)\frac{\alpha_\parallel}{\alpha_\parallel E+1}+\left(\frac{\delta}{2}-\left\{\frac{\Delta_\parallel^2-\Delta_\perp^2}{6\Delta_\parallel}\right\}\right)\frac{c_\parallel}{c_\parallel E+1}\right]\right\} \quad (1.3)$$

where, $b_\parallel \equiv 1/(E_g+\Delta_\parallel)$, $c_\perp \equiv 1/(E_g+\frac{2}{3}\Delta_\perp)$, $b_\perp \equiv 1/(E_g+\Delta_\perp)$, $c_\parallel \equiv 1/(E_g+\frac{2}{3}\Delta_\parallel)$ and $\alpha \equiv 1/E_g$

The Gaussian distribution *F(V)* of the impurity potential is given by [122, 123]

$$F(V) = \left(\pi\eta_g^2\right)^{-1/2}\exp\left(-V^2/\eta_g^2\right) \quad (1.4)$$

where, η_g is the impurity scattering potential. It appears from (1.4) that the variance parameter η_g is not equal to zero, but the mean value is zero. Further, the impurities are assumed to be uncorrelated and the band mixing effect has been neglected in this simplified theoretical formalism.

We have to average the kinetic energy in the order to obtain the E-k dispersion relation in nonlinear optical materials in the presence of band tails. Using the (1.3) and (1.4), we get

$$\left[\frac{\hbar^2 k_z^2}{2m_\parallel^*}\int_{-\infty}^{E}F(V)dV\right]+\left[\left(\frac{b_\parallel\, c_\perp}{b_\perp\, c_\parallel}\right)\frac{\hbar^2 k_s^2}{2m_\perp^*}\int_{-\infty}^{E}F(V)dV\right]=\left\{\int_{-\infty}^{E}\frac{(E-V)[\alpha(E-V)+1][b_\parallel(E-V)+1]}{c_\parallel(E-V)+1}F(V)dV\right.$$
$$+\frac{\alpha b_\parallel}{c_\parallel}\left[\delta\int_{-\infty}^{E}(E-V)F(V)dV+\frac{2}{9}\left(\Delta_\parallel^2-\Delta_\perp^2\right)\int_{-\infty}^{E}F(V)dV\right]-\left(\frac{2}{9}\right)\frac{\alpha b_\parallel}{c_\parallel}\left(\Delta_\parallel^2-\Delta_\perp^2\right)\left.\int_{-\infty}^{E}\frac{F(V)dV}{[c_\parallel(E-V)+1]}\right\}$$
$$-\left(\frac{\hbar^2 k_s^2}{2m_\perp^*}\right)\left\{\left(\frac{b_\parallel\, c_\perp}{b_\perp\, c_\parallel}\right)\left[\left(\frac{\delta}{2}+\frac{\Delta_\parallel^2-\Delta_\perp^2}{6\Delta_\parallel}\right)\alpha\int_{-\infty}^{E}\frac{F(V)dV}{\alpha(E-V)+1}+\left(\frac{\delta}{2}-\frac{\Delta_\parallel^2-\Delta_\perp^2}{6\Delta_\parallel}\right)c_\parallel\int_{-\infty}^{E}\frac{F(V)dV}{c_\parallel(E-V)+1}\right]\right\}$$
$$(1.5)$$

The (1.5) can be rewritten as [124–128]

$$\left(\frac{\hbar^2 k_z^2}{2m_\perp^*}\right) I(1) + \left(\frac{b_\parallel}{b_\perp}\frac{c_\perp}{c_\parallel}\right)\frac{\hbar^2 k_s^2}{2m_\perp^*} I(1) = \left\{ I_3(c_\parallel) + \frac{\alpha b_\parallel}{c_\parallel}\left[\delta I(4) + \frac{2}{9}\left(\Delta_\parallel^2 - \Delta_\perp^2\right) I(1)\right] - \left(\frac{2}{9}\right)\frac{\alpha b_\parallel}{c_\parallel}\left(\Delta_\parallel^2 - \Delta_\perp^2\right) I_6(c_\parallel)\right\}$$
$$- \frac{\hbar^2 k_s^2}{2m_\perp^*}\left\{\left(\frac{b_\parallel}{b_\perp}\frac{c_\perp}{c_\parallel}\right)\left[\left(\frac{\delta}{2} + \frac{\Delta_\parallel^2 - \Delta_\perp^2}{6\Delta_\parallel}\right)\alpha I(\alpha)\left(\frac{\delta}{2} - \left\{\frac{\Delta_\parallel^2 - \Delta_\perp^2}{6\Delta_\parallel}\right\}\right) c_\parallel I(c_\parallel)\right]\right\} \tag{1.6}$$

where,

$$I(1) \equiv \int_{-\infty}^{E} F(V)dV \tag{1.7}$$

$$I_3(c_\parallel) \equiv \int_{-\infty}^{E} \frac{(E-V)[\alpha(E-V)+1][b_\parallel(E-V)+1]}{[c_\parallel(E-V)+1]} F(V)dV \tag{1.8}$$

$$I(4) \equiv \int_{-\infty}^{E} (E-V)F(V)dV \tag{1.9}$$

$$I(\alpha) \equiv \int_{-\infty}^{E} \frac{F(V)dV}{[\alpha(E-V)+1]} \tag{1.10}$$

Substituting $E - V \equiv x$ and $x/\eta_g \equiv t_0$ we get from (1.7)

$$I(1) = \left(\exp(-E^2/\eta_g^2)/\sqrt{\pi}\right)\int_0^\infty \exp\left[-t_0^2 + (2Et_0/\eta_g)\right]dt_0$$

Thus,

$$I(1) = \left[\frac{1 + Erf(E/\eta_g)}{2}\right] \tag{1.11}$$

where, $Erf(E/\eta_g)$ is the error function of (E/η_g).

From (1.9), one can write

$$
\begin{aligned}
I(4) &= \left(1/\eta_g\sqrt{\pi}\right)\int_{-\infty}^{E}(E-V)\exp(-V^2/\eta_g^2)dV \\
&= \frac{E}{2}[1+Erf(E/\eta_g)] - \left\{\frac{1}{\sqrt{\pi\eta_g^2}}\int_{-\infty}^{E} V\exp(-V^2/\eta_g^2)dV\right\}
\end{aligned}
\tag{1.12}
$$

After computing this simple integration, one obtains
Thus,

$$
I(4) = \eta_g \exp(-E^2/\eta_g^2)\left(2\sqrt{\pi}\right)^{-1} + \frac{E}{2}(1+Erf(E/\eta_g)) = \gamma_0(E,\eta_g) \tag{1.13}
$$

From (1.10), we can write

$$
I(\alpha) = \frac{1}{\sqrt{\pi\eta_g^2}}\int_{-\infty}^{E}\frac{\exp\left(-V^2/\eta_g^2\right)dV}{[\alpha(E-V)+1]} \tag{1.14}
$$

When, $V \to \pm\infty$, $\frac{1}{[\alpha(E-V)+1]} \to 0$ and $\exp\left(-V^2/\eta_g^2\right) \to 0$
Thus (1.14) can be expressed as

$$
I(\alpha) = \left(1/\alpha\eta_g\sqrt{\pi}\right)\int_{-\infty}^{\infty}\exp(-t^2)(u-t)^{-1}dt \tag{1.15}
$$

where, $\frac{V}{\eta_g} \equiv t$ and $u \equiv \left(\frac{1+\alpha E}{\alpha\eta}\right)$.

It is well known that [129, 130]

$$
W(Z) = (i/\pi)\int_{-\infty}^{\infty}(Z-t)^{-1}\exp(-t^2)dt \tag{1.16}
$$

In which $i = \sqrt{-1}$ and Z, in general, is a complex number. We also know [129, 130],

$$
W(Z) = \exp(-Z^2)Erfc(-iZ) \tag{1.17}
$$

where, $Erfc(Z) \equiv 1 - Erf(Z)$.

Thus, $Erfc(-iu) = 1 - Erf(-iu)$
Since, $Erfc(-iu) = -Erf(iu)$
Therefore, $Erfc(-iu) = 1 + Erf(iu)$.

Thus,

$$I(\alpha) = \left[-i\sqrt{\pi}/\alpha\eta_g\right]\exp\left(-u^2\right)[1 + Erf(iu)] \tag{1.18}$$

We also know that [129, 130]

$$Erf(x+iy) = Erf(x) + \left(\frac{e^{-x^2}}{2\pi x}\right)\left[(1-\cos(2xy)) + i\sin(2xy) + \frac{2}{\pi}e^{-x^2}\sum_{p=1}^{\infty}\frac{\exp(-p^2/4)}{[p^2+4x^2]}\right]$$
$$\left[f_p(x,y) + ig_p(x,y) + \varepsilon(x,y)\right] \tag{1.19}$$

where,

$f_p(x,y) \equiv [2x - 2x\cosh(py)\cos(2xy) + p\sinh(py)\sin(2xy)]$,
$g_p(x,y) \equiv [2x\cosh(py)\sin(2xy) + p\sinh(py)\cos(2xy)]$, $|\varepsilon(x,y)| \approx 10^{-16}|Erf(x+iy)|$

Substituting $x = 0$ and $y = u$ in (1.19), one obtains,

$$Erf(iu) = \left(\frac{2i}{\pi}\right)\sum_{p=1}^{\infty}\left\{\frac{\exp(-p^2/4)}{p}\sinh(pu)\right\} \tag{1.20}$$

Therefore, one can write

$$I(\alpha) = C_{21}\left(\alpha, E, \eta_g\right) - iD_{21}\left(\alpha, E, \eta_g\right) \tag{1.21}$$

where,

$$c_{21}\left(\alpha, E, \eta_g\right) \equiv \left[\frac{2}{\alpha\eta_g\sqrt{\pi}}\right]\exp(-u^2)\left[\sum_{p=1}^{\infty}\left\{\frac{\exp(-p^2/4)}{p}\sinh(pu)\right\}\right]$$
$$\text{and } D_{21}\left(\alpha, E, \eta_g\right) \equiv \left[\frac{\sqrt{\pi}}{\alpha\eta_g}\exp\left(-u^2\right)\right].$$

The (1.21) consists of both real and imaginary parts and therefore, $I(\alpha)$ ***is complex, which can also be proved by using the method of analytic continuation of the subject Complex Analysis***.

The integral $I_3\left(c_{\|}\right)$ in (1.8) can be written as

$$I_3(c_{\|}) = \left(\frac{\alpha b_{\|}}{c_{\|}}\right)I(5) + \left(\frac{\alpha c_{\|} + b_{\|}c_{\|} - \alpha b_{\|}}{c_{\|}^2}\right)I(4) + \frac{1}{c_{\|}}\left(1 - \frac{\alpha}{c_{\|}}\right)\left(1 - \frac{b_{\|}}{c_{\|}}\right)I(1)$$
$$- \left\{\frac{1}{c_{\|}}\left(1 - \frac{\alpha}{c_{\|}}\right)\left(1 - \frac{b_{\|}}{c_{\|}}\right)I(c_{\|})\right\} \quad (1.22)$$

where

$$I(5) \equiv \int_{-\infty}^{E} (E - V)^2 F(V)dV \quad (1.23)$$

From (1.23) one can write

$$I(5) = \frac{1}{\sqrt{\pi\eta_g^2}}\left[E^2 \int_{-\infty}^{E} \exp\left(\frac{-V^2}{\eta_g^2}\right)dV - 2E \int_{-\infty}^{E} V\exp\left(\frac{-V^2}{\eta_g^2}\right)dV + \int_{-\infty}^{E} V^2 \exp\left(\frac{-V^2}{\eta_g^2}\right)dV\right]$$

The evaluations of the component integrals lead us to write

$$I(5) = \frac{\eta_g E}{2\sqrt{\pi}}\exp\left(\frac{-E^2}{\eta_g^2}\right) + \frac{1}{4}\left(\eta_g^2 + 2E^2\right)\left[1 + Erf\left(\frac{E}{\eta_g}\right)\right] = \theta_0(E, \eta_g) \quad (1.24)$$

Thus combining the aforementioned equations, $I_3(c_{\|})$ can be expressed as

$$I_3(c_{\|}) = A_{21}(E, \eta_g) + iB_{21}(E, \eta_g) \quad (1.25)$$

where,

$$A_{21}(E, \eta_g) \equiv \left[\frac{ab_{\|}}{c_{\|}}\left[\frac{\eta_g E}{2\sqrt{\pi}}\exp\left(\frac{-E^2}{\eta_g^2}\right) + \frac{1}{4}\left(\eta_g^2 + 2E^2\right)\left\{1 + Erf\left(\frac{E}{\eta_g}\right)\right\}\right]\right]$$
$$+ \left[\frac{\alpha c_{\|} + b_{\|}c_{\|} - \alpha b_{\|}}{c_{\|}^2}\right]\left\{\frac{E}{2}[1 + Erf(E/\eta_g)] + \frac{\eta_g \exp\left(-E^2/\eta_g^2\right)}{2\sqrt{\pi}}\right\}$$
$$+ \frac{1}{c_{\|}}\left(1 - \frac{\alpha}{c_{\|}}\right)\left(1 - \frac{b_{\|}}{c_{\|}}\right)\frac{1}{2}[1 + Erf(E/\eta_g)]$$
$$- \left\{\frac{2}{c_{\|}^2\eta_g\sqrt{\pi}}\left(1 - \frac{\alpha}{c_{\|}}\right)\left(1 - \frac{b_{\|}}{c_{\|}}\right)\exp\left(-u_1^2\right)\right\}\left[\sum_{p=1}^{\infty}\left\{\frac{\exp\left(-p^2/4\right)}{p} sinh(pu_1)\right\}\right],$$

$$u_1 \equiv \left[\frac{1 + c_{\|}E}{c_{\|}\eta_g}\right] \text{and } B_{21}(E, \eta_g) \equiv \frac{\sqrt{\pi}}{c_{\|}^2\eta_g}\left(1 - \frac{\alpha}{c_{\|}}\right)\left(1 - \frac{b_{\|}}{c_{\|}}\right)\exp\left(-u_1^2\right)\bigg].$$

Therefore, the combination of all the appropriate integrals together with algebraic manipulations leads to the expression of the dispersion relation of the conduction electrons of HD nonlinear optical materials forming Gaussian band tails as

$$\frac{\hbar^2 k_z^2}{2m_{\parallel}^* T_{21}(E,\eta_g)} + \frac{\hbar^2 k_s^2}{2m_{\perp}^* T_{22}(E,\eta_g)} = 1 \tag{1.26}$$

where, $T_{21}(E,\eta_g)$ and $T_{22}(E,\eta_g)$ have both real and complex parts and are given by

$$T_{21}(E,\eta_g) \equiv \left[T_{27}(E,\eta_g) + iT_{28}(E,\eta_g)\right],\ T_{27}(E,\eta_g) \equiv \left[\frac{T_{23}(E,\eta_g)}{T_5(E,\eta_g)}\right],$$

$$T_{23}(E,\eta_g) \equiv \left[A_{21}(E,\eta_g) + \frac{\alpha b_{\parallel}}{c_{\parallel}}\left[\delta\gamma_0(E,\eta_g) + \frac{1}{9}\left(\Delta_{\parallel}^2 - \Delta_{\perp}^2\right)\left[1 + Erf(E/\eta_g)\right]\right]\right.$$
$$\left. - \left\{\frac{2}{9}\left(\frac{\alpha b_{\parallel}}{c_{\parallel}}\right)\left(\Delta_{\parallel}^2 - \Delta_{\perp}^2\right)G_{21}(c_{\parallel}, E, \eta_g)\right\}\right],$$

$$G_{21}(E,\eta_g) \equiv \frac{2}{c_{\parallel}\eta_g\sqrt{\pi}}\exp(-u_1^2)\sum_{p=1}^{\infty}\left\{\frac{\exp(-p^2/4)}{p}sinh(pu_1)\right\},\ T_5(E,\eta_g) \equiv \frac{1}{2}\left[1 + Erf(E/\eta_g)\right],$$

$$T_{28}(E,\eta_g) \equiv \left[\frac{T_{24}(E,\eta_g)}{T_5(E,\eta_g)}\right],\ T_{24}(E,\eta_g) \equiv \left[B_{21}(E,\eta_g) + \frac{2}{9}\frac{ab_{\parallel}}{c_{\parallel}}\left(\Delta_{\parallel}^2 - \Delta_{\perp}^2\right)H_{21}(c_{\parallel}, E, \eta_g)\right],$$

$$H_{21}(c_{\parallel}, E,\eta_g) \equiv \left[\frac{\sqrt{\pi}}{\eta_g c_{\parallel}}\exp(-u_1^2)\right],\ T_{22}(E,\eta_g) \equiv \left[T_{29}(E,\eta_g) + iT_{30}(E,\eta_g)\right],$$

$$T_{29}(E,\eta_g) \equiv \frac{T_{23}(E,\eta_g)T_{25}(E,\eta_g) - T_{24}(E,\eta_g)T_{26}(E,\eta_g)}{\left[(T_{25}(E,\eta_g))^2 + (T_{26}(E,\eta_g))^2\right]}$$

$$T_{25}(E,\eta_g) \equiv \left[\left(\frac{b_{\parallel}}{b_{\perp}}\frac{c_{\perp}}{c_{\parallel}}\right)\frac{1}{2}\left[1 + Erf\left(\frac{E}{\eta_g}\right)\right] + \left(\frac{b_{\parallel}}{b_{\perp}}\frac{c_{\perp}}{c_{\parallel}}\right)\left(\frac{\delta}{2} + \left[\frac{\Delta_{\parallel}^2 - \Delta_{\perp}^2}{6\Delta_{\parallel}}\right]\right)\alpha_{\parallel}C_{21}(\alpha_{\parallel}, E,\eta_g)\right.$$
$$\left. + \left(\frac{b_{\parallel}c_{\perp}}{b_{\perp}}\right)\left(\frac{\delta}{2} + \left[\frac{\Delta_{\parallel}^2 - \Delta_{\perp}^2}{6\Delta_{\parallel}}\right]\right)G_{21}(\alpha_{\parallel}, E,\eta_g)\right],$$

$$C_{21}(\alpha, E,\eta_g) \equiv \left[\frac{2}{\alpha\sqrt{\pi}\eta_g}\exp(-u^2)\left[\sum_{p=1}^{\infty}\frac{\exp(-p^2/4)}{p}sinh(pu)\right]\right],$$

$$T_{26}(E,\eta_g) \equiv \left(\frac{b_{\parallel}}{b_{\perp}}\frac{c_{\perp}}{c_{\parallel}}\right)\left(\frac{\delta}{2} - \frac{\Delta_{\parallel}^2 - \Delta_{\perp}^2}{6\Delta_{\parallel}}\right)\alpha D_{21}(\alpha, E,\eta_g) + \frac{b_{\parallel}c_{\perp}}{b_{\perp}}\left(\frac{\delta}{2} - \frac{\Delta_{\parallel}^2 - \Delta_{\perp}^2}{6\Delta_{\parallel}}\right)H_{21}(c_{\parallel}, E,\eta_g)$$

$$\text{And } T_{30}(E,\eta_g) \equiv \frac{T_{24}(E,\eta_g)T_{25}(E,\eta_g) + T_{23}(E,\eta_g)T_{26}(E,\eta_g)}{\left[(T_{25}(E,\eta_g))^2 + (T_{26}(E,\eta_g))^2\right]}$$

From (1.26), it appears that the energy spectrum in HD nonlinear optical semiconductors is complex. **The complex nature of the electron dispersion law in HD semiconductors occurs from the existence of the essential poles in the corresponding electron energy spectrum in the absence of band tails.** It may be noted that the complex band structures have already been studied for bulk semiconductors and super lattices without heavy doping [131, 132] and bears no relationship with the complex electron dispersion law as indicated by (1.26). The physical picture behind the formulation of the complex energy spectrum in HDS is the interaction of the impurity atoms in the tails with the splitting constants of the valance bands. More is the interaction; more is the prominence of the complex part than the other case. In the absence of band tails, $\eta_g \to 0$, and there is no interaction of the impurity atoms in the tails with the spin orbit constants. As a result, there exist no complex energy spectrum and (1.26) gets converted into (1.2) when $\eta_g \to 0$. Besides, the complex spectra are not related to same evanescent modes in the band tails and the conduction bands.

It is interesting to note that the single important concept in the whole spectra of materials and allied sciences is the effective electron mass which is in disguise in the apparently simple (1.26), and can, briefly be described as follows:

Effective electron mass: The effective mass of the carriers in semiconductors, being connected with the mobility, is known to be one of the most important physical quantities, used for the analysis of electron devices under different operating conditions [133]. The carrier degeneracy in semiconductors influences the effective mass when it is energy dependent. Under degenerate conditions, only the electrons at the Fermi surface of n-type semiconductors participate in the conduction process and hence, the effective mass of the electrons corresponding to the Fermi level (EEM) would be of interest in electron transport under such conditions. The Fermi energy is again determined by the electron energy spectrum and the carrier statistics and therefore, these two features would determine the dependence of the effective electron mass in degenerate n-type semiconductors under the degree of carrier degeneracy. In recent years, various energy wave vector dispersion relations have been proposed [134–151] which have created the interest in studying the effective mass in such materials under external conditions. It has, therefore, different values in different materials and varies with electron concentration, with the magnitude of the reciprocal quantizing magnetic field under magnetic quantization, with the quantizing electric field as in inversion layers, with the nano-thickness as in UFs and nano wires and with superlattice period as in the quantum confined superlattices of small gap semiconductors with graded interfaces having various carrier energy spectra [152–184].

The transverse and the longitudinal EEMs at the Fermi energy (E_{F_h}) of HD nonlinear optical materials can, respectively, be expressed as

$$m_{\perp}^{*}(E_{F_h}, \eta_g) = m_{\perp}^{*}\{T_{29}(E, \eta_g)\}'\Big|_{E=E_{F_h}} \tag{1.27}$$

and

$$m_{\|}^{*}\left(E_{F_h},\eta_g\right)=m_{\|}^{*}\left\{T_{27}\left(E,\eta_g\right)\right\}'\Big|_{E=E_{F_h}} \tag{1.28}$$

where E_{F_h} is the Fermi energy of HDS in the presence of band tails as measured from the edge of the conduction band in the vertically upward direction in the absence of band tails and the primes denote the differentiations of the differentiable functions with respect to Fermi energy in the appropriate case.

In the absence of band tails $\eta_g \to 0$ and we get

$$m_{\perp}^{*}(E_F,0)=\frac{\hbar^2}{2}\left[\frac{\psi_2(E)\{\psi_1(E)\}'-\psi_1(E)\{\psi_2(E)\}'}{\{\psi_2(E)\}^2}\right]\Bigg|_{E=E_F} \tag{1.29}$$

and

$$m_{\|}^{*}(E_F,0)=\frac{\hbar^2}{2}\left[\frac{\psi_3(E)\{\psi_1(E)\}'-\{\psi_1(E)\}\{\psi_3(E)\}'}{\{\psi_3(E)\}^2}\right]\Bigg|_{E=E_F} \tag{1.30}$$

where E_F is the Fermi energy as measured from the edge of the conduction band in the vertically upward direction in the absence of any perturbation, $\psi_1(E)=\gamma(E)$, $\psi_2(E)=f_1(E)$ and $\psi_3(E)=f_2(E)$,

Comparing the aforementioned equations, one can infer that the **effective masses exist in the forbidden zone, which is impossible without the effect of band tailing. For semiconductors, in the absence of band tails the effective mass in the band gap is infinity** .

The DOS function is given by

$$N_{HD}\left(E,\eta_g\right)=\frac{2g_v m_{\perp}^{*}\sqrt{2m_{\|}^{*}}}{3\pi^2\hbar^3}R_{11}\left(E,\eta_g\right)\cos\left[\psi_{11}\left(E,\,\eta_g\right)\right] \tag{1.31a}$$

where,

$$R_{11}\left(E,\eta_g\right)\equiv\left[\left[\{T_{29}(E,\eta_g)\}'\sqrt{x(E,\eta_g)}+\frac{T_{29}(E,\eta_g)\{x(E,\eta_g)\}'}{2\sqrt{x(E,\eta_g)}}-\{T_{30}(E,\eta_g)\}'\sqrt{y(E,\eta_g)}-\frac{T_{30}(E,\eta_g)\{y(E,\eta_g)\}'}{2\sqrt{y(E,\eta_g)}}\right]^2\right.$$
$$\left.+\left[\{T_{29}(E,\eta_g)\}'\sqrt{y(E,\eta_g)}+\frac{T_{29}(E,\eta_g)\{y(E,\eta_g)\}'}{2\sqrt{y(E,\eta_g)}}+\{T_{30}(E,\eta_g)\}'\sqrt{x(E,\eta_g)}-\frac{T_{30}(E,\eta_g)\{x(E,\eta_g)\}'}{2\sqrt{x(E,\eta_g)}}\right]^2\right]^{1/2},$$

$$x(E,\eta_g)\equiv\frac{1}{2}\left[T_{27}(E,\eta_g)+\sqrt{\{T_{27}(E,\eta_g)\}^2+\{T_{28}(E,\eta_g)\}^2}\right],$$

$$y(E,\eta_g)\equiv\frac{1}{2}\left[\sqrt{\{T_{27}(E,\eta_g)\}^2+\{T_{28}(E,\eta_g)\}^2}-T_{27}(E,\eta_g)\right]$$

and

$$\psi_{11}(E,\eta_g) \equiv \tan^{-1}\left[\left[\{T_{29}(E,\eta_g)\}'\sqrt{y(E,\eta_g)} + \frac{T_{29}(E,\eta_g)}{2\sqrt{y(E,\eta_g)}} + \{T_{30}(E,\eta_g)\}'\sqrt{x(E,\eta_g)} + \frac{T_{30}\{x(E,\eta_g)\}'}{2\sqrt{x(E,\eta_g)}}\right]\right.$$
$$\left.\left[\{T_{29}(E,\eta_g)\}'\sqrt{x(E,\eta_g)} + \frac{T_{29}(E,\eta_g)\{x(E,\eta_g)\}'}{2\sqrt{x(E,\eta_g)}} - \{T_{30}(E,\eta_g)\}'\sqrt{y(E,\eta_g)} + \frac{T_{30}\{y(E,\eta_g)\}'}{2\sqrt{y(E,\eta_g)}}\right]^{-1}\right].$$

The oscillatory nature of the DOS for HD nonlinear optical materials is apparent from (1.31a). For,$\psi_{11}(E,\eta_g) \geq \pi$, the cosine function becomes negative leading to the negative values of the DOS. The electrons cannot exist for the negative values of the DOS and therefore, this region is forbidden for electrons, which indicates that in the band tail, **there appears a new forbidden zone in addition to the normal band gap of the semiconductor**.

The use of (1.31a) leads to the expression of the electron concentration as

$$n_0 = \frac{2g_v m_\perp^* \sqrt{2m_\parallel^*}}{3\pi^2\hbar^3}\left[I_{11}(E_{F_k},\eta_g) + \sum_{r=1}^{s} L(r)[I_{11}(E_{F_h},\eta_g)]\right] \tag{1.31b}$$

where, $I_{11}(E_{F_k},\eta_g) \equiv \left[T_{29}(E_{F_h},\eta_g)\sqrt{x(E_{F_h},\eta_g)} - T_{30}(E_{F_h},\eta_g)\sqrt{y(E_{F_h},\eta_g)}\right]$, $L(r) = 2(k_BT)^{2r}(1-2^{1-2r})\xi(2r)\frac{\partial^{2r}}{\partial(Fermi\ energy)^{2r}}$, r is the set of real positive integers whose upper s and $\xi(2r)$ is the Zeta function of order 2r [129, 130].

The ER for HD semiconductors is given by

$$\frac{D}{\mu} = \frac{n_0}{|e|}\text{Real part of }\left[\frac{\partial n_0}{\partial(E_{F_h} - \bar{E}_{hd})}\right]^{-1} \tag{1.31c}$$

where, $\bar{E}_{hd}$ is the electron energy within the band gap, as measured from $k = 0$ and should be obtained from the dispersion relation of the HD semiconductors under the conditions $E = \bar{E}_{hd}$ when $k = 0$.

For HD non-linear optical semiconductors, $\bar{E}_{hd}$ is the smallest negative root of the equation.

$$[T_{27}(\bar{E}_{hd},\eta_g)T_{29}(\bar{E}_{hd},\eta_g) - T_{28}(\bar{E}_{hd},\eta_g)T_{30}(\bar{E}_{hd},\eta_g)] = 0 \tag{1.31d}$$

Therefore, the ER can be numerically evaluated by using (1.31b), (1.31c), (1.31d) and the allied definitions.

For dimensional quantization along z- direction, the dispersion relation of the 2D electrons in this case can be written following (1.26) as

$$\frac{\hbar^2(\mathrm{n}_z\pi/\mathrm{d}_z)^2}{2m_\parallel^* T_{21}(E,\eta_g)} + \frac{\hbar^2 k_s^2}{2m_\perp^* T_{22}(E,\eta_g)} = 1 \tag{1.32}$$

where, $n_z(=1, 2, 3, \ldots)$ and d_z are the size quantum number and the nano-thickness along the z-direction respectively.

The general expression of the total 2D DOS $(N_{2DT}(E))$ can, in general, be expressed as

$$N_{2DT}(E) = \frac{2g_v}{(2\pi)^2} \sum_{n_z=1}^{n_{z\max}} \frac{\partial A(E, n_z)}{\partial E} H(E - E_{n_z}) \tag{1.33}$$

where, g_v is the valley degeneracy, $A(E, n_z)$ is the area of the constant energy 2D wave vector space and in this case it is for QWs, $H(E - E_{n_z})$ is the Heaviside step function and E_{n_z} is the corresponding sub-band energy. Using (1.32) and (1.33), the expression of the $N_{2DT}(E)$ for QWs of HD nonlinear optical semiconductors can be written as

$$N_{2DT}(E) = \frac{m_{\perp}^* g_v}{\pi \hbar^2} \sum_{n_z=1}^{n_{z\max}} T'_{1D}(E, \eta_g, \mathrm{n}_z) H(E - E_{n_z D1}) \tag{1.34}$$

where, $T_{1D}(E, \eta_g, \mathrm{n}_z) = \left[1 - \frac{\hbar^2 (n_z \pi/\mathrm{d}_z)^2}{2m_{\|}^* T_{21}(E, \eta_g)}\right] T_{22}(E, \eta_g)$ and the sub band energies $\mathrm{E}_{n_z D1}$ in this case is given by the following equation

$$\frac{\hbar^2 (n_z \pi/\mathrm{d}_z)^2}{2m_{\|}^* T_{21}(E_{n_z D1}, \eta_g)} = 1 \tag{1.35}$$

Thus we observe that both the total DOS and sub-band energies of QWs of HD nonlinear optical semiconductors are complex due to the presence of the pole in energy axis of the corresponding materials in the absence of band tails.

The EEM in this case is given by

$$m^*(E_{F1HD}, \eta_g, \mathrm{n}_z) = m_{\perp}^* [\text{Real part of } T'_{1D}(E_{F1HD}, \eta_g, \mathrm{n}_z)] \tag{1.36}$$

Thus, we observe that the EEM is the function of size quantum number and the Fermi energy due to the combined influence of the crystal filed splitting constant and the anisotropic spin-orbit splitting constants respectively. Besides it is a function of η_g due to which the EEM exists in the band gap, which is otherwise impossible.

Combining (1.34) with the Fermi-Dirac occupation probability factor, integrating between $E_{\eta_z D1}$ to infinity and applying the generalized Sommerfeld's lemma [185], the 2D carrier statistics in this case assumes the form

$$n_{2D} = \frac{m_{\perp}^* g_v}{\pi \hbar^2} \sum_{r=1}^{n_{z\max}} [\text{Real Part of } [T_{1D}(E_{F1HD}, \eta_g, \mathrm{n}_z) + T_{2D}(E_{F1HD}, \eta_g, \mathrm{n}_z)]] \tag{1.37}$$

where, $T_{2D}(E_{F1HD}, \eta_g, \mathrm{n}_z) = \sum_{r=1}^{s} L(r)[T_{1D}(E_{F1HD}, \eta_g, \mathrm{n}_z)]$, E_{F1HD} is the Fermi energy in the presence of size quantization of the QWs of HD non-linear optical materials as measured from the edge of the conduction band in the vertically upward direction in the absence of any perturbation.

The ER of QWs of HDS can, in general, be expressed as

$$\frac{D}{\mu} = \frac{n_{2D}}{|e|} \text{Real part of} \left[\frac{\partial n_{2D}}{\partial (E_{FHD} - E_{n_z D})} \right]^{-1} \tag{1.38}$$

where, E_{FHD} is the Fermi energy in HD sized quantized 2-D materials as measured from the edge of the conduction band in the vertically upward direction in the absence of any perturbation and $E_{n_z D}$ is the corresponding sub-band energy.

Therefore combining (1.37) and (1.38) we can study the ER in this case.

In the absence of heavy doping, the 2D dispersion relation the EEM in the x-y plane at the Fermi level, the total 2D DOS, the electron concentration and the ER for QWs of non-linear optical materials in the absence of band tails can, respectively, be written as

$$\psi_1(E) = \psi_2(E) k_s^2 + \psi_3(E)(n_z \pi / d_z)^2 \tag{1.39}$$

$$m^*(E_{Fs}, n_z) = \left(\frac{\hbar^2}{2}\right) [\psi_2(E_{Fs})]^{-2} \left[\{\psi_2(E_{Fs})\} \left\{ \{\psi_1(E_{Fs})\}' - \{\psi_3(E_{Fs})\}' \left(\frac{n_z \pi}{d_z}\right)^2 \right\} - \left\{ \psi_1(E_{Fs}) - \psi_3(E_{Fs}) \left(\frac{n_z \pi}{d_z}\right)^2 \right\} \{\psi_2(E_{Fs})\}' \right] \tag{1.40}$$

$$N_{2DT}(E) = \left(\frac{g_v}{2\pi}\right) \sum_{n_z=1}^{n_{z\max}} [\psi_2(E)]^2 \left[\psi_2(E) \left\{ \{\psi_1(E)\}' - \{\psi_3(E)\}' \left(\frac{n_z \pi}{d_z}\right)^2 \right\} - \left\{ \psi_1(E) - \psi_3(E) \left(\frac{n_z \pi}{d_z}\right)^2 \right\} \{\psi_2(E)\}' \right] H(E - E_{n_{z1}}) \tag{1.41}$$

$$\psi_1(E_{n_{z1}}) = \psi_2(E_{n_{z1}})(n_2 \pi / d_z)^2 \tag{1.42}$$

$$n_{2D} = \frac{g_v}{2\pi} \sum_{n_x=1}^{n_{x\max}} [T_{51}(E_{Fs}, n_z) + T_{52}(E_{Fs}, n_z)] \tag{1.43}$$

$$\frac{D}{\mu} = \frac{1}{|e|} \frac{\sum_{n_z=1}^{n_{z\max}} [T_{51}(E_{Fs}, n_z) + T_{52}(E_{Fs}, n_z)]}{\sum_{n_z=1}^{n_{z\max}} [\{T_{51}(E_{Fs}, n_z)\}' + \{T_{52}(E_{Fs}, n_z)\}']} \tag{1.44}$$

where, $\psi_1(E) = \gamma(E), \psi_2(E) = f_1(E), \psi_3(E) = f_2(E)$, $E_{n_{z1}}$ are the sub band energies, E_{Fs} is the Fermi energy in the 2-D sized quantized material in the presence of size quantization and in the absence of heavy doping as measured from the edge of the conduction band in the vertically upward direction in the absence of any quantization,

$$T_{51}(E_{Fs}, n_z) \equiv \left[\frac{\psi_1(E_{Fs}) - \psi_3(E_{Fs})(n_z\pi/d_z)^2}{\psi_2(E_{Fs})}\right] \text{ and}$$

$$T_{52}(E_{Fs}, n_z) \equiv \sum_{r=1}^{s} L(r)[T_{51}(E_{Fs}, n_z)].$$

It may be noted that the expanded form of (1.44) was derived for the first time by A.N. Chakravarti et al. [122, 123].

In the absence of heavy doping, the DOS for bulk specimens of non-linear optical semiconductors is given by

$$D_0(E) = g_v(3\pi^2)^{-1}\psi_4(E) \tag{1.45}$$

$$\psi_4(E) \equiv \left[\frac{3}{2}\frac{\sqrt{\psi_1(E)}[\psi_1(E)]'}{\psi_2(E)\sqrt{\psi_3(E)}} - \frac{[\psi_2(E)]'[\psi_1(E)]^{3/2}}{[\psi_2(E)]^2\sqrt{\psi_3(E)}} - \frac{1}{2}\frac{[\psi_3(E)]'[\psi_1(E)]^{3/2}}{\psi_2(E)[\psi_3(E)]^{3/2}}\right],$$

$$[\psi_1(E)]' \equiv \left[(2E+E_g)\psi_1(E)\left[E(E+E_g)\right]^{-1} + E(E+E_g)(2E+2E_g+\delta+\Delta_{\|})\right],$$

$$[\psi_2(E)]' \equiv \left[2m_\perp^*\left(E_g+\frac{2}{3}\Delta_\perp\right)\right]^{-1}\left[\hbar^2E_g(E_g+\Delta_\perp)\right]\left[\delta+2E+2E_g+\frac{2}{3}\Delta_{\|}\right]$$

$$\text{and } [\psi_3(E)]' \equiv \left[2m_{\|}^*\left(E_g+\frac{2}{3}\Delta_{\|}\right)\right]^{-1}\left[\hbar^2E_g(E_g+\Delta_{\|})\right]\left[2E+2E_g+\frac{2}{3}\Delta_{\|}\right]$$

Combining (1.45) with the Fermi-Dirac occupation probability factor and using the generalized Sommerfeld's lemma, the electron concentration can be written as

$$n_0 = g_v(3\pi^2)^{-1}[M(E_F) + N(E_F)] \tag{1.46a}$$

where, $M(E_F) \equiv \left[\frac{[\psi_1(E_F)]^{\frac{3}{2}}}{\psi_2(E_F)\sqrt{\psi_3(E_F)}}\right]$, E_F is the Fermi energy of the bulk specimen in the absence of band tails as measured from the edge of the conduction band in the vertically upward direction and

$$N(E_F) \equiv \sum_{r=1}^{s} L(r)M(E_F)$$

The ER in this case can, in general, be expressed as

$$\frac{D}{\mu} = \frac{n_0}{|e|}\left[\frac{\partial n_0}{\partial E_F}\right]^{-1} \tag{1.46b}$$

Thus using (1.46a) and (1.46b), the ER assumes the form

$$\frac{D}{\mu} = \frac{1}{|e|} \frac{[M(E_F) + N(E_F)]}{[\{M(E_F)\}' + \{N(E_F)\}']} \tag{1.47}$$

It may be noted that the expanded form of (1.47) was derived for the first time by A.N. Chakravarti et al. [152–154].

1.2.2 The ER in QWs of HD III-V Semiconductors

The dispersion relation of the conduction electrons of III-V compounds are described by the models of Kane (both three and two bands) [37, 38, 161–163], Stillman et al. [39] and Palik et al. [41] respectively. For the purpose of complete and coherent presentation and relative comparison, the ER in QWs of HD III-V semiconductors have also been investigated in accordance with the aforementioned different dispersion relations as follows:

(a) **The Three Band Model of Kane**

Under the conditions, $\delta = 0$, $\Delta_\| = \Delta_\perp = \Delta$ (isotropic spin orbit splitting constant) and $m_\|^* = m_\perp^* = m_c$ (isotropic effective electron mass at the edge of the conduction band), (1.2) gets simplified as

$$\frac{\hbar^2 k^2}{2m_c} = I_{11}(E),\ I_{11}(E) \equiv \frac{E(E + E_{g_0})(E + E_{g_0} + \Delta)(E_{g_0} + \frac{2}{3}\Delta)}{E_{g_0}(E_{g_0} + \Delta)(E + E_{g_0} + \frac{2}{3}\Delta)} \tag{1.48}$$

which is known as the three band model of Kane [37, 38] and is often used to investigate the physical properties of III-V materials.

Under the said conditions, the HD electron dispersion law in this case can be written from (1.26) as

$$\frac{\hbar^2 k^2}{2m_c} = T_{31}(E, \eta_g) + iT_{32}(E, \eta_g) \tag{1.49}$$

where,

$$T_{31}(E, \eta_g) \equiv \left(\frac{2}{1 + Erf(E/\eta_g)}\right)\left[\frac{\alpha b}{c}\theta_0(E, \eta_g) + \left[\frac{\alpha c + bc - \alpha b}{c^2}\right]\gamma_0(E, \eta_g) + \frac{1}{c}\left(1 - \frac{\alpha}{c}\right)\left(1 - \frac{b}{c}\right)\frac{1}{2}\left[1 + Erf\left(\frac{E}{\eta_g}\right)\right]\right.$$
$$\left. - \frac{1}{c}\left(1 - \frac{\alpha}{c}\right)\left(1 - \frac{b}{c}\right)\frac{2}{c\eta_g\sqrt{\pi}}\exp(-u_2^2)\left[\sum_{p=1}^{\infty}\frac{\exp(-p^2/4)}{p}\sinh(pu_2)\right]\right],\ b \equiv (E_g + \Delta)^{-1},\ c \equiv \left(E_g + \frac{2}{3}\Delta\right)^{-1},$$
$$u_2 \equiv \frac{1 + cE}{c\eta_g} \text{ and } T_{32}(E, \eta_g) \equiv \left(\frac{2}{1 + Erf(E/\eta_g)}\right)\frac{1}{c}\left(1 - \frac{\alpha}{c}\right)\left(1 - \frac{b}{c}\right)\frac{\sqrt{\pi}}{c\eta_g}\exp(-u_2^2).$$

Thus, the complex energy spectrum occurs due to the term $T_{32}(E, \eta_g)$ and this imaginary band is quite different from the forbidden energy band.

The EEM at the Fermi level is given by

$$m^*(E_{F_h}, \eta_g) = m_c\{T_{31}(E, \eta_g)\}'\Big|_{E=E_{F_h}} \tag{1.50}$$

Thus, the EEM in HD III-V, ternary and quaternary materials exists in the band gap, which is the new attribute of the theory of band tailing.

In the absence of band tails, $\eta_g \to 0$ and the EEM assumes the form

$$m^*(E_F) = m_c\{I_{11}(E)\}'|_{E=E_F} \tag{1.51}$$

The DOS function in this case can be written as

$$N_{HD}(E, \eta_g) = \frac{g_v}{3\pi^2}\left(\frac{2m_c}{\hbar^2}\right)^{3/2} R_{21}(E, \eta_g)\cos\left[\vartheta_{21}(E, \eta_g)\right] \tag{1.52}$$

where,

$$R_{21}(E, \eta_g) \equiv \left[\left[\frac{\left[\{\alpha_{11}(E, \eta_g)\}'\right]^2}{4\alpha_{11}(E, \eta_g)} + \frac{\left[\{\beta_{11}(E, \eta_g)\}'\right]^2}{4\beta_{11}(E, \eta_g)}\right]^2\right]^{1/2}$$

$$\alpha_{11}(E, \eta_g) \equiv \frac{1}{2}\left[T_{33}(E, \eta_g) + \sqrt{\{T_{33}(E, \eta_g)\}^2 + \{T_{34}(E, \eta_g)\}^2}\right],$$

$$T_{33}(E, \eta_g) \equiv \left[\{T_{31}(E, \eta_g)\}^3 - 3T_{31}(E, \eta_g)\{T_{32}(E, \eta_g)\}^2\right],$$

$$T_{34}(E, \eta_g) \equiv \left[3T_{32}(E, \eta_g)\{T_{31}(E, \eta_g)\}^2 - \{T_{32}(E, \eta_g)\}^3\right],$$

$$\beta_{11}(E, \eta_g) \equiv \frac{1}{2}\left[\sqrt{\{T_{33}(E, \eta_g)\}^2 + \{T_{34}(E, \eta_g)\}^2} - T_{33}(E, \eta_g)\right] \text{ and}$$

$$\vartheta_{21}(E, \eta_g) \equiv \tan^{-1}\left[\frac{\{\beta_{11}(E, \eta_g)\}'}{\{\alpha_{11}(E, \eta_g)\}'}\sqrt{\frac{\alpha_{11}(E, \eta_g)}{\beta_{11}(E, \eta_g)}}\right].$$

Thus, the oscillatory DOS function becomes negative for $\vartheta_{21}(E, \eta_g) \geq \pi$ and a new forbidden zone will appear in addition to the normal band gap.

The electron concentration can be expressed as

$$n_0 = \frac{g_v}{3\pi^2}\left(\frac{2m_c}{\hbar^2}\right)^{3/2}\left[\bar{I}_{111e}(E_{F_h}, \eta_g) + \sum_{r=1}^{s} L(r)[\bar{I}_{111e}(E_{F_h}, \eta_g)]\right] \tag{1.53}$$

where

$$\bar{I}_{111e}(E_{F_h}, \eta_g) = \{\gamma_2(E_{F_h}, \eta_g)\}^{3/2}$$

In this case, $\bar{E}_{hd}$ is given by

$$T_{31}\left(\bar{E}_{hd}, \eta_g\right) = 0 \tag{1.54}$$

The numerical evaluation of the ER has been done by using (1.53), (1.31c), (1.54) and the allied definitions.

For dimensional quantization along z-direction, the dispersion relation of the 2D electrons in this case can be written following (1.49) as

$$\frac{\hbar^2 (n_z \pi/d_z)^2}{2m_c} + \frac{\hbar^2 (k_s)^2}{2m_c} = T_{31}(E, \eta_g) + iT_{32}(E, \eta_g) \tag{1.55}$$

The expression of the $N_{2DT}(E)$ in this case assumes the form

$$N_{2DT}(E) = \frac{m_c g_v}{\pi \hbar^2} \sum_{n_z=1}^{n_{z\max}} T'_{5D}(E, \eta_g, n_z) H(E - E_{n_z D5}) \tag{1.56}$$

where, $T_{5D}(E, \eta_g, n_z) = \left[T_{31}(E, \eta_g) + iT_{32}(E, \eta_g) - \hbar^2 (n_z \pi/d_z)^2 (2m_c)^{-1}\right]$ and the sub band energies $E_{n_z D5}$ in this case given by

$$\{\hbar^2 (n_z/d_z)^2\}(2m_c)^{-1} = T_{31}(E_{n_z D5}, \eta_g) \tag{1.57}$$

Thus we observe that both the total DOS in QWs of HD III-V compounds and the sub band energies are complex due to the presence of the pole in energy axis of the corresponding materials in the absence of band tails.

The EEM in this case is given by

$$m^*\left(E_{F1HD}, \eta_g, n_z\right) = m_c\left[T'_{31}(E_{F1HD}, \eta_g, n_z)\right] \tag{1.58}$$

Therefore under the same conditions as used in obtaining (1.48) from (1.2), the 2D carrier statistics in this case can be written by using the same conditions from (1.37) as

$$n_{2D} = \frac{m_c g_v}{\pi \hbar^2} \sum_{n_z=1}^{n_{z\max}} [\text{Real part of } [T_{5D}\left(E_{F1HD}, \eta_g, n_z\right) + T_{6D}\left(E_{F1HD}, \eta_g, n_z\right)]] \tag{1.59}$$

where, $T_{6D}\left(E_{F1HD}, \eta_g, n_z\right) = \sum_{r=1}^{s} L(r)[T_{5D}\left(E_{F1HD}, \eta_g, n_z\right)]$,

Therefore combining (1.38) and (1.59) we can study the ER in this case.

In the absence of band tails, the 2D dispersion relation, EEM in the x-y plane at the Fermi level, the total 2D DOS, the sub-band energy, the electron concentration and the ER for QWs of III-V materials assume the following forms

$$\frac{\hbar^2 k_s^2}{2m_c} + \frac{\hbar^2}{2m_c}(n_z\pi/d_z)^2 = I_{11}(E) \tag{1.60}$$

$$m^*(E_{Fs}) = m\{I_{11}(E_{Fs})\}' \tag{1.61}$$

It is worth noting that the EEM in this case is a function of Fermi energy alone and is independent of size quantum number.

$$N_{2DT}(E) = \left(\frac{m_c g_v}{\pi\hbar^2}\right)\sum_{n_z=1}^{n_{z\max}}\left\{[I_{11}(E)]' H(E - E_{n_{z_2}})\right\} \tag{1.62}$$

where, the sub-band energies $E_{n_{z_2}}$ can be expressed as

$$I_{11}\left(E_{n_{z_2}}\right) = \frac{\hbar^2}{2m_c}(n_z\pi/d_z)^2 \tag{1.63}$$

$$n_{2D} = \frac{m_c g_v}{\pi\hbar^2}\sum_{n_z=1}^{n_{z\max}}[T_{53}(E_{Fs}, n_z) + T_{54}(E_{Fs}, n_z)] \tag{1.64}$$

$$\frac{D}{\mu} = \frac{1}{|e|}\frac{\sum_{n_z=1}^{n_{z\max}}[T_{53}(E_{Fs}, n_z) + T_{54}(E_{Fs}, n_z)]}{\sum_{n_z=1}^{n_{z\max}}[\{T_{53}(E_{Fs}, n_z)\}' + \{T_{54}(E_{Fs}, n_z)\}']} \tag{1.65}$$

where $T_{53}(E_{Fs}, n_z) \equiv \left[I_{11}(E_{Fs}) - \frac{\hbar^2}{2m_c}\left(\frac{n_z\pi}{d_z}\right)^2\right]$ and $T_{54}(E_{Fs}, n_z) \equiv \sum_{r=1}^{s} L(r) T_{53}(E_{Fs}, n_z)$.

The expanded form of (1.65) was formulated for the first time by Chakravarti et al. [157, 158].

In the absence of band tails, the DOS function, the electron concentration, and the ER in bulk III-V, ternary and quaternary materials in accordance with the unperturbed three band model of Kane assume the following forms

$$D_0(E) = 4\pi g_v\left(\frac{2m_c}{h^2}\right)^{3/2}\sqrt{I_{11}(E)}\,[I'_{11}(E)] \tag{1.66}$$

$$n_0 = \frac{g_v}{3\pi^2}\left(\frac{2m_c}{\hbar^2}\right)^{3/2}[M_1(E_F) + N_1(E_F)] \tag{1.67}$$

and

$$\frac{D}{\mu} = \frac{1}{e}[M_1(E_F) + N_1(E_F)]\left[\{M_1(E_F)\}' + \{N_1(E_F)\}'\right]^{-1} \tag{1.68}$$

where,

$$I'_{11}(E) \equiv I_{11}(E)\left[\frac{1}{E} + \frac{1}{E + E_{g_0}} + \frac{1}{E + E_{g_0} + \Delta} + \frac{1}{E + E_{g_0} + \frac{2}{3}\Delta}\right],$$

$$M_1(E_F) \equiv [I_{11}(E_F)]^{3/2}$$

and $N_1(E_F) \equiv \sum_{r=1}^{s} L(r)M_1(E_F)$

The expanded form of (1.68) was formulated for the first time by Chakravarti et al. [152–154].

Under the inequalities $\Delta \gg E_{g_0}$ or $\Delta \ll E_{g_0}$, (1.48) can be expressed as

$$E(1 + \alpha E) = \frac{\hbar^2 k^2}{2m_c} \tag{1.69}$$

where $\alpha \equiv \left(E_{g_0}\right)^{-1}$ and is known as band non-parabolicity.

It may be noted that (1.69) is the well-known two band model of Kane and is used in the literature to study the physical properties of those III-V and opto-electronic materials whose energy band structures obey the aforementioned inequalities.

The dispersion relation in HD III-V, ternary and quaternary materials whose energy spectrum in the absence of band tails obeys the two band model of Kane as defined by (1.69), can be written as

$$\frac{\hbar^2 k^2}{2m_c} = \gamma_2(E, \eta_g) \tag{1.70}$$

where, $\gamma_2(E, \eta_g) \equiv \left[\frac{2}{1 + Erf(E/\eta_g)}\right]\left[\gamma_0(E, \eta_g) + \alpha\theta_0(E, \eta_g)\right]$.

The EEM in this case can be written as

$$m^*\left(E_{F_h}, \eta_g\right) = m_c\{\gamma_2(E, \eta_g)\}'\Big|_{E=E_{F_h}} \tag{1.71}$$

Thus, one again observes that the EEM in this case exists in the band gap.

In the absence of band tails, $\eta_g \to 0$and the EEM assumes the well-known form

$$m^*(E_F) = m_c\{1 + 2\alpha E\}|_{E=E_F} \tag{1.72}$$

The DOS function in this case can be written as

$$N_{HD}(E,\eta_g) = \frac{g_v}{2\pi^2}\left(\frac{2m_c}{\hbar^2}\right)^{3/2}\sqrt{\gamma_2(E,\eta_g)}\{\gamma_2(E,\eta_g)\}' \tag{1.73}$$

Since, the original two band Kane model is an all zero and no pole function with respect to energy, therefore the HD counterpart will be totally real and the complex band vanishes.

The electron concentration at low temperatures is given by

$$n_0 = \frac{g_v}{3\pi^2}\left(\frac{2m_c}{\hbar^2}\right)^{3/2}\left[\bar{I}_{111}(E_{F_h},\eta_g) + \sum_{r=1}^{s} L(r)\left[\bar{I}_{111}(E_{F_h},\eta_g)\right]\right] \tag{1.74}$$

where,

$$\bar{I}_{111}(E_{F_h},\eta_g) = \{\gamma_2(E_{F_h},\eta_g)\}^{3/2}$$

In this case, $\bar{E}_{hd}$ is given by

$$\gamma_2(\bar{E}_{hd},\eta_g) = 0 \tag{1.75}$$

One can numerically compute the ER by using (1.74), (1.75), (1.31c) and the allied definitions in this case.

For dimensional quantization along z-direction, the dispersion relation of the 2D electrons in this case can be written following (1.70) as

$$\frac{\hbar^2(\mathrm{n}_z\pi/d_z)^2}{2m_c} + \frac{\hbar^2(k_s)^2}{2m_c} = \gamma_2(E,\eta_g) \tag{1.76}$$

the expression of the $N_{2DT}(E)$ in this case can be written as

$$N_{2DT}(E) = \frac{m_c g_v}{\pi\hbar^2}\sum_{n_z=1}^{n_{z\max}} T'_{7D}(E,\ \eta_g,\mathrm{n}_z)H(\mathrm{E}-\mathrm{E}_{n_zD7}) \tag{1.77}$$

where, $T_{7D}(E,\ \eta_g,\mathrm{n}_z) = [\gamma_2(E,\eta_g) - \hbar^2(n_z\pi/\mathrm{d}_z)^2(2m_c)^{-1}]$,

The sub-band energies E_{n_zD7} in this case given by

$$\left\{\hbar^2(\mathrm{n}_z\pi/d_z)^2\right\}(2m_c)^{-1} = \gamma_2(\mathrm{E}_{n_zD7},\eta_g) \tag{1.78}$$

Thus, we observe that both the total DOS and sub-band energies of QWs of HD III-V compounds in accordance with two band model of Kane are not at all complex since the dispersion relation in accordance with the said model is an all zero function with no pole in the finite complex plane.

The EEM in this case is given by

$$m^*(E_{F1HD}, \eta_g, \mathrm{n}_z) = m_c[\gamma_2'(E_{F1HD}, \eta_g, \mathrm{n}_z)] \tag{1.79}$$

Therefore under the same conditions as used in obtaining (1.48) from (1.2), the 2D carrier statistics in this case can be written by using the same conditions from (1.77) as

$$n_{2D} = \frac{m_c g_v}{\pi\hbar^2} \sum_{n_z=1}^{n_{z\max}} \left[T_{7D}(E_{F1HD}, \eta_g, \mathrm{n}_z) + T_{8D}(E_{F1HD}, \eta_g, \mathrm{n}_z)\right] \tag{1.80}$$

where, $T_{8D}(E_{F1HD}, \eta_g, \mathrm{n}_z) = \sum_{r=1}^{s} L(r)\left[T_{7D}(E_{F1HD}, \eta_g, \mathrm{n}_z)\right]$,

Therefore, combining (1.38) and (1.80) we can get the ER in this case.
Under the inequalities $\Delta \gg E_{g_0}$ or $\Delta \ll E_{g_0}$, (1.60) assumes the form

$$E(1 + \alpha E) = \frac{\hbar^2 k_s^2}{2m_c} + \frac{\hbar^2}{2m_c}\left(\frac{n_z\pi}{d_z}\right)^2 \tag{1.81a}$$

The EEM can be written from (1.81a) as

$$m^*(E_{F_S}) = m_c(1 + 2\alpha E_{F_S}) \tag{1.81b}$$

The total 2D DOS function assumes the form

$$N_{2DT}(E) = \frac{m_c g_v}{\pi\hbar^2} \sum_{n_z=1}^{n_{z\max}} (1 + 2\alpha E) H(E - E_{n_{z3}}) \tag{1.82}$$

where, the sub-band energy ($E_{n_{z3}}$) can be expressed as

$$\frac{\hbar^2}{2m_c}(n_z\pi/d_z)^2 = E_{n_{z3}}(1 + \alpha E_{n_{z3}}) \tag{1.83}$$

The 2D electron statistics can be written as

$$\begin{aligned} n_{2D} &= \frac{m_c g_v}{\pi\hbar^2} \sum_{n_z=1}^{n_{z\max}} \int_{E_{n_{z3}}}^{\infty} \frac{(1 + 2\alpha E)dE}{1 + \exp\left(\frac{E - E_{F_S}}{k_B T}\right)} \\ &= \frac{m_c k_B T g_v}{\pi\hbar^2} \sum_{n_z=1}^{n_{z\max}} \left[(1 + 2\alpha E_{n_{z3}}) F_0(\eta_{n_1}) + 2\alpha k_B T F_1(\eta_{n_1})\right] \end{aligned} \tag{1.84}$$

where, $\eta_{n_1} \equiv (E_{Fs} - E_{n_{x3}})/k_B T$ and $F_j(\eta)$ is the one parameter Fermi-Dirac integral of order j which can be written [186] as

$$F_j(\eta) = \left(\frac{1}{\Gamma(j+1)}\right)\int\limits_0^{\infty}\frac{x^j\,dx}{1+\exp(x-\eta)}, \qquad j > -1 \tag{1.85}$$

or for all j, analytically continued as a complex contour integral around the negative x-axis

$$F_j(\eta) = \left(\frac{\Gamma(-j)}{2\pi\sqrt{-1}}\right)\int\limits_{-\infty}^{+0}\frac{x^j\,dx}{1+\exp(-x-\eta)} \tag{1.86}$$

where η is the dimensionless parameter and x is independent variable,

Therefore in this case the ER can be written as

$$\frac{D}{\mu} = \frac{k_BT}{|e|}\frac{\sum_{n_z=1}^{n_{z\max}}\left[\left(1+2\alpha E_{n_{z3}}\right)F_0\left(\eta_{n_1}\right)+2\alpha k_BTF_1\left(\eta_{n_1}\right)\right]}{\sum_{n_z=1}^{n_{z\max}}\left[\left(1+2\alpha E_{n_{z3}}\right)F_{-1}\left(\eta_{n_1}\right)+2\alpha k_BTF_0\left(\eta_{n_1}\right)\right]} \tag{1.87}$$

The (1.87) was formulated for the first time by Chakravarti et al. [157, 158].

The forms of the DOS, the electron statistics and the ER for bulk specimens of III-V materials in the absence of band tails whose energy band structures are defined by the two-band model of Kane can, respectively, be written as

$$D_0(E) = 4\pi g_v\left(\frac{2m_c}{\hbar^2}\right)^{3/2}\sqrt{I_{11e}(E)}\left[I'_{11e}(E)\right] \tag{1.88}$$

$$n_0 = \frac{g_v}{3\pi^2}\left(\frac{2m_c}{\hbar^2}\right)^{3/2}\left[M_2(E_F)+N_2(E_F)\right] \tag{1.89}$$

and

$$\frac{D}{\mu} = \frac{1}{|e|}\left[M_2(E_F)+N_2(E_F)\right]\left[\{M_2(E_F)\}'+\{N_2(E_F)\}'\right]^{-1} \tag{1.90}$$

where, $I_{11e}(E) \equiv E(1+\alpha E)$, $I'_{11e}(E) \equiv (1+2\alpha E)$, $M_2(E_F) \equiv \left[I_{11e}(E_F)\right]^{3/2}$ and $N_2(E_F) \equiv \sum\limits_{r=1}^{s} L(r)M_2(E_F)$

(c) Under the constraints $\Delta \gg E_{g_0}$ or $\Delta \ll E_{g_0}$ together with the inequality $\alpha E_F \ll 1$, the (1.89) and (1.90) assumes the forms as

$$n_0 = g_vN_c\left[F_{1/2}(\eta)+\left(\frac{15\alpha k_BT}{4}\right)F_{3/2}(\eta)\right] \tag{1.91}$$

and

$$\frac{D}{\mu} = \left[\frac{k_B T}{|e|}\right] \left[\frac{F_{1/2}(\eta) + \left(\frac{15\alpha k_B T}{4}\right) F_{3/2}(\eta)}{F_{-1/2}(\eta) + \left(\frac{15\alpha k_B T}{4}\right) F_{1/2}(\eta)}\right] \tag{1.92}$$

where, $N_c \equiv 2\left(\frac{2\pi m_c k_B T}{h^2}\right)^{3/2}$ and $\eta \equiv \frac{E_F}{k_B T}$

It may be noted that (1.91) and (1.92) were derived for the first time by Nag and Chakravarti [187, 188].

The dispersion relation in HDS whose energy spectrum in the absence of band tails obeys the parabolic energy bands is given by

$$\frac{\hbar^2 k^2}{2m_c} = \gamma_3(E, \eta_g) \tag{1.93}$$

where, $\gamma_3(E, \eta_g) \equiv \left[\frac{2}{(1+Erf(E/\eta_g))}\right] \gamma_0(E, \eta_g)$.

Since the dispersion relation in accordance with the said model is an all zero function with no pole in the finite complex plane, therefore the HD counterpart will be totally real, which is also apparent form the expression (1.93).

The EEM in this case can be written as

$$m^*(E_{F_h}, \eta_g) = m_c\{\gamma_3(E_{F_h}, \eta_g)\}' \tag{1.94}$$

In the absence of band tails, $n_g \to 0$ and the EEM assumes the form

$$m^*(E_F) = m_c \tag{1.95}$$

It is well-known that the EEM in unperturbed parabolic energy bands is a constant quantity in general excluding cross-fields configuration. However, the same mass in the corresponding HD bulk counterpart becomes a complicated function of Fermi energy and the impurity potential together with the fact that the EEM also exists in the band gap solely due to the presence of finite η_g.

The DOS function in this case can be written as

$$N_{HD}(E, \eta_g) = \frac{g_v}{2\pi^2}\left(\frac{2m_c}{\hbar^2}\right)^{3/2} \sqrt{\gamma_3(E, \eta_g)}\{\gamma_3(E, \eta_g)\}' \tag{1.96}$$

The electron concentration is given by

$$n_0 = \frac{g_v}{3\pi^2}\left(\frac{2m_c}{\hbar^2}\right)^{3/2}\left[\bar{I}_{113}(E_{F_h}, \eta_g) + \sum_{r=1}^{s} L(r)[\bar{I}_{113}(E_{F_h}, \eta_g)]\right] \tag{1.97}$$

where

$$\bar{I}_{113}(E_{F_h},\eta_g)=\{\gamma_3(E_{F_h},\eta_g)\}^{3/2}$$

In this case, $\bar{E}_{hd}$ is given by

$$\gamma_3(\bar{E}_{hd},\eta_g)=0 \tag{1.98}$$

One can numerically compute the ER by using (1.97), (1.98), (1.31c) and the allied definitions in this case.

For dimensional quantization along z-direction, the dispersion relation of the 2D electrons in this case can be written following (1.93) as

$$\frac{\hbar^2(\mathrm{n}_z\pi/d_z)^2}{2m_c}+\frac{\hbar^2(k_s)^2}{2m_c}=\gamma_3(E,\eta_g) \tag{1.99}$$

the expression of the $N_{2DT}(E)$ in this case can be written as

$$N_{2DT}(E)=\frac{m_c g_v}{\pi\hbar^2}\sum_{n_z=1}^{n_{z\max}}T'_{9D}(E,\,\eta_g,\mathrm{n}_z)H(E-E_{n_zD9}) \tag{1.100}$$

where, $T_{9D}(E,\eta_g,\mathrm{n}_z)=[\gamma_3(E,\eta_g)-\hbar^2(\mathrm{n}_z\pi/\mathrm{d}_z)^2(2m_c)^{-1}]$.

The sub band energies E_{n_zD9} in this case given by

$$\{\hbar(n_z\pi/d_z)^2\}(2m_c)^{-1}=\gamma_3(E_{n_zD9},\eta_g) \tag{1.101}$$

The EEM in this case can be written as

$$m^*(E_{F1HD},\eta_g,n_z)=m_c[\gamma'_3(E_{F1HD},\eta_g)] \tag{1.102}$$

Therefore under the same conditions as used in obtaining (1.48) from (1.2), the 2D carrier statistics in this case can be written by using the same conditions from (1.77) as

$$n_{2D}=\frac{m_c g_v}{\pi\hbar^2}\sum_{n_z=1}^{n_{z\max}}\left[T_{9D}(E_{F1HD},\eta_g,\mathrm{n}_z)+T_{10D}(E_{F1HD},\eta_g,\mathrm{n}_z)\right] \tag{1.103}$$

where, $T_{10D}(E_{F1HD},\eta_g,\mathrm{n}_z)=\sum_{r=1}^{s}L(r)\left[T_{9D}(E_{F1HD},\eta_g,\mathrm{n}_z)\right]$,

Therefore combining (1.38) and (1.80) we can get the ER in this case.

Under the condition $\alpha\to 0$, the expressions of total 2D DOS, for semiconductors without forming band tails whose bulk electrons are defined by the isotropic parabolic energy bands can, be written from (1.82) as

$$N_{2DT}(E) = \frac{m_c g_v}{\pi\hbar^2} \sum_{n_z=1}^{n_{z\max}} H(E - E_{n_{zp}}) \tag{1.104}$$

The sub-band energy $\left(E_{n_{zp}}\right)$, the n_{2D} and the ER can, respectively, be expressed as

$$E_{n_{zp}} = \frac{\hbar^2}{2m_c}\left(\frac{n_z\pi}{d_z}\right)^2 \tag{1.105}$$

$$n_{2D} = \frac{m_c k_B T g_v}{\pi\hbar^2} \sum_{n_z=1}^{n_{z\max}} F_0(\eta_{n_2}) \tag{1.106a}$$

$$\frac{D}{\mu} = \frac{k_B T}{|e|}\left(\sum_{n_z=1}^{n_{z\max}} F_0(\eta_{n_2})\right)\left(\sum_{n_z=1}^{n_{z\max}} F_{-1}(\eta_{n_2})\right)^{-1}, \text{ where,}$$
$$\eta_{n_2} \equiv \frac{1}{k_B T}\left[E_{Fs} - \frac{\hbar^2}{2m_c}\left(\frac{n_z\pi}{d_z}\right)^2\right] \tag{1.106b}$$

It may be noted that (1.106b) was derived for the first time by Chakravarti et al. [37, 38].

Converting the summation over n_z to the integration over n_z, (1.106b) gets transformed to the well-known relation as [37, 38]

$$\frac{D}{\mu} = \frac{k_B T}{|e|}\left[F_{1/2}(\eta)/F_{-1/2}(\eta)\right] \tag{1.106c}$$

This indirect test not only exhibits the mathematical compatibility of our formulation but also shows the fact that our simple analysis is a more generalized one, since one can obtain the corresponding results for relatively wide gap 2D materials having parabolic energy bands under certain limiting conditions from our present derivation.

(b) **The Model of Stillman et al.**

In accordance with the model of Stillman et al. [39], The dispersion relation of III-V materials assumes the form

$$E = \bar{t}_{11}k^2 - \bar{t}_{12}k^4 \tag{1.107}$$

where, $\bar{t}_{11} \equiv \frac{\hbar^2}{2m_c}$; $\bar{t}_{12} \equiv \left(1 - \frac{m_c}{m_0}\right)^2\left(\frac{\hbar^2}{2m_c}\right)^2\left[\left(3E_{g_0} + 4\Delta + \frac{2\Delta^2}{E_{g0}}\right)\left\{\left(E_{g_0} + \Delta\right)(2\Delta + 3E_{g_0})\right\}^{-1}\right]$ and m_0 is the free electron mass

In the presence of band tails, (1.107) gets transformed as

$$\frac{\hbar^2 k^2}{2m_c} = I_{12}(E, \eta_g) \quad (1.108)$$

where, $I_{12}(E, \eta_g) = a_{11}[1 - (1 - a_{12}\gamma_3(E, \eta_g))^{\frac{1}{2}}]$, $a_{11} \equiv \left(\frac{\hbar^2 \bar{t}_{11}}{4m_c \bar{t}_{12}}\right)$ *and* $a_{12} \equiv \frac{4\bar{t}_{12}}{\bar{t}_{11}{}^2}$

The EEM can be written as

$$m^*(E_{F_h}, \eta_g) = m_c\{I_{12}(E_{F_h}, \eta_g)\}' \quad (1.109)$$

The DOS function in this case can be written as

$$N_{HD}(E, \eta_g) = \frac{g_v}{2\pi^2}\left(\frac{2m_c}{\hbar^2}\right)^{3/2} \sqrt{I_{12}(E, \eta_g)}\{I_{12}(E, \eta_g)\}' \quad (1.110)$$

The electron concentration is given by

$$n_0 = \frac{g_v}{3\pi^2}\left(\frac{2m_c}{\hbar^2}\right)^{3/2}\left[\bar{I}_{121}(E_{F_h}, \eta_g) + \sum_{r=1}^{s} L(r)[\bar{I}_{121}(E_{F_h}, \eta_g)]\right] \quad (1.111)$$

where

$$\bar{I}_{121}(E_{F_h}, \eta_g) = \{I_{12}(E_{F_h}, \eta_g)\}^{3/2}$$

In this case, $\bar{E}_{hd}$ is expressed through the equation

$$\gamma_3(\bar{E}_{hd}, \eta_g) = 0 \quad (1.112)$$

One can numerically compute the ER by using (1.111), (1.112), (1.31c) and the allied definitions in this case.

For dimensional quantization along z-direction, the dispersion relation of the 2D electrons in this case can be written following (1.108) as

$$\frac{\hbar^2(n_z\pi/d_z)^2}{2m_c} + \frac{\hbar^2(k_s)^2}{2m_c} = I_{12}(E, \eta_g) \quad (1.113)$$

the expression of the $N_{2DT}(E)$ in this case can be written as

$$N_{2DT}(E) = \frac{m_c g_v}{\pi\hbar^2}\sum_{n_z=1}^{n_{z\max}} T'_{11\,D}(E, \eta_g, n_z)H(E - E_{n_z D11}) \quad (1.114)$$

where, $T_{11\,D}(E, \eta_g, n_z) = [I_{12}(E, \eta_g) - \hbar^2(n_z\pi/d_z)^2(2m_c)^{-1}]$,

The sub band energies E_{n_zD11} in this case given by

$$\{(\hbar n_z\pi/d_z)^2\}(2m_c)^{-1} = I_{12}(E_{n_zD11}, \eta_g) \tag{1.115}$$

The EEM in this case assumes the form

$$m^*(E_{F1HD}, \eta_g, n_z) = m_c[I'_{12}(E_{F1HD}, \eta_g, \mathrm{n}_z)] \tag{1.116}$$

The 2-D electron statistics in this case can be written as

$$n_{2D} = \frac{m_c g_v}{\pi\hbar^2}\sum_{n_z=1}^{n_{z\max}}\left[T_{11D}\left(E_{F1HD}, \eta_g, \mathrm{n}_z\right) + T_{12D}\left(E_{F1HD}, \eta_g, \mathrm{n}_z\right)\right] \tag{1.117}$$

where, $T_{12D}\left(E_{F1HD}, \eta_g, \mathrm{n}_z\right) = \sum_{r=1}^{s} L(r)\left[T_{11D}\left(E_{F1HD}, \eta_g, \mathrm{n}_z\right)\right]$,

Therefore combining (1.117) and (1.38) we can get the ER in this case.

For unperturbed material, the 2-D EEM can be expressed as

$$m^*(E_{Fs}) = m_c\{I_{12}(E_{Fs})\}' \tag{1.118}$$

where $I_{12}(E) \equiv a_{11}[1 - (1 - a_{12}(E))^{\frac{1}{2}}]$

It appears that the EEM in this case is a function of Fermi energy alone and is independent of size quantum number.

The total 2D DOS function in the absence of band tails in this case can be written as

$$N_{2DT}(E) = \left(\frac{m_c g_v}{\pi\hbar^2}\right)\sum_{n_z=1}^{n_{z\max}}\left\{[I_{12}(E)]' H(E - E_{n_{z3}})\right\} \tag{1.119}$$

where, the sub band energies $E_{n_{z3}}$ can be expressed as

$$I_{12}\left(E_{n_{z3}}\right) = \frac{\hbar^2}{2m_c}(n_z\pi/d_z)^2 \tag{1.120}$$

The 2D electron concentration assumes the form

$$n_{2D} = \frac{m_c g_v}{\pi\hbar^2}\sum_{n_z=1}^{n_{z\max}}[T_{55}(E_{Fs}, n_z) + T_{56}(E_{Fs}, n_z)] \tag{1.121}$$

where $T_{55}(E_{Fs}, n_z) \equiv \left[I_{12}(E_{Fs}) - \frac{\hbar^2}{2m_c}\left(\frac{n_z\pi}{d_z}\right)^2\right]$ and $T_{56}(E_{Fs}, n_z) \equiv \sum_{r=1}^{s} L(r)T_{55}(E_{Fs}, n_z)$

Using (1.121), the ER in this case is given by

$$\frac{D}{\mu} = \frac{1}{|e|} \frac{\sum_{n_z=1}^{n_{z\max}} [T_{55}(E_{Fs}, n_z) + T_{56}(E_{Fs}, n_z)]}{\sum_{n_z=1}^{n_{z\max}} [\{T_{55}(E_{Fs}, n_z)\}' + \{T_{56}(E_{Fs}, n_z)\}']} \quad (1.122a)$$

The expression of electron concentration for bulk specimens of III-V semiconductors (in the absence of band tails) can be written in accordance with the model of Stillman et al. as

$$n_0 = \frac{g_v}{3\pi^2} \left(\frac{2m_c}{\hbar^2} \right)^{3/2} [M_{A_{10}}(E_F) + N_{A_{10}}(E_F)] \quad (1.122b)$$

where, $M_{A_{10}}(E_F) = [I_{12}(E_F)]^{3/2}$ and $N_{A_{10}}(E_F) = \sum_{r=1}^{s} L(r)[(M_{A_{10}} E_F)]$

The ER in this case can be expressed as

$$\frac{D}{\mu} = \frac{1}{e} [M'_{A_{10}}(E_F) + N'_{A_{10}}(E_F)]^{-1} [M_{A_{10}}(E_F) + N_{A_{10}}(E_F)] \quad (1.122c)$$

(c) **Model of Palik et al.**

The energy spectrum of the conduction electrons in III-V semiconductors up to the fourth order in effective mass theory, taking into account the interactions of heavy hole, light hole and the split-off holes can be expressed in accordance with the model of Palik et al. [41] as

$$E = \frac{\hbar^2 k^2}{2m_c} - \bar{B}_{11} k^4 \quad (1.123)$$

where $\bar{B}_{11} = \left[\frac{\hbar^4}{4E_{g_0}(m_c)^2} \right] \left[\frac{1 + \frac{x_{11}^2}{2}}{1 + \frac{x_{11}}{2}} \right] (1 - y_{11})^2$, $x_{11} = \left[1 + \left(\frac{\Delta}{E_{g_0}} \right) \right]^{-1}$ *and* $y_{11} = \frac{m_c}{m_o}$

The (1.123) gets simplified as

$$\frac{\hbar^2 k^2}{2m_c} = I_{13}(E) \quad (1.124)$$

where $I_{13}(E) = \bar{b}_{12} \left[\bar{a}_{12} - \left((\bar{a}_{12})^2 - 4E\bar{B}_{11} \right)^{1/2} \right]$, $\bar{a}_{12} = \left(\frac{\hbar^2}{2m_c} \right)$ *and* $\bar{b}_{12} \left[\frac{\bar{a}_{12}}{2\bar{B}_{11}} \right]$

Under the condition of heavy doping forming Gaussian band tails, (1.124) assumes the form

$$\frac{\hbar^2 k^2}{2m_c} = I_{13}(E, \eta_g) \quad (1.125)$$

where, $I_{13}(E,\eta_g) = \bar{b}_{12}[\bar{a}_{12} - ((\bar{a}_{12})^2 - 4\bar{B}_{11}\gamma_3(E,\eta_g))^{1/2}]$

The EEM can be written as

$$m^*\left(E_{F_h},\eta_g\right) = m_c\left\{I_{13}\left(E_{F_h,\eta_g}\right)\right\}' \tag{1.126}$$

The DOS function in this case can be expressed as

$$N_{HD}(E,\eta_g) = \frac{g_v}{2\pi^2}\left(\frac{2m_c}{\hbar^2}\right)^{3/2}\sqrt{I_{13}(E,\eta_g)}\{I_{13}(E,\eta_g)\}' \tag{1.127}$$

Since, the original band model in this case is a no pole function, in the finite complex plane therefore, the HD counterpart will be totally real and the complex band vanishes.

The electron concentration is given by

$$n_0 = \frac{g_v}{3\pi^2}\left(\frac{2m_c}{\hbar^2}\right)^{3/2}\left[\bar{I}_{123}\left(E_{F_h},\eta_g\right) + \sum_{r=1}^{s} L(r)[\bar{I}_{123}(E_{F_h},\eta_g)]\right] \tag{1.128}$$

where,

$$\bar{I}_{123}(E_{F_h},\eta_g) = \{I_{13}(E_{F_h},\eta_g)\}^{3/2}$$

In this case, $\bar{E}_{hd}$ is given by

$$\gamma_3\left(\bar{E}_{hd},\eta_g\right) = 0 \tag{1.129}$$

One can numerically compute the ER by using (1.128), (1.129), (1.31c) and the allied definitions in this case.

For dimensional quantization along z-direction, the dispersion relation of the 2D electrons in this case can be written following (1.108) as

$$\frac{\hbar^2(\mathrm{n}_z\pi/d_z)^2}{2m_c} + \frac{\hbar^2(k_s)^2}{2m_c} = I_{13}(E,\eta_g) \tag{1.130}$$

the expression of the $N_{2DT}(E)$ in this case can be written as

$$N_{2DT}(E) = \frac{m_c g_v}{\pi\hbar^2}\sum_{n_z}^{n_{zmax}} T'_{13D}(E,\eta_g,n_z)H(E - E_{n_zD13}) \tag{1.131}$$

where, $T_{13D}(E,\ \eta_g,\mathrm{n}_z) = [I_{13}(E,\eta_g) - \hbar^2(n_z\pi/\mathrm{d}_z)^2(2m_c)^{-1}]$,

The sub band energies E_{n_zD13} in this case given by

$$\{\hbar(n_z\pi/d_z)^2\}(2m_c)^{-1} = I_{13}(E_{n_zD13}, \eta_g) \tag{1.132}$$

The EEM in this case can be expressed as

$$m^*(E_{F1HD}, \eta_g, n_z) = m_c[I'_{13}(E_{F1HD}, \eta_g, n_z)] \tag{1.133}$$

The 2-D electron statistics in this case can be written as

$$n_{2D} = \frac{m_c g_v}{\pi\hbar^2}\sum_{n_z=1}^{n_{z\max}}\left[T_{13D}(E_{F1HD}, \eta_g, \mathrm{n}_z) + T_{14D}(E_{F1HD}, \eta_g, \mathrm{n}_z)\right] \tag{1.134}$$

where, $T_{14D}(E_{F1HD}, \eta_g, \mathrm{n}_z) = \sum_{r=1}^{s} L(r)\left[T_{13D}(E_{F1HD}, \eta_g, \mathrm{n}_z)\right]$,

Therefore combining (1.134) and (1.38) we can get the ER in this case.

The 2D electron dispersion relation in the absence of band tails this case assumes the form

$$\frac{\hbar^2 k_s^2}{2m_c} + \frac{\hbar^2}{2m_c}(n_z\pi/d_z)^2 = I_{13}(E) \tag{1.135a}$$

The EEM in this case can be written from (1.135a) as

$$m^*(E_{F_S}) = m_c[I_{13}(E_{F_S})]' \tag{1.135b}$$

The total 2D DOS function can be written as

$$N_{2DT}(E) = \left(\frac{m_c g_v}{\pi\hbar^2}\right)\sum_{n_z=1}^{n_{z\max}}\{[I_{13}(E)]' H(E - E_{n_{z4}})\} \tag{1.136}$$

where, the sub band energies$E_{n_{z4}}$ can be expressed as

$$I_{13}\left(E_{n_{z4}}\right) = \frac{\hbar^2}{2m_c}(n_z\pi/d_z)^2 \tag{1.137}$$

The 2D electron concentration assumes the form

$$n_{2D} = \frac{m_c g_v}{\pi\hbar^2}\sum_{n_z=1}^{n_{z\max}}[T_{57}(E_{Fs}, n_z) + T_{58}(E_{Fs}, n_z)] \tag{1.138}$$

where $T_{57}(E_{Fs}, n_z) \equiv \left[I_{13}(E_{Fs}) - \frac{\hbar^2}{2m_c}\left(\frac{n_z\pi}{d_z}\right)^2\right]$ and $T_{58}(E_{Fs}, n_z) \equiv \sum_{r=1}^{s} L(r) T_{57}(E_{Fs}, n_z)$

Using the appropriate equations, the ER in this case can be written as

$$\frac{D}{\mu}=\frac{1}{|e|}\frac{\sum\limits_{n_z=1}^{n_{z\max}}[T_{37}(E_{Fs},n_z)+T_{38}(E_{Fs},n_z)]}{\sum\limits_{n_z=1}^{n_{z\max}}[\{T_{37}(E_{Fs},n_z)\}'+\{T_{38}(E_{Fs},n_z)\}']} \quad (1.139)$$

1.2.3 The ER in QWs of HD II-VI Semiconductors

The carrier energy spectra in bulk specimens of II-VI compounds in accordance with Hopfield model [58] can be written as

$$E=a_o'k_s^2+b_o'k_z^2\pm\bar{\lambda}_o k_s \quad (1.140)$$

where $a_o'\equiv\hbar^2/2m_\perp^*$, $b_o'\equiv\hbar^2/2m_\parallel^*$, and $\bar{\lambda}_o$ represents the splitting of the two-spin states by the spin orbit coupling and the crystalline field.

Therefore the dispersion relation of the carriers in HD II-VI materials in the presence of Gaussian band tails can be expressed as

$$\gamma_3(E,\eta_g)=a_o'k_s^2+b_o'k_z^2\pm\bar{\lambda}_o k_s \quad (1.141)$$

Thus, the energy spectrum in this case is real since the corresponding E-k relation in the absence of band tails as given by (1.141) is a no pole function in the finite complex plane.

The transverse and the longitudinal EEMs masses are, respectively, given by

$$m_\perp^*(E_{F_h},\eta_g)=m_\perp^*\{\gamma_3(E,\eta_g)\}'\left[1\pm\left(\frac{\bar{\lambda}_o}{\sqrt{(\bar{\lambda}_o)+4a_o'\gamma_3(E,\eta_g)}}\right)\right]\Bigg|_{E=E_{F_h}} \quad (1.142)$$

and

$$m_\parallel^*(E_{F_h},\eta_g)=m_\parallel^*\{\gamma_3(E,\eta_g)\}'\Big|_{E=E_{F_h}} \quad (1.143)$$

Thus the transverse EEM in HD II-VI semiconductors is a function of electron energy and is double valued due to the presence of $\bar{\lambda}_o$ and due to heavy doping the same mass exists in the band gap.

In the absence of band tails, $n_g\to 0$, we get

$$m_{\perp}^{*}(E_{F_h}) = m_{\perp}^{*}\left[1 \pm \left(\frac{\bar{\lambda}_o}{\sqrt{(\bar{\lambda}_o)^2 + 4a_o' E}}\right)\right]\Bigg|_{E=E_F} \tag{1.144}$$

and

$$m_{\parallel}^{*}(E_F) = m_{\parallel}^{*} \tag{1.145}$$

The volume in k-space as enclosed (1.141) can be expressed as

$$V(E,\eta_g) = \frac{4\pi}{3a_0'\sqrt{b_0'}}\left[\{\gamma_3(E,\eta_g)\}^{3/2} + \frac{3}{8}\frac{(\bar{\lambda}_0)^2\sqrt{\gamma_3(E,\eta_g)}}{a_0'} \pm \left(\frac{3}{4}\frac{\bar{\lambda}_0}{\sqrt{a_0'}}\right)\sqrt{\gamma_3(E,\eta_g) + \frac{(\bar{\lambda}_0)^2}{4a_0'}}\sin^{-1}\left[\frac{\sqrt{\gamma_3(E,\eta_g)}}{\gamma_3(E,\eta_g) + \frac{(\bar{\lambda}_0)^2}{4a_0'}}\right]\right] \tag{1.146}$$

Therefore, the electron concentration can be written as

$$n_0 = \frac{g_v}{3\pi^2 a_0'\sqrt{b_0'}}\left[\bar{I}_{124}(E_{F_h},\eta_g) + \sum_{r=1}^{s} L(r)\left[\bar{I}_{124}(E_{F_h},\eta_g)\right]\right] \tag{1.147}$$

where,

$$\bar{I}_{124}(E_{F_h},\eta_g) = \left[\{\gamma_3(E_{F_k},\eta_g)\}^{3/2} + \frac{3}{8}\frac{(\bar{\lambda}_0)^2\sqrt{\gamma_3(E_{F_k},\eta_g)}}{a_0'}\right]$$

In this case, $\bar{E}_{hd}$ is given by

$$\{\gamma_3(\bar{E}_{hd},\eta_g)\} = 0 \tag{1.148}$$

Thus, one can numerically evaluate the ER by using (1.147), (1.31c), (1.148) and the allied definitions in this case.

The dispersion relation of the conduction electrons of QWs of HD II-VI materials for dimensional quantization along z-direction can be written following (1.141) as

$$\gamma_3(E,\eta_g) = a_0' k_s^2 + b_0'\left(\frac{\pi n_z}{d_z}\right)^2 \pm \bar{\lambda}_0 k_s \tag{1.149}$$

The EEM can be expressed following (1.149) as

$$m^*(E_{F1HD}, n_z, \eta_g) = m_\perp^*[1 \mp \frac{(\bar{\lambda}_0)\gamma_3'(E_{F1HD}, \eta_g)}{[(\bar{\lambda}_0)^2 - 4a_0'b_0'\left(\frac{n_z\pi}{d_z}\right)^2 + 4a_0'\gamma_3(E_{F1HD}, \eta_g)]^{1/2}} \tag{1.150}$$

Thus we observe that the doubled valued effective mass in 2-D QWs of HD II-VI materials is a function of Fermi energy, size quantum number and the screening potential respectively together with the fact that the same mass exists in the band gap due to the sole presence of the splitting of the two-spin states by the spin orbit coupling and the crystalline field.

The sub-band energy in this case is given by

$$\gamma_3(E_{n_zD14}, \eta_g) = b_0'\left(\frac{\pi n_z}{d_z}\right)^2 \tag{1.151}$$

The surface electron concentration at low temperatures assumes the form

$$n_{2D} = \frac{g_v m_\perp^*}{\pi\hbar^2} \sum_{n_z=1}^{n_{z\max}} \left(\gamma_3(E_{F1HD}, \eta_g) - E_{n_zD14} + (\bar{\lambda}_0)^2 m_\perp^* \hbar^{-2}\right) \tag{1.152}$$

Therefore combining (1.152) and (1.38) we can get the ER in this case.

The dispersion relation of the conduction electrons of QWs of II-VI materials for dimensional quantization along z-direction in the absence of band tails can be written following (1.140) as

$$E = a_0'k_s^2 + b_0'\left(\frac{n_z k}{d_z}\right)^2 \pm \bar{\lambda}_0 k_s \tag{1.153}$$

Using (1.153), the EEM in this case can be written as

$$m^*(E_{Fs}, n_z) = m_\perp^*\left[1 \mp \frac{(\bar{\lambda}_0)}{\left[(\bar{\lambda}_0)^2 - 4a_0'b_0'\left(\frac{n_z\pi}{d_z}\right)^2 + 4a_0'E_{Fs}\right]^{1/2}}\right] \tag{1.154}$$

The sub-band energy $E_{n_{z4}}$ assumes the form

$$E_{n_{z4}} = b_0'(n_z\pi/d_z)^2 \tag{1.155}$$

The area of constant energy 2D quantized surface in this case is given by where

$$A_\pm(E, n_z) = \left[\frac{\pi}{2(a_0')^2}\left[(\bar{\lambda}_0)^2 + 2a_0'\left(E - E_{n_{z5}}\right) \pm \bar{\lambda}_0\left[(\bar{\lambda}_0)^2 + 4a_0'\left(E - E_{n_{z5}}\right)\right]^{\frac{1}{2}}\right]\right]$$

The surface electron concentration can be expressed in this case as

$$n_{2D} = \frac{-2g_v}{2(2\pi)^2} \sum_{n_z=1}^{n_z \max} \int_{E_{n_{z5}}}^{\infty} [A_+(E_{Fs}, n_z) + A_-(E_{Fs}, n_z)] \frac{\partial}{\partial E} \{f_0(E)\} dE \quad (1.156)$$

where $f_0(E)$ is the Fermi-Dirac occupation probability factor.

From (1.156) we get

$$n_{2D} = \frac{g_v m_\perp^* k_B T}{\pi \hbar^2} \sum_{n_z=1}^{n_z \max} F_0(\eta_{n_{z8}}) \quad (1.157)$$

where $\eta_{n_{z8}} = \left(E_{Fs} - E_{F_{z5}} + (\bar{\lambda}_0)^2 m_\perp^* \hbar^{-2} \right) (k_B T)^{-1}$

Therefore the ER is given by

$$\frac{D}{\mu} = \frac{k_B T}{e} \left(\sum_{n_z=1}^{n_z \max} F_0\left(\eta_{n_{z8}}\right) \right) \left(\sum_{n_z=1}^{n_z \max} F_{-1}\left(\eta_{n_{z8}}\right) \right)^{-1} \quad (1.158)$$

1.2.4 The ER from QWs of HD IV-VI Semiconductors

The dispersion relation of the conduction electrons in IV-VI semiconductors can be expressed in accordance with Dimmock [189] as

$$\left[\bar{\varepsilon} - \frac{E_{g_0}}{2} - \frac{\hbar^2 k_s^2}{2m_t^-} - \frac{\hbar^2 k_s^2}{2m_l^-} \right] \left[\bar{\varepsilon} + \frac{E_{g_0}}{2} + \frac{\hbar^2 k_s^2}{2m_t^-} + \frac{\hbar^2 k_s^2}{2m_l^-} \right] = P_\perp^2 k_s^2 + P_\parallel^2 k_z^2 \quad (1.159)$$

where, $\bar{\varepsilon}$ is the energy as measured from the center of the band gap E_{g_0}, $m_t^\pm$ and $m_l^\pm$ represent the contributions to the transverse and longitudinal effective masses of the external L_6^+ and L_6^- bands arising from the $\vec{k}.\vec{p}$ perturbations with the other bands taken to the second order.

Substituting, $P_\perp^2 \equiv (\hbar^2 E_g / 2m_t^*)$, $P_\parallel^2 \equiv \left(\frac{\hbar^2 E_g}{2m_l^*} \right)$ and $\bar{\varepsilon} \equiv [E + (\frac{E_g}{2})]$ (where, m_t^* and m_l^* are the transverse and the longitudinal effective masses at k = 0), (1.159) gets transformed as

$$\left[E - \frac{\hbar^2 k_s^2}{2m_t^-} - \frac{\hbar^2 k_z^2}{2m_l^-} \right] \left[1 + \alpha E + \alpha \frac{\hbar^2 k_s^2}{2m_t^+} + \alpha \frac{\hbar^2 k_z^2}{2m_l^+} \right] = \frac{\hbar^2 k_s^2}{2m_t^*} + \frac{\hbar^2 k_z^2}{2m_l^*} \quad (1.160)$$

From (1.160), we can write

$$\frac{\alpha\hbar^4 k_s^4}{4m_t^+ m_t^-} + \hbar^2 k_s^2 \left[\left(\frac{1}{2m_t^*} - \frac{1}{2m_t^-} \right) + \alpha E \left(\frac{1}{2m_t^-} - \frac{1}{2m_t^+} \right) + \frac{\alpha\hbar^2 k_z^2}{4m_l^- m_t^+} \right] + \left[\left(\frac{\hbar^2 k_z^2}{2m_l^*} + \frac{\hbar^2 k_z^2}{2m_l^-} \right) + \frac{\alpha E}{2} \hbar^2 k_z^2 \left(\frac{1}{m_l^-} - \frac{1}{m_l^+} \right) + \frac{\alpha\hbar^4 k_s^4}{4m_l^+ m_l^-} - E(1+\alpha E) \right] = 0 \tag{1.161}$$

Using (1.161), the dispersion relation of the conduction electrons in HD IV-VI materials can be expressed as

$$\frac{\alpha\hbar^4 k_s^4}{4m_t^+ m_t^-} Z_0(E,\eta_g) + \hbar^2 k_s^2 [\lambda_{71}(E,\eta_g) k_z^2 + \lambda_{72}(E,\eta_g)] + [\lambda_{73}(E,\eta_g) k_z^2 + \lambda_{74}(E,\eta_g) k_z^2 - \lambda_{75}(E,\eta_g)] = 0 \tag{1.162}$$

where, $Z_0(E,\eta_g) \equiv \frac{1}{2}\left[1 + Erf\left(\frac{E}{\eta_g}\right)\right]$, $\lambda_{70}(E,\eta_g) \equiv \frac{\alpha}{4m_t^+ m_t^-} Z_0(E,\eta_g)$

$$\lambda_{71}(E,\eta_g) \equiv \left[\frac{\alpha\hbar^2}{4m_t^- m_l^+} Z_0(E,\eta_g) + \frac{\alpha\hbar^2}{4m_l^- m_t^+} Z_0(E,\eta_g) \right],$$

$$\lambda_{72}(E,\eta_g) \equiv \left[\left(\frac{1}{2m_t^*} - \frac{1}{2m_t^-} \right) Z_0(E,\eta_g) + \alpha \left(\frac{1}{2m_t^-} - \frac{1}{2m_t^+} \right) \gamma_0(E,\eta_g) \right],$$

$$\lambda_{73}(E,\eta_g) \equiv \left[\left(\frac{\hbar^2}{2m_l^*} + \frac{\hbar^2}{2m_l^-} \right) Z_0(E,\eta_g) + \frac{\alpha\hbar^2}{2} \left(\frac{1}{m_l^-} - \frac{1}{2m_l^+} \right) \gamma_0(E,\eta_g) \right],$$

$$\lambda_{74}(E,\eta_g) \equiv \frac{\alpha\hbar^4 Z_0(E,\eta_g)}{4m_l^+ m_l^-} \text{ and } \lambda_{75}(E,\eta_g) \equiv [\gamma_0(E,\eta_g) + \alpha\theta(E,\eta_g)]$$

Thus, the energy spectrum in this case is real since the corresponding dispersion relation in the absence of band tails as given by (1.162) is a pole-less function with respect to energy axis in the finite complex plane.

The respective transverse and the longitudinal EEMs' in this case can be written as

$$m_\perp^*(E_{F_h},\eta_g) = \{2Z_0(E,\eta_g)\}^{-2} \left[Z_0(E,\eta_g) \left[-\{\lambda_{72}(E,\eta_g)\}' + \frac{\{\lambda_{78}(E,\eta_g)\}'}{2\sqrt{\lambda_{78}(E,\eta_g)}} \right] - \{Z_0(E,\eta_g)\}' \left[-\lambda_{72}(E,\eta_g) + \sqrt{\lambda_{78}(E,\eta_g)} \right] \right] \Bigg|_{E=E_{F_h}} \tag{1.163}$$

where,

$$\lambda_{78}(E,\eta_g) \equiv [4\lambda_{70}(E,\eta_g)\,\lambda_{75}(E,\eta_g)]$$

and

$$m_{\|}^{*}(E_{F_h},\eta_g)=\frac{\hbar^2}{4}\left[-\{\lambda_{84}(E,\eta_g)\}'+\frac{\{\lambda_{84}(E,\eta_g)\}'\lambda_{84}(E,\eta_g)+2\{\lambda_{85}(E,\eta_g)\}'}{\sqrt{(\lambda_{84}(E,\eta_g))^2+4\lambda_{85}(E,\eta_g)}}\right]\Bigg|_{E=E_{F_h}} \quad (1.164)$$

in which, $\lambda_{84}(E,\eta_g)\equiv\frac{\lambda_{73}(E,\eta_g)}{\lambda_{74}(E,\eta_g)}$ and $\lambda_{85}(E,\eta_g)\equiv\frac{\lambda_{75}(E,\eta_g)}{\lambda_{74}(E,\eta_g)}$

Thus, we can see that the both the EEMs' in this case exist in the band gap. In the absence of band tails, $\eta_g\to 0$, we get

$$m_{\perp}^{*}(E_F)=\frac{\hbar^2}{2}\left[-\{\alpha_{11}(E)\}'+\frac{\alpha_{511}\{T_{311}(E)\}'}{2\sqrt{T_{311}(E)}}\right]\Bigg|_{E=E_F} \quad (1.165)$$

where

$$\alpha_{11}(E)\equiv\frac{2m_t^+m_t^-}{\alpha\hbar^2}\alpha_{211}(E),\ \alpha_{211}(E)\equiv\left[\frac{1}{2m_t^*}-\frac{\alpha E}{2m_t^+}+\frac{1+\alpha E}{2m_t^-}\right],\ \alpha_{511}\equiv\frac{2m_t^+m_t^-}{\alpha\hbar^2}\omega_{11}$$

$$(\omega_{11})\equiv\left[\frac{\alpha^2}{16}\left[\frac{1}{m_t^-m_t^+}+\frac{1}{m_l^-m_t^+}\right]^2-\frac{\alpha^2}{4m_t^-m_t^+m_l^-m_l^+}\right]^{1/2},\ T_{311}(E)\equiv\frac{\omega_{311}(E)}{(\omega_{11})^2},$$

$$\omega_{311}(E)\equiv\left[\frac{\alpha E(1+\alpha E)}{m_t^+m_t^-}+\left[\frac{1}{2m_t^*}-\left(\frac{\alpha E}{2m_t^+}\right)+\frac{(1+\alpha E)}{2m_t^-}\right]^2\right]$$

and

$$m_{\|}^{*}(E_F)=\left(\frac{m_t^+m_l^-}{\alpha}\right)\left[\left(\frac{\alpha}{2m_l^+}-\frac{\alpha}{2m_l^-}\right)+\frac{1}{2}\left\{\frac{2\left[\frac{1}{2m_l^*}+\frac{1+\alpha E}{2m_l^-}-\frac{\alpha E}{2m_l^+}\right]\left(\frac{\alpha}{2m_l^-}-\frac{\alpha}{2m_l^+}\right)+\frac{\alpha(1+2\alpha E)}{m_l^-m_l^+}}{\left[\left[\frac{1}{2m_l^*}+\frac{1+\alpha E}{2m_l^-}-\frac{\alpha E}{2m_l^+}\right]^2+\frac{\alpha E(1+\alpha E)}{m_l^-m_l^+}\right]^{1/2}}\right\}\right]\Bigg|_{E=E_F} \quad (1.166)$$

The volume in k-space as enclosed by Eq. (1.162) can be written through the integral as

$$V(E,\eta_g)=2\pi\int_0^{\lambda_{86}(E,\eta_g)}\left[-[\lambda_{79}(E,\eta_g)k_z^2+\lambda_{80}(E,\eta_g)]+\sqrt{\lambda_{81}(E,\eta_g)k_z^4+\lambda_{82}(E,\eta_g)k_z^2+\lambda_{83}(E,\eta_g)}\right]dk_z \quad (1.167)$$

where,

$$\lambda_{86}(E,\eta_g) \equiv \left[\frac{\sqrt{[\lambda_{84}(E,\eta_g)]^2+4\lambda_{85}(E,\eta_g)} - \lambda_{84}(E,\eta_g)}{2}\right]^{1/2}, \quad \lambda_{79}(E,\eta_g) \equiv \frac{\lambda_{71}(E,\eta_g)}{2\hbar^2 Z_0(E,\eta_g)}$$

$$\lambda_{81}(E,\eta_g) \equiv \frac{\lambda_{76}(E,\eta_g)}{4\hbar^4 [Z_0(E,\eta_g)]^2}, \quad \lambda_{76}(E,\eta_g) \equiv [\lambda_{71}(E,\eta_g)]^2, \quad \lambda_{76}(E,\eta_g) \equiv [\lambda_{71}(E,\eta_g)]^2$$

$$\lambda_{77}(E,\eta_g) \equiv [2\lambda_{71}(E,\eta_g)\lambda_{72}(E,\eta_g) - 4\lambda_{70}(E,\eta_g)\lambda_{73}(E,\eta_g) - 4\lambda_{70}(E,\eta_g)\lambda_{74}(E,\eta_g)],$$

$$\lambda_{83}(E,\eta_g) \equiv \frac{\lambda_{78}(E,\eta_g)}{9\hbar^4 [Z_0(E,\eta_g)]^2} \quad \text{and } \lambda_{78}(E,\eta_g) \equiv [4\lambda_{70}(E,\eta_g)\lambda_{75}(E,\eta_g)]$$

Thus,

$$V(E,\eta_g) = [\lambda_{87}(E,\eta_g)] \int_0^{\lambda_{86}(E,\eta_g)} \left[\sqrt{k_z^4 + \lambda_{88}(E,\eta_g)k_z^2 + \lambda_{89}(E,\eta_g)} - \lambda_{90}(E,\eta_g)\right] dk_z \tag{1.168}$$

where,

$$\lambda_{87}(E,\eta_g) \equiv 2\pi\sqrt{\lambda_{81}(E,\eta_g)},$$

$$\lambda_{88}(E,\eta_g) \equiv \frac{\lambda_{82}(E,\eta_g)}{\lambda_{81}(E,\eta_g)},$$

$$\lambda_{89}(E,\eta_g) \equiv \frac{\lambda_{83}(E,\eta_g)}{\lambda_{81}(E,\eta_g)}$$

and

$$\lambda_{90}(E,\eta_g) \equiv 2\pi\left[\frac{\lambda_{79}(E,\eta_g)\{\lambda_{86}(E,\eta_g)\}^3}{3} + \lambda_{80}(E,\eta_g)\lambda_{89}(E,\eta_g)\right].$$

The (1.168) can be written as

$$V(E,\eta_g) = [\lambda_{87}(E,\eta_g)\lambda_{95}(E,\eta_g) - \lambda_{90}(E,\eta_g)] \tag{1.169}$$

in which,

$$\lambda_{95}(E,\eta_g) \equiv \left[\frac{\lambda_{91}(E,\eta_g)}{3}\left[-E_i\left[\lambda_{93}(E,\eta_g),\lambda_{94}(E,\eta_g)\right]\right.\right.$$
$$\left[\{\lambda_{91}(E,\eta_g)\}^2+\{\lambda_{92}(E,\eta_g)\}^2+2\{\lambda_{92}(E,\eta_g)\}^2F_i\left[\lambda_{93}(E,\eta_g),\lambda_{94}(E,\eta_g)\right]\right]$$
$$+\left(\frac{\{\lambda_{86}(E,\eta_g)\}}{3}\right)\left[\{\lambda_{86}(E,\eta_g)\}^2+\{\lambda_{91}(E,\eta_g)\}^2+2\{\lambda_{92}(E,\eta_g)\}^2\right]$$
$$\left.\left.\left[\left[\{\lambda_{91}(E,\eta_g)\}^2+\{\lambda_{86}(E,\eta_g)\}^2\right]^{1/2}\left[\{\lambda_{92}(E,\eta_g)\}^2+\{\lambda_{86}(E,\eta_g)\}^2\right]^{-1/2}\right]\right]\right],$$
$$\{\lambda_{91}(E,\eta_g)\}^2\equiv\frac{1}{2}\left[\sqrt{\{\lambda_{88}(E,\eta_g)\}^2-4\lambda_{89}(E,\eta_g)}+\lambda_{88}(E,\eta_g)\right],\quad E_i\left[\lambda_{93}(E,\eta_g),\lambda_{94}(E,\eta_g)\right]$$

is the incomplete elliptic integral of the 2nd kind and is given by Abramowitz and Stegun [129] and Gradshteyn and Ryzhik [130],

$$E_i\left[\lambda_{93}(E,\eta_g),\lambda_{94}(E,\eta_g)\right] \equiv \int_0^{\lambda_{93}(E,\eta_g)}\left[\left\{1-\{\lambda_{94}(E,\eta_g)\}^2\sin^2\xi\right\}^{1/2}\right]d\xi,$$

ξ is the variable of integration in this case,

$$\lambda_{93}(E,\eta_g)\equiv\tan^{-1}\left[\frac{\lambda_{86}(E,\eta_g)}{\lambda_{92}(E,\eta_g)}\right],\ \{\lambda_{92}(E,\eta_g)\}^2\equiv\frac{1}{2}\left[\lambda_{88}(E,\eta_g)-\sqrt{\{\lambda_{88}(E,\eta_g)\}^2-4\lambda_{89}(E,\eta_g)}\right],$$
$$\lambda_{94}(E,\eta_g)\equiv\frac{\sqrt{\{\lambda_{91}(E,\eta_g)\}^2-\{\lambda_{92}(E,\eta_g)\}^2}}{\lambda_{91}(E,\eta_g)},\ F_i\left[\lambda_{93}(E,\eta_g),\lambda_{94}(E,\eta_g)\right]$$

is the incomplete elliptic integral of the 1st kind and is given by Abramowitz and Stegun [129] and Gradshteyn and Ryzhik [130],

$$F_i\left[\lambda_{93}(E,\eta_g),\lambda_{94}(E,\eta_g)\right] \equiv \int_0^{\lambda_{93}(E,\eta_g)}\left[\left\{1-\{\lambda_{94}(E,\eta_g)\}^2\sin^2\xi\right\}^{-1/2}\right]d\xi.$$

The DOS function in this case is given by

$$N_{HD}(E,\eta_g)=\frac{g_v}{4\pi^3}\left[\{\lambda_{87}(E,\eta_g)\}'\lambda_{95}(E,\eta_g)+\{\lambda_{95}(E,\eta_g)\}'\lambda_{87}(E,\eta_g)-\{\lambda_{90}(E,\eta_g)\}'\right] \tag{1.170}$$

Therefore the electron concentration can be expressed as

$$n_0 = \frac{g_v}{4\pi^3}\left[\bar{I}_{125}(E_{F_h}, \eta_g) + \sum_{r=1}^{s} L(r)[\bar{I}_{125}(E_{F_h}, \eta_g)]\right] \tag{1.171}$$

where,

$$\bar{I}_{125}(E_{F_h}, \eta_g) = [\{\lambda_{87}(E_{F_h}, \eta_g)\}\lambda_{95}(E_{F_h}, \eta_g) - \{\lambda_{90}(E_{F_h}, \eta_g)\}]$$

In this case, $\overline{E}_{hd}$ is given by

$$\{\lambda_{75}(\overline{E}_{hd}, \eta_g)\} = 0 \tag{1.172}$$

Thus, one can numerically evaluate the ER by using (1.31c), (1.171) and (1.172) and the allied definitions in this case.

The 2D dispersion relation of the conduction electrons in QWs of IV-VI materials in the absence of band tails for the dimensional quantization along z direction can be expressed as

$$E(1+\alpha E) + \alpha E\left(\frac{\hbar^2 k_x^2}{2x_4} + \frac{\hbar^2 k_y^2}{2x_5}\right) + \alpha E\frac{\hbar^2}{2x_6}\left(\frac{n_z\pi}{d_z}\right)^2 - (1+\alpha E)\left(\frac{\hbar^2 k_x^2}{2x_1} + \frac{\hbar^2 k_y^2}{2x_2}\right)$$
$$- \alpha\left(\frac{\hbar^2 k_x^2}{2x_1} + \frac{\hbar^2 k_y^2}{2x_2}\right)\left(\frac{\hbar^2 k_x^2}{2x_4} + \frac{\hbar^2 k_y^2}{2x_5}\right) - \alpha\left(\frac{\hbar^2 k_x^2}{2x_1} + \frac{\hbar^2 k_y^2}{2x_2}\right)\frac{\hbar^2}{2x_6}\left(\frac{n_z\pi}{d_z}\right)^2 - (1+\alpha E)\frac{\hbar^2}{2x_3}\left(\frac{n_z\pi}{d_z}\right)^2$$
$$- \alpha\frac{\hbar^2}{2x_3}\left(\frac{n_z\pi}{d_z}\right)^2\left(\frac{\hbar^2 k_x^2}{2x_4} + \frac{\hbar^2 k_y^2}{2x_5}\right) - \alpha\frac{\hbar^2}{2x_3}\left(\frac{n_z\pi}{d_z}\right)^2\frac{\hbar^2}{2x_6}\left(\frac{n_z\pi}{d_z}\right)^2 = \frac{\hbar^2 k_x^2}{2m_1} + \frac{\hbar^2 k_y^2}{2m_2} + \frac{\hbar^2}{2m_3}\left(\frac{n_z\pi}{d_z}\right)^2 \tag{1.173}$$

where $x_4 = m_t^+$, $x_5 = \frac{m_t^+ + 2m_l^+}{3}$, $x_6 = \frac{3m_t^+ m_l^+}{2m_l^+ + m_t^+}$, $x_1 = m_t^-$, $x_2 = \frac{m_t^- + 2m_l^-}{3}$, $x_3 = \frac{3m_t^- m_l^-}{2m_l^- + m_t^-}$, $m_1 = m_t^*, m_2 = \frac{m_t^* + 2m_l^*}{3}$ and $m_3 = \frac{3m_l^* m_t^*}{m_t^* + 2m_l^*}$.

Therefore, the HD 2-D dispersion relation In this case assumes the form

$$\gamma_2(E,\eta_g) + \alpha\gamma_3(E,\eta_g)\left(\frac{\hbar^2 k_x^2}{2x_4} + \frac{\hbar^2 k_y^2}{2x_5}\right) + \alpha\gamma_3(E,\eta_g)\frac{\hbar^2}{2x_6}\left(\frac{n_z\pi}{d_z}\right)^2 - (1+\alpha\gamma_3(E,\eta_g))\left(\frac{\hbar^2 k_x^2}{2x_1} + \frac{\hbar^2 k_y^2}{2x_2}\right)$$
$$- \alpha\left(\frac{\hbar^2 k_x^2}{2x_1} + \frac{\hbar^2 k_y^2}{2x_2}\right)\left(\frac{\hbar^2 k_x^2}{2x_4} + \frac{\hbar^2 k_y^2}{2x_5}\right) - \alpha\left(\frac{\hbar^2 k_x^2}{2x_1} + \frac{\hbar^2 k_y^2}{2x_2}\right)\frac{\hbar^2}{2x_6}\left(\frac{n_z\pi}{d_z}\right)^2 - (1+\alpha\gamma_3(E,\eta_g))\frac{\hbar^2}{2x_3}\left(\frac{n_z\pi}{d_z}\right)^2$$
$$- \alpha\frac{\hbar^2}{2x_3}\left(\frac{n_z\pi}{d_z}\right)^2\left(\frac{\hbar^2 k_x^2}{2x_4} + \frac{\hbar^2 k_y^2}{2x_5}\right) - \alpha\frac{\hbar^2}{2x_3}\left(\frac{n_z\pi}{d_z}\right)^2\frac{\hbar^2}{2x_6}\left(\frac{n_z\pi}{d_z}\right)^2 = \frac{\hbar^2 k_x^2}{2m_1} + \frac{\hbar^2 k_y^2}{2m_1} + \frac{\hbar^2}{2m_3}\left(\frac{n_z\pi}{d_z}\right)^2, \tag{1.174}$$

Substituting, $k_x = r\text{Cos}\,\theta$ and $k_y = r\text{Sin}\theta$ (where r and θ are 2D polar coordinates in 2D wave vector space) in (1.174), we can write

$$r^4\left[\alpha\frac{1}{4}\left(\frac{\hbar^2\mathrm{Cos}^2\theta}{x_1}+\frac{\hbar^2\mathrm{Sin}^2\theta}{x_2}\right)\left(\frac{\hbar^2\mathrm{Cos}^2\theta}{x_4}+\frac{\hbar^2\mathrm{Sin}^2\theta}{x_5}\right)\right]+r^2\frac{1}{2}\left[\left(\frac{\hbar^2\mathrm{Cos}^2\theta}{m_1}+\frac{\hbar^2\mathrm{Sin}^2\theta}{m_2}\right)\right.$$
$$+\alpha\frac{\hbar^2}{2x_3}\left(\frac{n_z\pi}{d_z}\right)^2\left(\frac{\hbar^2\mathrm{Cos}^2\theta}{x_4}+\frac{\hbar^2\mathrm{Sin}^2\theta}{x_5}\right)+\alpha\left(\frac{\hbar^2\mathrm{Cos}^2\theta}{x_1}+\frac{\hbar^2\mathrm{Sin}^2\theta}{x_2}\right)\frac{\hbar^2}{2x_6}\left(\frac{n_z\pi}{d_z}\right)^2$$
$$\left.+\hbar^2(1+\alpha\gamma_3(E,\eta_g))\left(\frac{\mathrm{Cos}^2\theta}{x_1}+\frac{\mathrm{Sin}^2\theta}{x_2}\right)-\hbar^2\alpha\gamma_3(E,\eta_g)\left(\frac{\mathrm{Cos}^2\theta}{x_4}+\frac{\mathrm{Sin}^2\theta}{x_5}\right)\right].$$
$$-\left[\gamma_2(E,\eta_g)+\alpha\gamma_3(E,\eta_g)\frac{\hbar^2}{2x_6}\left(\frac{n_z\pi}{d_z}\right)^2-(1+\alpha\gamma_3(E,\eta_g))\frac{\hbar^2}{2x_3}\left(\frac{n_z\pi}{d_z}\right)^2-\alpha\left(\frac{\hbar^2}{4x_3x_6}\left(\frac{n_z\pi}{d_z}\right)^4\right)\right]=0 \tag{1.175}$$

The area $A(E,n_z)$ of the 2D wave vector space can be expressed as

$$A(E,n_z)=\bar{J}_1-\bar{J}_2 \tag{1.176}$$

where

$$\bar{J}_1\equiv 2\int_0^{\pi/2}\frac{c_1}{b_1}d\theta \tag{1.177}$$

and

$$\bar{J}_2\equiv 2\int_0^{\pi/2}\frac{ac_1^2}{b_1^3}d\theta \tag{1.178}$$

in which

$$a\equiv\left[\alpha\left(\frac{\hbar^4}{4}\right)\left(\frac{\mathrm{Cos}^2\theta}{x_1}+\frac{\mathrm{Sin}^2\theta}{x_2}\right)\left(\frac{\mathrm{Cos}^2\theta}{x_4}+\frac{\mathrm{Sin}^2\theta}{x_5}\right)\right],$$
$$b_1\equiv\left(\frac{\hbar^2}{4}\right)\left[\left(\frac{\mathrm{Cos}^2\theta}{m_1}+\frac{\mathrm{Sin}^2\theta}{m_2}\right)+\alpha\left(\frac{\hbar^4}{2x_3}\right)\left(\frac{n_z\pi}{d_z}\right)^2\left(\frac{\mathrm{Cos}^2\theta}{x_4}+\frac{\mathrm{Sin}^2\theta}{x_5}\right)\right.$$
$$+\alpha\left(\frac{\hbar^4}{2x_6}\right)\left(\frac{n_z\pi}{d_z}\right)^2\left(\frac{\mathrm{Cos}^2\theta}{m_1}+\frac{\mathrm{Sin}^2\theta}{x_2}\right)$$
$$\left.+(1+\alpha\gamma_3(E,\eta_g))\left(\frac{\mathrm{Cos}^2\theta}{x_1}+\frac{\mathrm{Sin}^2\theta}{x_2}\right)-\alpha\gamma_3(E,\eta_g)\left(\frac{\mathrm{Cos}^2\theta}{x_4}+\frac{\mathrm{Sin}^2\theta}{x_5}\right)\right]$$

and

$$c_1 \equiv \left[\gamma_2(E,\eta_g) + \alpha\gamma_3(E,\eta_g)\left(\frac{\hbar^2}{2x_6}\right)\left(\frac{n_z\pi}{d_z}\right)^2 - (1+\alpha\gamma_3(E,\eta_g))\left(\frac{\hbar^2}{2x_3}\right)\left(\frac{n_z\pi}{d_z}\right)^2 - \alpha\left(\frac{\hbar^4}{4x_3x_6}\right)\left(\frac{n_z\pi}{d_z}\right)^4\right]$$

The (1.177) can be expressed as

$$\bar{J}_1 \equiv 2\int_0^{\pi/2} \frac{t_{31}(E,n_z)d\theta}{A_{11}(E,n_z)\cos^2\theta + B_{11}(E,n_z)\sin^2\theta}$$

where,

$$t_{31}(E,n_z) \equiv c_1,\ A_{11}(E,n_z) \equiv \frac{\hbar^2}{2m_1}t_{11}(E,n_z),$$

$$t_{11}(E,n_z) \equiv \left[1 + m_1\left[\frac{1}{x_4}\frac{\alpha\hbar^2}{2x_3}\left(\frac{n_z\pi}{d_z}\right)^2 + \frac{\alpha\hbar^2}{2x_1x_6}\left(\frac{n_z\pi}{d_z}\right)^2 + \frac{1+\alpha\gamma_2(E,\eta_g)}{x_1} - \frac{\alpha\gamma_3(E,\eta_g)}{x_4}\right]\right]$$

$$B_{11}(E,n_z) \equiv \frac{\hbar^2}{2m_2}t_{21}(E,n_z)\ \text{and}$$

$$t_{21}(E,n_z) \equiv \left[1 + m_2\left[\frac{\alpha\hbar^2}{2x_3x_5}\left(\frac{n_z\pi}{d_z}\right)^2 + \frac{\alpha\hbar^2}{2x_2x_6}\left(\frac{n_z\pi}{d_z}\right)^2 + \frac{1+\alpha\gamma_3(E,\eta_g)}{x_2} - \frac{\alpha\gamma_3(E,\eta_g)}{x_5}\right]\right].$$

Performing the integration, we get

$$\bar{J}_1 = \pi t_{31}(E,n_z)[A_{11}(E,n_z)B_{11}(E,n_z)]^{-1/2} \tag{1.179}$$

From (1.178) we can write

$$\bar{J}_2 = \frac{\alpha t_{31}^2(E,n_z)\hbar^4}{2B_{11}^3(E,n_z)}I \tag{1.180}$$

where,

$$I \equiv \int_0^{\infty} \frac{(a_1 + a_2z^2)(a_3 + a_4z^2)dz}{[(\bar{a})^2 + z^2]^3},\quad (\bar{a})^2 = \left(\frac{A_{11}(E,n_z)}{B_{11}(E,n_z)}\right), \tag{1.181}$$

in which $a_1 \equiv \frac{1}{x_1}$, $a_2 \equiv \frac{1}{x_2}$, $z = \tan\theta, \theta$ is a new variable, $a_3 \equiv \frac{1}{x_4}$, $a_4 \equiv \frac{1}{x_5}$ and $(\bar{a})^2 \equiv \left(\frac{A_1(E,n_z)}{B_1(E,n_z)}\right)$.

The use of the Residue theorem leads to the evaluation of the integral in (1.181) as

$$I = \frac{\pi}{4\bar{a}}[a_1 a_4 + 3a_2 a_4], \tag{1.182}$$

Therefore, the 2D area of the 2D wave vector space can be written as

$$A_{HD}(E, n_z) = \frac{\pi t_{31}(E, n_z)}{\sqrt{A_{11}(E, n_z) B_{11}(E, n_z)}} \left[1 - \frac{1}{x_5}\left(\frac{1}{x_1} + \frac{3}{x_2}\right) \frac{\alpha t_{31}(E, n_z)\hbar^4}{8B_{11}^2(E, n_z)}\right] \tag{1.183}$$

The EEM for the HD QWs of IV-VI materials can thus be written as

$$m^*(E, n_z) = \frac{\hbar^2}{2}[\theta_{5HD}(E, n_z)]\Big|_{E=E_{F1HD}} \tag{1.184}$$

where,

$$\begin{aligned}
\theta_{5HD}(E, n_z) \equiv & \left[1 - \frac{1}{x_5}\left(\frac{1}{x_1} + \frac{3}{x_2}\right) \frac{\alpha t_{31}(E, n_z)\hbar^4}{8[B_{11}(E, n_z)]^2}\right] [A_{11}(E, n_z) B_{11}(E, n_z)]^{-1} \\
& \left[\sqrt{A_{11}(E, n_z) B_{11}(E, n_z)} \{t_{31}(E, n_z)\}' - t_{31}(E, n_z) \left\{\frac{1}{2}\{A_{11}(E, n_z)\}' \left[\frac{B_{11}(E, n_z)}{A_{11}(E, n_z)}\right]^{1/2}\right.\right. \\
& \left.\left. + \frac{1}{2}\{B_{11}(E, n_z)\}' \left[\frac{A_{11}(E, n_z)}{B_{11}(E, n_z)}\right]^{1/2}\right\}\right] \\
& - \frac{1}{8} \frac{t_{31}(E, n_z)\alpha\hbar^4}{\sqrt{A_{11}(E, n_z) B_{11}(E, n_z)}} \frac{1}{x_5}\left(\frac{1}{x_1} + \frac{3}{x_2}\right) [B_{11}(E, n_z)]^{-4} \\
& \left[\{B_{11}(E, n_z)\}^2 \{t_{31}(E, n_z)\}' - 2B_{11}(E, n_z)\{B_{11}(E, n_z)\}' t_{31}(E, n_z)\right].
\end{aligned}$$

Thus, the EEM is a function of Fermi energy and the quantum number due to the band non-parabolicity.

The total DOS function can be written as

$$N_{2DT}(E) = \left(\frac{g_v}{2\pi}\right) \sum_{n_z=1}^{n_{z\max}} \theta_{5HD}(E, n_z) H(E - E_{n_{z7HD}}) \tag{1.185}$$

where the sub-band energy $\left(E_{n_{z7}}\right)$ in this case can be written as

$$\begin{aligned}
& \gamma_2\left(E_{n_{z7}HD}, \eta_g\right) + \alpha\gamma_3\left(E_{n_{z7}HD}, \eta_g\right) \frac{\hbar^2}{2x_6}\left(\frac{n_z\pi}{d_z}\right)^2 - \left(1 + \alpha\gamma_3\left(E_{n_{z7}HD}, \eta_g\right)\right) \frac{\hbar^2}{2x_3}\left(\frac{n_z\pi}{d_z}\right)^2 \\
& - \alpha\frac{\hbar^2}{2x_3}\left(\frac{n_z\pi}{d_z}\right)^2 \frac{\hbar^2}{2x_6}\left(\frac{n_z\pi}{d_z}\right)^2 - \left[\frac{\hbar^2}{2m_3}\left(\frac{n_z\pi}{d_z}\right)^2\right] = 0
\end{aligned} \tag{1.186}$$

The use (5.61) leads to the expression of 2D electron statistics as

$$n_{2D} = \frac{g_v}{2\pi} \sum_{n_z=1}^{n_{z\max}} [T_{55\,HD}(E_{F1HD}, n_z) + T_{56\,HD}(E_{F1HD}, n_z)] \tag{1.187}$$

where $T_{55\,HD}(E_{F1HD}, n_z) \equiv \frac{A_{HD}(E_{F1\,HD}, n_z)}{\pi}$ and $T_{56\,HD}(E_{F1HD}, n_z) \equiv \sum_{r=1}^{s} L(r) T_{55\,HD}(E_{F1\,HD}, n_z)$.

In the absence of heavy doping the EEM in QWs of IV-VI materials can be written as

$$m^*(E, n_z) = \frac{\hbar^2}{2} [\theta_5(E, n_z)] \Big|_{E=E_{Fs}} \tag{1.188}$$

where,

$$\begin{aligned}
\theta_5(E, n_z) \equiv & \left[1 - \frac{1}{x_5}\left(\frac{1}{x_1} + \frac{3}{x_2}\right) \frac{\alpha t_{30}(E, n_z)\hbar^4}{8[B_{10}(E, n_z)]^2}\right] [A_{10}(E, n_z) B_{10}(E, n_z)]^{-1} \\
& \left[\sqrt{A_{10}(E, n_z) B_{10}(E, n_z)} \{t_{30}(E, n_z)\}' - t_{30}(E, n_z) \left\{\frac{1}{2}\{A_{10}(E, n_z)\}' \left[\frac{B_{10}(E, n_z)}{A_{10}(E, n_z)}\right]^{1/2}\right.\right. \\
& \left.\left. + \frac{1}{2}\{B_{10}(E, n_z)\}' \left[\frac{A_{10}(E, n_z)}{B_{10}(E, n_z)}\right]^{1/2}\right\}\right] \\
& - \frac{1}{8} \frac{t_{30}(E, n_z)\alpha\hbar^4}{\sqrt{A_{10}(E, n_z) B_{10}(E, n_z)}} \frac{1}{x_5} \left(\frac{1}{x_1} + \frac{3}{x_2}\right) [B_{10}(E, n_z)]^{-4} \\
& \left[\{B_{10}(E, n_z)\}^2 \{t_{30}(E, n_z)\}' - 2B_{10}(E, n_z)\{B_{10}(E, n_z)\}' t_{30}(E, n_z)\right]
\end{aligned}$$

where

$$t_{30}(E, n_z) \equiv c_0,$$

$$\begin{aligned}
c_0 \equiv & \left[E(1 + \alpha E) + \alpha E\left(\frac{\hbar^2}{2x_6}\right)\left(\frac{n_z \pi}{d_z}\right)^2\right. \\
& \left. - (1 + \alpha E)\left(\frac{\hbar^2}{2x_3}\right)\left(\frac{n_z \pi}{d_z}\right)^2 - \alpha\left(\frac{\hbar^4}{4x_3 x_6}\right)\left(\frac{n_z \pi}{d_z}\right)^4\right],
\end{aligned}$$

$$\begin{aligned}
A_{10}(E, n_z) \equiv & \frac{\hbar^2}{2m_1} t_{10}(E, n_z), t_{10}(E, n_z) \\
\equiv & \left[1 + m_1\left[\frac{1}{x_4}\frac{\alpha\hbar^2}{2x_3}\left(\frac{n_z \pi}{d_z}\right)^2 + \frac{\alpha\hbar^2}{2x_1 x_6}\left(\frac{n_z \pi}{d_z}\right)^2 + \frac{1 + \alpha E}{x_1} - \frac{\alpha E}{x_4}\right]\right],
\end{aligned}$$

$$B_{10}(E,n_z) \equiv \frac{\hbar^2}{2m_2} t_{20}(E,n_z) \; and \; t_{20}(E,n_z) \equiv \left[1 + m_2\left[\frac{\alpha\hbar^2}{2x_3x_5}\left(\frac{n_z\pi}{d_z}\right)^2 + \frac{\alpha\hbar^2}{2x_2x_6}\left(\frac{n_z\pi}{d_z}\right)^2 + \frac{1+\alpha E}{x_2} - \frac{\alpha E}{x_5}\right]\right]$$

Thus, the EEM is a function of Fermi energy and the quantum number due to the band non-parabolicity.

The total DOS function can be written as

$$N_{2DT}(E) = \left(\frac{g_v}{2\pi}\right)\sum_{n_z=1}^{n_{z\max}} \theta_5(E,\mathrm{n}_z)H(\mathrm{E}-\mathrm{E}_{n_{z7}}) \tag{1.189}$$

where the sub-band energy $\left(\mathrm{E}_{n_{z7}}\right)$ in this case can be written as

$$\begin{aligned} &E_{n_{z7}}\left(1+\alpha E_{n_{z7}}\right) + \alpha E_{n_{z7}}\frac{\hbar^2}{2x_6}\left(\frac{n_z\pi}{d_z}\right)^2 - \left(1+\alpha E_{n_{z7}}\right)\frac{\hbar^2}{2x_3}\left(\frac{n_z\pi}{d_z}\right)^2 \\ &-\alpha\frac{\hbar^2}{2x_3}\left(\frac{n_z\pi}{d_z}\right)^2\frac{\hbar^2}{2x_6}\left(\frac{n_z\pi}{d_z}\right)^2 - \left[\frac{\hbar^2}{2m_3}\left(\frac{n_z\pi}{d_z}\right)^2\right] = 0 \end{aligned} \tag{1.190}$$

In the absence of heavy doping, the expression of 2D electron statistics can be written as

$$n_{2D} = \frac{g_v}{2\pi}\sum_{n_z=1}^{n_{z\max}}\left[T_{550}(E_{Fs},n_z) + T_{560}(E_{Fs},n_z)\right] \tag{1.191}$$

where, $T_{550}(E_{Fs},\, n_z) \equiv \frac{A_0(E_{Fs},n_z)}{\pi}$, $A_0(E_F,\, n_z) = \frac{\pi t_{30}(E,n_z)}{\sqrt{A_{10}(E,n_z)B_{10}(E,n_z)}}\left[1 - \frac{1}{x_5}\left(\frac{1}{x_1}+\frac{3}{x_2}\right)\frac{\alpha t_{30}(E,n_z)\hbar^4}{8B_{10}^2(E,n_z)}\right]$, and $T_{560}(E_{Fs},n_z) = \sum_{r=1}^{s} L(r)T_{550}(E_{Fs},\mathrm{n}_z)$

Thus, the expression for the 2D DMR in QWs of IV-VI compounds assumes the form

$$\frac{D}{\mu} = \frac{1}{|e|}\frac{\sum_{n_z=1}^{n_{z\max}}\left[T_{55}(E_{Fs},n_z) + T_{56}(E_{Fs},n_z)\right]}{\sum_{n_z=1}^{n_{z\max}}\left[(T_{55}(E_{Fs},n_z))' + (T_{56}(E_{Fs},n_z))'\right]} \tag{1.192}$$

For bulk specimens of IV-VI materials, the expressions of electron concentration and the ER assume the forms

$$n_0 = \left(\frac{g_v}{2\pi^2}\right)[M_{A_4}(E_{F_b}) + N_{A_4}(E_{F_b})] \tag{1.193}$$

$$\frac{D}{\mu} = \frac{1}{e}[M_{A_4}(E_{F_b}) + N_{A_4}(E_{F_b})]\left[M'_{A_4}(E_{F_b}) + N'_{A_4}(E_{F_b})\right]^{-1} \tag{1.194}$$

where

$$M_{A_4}(E_{F_b}) = \left[\alpha_5 J_{A_1}(E_{F_b}) - \alpha_3(E_{F_b})\bar{\tau}_{A_1}(E_{F_b}) - \frac{\alpha_4}{3}\left[\bar{\tau}_{A_1}(E_{F_b})\right]^3\right],$$

$$\alpha_5 = \left[\frac{2m_t^+ m_t^-}{\alpha\hbar^2}\omega_{A_1}\right],$$

$$\omega_{A_1} = \left[\frac{\alpha^2}{16}\left[\frac{1}{m_t^- m_l^+} + \frac{1}{m_l^- m_t^+}\right]^2 - \frac{\alpha^2}{4m_l^+ m_t^- m_l^- m_t^+}\right],$$

$$J_{A_1}(E_{F_b}) = \frac{A_A(E_{F_b})}{3}\left[-(A_A^2(E_{F_b}) + B_A^2(E_{F_b}))E(\lambda, q) + 2B_A^2(E_{F_b})F(\lambda, q)\right] + \frac{\bar{\tau}_{A_1}(E_{F_b})}{3}\left[(\bar{\tau}_{A_1}(E_{F_b}))^2 + A_A^2(E_{F_b}) + 2B_A^2(E_{F_b})\right]$$
$$\left[A_A^2(E_{F_b}) + \bar{\tau}_{A_1}^2(E_{F_b})\right]^{1/2}\left[B_A^2(E_{F_b}) + \bar{\tau}_{A_1}^2(E_{F_b})\right]^{-1/2}$$

$$\lambda = \tan^{-1}\frac{\bar{\tau}_{A_1}(E_{F_b})}{B_A(E_{F_b})}, q = \left[\frac{\sqrt{A_A^2(E_{F_b}) - B_A^2(E_{F_b})}}{A_A(E_{F_b})}\right], A_A(E_{F_b}) = \left[\tau_{A_2}(E_{F_b}) + \sqrt{\tau_{A_2}^2(E_{F_b}) - 4\tau_{A_3}^2(E_{F_b})}\right]^{1/2}/\sqrt{2},$$

$$B_A(E_{F_b}) = \left[\tau_{A_2}(E_{F_b}) - \sqrt{\tau_{A_2}^2(E_{F_b}) - 4\tau_{A_3}(E_{F_b})}\right]^{1/2}/\sqrt{2}, \tau_{A_2}(E_{F_b}) = \frac{\omega_{A_2}(E_{F_b})}{\omega_{A_1}^2}, \tau_{A_3}(E_{F_b}) = \frac{\omega_{A_3}(E_{F_b})}{\omega_{A_1}^2},$$

$$\omega_{A_2}(E_{F_b}) = \left[\frac{\alpha}{2}\left[\frac{1}{2m_t^*} - \frac{\alpha.E_{F_b}}{2m_t^+} + \frac{1+\alpha.E_{F_b}}{2m_t^-}\right]\cdot\left[\frac{1}{m_t^- m_l^+} + \frac{1}{m_l^- m_t^+}\right] - \frac{\alpha}{m_t^+ m_t^-}\left[\frac{1}{2m_l^*} + \frac{\alpha.E_{F_b}}{2m_l^+} + \frac{1+\alpha.E_{F_b}}{2m_l^-}\right]\right]$$

$$\omega_{A_3}(E_{F_b}) = \left[\frac{\alpha E_{F_b}(1+\alpha E_{F_b})}{m_t^+ m_t^-} + \left[\frac{1}{2m_t^*} - \frac{\alpha E_{F_b}}{2m_t^+} + \frac{1+\alpha E_{F_b}}{2m_t^-}\right]^2\right], \alpha_2(E_{F_b}) = \left[\frac{1}{2m_t^*} - \frac{\alpha E_{F_b}}{2m_t^+} + \frac{1+\alpha E_{F_b}}{2m_t^-}\right],$$

$$\alpha_3 = \frac{\alpha\hbar^2}{4}\left[\frac{1}{m_t^- m_l^+} + \frac{1}{m_l^- m_t^+}\right],$$

$$\tau_{A_1}(E_{F_b}) = \left[\frac{2m_l^+ m_l^-}{\alpha\hbar^2}\right]^{\frac{1}{2}}\left[-\left[\frac{1}{2m_l^*} + \frac{1+\alpha E_{F_b}}{m_l^-} - \frac{\alpha E_{F_b}}{2m_l^+}\right] + \left[\left[\frac{1}{2m_l^*} + \frac{1+\alpha E_{F_b}}{m_l^-} - \frac{\alpha E_{F_b}}{2m_l^+}\right]^2 + \frac{\alpha E_{F_b}(1+\alpha E_{F_b})}{m_l^+ m_l^-}\right]^{1/2}\right]^{1/2}$$

$E(\lambda, q)$ is the in complete Elliptic integral of second kind, $F(\lambda, q)$ is the incomplete Elliptic integral of first kind and $N_{A_4}(E_{F_b}) = \sum_{r=1}^{s} L(r)[M_{A_4}(E_{F_b})]$.

1.2.5 The ER from QWs of HD Stressed Kane Type Semiconductors

The electron energy spectrum in stressed Kane type semiconductors can be written [190–192] as

$$\left(\frac{k_x}{\bar{a}_0(E)}\right)^2+\left(\frac{k_y}{\bar{b}_0(E)}\right)^2+\left(\frac{k_z}{\bar{c}_0(E)}\right)^2=1 \tag{1.195}$$

where, $[\bar{a}_0(E)]^2 \equiv \frac{\bar{K}_0(E)}{\bar{A}_0(E)+\frac{1}{2}\bar{D}_0(E)}$, $\bar{K}_0(E) \equiv \left[E - C_1\varepsilon - \frac{2C_2^2\varepsilon_{xy}^2}{3E_g'}\right]\left(\frac{3E_g'}{2B_2^2}\right)$, C_1 is the conduction band deformation potential, ε is the trace of the strain tensor $\hat{\varepsilon}$ which can be written as

$$\hat{\varepsilon}=\begin{bmatrix}\varepsilon_{xx} & \varepsilon_{xy} & 0\\ \varepsilon_{xy} & \varepsilon_{yy} & 0\\ 0 & 0 & \varepsilon_{zz}\end{bmatrix},$$

C_2 is a constant which describes the strain interaction between the conduction and valance bands, $E_g' \equiv E_g + E - C_1\varepsilon, B_2$ is the momentum matrix element,

$$\bar{A}_0(E) \equiv \left[1-\frac{(\bar{a}_0+C_1)}{E_g'}+\frac{3\bar{b}_0\varepsilon_{xx}}{2E_g'}-\frac{\bar{b}_0\varepsilon}{2E_g'}\right],$$

$$\bar{a}_0 \equiv -\frac{1}{3}(\bar{b}_0+2\bar{m}),\ \bar{b}_0 \equiv \frac{1}{3}(\bar{l}-\bar{m}),\ \bar{d}_0 \equiv \frac{2\bar{n}}{\sqrt{3}},$$

$\bar{l}, \bar{m}, \bar{n}$ are the matrix elements of the strain perturbation operator, $\bar{D}_0(E) \equiv \left(\bar{d}_0\sqrt{3}\frac{\varepsilon_{xy}}{E_g'}\right)$, $[\bar{b}_0(E)]^2 \equiv \frac{\bar{K}_0(E)}{\bar{A}_0(E)-\frac{1}{2}\bar{D}_0(E)}$, $[\bar{c}_0(E)]^2 \equiv \frac{\bar{K}_0(E)}{\bar{L}_0(E)}$ and $\bar{L}_0(E) \equiv \left[1-\frac{(\bar{a}_0+C_1)}{E_g'}+\frac{3\bar{b}_0\varepsilon_{zz}}{E_g'}-\frac{\bar{b}_0\varepsilon}{2E_g'}\right]$.

The use of (1.195) can be written as

$$(E-\alpha_1)k_x^2+(E-\alpha_2)k_y^2+(E-\alpha_3)k_z^2=t_1E^3-t_2E^2+t_3E+t_4 \tag{1.196a}$$

where

$$\alpha_1 \equiv \left[E_g - C_1\varepsilon - (\bar{a}_0 + C_1)\varepsilon + \frac{3}{2}\bar{b}_0\varepsilon_{xx} - \frac{\bar{b}_0}{2}\varepsilon + \left(\sqrt{3}/2\right)\varepsilon_{xy}\bar{d}_0\right],$$

$$\alpha_2 \equiv \left[E_g - C_1\varepsilon - (\bar{a}_0 + C_1)\varepsilon + \frac{3}{2}\bar{b}_0\varepsilon_{xx} - \frac{\bar{b}_0}{2}\varepsilon - \left(\sqrt{3}/2\right)\varepsilon_{xy}\bar{d}_0\right],$$

$$\alpha_3 \equiv \left[E_g - C_1\varepsilon - (\bar{a}_0 + C_1)\varepsilon + \frac{3}{2}\bar{b}_0\varepsilon_{zz} - \frac{\bar{b}_0}{2}\varepsilon\right],$$

$$t_1 \equiv \left(3/2B_2^2\right),\ t_2 \equiv \left(1/2B_2^2\right)[6(E_g - C_1\varepsilon) + 3C_1\varepsilon],$$

$$t_3 \equiv \left(1/2B_2^2\right)[3(E_g - C_1\varepsilon)^2 + 6C_1\varepsilon(E_g - C_1\varepsilon) - 2C_2^2\varepsilon_{xy}^2] \text{ and}$$

$$t_4 \equiv \left(1/2B_2^2\right)[-3C_1\varepsilon(E_g - C_1\varepsilon)^2 + 2C_2^2\varepsilon_{xy}^2].$$

The (1.196a) can be written as

$$Ek^2 - T_{17}k_x^2 - T_{27}k_y^2 - T_{37}k_z^2 = [q_{67}E^3 - R_{67}E^2 + V_{67}E + \rho_{67}] \tag{1.196b}$$

where, $T_{17} = \alpha_1$, $T_{27} = \alpha_2$, $T_{37} = \alpha_3$, $t_1 = q_{67}$, $t_2 = R_{67}$, $t_3 = V_{67}$ and $t_4 = \rho_{67}$

Under the condition of heavy doping, (1.196b) can be written as

$$\begin{aligned} I(4)k^2 - T_{17}I(1)k_x^2 - T_{27}I(1)k_y^2 - T_{37}k_z^2I(1) \\ = [q_{67}I(6) - R_{67}I(5) + V_{67}I(4) + \rho_{67}I(1)] \end{aligned} \tag{1.196c}$$

where,

$$I(6) = \int_{-\infty}^{E} (E - V)^3 F(V)dV \tag{1.197}$$

The (1.197) can be written as

$$I(6) = E^3 I(1) - 3E^2 I(7) + 3EI(8) - I(9) \tag{1.198}$$

In which,

$$I(7) = \int_{-\infty}^{E} VF(V)dV \tag{1.199}$$

$$I(8) = \int_{-\infty}^{E} V^2F(V)dV \tag{1.200}$$

$$I(9) = \int_{-\infty}^{E} V^3F(V)dV \tag{1.201}$$

Using (1.4), together with simple algebraic manipulations, one obtains

$$I(7) = \frac{-\eta_g}{2\sqrt{\pi}} \exp\left(\frac{-E^2}{\eta_g^2}\right) \tag{1.202}$$

$$I(8) = \frac{\eta_g^2}{4}\left[1 + Erf\left(\frac{E}{\eta_g}\right)\right] \tag{1.203}$$

and

$$I(9) = \frac{-\eta_g^3}{2\sqrt{\pi}} \exp\left(\frac{-E^2}{\eta_g^2}\right)\left[1 + \frac{E^2}{\eta_g^2}\right] \tag{1.204}$$

Thus (1.197) can be written as

$$I(6) = \left[\frac{E}{2}\left[1 + Erf\left(\frac{E}{\eta_g}\right)\right]\left[E^2 + \frac{3}{2}\eta_g^2\right] + \frac{\eta_g}{2\sqrt{\pi}} \exp\left(\frac{-E^2}{\eta_g^2}\right)[4E^2 + \eta_g^2]\right] \tag{1.205}$$

Thus, combining the appropriate equations, the dispersion relations of the conduction electrons in HD stressed materials can be expressed as

$$P_{11}(E,\eta_g)k_x^2 + Q_{11}(E,\eta_g)k_y^2 + S_{11}(E,\eta_g)k_z^2 = 1 \tag{1.206}$$

where,

$$P_{11}(E,\eta_g) \equiv \left[\frac{\gamma_0(E,\eta_g) - (T_{17}/2)[1 + Erf(E/\eta_g)]}{\Delta_{14}(E,\eta_g)}\right],$$

$$\Delta_{14}(E,\eta_g) \equiv \left[q_{67}\left\{\frac{E}{2}\left[1 + Erf\left(\frac{E}{\eta_g}\right)\right]\left[E^2 + \frac{3}{2}\eta_g^2\right] + \frac{\eta_g}{2\sqrt{\pi}} \exp\left(\frac{-E^2}{\eta_g^2}\right)[4E^2 + \eta_g^2]\right\}\right.$$
$$\left. - R_{67}\theta_0(E,\eta_g) + V_{67}\gamma_0(E,\eta_g) + \frac{\rho_{67}}{2}[1 + Erf(E/\eta_g)]\right],$$

$$Q_{11}(E,\eta_g) \equiv \left[\frac{\gamma_0(E,\eta_g) - (T_{27}/2)[1 + Erf(E/\eta_g)]}{\Delta_{14}(E,\eta_g)}\right] \text{ and}$$

$$S_{11}(E,\eta_g) \equiv \left[\frac{\gamma_0(E,\eta_g) - (T_{37}/2)[1 + Erf(E/\eta_g)]}{\Delta_{14}(E,\eta_g)}\right].$$

Thus, the energy spectrum in this case is real since the dispersion relation of the corresponding materials in the absence of band tails as given by (1.195) is a pole-less function in the finite complex plane.

The EEMs along x, y and z directions in this case can be written as

$$m_{xx}^*(E_{F_h},\eta_g) = \frac{\hbar^2}{2}\left[\left[\gamma_0(E_{F_h},\eta_g) - (T_{17}/2)\left[1 + Erf(E_{F_h}/\eta_g)\right]\right]^{-2}. \right.$$
$$\left[\{\Delta_{14}(E_{F_h},\eta_g)\}'\left[\gamma_0(E_{F_h},\eta_g) - (T_{17}/2)\left[1 + Erf(E_{F_h}/\eta_g)\right]\right]\right.$$
$$\left.\left. -\Delta_{14}(E_{F_h},\eta_g)\left[\frac{1}{2}\left[1 + Erf\left(\frac{E_{F_h}}{\eta_g}\right)\right] - \left\{\frac{T_{17}}{\eta_g\sqrt{\pi}}\exp\left(\frac{-E_{F_h}^2}{\eta_g^2}\right)\right\}\right]\right]\right] \tag{1.207}$$

$$m_{yy}^*(E_{F_h},\eta_g) = \frac{\hbar^2}{2}\left[\left[\gamma_0(E_{F_h},\eta_g) - (T_{27}/2)\left[1 + Erf(E_{F_h}/\eta_g)\right]\right]^{-2} \right.$$
$$\left[\{\Delta_{14}(E_{F_h},\eta_g)\}'\left[\gamma_0(E_{F_h},\eta_g) - (T_{27}/2)\left[1 + Erf(E_{F_h}/\eta_g)\right]\right]\right.$$
$$\left.\left. -\Delta_{14}(E_{F_h},\eta_g)\left[\frac{1}{2}\left[1 + Erf\left(\frac{E_{F_h}}{\eta_g}\right)\right] - \left\{\frac{T_{27}}{\eta_g\sqrt{\pi}}\exp\left(\frac{-E_{F_h}^2}{\eta_g^2}\right)\right\}\right]\right]\right] \tag{1.208}$$

and

$$m_{zz}^*(E_{F_h},\eta_g) = \frac{\hbar^2}{2}\left[\left[\gamma_0(E_{F_h},\eta_g) - (T_{37}/2)\left[1 + Erf(E_{F_h}/\eta_g)\right]\right]^{-2}. \right.$$
$$\left[\{\Delta_{14}(E_{F_h},\eta_g)\}'\left[\gamma_0(E_{F_h},\eta_g) - (T_{37}/2)\left[1 + Erf(E_{F_h}/\eta_g)\right]\right]\right.$$
$$\left.\left. -\Delta_{14}(E_{F_h},\eta_g)\left[\frac{1}{2}\left[1 + Erf\left(\frac{E_{F_h}}{\eta_g}\right)\right] - \left\{\frac{T_{37}}{\eta_g\sqrt{\pi}}\exp\left(\frac{-E_{F_h}^2}{\eta_g^2}\right)\right\}\right]\right]\right] \tag{1.209}$$

Thus, we can see that the EEMs in this case exist within the band gap.
In the absence of band tails, $\eta_g \to 0$ we get

$$m_{xx}^*(E_F) = \hbar^2\bar{a}_0(E_F)\{\bar{a}_0(E_F)\}' \tag{1.210}$$

$$m_{xx}^*(E_F) = \hbar^2\bar{b}_0(E_F)\{\bar{b}_0(E_F)\}' \tag{1.211}$$

and

$$m_{xx}^*(E_F) = \hbar^2\bar{c}_0(E_F)\{\bar{c}_0(E_F)\}' \tag{1.212}$$

The DOS function in this case can be written as

$$N_{HD}(E,\eta_g)=\frac{g_v}{3\pi^2}\{\Delta_{15}(E,\eta_g)\}^{-2}\left[\frac{3}{2}\{\Delta_{15}(E,\eta_g)\}\sqrt{\Delta_{14}(E,\eta_g)}\{\Delta_{14}(E,\eta_g)\}'-\{\Delta_{14}(E,\eta_g)\}^{3/2}\{\Delta_{15}(E,\eta_g)\}'\right] \tag{1.213}$$

where,

$$\Delta_{15}(E,\eta_g)\equiv\left[[\gamma_0(E,\eta_g)-(T_{17}/2)[1+Erf(E/\eta_g)]]\,[\gamma_0(E,\eta_g)-(T_{27}/2)[1+Erf(E/\eta_g)]]\right.$$
$$\left.[\gamma_0(E,\eta_g)-(T_{37}/2)[1+Erf(E/\eta_g)]]\right]^{1/2}.$$

Using (1.213), the electron concentration at can be written as

$$n_0=\frac{g_v}{3\pi^2}\left[\bar{I}_{126}(E_{F_k},\eta_g)+\sum_{r=1}^{s}L(r)\left[\bar{I}_{126}(E_{F_k},\eta_g)\right]\right] \tag{1.214}$$

where,

$$\bar{I}_{126}(E_{F_k},\eta_g)=\left[\frac{\{\Delta_{14}(E_{F_k},\eta_g)\}^{3/2}}{\Delta_{15}(E_{F_k},\eta_g)}\right]$$

In this case, $\bar{E}_{hd}$ is given by

$$\{\Delta_{14}(\bar{E}_{hd},\eta_g)\}=0 \tag{1.215}$$

Thus, one can numerically evaluate the ER by using (1.214), (1.215), (1.31c) and the allied definitions in this case.

The dispersion relation of the conduction electrons in HD QWs of stressed Kane type semiconductors can be written as

$$P_{11}(E,\eta_g)k_x^2+Q_{11}(E,\eta_g)k_y^2+S_{11}(E,\eta_g)(\pi n_z/d_z)^2=1 \tag{1.216}$$

The EEM can be expressed as

$$m^*(E_{F1HD},\eta_g,n_z)=\frac{\hbar^2}{2}A'_{56}(E_{F1HD},\eta_g,n_z)$$

where,

$$A_{56}(E,\eta_g,n_z)=\frac{\pi\left[1-S_{11}(E,\eta_g)(\eta_z\pi/d_z)^2\right]}{\sqrt{P_{11}(E,\eta_g)Q_{11}(E,\eta_g)}} \tag{1.217}$$

From (1.217), it appears that the EEM is a function of Fermi energy, and size quantum number and the same mass exists in the band gap.

Thus, the total 2D DOS function can be expressed as

$$N_{2DT}(E) = \left(\frac{g_v}{2\pi}\right) \sum_{n_z=1}^{n_{z\max}} A'_{56}\left(E_{F1HD}, \eta_g, n_z\right) \tag{1.218}$$

The sub band energies ($E_{n_{z8}HD}$) are given by

$$S_{11}(E_{n_{z8}HD}, \eta_g)(\pi n_z/d_z)^2 = 1 \tag{1.219}$$

The 2D surface electron concentration per unit area for QWs of stressed HD Kane type compounds can be written as

$$n_{2D} = \frac{g_v}{2\pi} \sum_{n_z=1}^{n_{z\max}} \left[T_{57HD}\left(E_{Fs1HD}, \eta_g, n_z\right) + T_{58HD}\left(E_{Fs1HD}, \eta_g, n_z\right)\right] \tag{1.220}$$

where,

$$T_{57}\left(E_{F1HD}, \eta_g, n_z\right) \equiv A_{56}\left(E_{F1HD}, \eta_g, n_z\right)$$

and

$$T_{58HD}\left(E_{F1HD}, \eta_g, n_z\right) \equiv \sum_{r=1}^{s} L(r) T_{57HD}\left(E_{F1HD}, \eta_g, n_z\right).$$

In the absence of band tails, the 2D electron energy spectrum in QWs of stressed materials assumes the form

$$\frac{k_x^2}{[\bar{a}_0(E)]^2} + \frac{k_y^2}{[\bar{b}_0(E)]^2} + \frac{1}{[\bar{c}_0(E)]^2}(n_z\pi/d_z)^2 = 1 \tag{1.221}$$

The area of 2D wave vector space enclosed by (1.221) can be written as

$$A(E, n_z) = \pi P^2(E, n_z)\bar{a}_0(E)\bar{b}_0(E)$$

where $P^2(E, n_z) = \left[1 - [n_z\pi/d_z\bar{c}_0(E)]^2\right]$.

From (1.221), the EEM can be written as

$$m^*(E_{F_s}, n_z) = \frac{\hbar^2}{2}[P^2(E_{F_s}, n_z)\bar{a}_0(E_{F_s})\bar{b}_0(E_{F_s})]' \tag{1.222}$$

Thus, the total 2D DOS function can be expressed as

$$N_{2DT}(E) = \left(\frac{g_v}{2\pi}\right) \sum_{n_z=1}^{n_{z\max}} \theta_6(E, n_z) H(E - E_{n_{z11}}) \tag{1.223}$$

in which, $\theta_6(E,n_z) = \left[2P(E,n_z)\{P(E,n_z)\}'\bar{a}_0(E)\bar{b}_0(E) + \{P(E,n_z)\}^2\{\bar{a}_0(E)\}'\bar{b}_0(E) + \{P(E,n_z)\}^2\{\bar{b}_0(E)\}'\bar{a}_0(E)\right]$

The sub band energies ($E_{n_{z11}}$) are given by

$$\bar{c}_0(E_{n_{z11}}) = n_z\pi/d_z \tag{1.224}$$

The 2D surface electron concentration per unit area for QWs of stressed Kane type compounds can be written as

$$n_{2D} = \frac{g_v}{2\pi}\sum_{n_z=1}^{n_{z\max}}\left[T_{61}(E_{Fs},n_z) + T_{62}(E_{Fs},n_z)\right] \tag{1.225}$$

where

$$T_{61}(E_{Fs},n_z) \equiv \left[P^2(E_{Fs},n_z)\bar{a}_0(E_{Fs})\bar{b}_0(E_{Fs})\right]$$

and

$$T_{62}(E_{Fs},n_z) \equiv \sum\nolimits_{r=1}^{s} L(r)T_{61}(E_{Fs},n_z)$$

.

The ER in this case assumes the form

$$\frac{D}{\mu} = \frac{1}{|e|}\frac{\sum_{n_z=1}^{n_{z\max}}\left[T_{61}(E_{Fs},n_z) + T_{62}(E_{Fs},n_z)\right]}{\sum_{n_z=1}^{n_{z\max}}\left[(T_{61}(E_{Fs},n_z))' + (T_{62}(E_{Fs},n_z))'\right]} \tag{1.226}$$

The DOS function for bulk specimens of stressed Kane type semiconductors in the absence of band tail can be written as

$$D_0(E) = g_v(3\pi^2)^{-1}\left[\bar{a}_0(E)\bar{b}_0(E)[\bar{c}_0(E)]' + \bar{a}_0(E)[\bar{b}_0(E)]'\bar{c}_0(E) + [\bar{a}_0(E)]'\bar{b}_0(E)\bar{c}_0(E)\right] \tag{1.227}$$

Combining (1.227) with the Fermi-Dirac occupation probability factor and using the generalized Sommerfeld lemma the electron concentration in this case can be expressed as

$$n_0 = g_v(3\pi^2)^{-1}\left[M_4(E_F) + N_4(E_F)\right] \tag{1.228}$$

where, $M_4(E_F) \equiv \left[\bar{a}_0(E_F)\bar{b}_0(E_F)\bar{c}_0(E_F)\right]$ and $N_4(E_F) \equiv \sum_{r=1}^{s} L(r)M_4(E_F)$.

The ER in this case is given by

$$\frac{D}{\mu} = \frac{1}{|e|}\frac{\left[M_4(E_F) + N_4(E_F)\right]}{\left[\{M_4(E_F)\}' + \{N_4(E_F)\}'\right]} \tag{1.229}$$

1.2.6 The ER from QWs of HD Te

The dispersion relation of the conduction electrons in Te can be expressed as [193]

$$E = \psi_1 k_z^2 + \psi_2 k_s^2 \pm [\psi_3^2 k_z^2 + \psi_4^2 k_s^2]^{1/2} \tag{1.230}$$

where, the values of the system constants are given in Table 1.1.

The carrier energy spectrum in HD Te can be written as

$$\gamma_3(E, \eta_g) = \psi_1 k_z^2 + \psi_2 k_s^2 \pm [\psi_3^2 k_z^2 + \psi_4^2 k_s^2]^{1/2} \tag{1.231}$$

The EEMs along k_z and k_s directions assume the forms

$$m_z^*(E_{F_h}, \eta_g) = \frac{\hbar^2}{2\psi_1}\left[1 - \frac{\psi_3}{\sqrt{\psi_3^2 + 4\psi_1\gamma_3(E_{F_h}, \eta_g)}}\right]\gamma_3'(E_{F_h}, \eta_g) \tag{1.232}$$

and

$$m_s^*(E_{F_h}, \eta_g) = \frac{\hbar^2}{2\psi_2}\left[1 - \frac{\psi_4}{\sqrt{\psi_4^2 + 4\psi_2\gamma_3(E_{F_h}, \eta_g)}}\right]\gamma_3'(E_{F_h}, \eta_g) \tag{1.233}$$

The investigations of EEMs require the expression of electron concentration, which can be written from (1.231) as

$$n_0 = \frac{g_v}{3\pi^2}[t_{1HD}(E_{F_h}, \eta_g) + t_{2HD}(E_{F_h}, \eta_g)] \tag{1.234a}$$

where, $t_{1HD}(E_{F_h}, \eta_g) = \left[3\psi_{5HD}(E_{F_h}, \eta_g)\Gamma_{3HD}(E_{F_h}, \eta_g) - \psi_6\Gamma_{3HD}^3(E_{F_h}, \eta_g)\right]$, $\psi_{5HD}(E_{F_h}, \eta_g) = \left[\frac{\gamma_3(E_{F_h}, \eta_g)}{\psi_2} + \frac{\psi_4^2}{2\psi_2^2}\right]$, $\Gamma_{3HD}(E_{F_h}, \eta_g) = \frac{\sqrt{\psi_3^2 + 4\psi_1\gamma_3(E_{F_h}, \eta_g)}}{2\psi_1}$, $\psi_6 = \frac{\psi_1}{\psi_2}$ and $t_{2HD}(E_{F_h}, \eta_g) = \sum_{r=1}^{s} L(r) t_{1HD}(E_{F_h}, \eta_g)$

In this case $\bar{E}_{hd}$ is given by

$$\{\gamma_3(\bar{E}_{hd}, \eta_g)\} = 0 \tag{1.234b}$$

Therefore by using (1.38), (1.234a) and (1.234b) we can study the ER in this case The 2D electron energy spectrum in HD QW of Te can be written using (1.230) as

$$k_s^2 = \psi_{5HD}(E, \eta_g) - \psi_6\left(\frac{\pi n_z}{d_z}\right)^2 \pm \psi_7\left[\psi_{8HD}^2(E, \eta_g) - \left(\frac{\pi n_z}{d_z}\right)^2\right]^{1/2} \tag{1.235}$$

Table 1.1 The numerical values of the energy band constants of few materials [202]

1	(a) The conduction electrons of n-Cadmium Germanium Arsenide can be described by three types of band models	1. The values of the energy band constants in accordance with the generalized electron dispersion relation of nonlinear optical materials are as follows $E_{g_0} = 0.57\,\text{eV}$, $\Delta_{\|} = 0.30\,\text{eV}$, $\Delta_{\|} = 0.36\,\text{eV}$, $m_{\|}^* = 0.034m_0$, $m_{\|}^* = 0.039m_0$, $T = 4\ K$, $\delta = -0.21\,\text{eV}$, $g_v = 1$ and $\varepsilon_{sc} = 18.4\varepsilon_0$ (ε_{sc} and ε_0 are the permittivity of the semiconductor material and free space respectively. 2. In accordance with the three band model of Kane the spectrum constants are given by $\Delta = (\Delta_{\|} + \Delta_{\perp})/2 = 0.33\,\text{eV}$, $E_{g_0} = 0.57\,\text{eV}$, $m_c = (m_{\|}^* + m_{\perp}^*)/2 = 0.0365m_0$ and $\delta = 0\,\text{eV}$. 3. In accordance with two band model of Kane, the spectrum constants are given by $E_{g_0} = 0.57\,\text{eV}$ and $m_c = (m_{\|}^* + m_{\perp}^*)/2 = 0.0365m_0$.
	(b) The conduction electrons of n-Cadmium Arsenide can be described by three types of band models	1. The values of the energy band constants in accordance with the generalized electron dispersion relation of nonlinear optical materials are as follows $\lvert E_{g_0}\rvert = 0.095\,\text{eV}$, $\Delta_{\|} = 0.27\,\text{eV}$, $\Delta_{\perp} = 0.25\,\text{eV}$, $m_{\|}^* = 0.00697m_0$, $m_{\perp}^* = 0.013933m_0$, $T = 4\ K$, $\delta = 0.085\,\text{eV}$, $g_v = 1$ and $\varepsilon_{sc} = 16\varepsilon_0$. 2. In accordance with the three band model of Kane, the spectrum constants are given by $\Delta = (\Delta_{\|} + \Delta_{\perp})/2 = 0.26\,\text{eV}$, $\lvert E_{g_0}\rvert = 0.095\,\text{eV}$, $m_c = (m_{\|}^* + m_{\perp}^*)/2 = 0.0105m_0$ and $\delta = 0\,\text{eV}$. 3. In accordance with two band model of Kane, the spectrum constants are given by $\lvert E_{g_0}\rvert = 0.095\,\text{eV}$, and $m_c = (m_{\|}^* + m_{\perp}^*)/2 = 0.0105m_0$
2	n-Indium Arsenide	The values $E_{g_0} = 0.36\,\text{eV}$, $\Delta = 0.43\,\text{eV}$, $m_c = 0.026m_0$, $g_v = 1$ and $\varepsilon_{sc} = 12.25\varepsilon_0$ are valid for three band model of Kane
3	n-Gallium Arsenide	The values $E_{g_0} = 1.55\,\text{eV}$, $\Delta = 0.35\,\text{eV}$, $m_c = 0.066m_0$, $g_v = 1$ and $\varepsilon_{sc} = 12.9\varepsilon_0$ are valid for three band model of Kane. The values $a_{13} = -1.97 \times 10^{-37}\ \text{eVm}^4$ and $a_{15} = -2.3 \times 10^{-34}\ \text{eVm}^4$ are valid for the Newson and Kurobe model [40]
4	n-Gallium Aluminium Arsenide	$E_{g_0} = (1.424 + 1.266x + 0.26x^2)\text{eV}$, $\Delta = (0.34 - 0.5x)\text{eV}$, $g_v = 1$, $m_c = (0.066 + 0.088x)m_0$ and $\varepsilon_{sc} = [13.18\text{-}3.12x]\varepsilon_0$

(continued)

Table 1.1 (continued)

5	n-Mercury Cadmium Telluride	$E_{g_0} = (-0.302 + 1.93x + 5.35 \times 10^{-4}(1-2x)T - 0.810x^2 + 0.832x^3)\text{eV}$, $\Delta = (0.63 + 0.24x - 0.27x^2)\text{eV}$, $m_c = 0.1m_0E_{g_0}(\text{eV})^{-1}$, $g_v = 1$ and $\varepsilon_{sc} = [20.262\text{-}14.812x + 5.22795x^2]\varepsilon_0$
6	n-Indium Gallium Arsenide Phosphide lattice matched to Indium Phosphide	$E_{g_0} = (1.337 - 0.73y + 0.13y^2)\text{eV}$, $\Delta = (0.114 + 0.26y - 0.22y^2)\text{eV}$, $y = (0.1896\text{-}0.4052x)/(0.1896\text{-}0.0123x)$, $m_c = (0.08\text{-}0.039y)m_0$ and $g_v = 1$, $\varepsilon_{sc} = [10.65 + 0.1320y]\varepsilon_0$
7	n-Indium Antimonide	$E_{g_0} = 0.2352\,\text{eV}$, $\Delta = 0.81\,\text{eV}$, $m_c = 0.01359m_0$, $g_v = 1$ and $\varepsilon_{sc} = 15.56\varepsilon_0$
8	n-Gallium Antimonide	The values of $E_{g_0} = 0.81\,\text{eV}$, $\Delta = 0.81\,\text{eV}$, $P = 9.48 \times 10^{-10}\text{eVm}$, $\bar{\varsigma}_0 = -2.1$, $\bar{v}_0 = -1.49$, $\bar{\omega}_0 = 0.42$, $g_v = 1$ and $\varepsilon_{sc} = 15.85\varepsilon_0$ are valid for the model of Seiler et al.
9	n-Cadmium Sulphide	$m^*_{\parallel} = 0.7m_0$, $m^*_{\perp} = 1.5m_0$, $\bar{\lambda}_0 = 1.4 \times 10^{-8}\text{eVm}$, $g_v = 1$ and $\varepsilon_{sc} = 15.5\varepsilon_0$
10	n-Lead Telluride	The values $m_t^- = 0.070m_0$, $m_l^- = 0.54m_0$, $m_t^+ = 0.010m_0$, $m_l^+ = 1.4m_0$, $P_{\parallel} = 141\,\text{meVnm}$, $P_{\perp} = 486\,\text{meVnm}$, $E_{g_0} = 190\,\text{meV}$, $g_v = 4$ and $\varepsilon_{sc} = 33\varepsilon_0$ are valid for the Dimmock model The values $m_1 = 0.0239m_0$, $m_2 = 0.024m_0$, $m_2' = 0.31m_0$, $m_3 = 0.24m_0$ are valid for the Cohen model
11	Stressed n-Indium Antimonide	The values $m_c = 0.01359m_0$, $E_{g_0} = 0.081\,\text{eV}$, $B_2 = 9{\times}10^{-10}\backslash text\{eVm\}$, $C_1 = 3\ \backslash text\{eV\}$, $C_2 = 2\ \backslash text\{eV\}$, $\bar{a}_0 = -10\,\text{eV}$, $\bar{b}_0 = -1.7\,\text{eV}$, $\bar{d}_0 = -4.4\,\text{eV}$, $S_{xx} = 0.6 \times 10^{-3}(kbar)^{-1}$, $S_{yy} = 0.42 \times 10^{-3}(kbar)^{-1}$, $S_{zz} = 0.39 \times 10^{-3}(kbar)^{-1}$, $S_{xy} = 0.5 \times 10^{-3}(kbar)^{-1}$, $\varepsilon_{xx} = \sigma S_{xx}$, $\varepsilon_{yy} = \sigma S_{yy}$, $\varepsilon_{zz} = \sigma S_{zz}$, $\varepsilon_{xy} = \sigma S_{xy}$, σ is the stress in kilobar, $g_v = 1$ are valid for the model of Seiler et al. [183]
12	Bismuth	$E_{g_0} = 0.0153\,\text{eV}$, $m_1 = 0.00194m_0$, $m_2 = 0.313m_0$, $m_3 = 0.00246m_0$, $m_2' = 0.36m_0$, $g_v = 3$, $M_2 = 1.25m_0$ and $M_2' = 0.36m_0$
13	Mercury Telluride	$m_v^* = 0.028m_0$, $g_v = 1$ and $\varepsilon_\infty = 15.2\varepsilon_0$
14	Platinum Antimonide	For valence bands, along $< 100 >$ direction, $\bar{\lambda}_0 = (0.02/4)\text{eV}$, $\bar{l} = (-0.32/4)\text{eV}$, $\bar{v} = (0.39/4)\text{eV}$, $\bar{n} = (-0.65/4)\text{eV}$, $\bar{a} = 0.643nm$, $I = 0.30(\text{eV})^2$, $\bar{\delta}_0 = 0.02\,\text{eV}$, $g_v = 6$, $\varepsilon_{sc} = 30\varepsilon_0$ and $\phi_w \approx 3.0\,\text{eV}$. For conduction bands, along $< 111 >$ direction, $g_v = 8$ $\bar{\lambda}_0 = (0.33/4)\text{eV}$, $\bar{l} = (1.09/4)\text{eV}$, $\bar{v} = (0.17/4)\text{eV}$ and $\bar{n} = (0.22/4)\text{eV}$

(continued)

Table 1.1 (continued)

15	n-Gallium Phosphide	$m_{\parallel}^{*} = 0.92m_0$, $m_{\perp}^{*} = 0.25m_0$, $k_0 = 1.7 \times 10^{15} m^{-1}$, $\lvert V_G \rvert = 0.21\,\text{eV}$ and $g_v = 6$
16	Germanium	$E_{g_0} = 0.785\,\text{eV}$, $m_{\parallel}^{*} = 1.57m_0$, $m_{\perp}^{*} = 0.0807m_0$ and $g_v = 4$
17	Tellurium	The values $\psi_1 = 6.7 \times 10^{-16}\text{meVm}^2$, $\psi_2 = 4.2 \times 10^{-16}\text{meVm}^2$, $\psi_3 = 6 \times 10^{-8}\text{meVm}$ and $\psi_4 = (3.6 \times 10^{-8}\text{meVm})$ are valid for the model of Bouat et al. [184]
18	Lead Germanium Telluride	The values $g_v = 4$ and $\phi_w \approx 6\,\text{eV}$ are valid for the model of Vassilev as given by [208]
19	Cadmium Antimonide	The values $a_1 = -32.3 \times 10^{-20}\text{eVm}^2$, $b_1 = -60.7 \times 10^{-20}\text{eVm}^2$, $a_2 = -16.3 \times 10^{-20}\text{eVm}^2$, $b_2 = -24.4 \times 10^{-20}\text{eVm}^2$, $a_3 = -91.9 \times 10^{-20}\text{eVm}^2$, $b_3 = -105 \times 10^{-20}\text{eVm}^2$, $A = 2.92 \times 10^{-10}\text{eVm}$, $B = -3.47 \times 10^{-10}\text{eVm}$, $G_3 = 1.3 \times 10^{-10}\text{eVm}$, and $\Delta_3 = 0.070\,\text{eV}$
20	Cadmium Diphosphide	The values $\beta_1 = 8.6 \times 10^{-21}\text{eVm}^2$, $\beta_2 = 1.8 \times 10^{-21}(\text{eVm})^2$, $\beta_4 = 0.0825\,\text{eV}$ and $\beta_5 = -1.9 \times 10^{-19}\text{eVm}^2$ are valid for the model of Chuiko [215]
21	Zinc Diphosphide	The values $\beta_1 = 8.7 \times 10^{-21}\text{eVm}^2$, $\beta_2 = 1.9 \times 10^{-21}(\text{eVm})^2$, $\beta_4 = 0.0875\,\text{eV}$ and $\beta_5 = -1.9 \times 10^{-19}\text{eVm}^2$ are valid for the model of Chuiko [215]
22	Bismuth Telluride	The values $E_{g_0} = 0.145\,\text{eV}$, $\bar{\alpha}_{11} = 4.9$, $\bar{\alpha}_{22} = 5.92$, $\bar{\alpha}_{33} = 9.5$, $\bar{\alpha}_{23} = 4.22$ and $g_v = 6$
23	Antimony	The values $\alpha_{11} = 16.7$, $\alpha_{22} = 5.98$, $\alpha_{33} = 11.61$ and $\alpha_{23} = 7.54$ are valid for the model of Ketterson [211]
24	Zinc Selenide	$m_{c2} = 0.16m_0$, $\Delta_2 = 0.42\,\text{eV}$ and $E_{g_{02}} = 2.82\,\text{eV}$
25	Lead Selenide	$m_t^{-} = 0.23m_0$, $m_l^{-} = 0.32m_0$, $m_t^{+} = 0.115m_0$, $m_l^{+} = 0.303m_0$, $P_{\parallel} \approx 138\,\text{meVnm}$, $P_{\perp} \approx 471\,\text{meVnm}$, $E_{g_0} = 0.28\text{eV}$ and $\varepsilon_{sc} = 21.0\varepsilon_0$

where, $\psi_7 = \frac{\psi_4\sqrt{\psi_1}}{\psi_2^{3/2}}$ and $\psi_{8HD}^2(E,\eta_g) = \left[\frac{\psi_4^4+4\gamma_3(E,\eta_g)\psi_2\psi_4^2+4\psi_2^2\psi_3^2}{4\psi_1\psi_2\psi_4^2}\right]$

The EEM in this case is given by

$$m^*(E_{F1HD},\eta_g,n_z) = \frac{\hbar^2}{2}\left[\psi'_{5HD}(E_{F1HD},\eta_g) \pm \frac{\psi_{8HD}(E_{F1HD},\eta_g)\psi'_{8HD}(E_{F1HD},\eta_g)}{\sqrt{\psi_{8HD}^2(E_{F1HD},\eta_g) - (\pi n_z/d_z)^2}}\right] \tag{1.236}$$

The total DOS function in this case can be expressed as

$$N_{2DT}(E) = \frac{g_v}{\pi}\sum_{n_z=1}^{n_{z\max}} \psi'_{5HD}(E,\eta_g)H(E - E_{n_{z59HD}}) \tag{1.237}$$

where $E_{n_{z59HD}}$ is the lowest positive root of the equation

$$\psi_{5HD}(E_{n_{z59HD}},\eta_g) - \psi_6\left(\frac{\pi n_z}{d_z}\right)^2 \pm \psi_7\left[\psi_{8HD}^2(E_{n_{z59HD}},\eta_g) - \left(\frac{\pi n_z}{d_z}\right)^2\right]^{1/2} = 0 \tag{1.238}$$

The surface electron concentration is given by

$$n_{2D} = \frac{g_v}{\pi}\sum_{n_z=1}^{n_{z\max}} \left[t_{1HDTe}(E_{F1HD},\eta_g,n_z) + t_{2HDTe}(E_{F1HD},\eta_g,n_z)\right] \tag{1.239}$$

where,

$$t_{1HDTe}(E_{F1HD},\eta_g,n_z) = \left[\psi_{5HD}(E_{F1HD},\eta_g,n_z) - \psi_6\left(\frac{\pi n_z}{d_z}\right)^2\right]$$

and

$$t_{2HDTe}(E_{F1HD},\eta_g,n_z) = \sum\nolimits_{r=1}^{s} L(r)\left[t_{1HDTe}(E_{F1HD},\eta_g,n_z)\right]$$

Thus using (1.239) and (1.38) we can study the ER in this case.

The 2D electron energy spectrum in QWs of Te in the absence of band tails assumes the form

$$k_s^2 = \psi_5(E) - \psi_6\left(\frac{\pi n_z}{d_z}\right)^2 \pm \psi_7\left[\psi_8^2(E) - \left(\frac{\pi n_z}{d_z}\right)^2\right]^{1/2} \tag{1.240}$$

where, $\psi_5(E) = \left[\frac{E}{\psi_2} + \frac{\psi_4^2}{2\psi_2^2}\right]$ and $\psi_8^2(E) = \left[\frac{\psi_4^4 + 4E\psi_2\psi_4^2 + 4\psi_2^2\psi_3^2}{4\psi_1\psi_2\psi_4^2}\right]$

Thus, the total 2D DOS function can be expressed as

$$N_{2DT}(E) = \left(\frac{g_v}{\pi}\right) \sum_{n_z=1}^{n_{z\max}} t'_{40}(E, n_z) H(E - E_{nz12}) \tag{1.241}$$

where, $t_{40}(E, n_z) = \left[\psi_5(E) - \psi_6\left(\frac{\pi n_z}{d_z}\right)^2 \pm \psi_7\left[\psi_8^2(E) - \left(\frac{\pi n_z}{d_z}\right)^2\right]^{1/2}\right]^{1/2}$

The sub-band energies$(E_{n_{z12}})$ are given by

$$E_{n_{z12}} = \psi_1 (n_z \pi / d_z)^2 \pm \psi_3 (n_z \pi / d_z) \tag{1.242a}$$

Using (1.240) the EEM can be expressed as

$$m^*(E_{F_s}, n_z) = \frac{\hbar^2}{2} t'_{40}(E_{F_s}, n_z) \tag{1.242b}$$

The 2D surface electron concentration per unit area for QWs of Te can be written as

$$n_{2D} = \frac{g_v}{\pi} \sum_{n_z=1}^{n_{z\max}} \left[t_{40}(E_{Fs}, n_z) + t_{41}(E_{Fs}, n_z)\right] \tag{1.243}$$

where $t_{41}(E_{Fs}, n_z) \equiv \sum_{r=1}^{s} L(r) t_{40}(E_{Fs}, n_z)$.

The ER in this case is given by

$$\frac{D}{\mu} = \frac{1}{e}\left[\sum_{n_z=1}^{n_{z\max}} \left[t_{40}(E_{F_s}, n_z) + t_{41}(E_{F_s}, n_z)\right]\right]\left[\sum_{n_z=1}^{n_{z\max}} \left[t'_{40}(E_{F_s}, n_z) + t'_{41}(E_{F_s}, n_z)\right]\right]^{-1} \tag{1.244}$$

The electron concentration and the ER for bulk specimens of Te in the absence of band tails can, respectively, be expressed as

$$n_0 = \frac{1}{3\pi^2}\left[M_9(E_F) + N_9(E_F)\right] \tag{1.245}$$

and

$$\frac{D}{\mu} = \frac{1}{e}\left[\frac{M_9(E_F) + N_9(E_F)}{M'_9(E_F) + N'_9(E_F)}\right] \tag{1.246}$$

where, $M_9(E_F) = \left[3\psi_5(E_F)\Gamma_3(E_F) - \psi_6\Gamma_3^3(E_F)\right], \psi_5(E_F) = \left[\frac{E_F}{\psi_2} + \frac{\psi_4^2}{2\psi_2^2}\right]$ and $\Gamma_3(E_F) = [2\psi_1]^{-1}\left[\sqrt{\psi_3^2 + 4\psi_1 E_F} - \psi_3\right]$ and $N_9(E_F) \equiv \sum_{r=1}^{s} L(r)M_9(E_F)$

1.2.7 The ER from QWs of HD Gallium Phosphide

The energy spectrum of the conduction electrons in n-GaP can be written as [194]

$$E = \frac{\hbar^2 k_s^2}{2m_\perp^*} + \frac{\hbar^2}{2m_\parallel^*}[\bar{A}' k_s^2 + k_z^2] - \left[\frac{\hbar^4 k_0^2}{m_\parallel^{*2}}(k_s^2 + k_z^2) + |V_G|^2\right]^{1/2} + |V_G| \quad (1.247)$$

where, k_0 and $|V_G|$ are constants of the energy spectrum and $\bar{A}' = 1$.

The dispersion relation of the conduction electrons in HD n-GaP can be expressed as

$$\gamma_3(E, \eta_g) = \frac{\hbar^2 k_s^2}{2m_\perp^*} + \frac{\hbar^2}{2m_\parallel^*}[\bar{A}' k_s^2 + k_z^2] - \left[\frac{\hbar^4 k_0^2}{m_\parallel^{*2}}(k_s^2 + k_z^2) + |V_G|^2\right]^{1/2} - |V_G| \quad (1.248)$$

The EEMs assume the forms as

$$m_z^*(E_{F_h}, \eta_g) = \frac{\hbar^2 \gamma_3'(E_{F_h}, \eta_g)}{b}[1 \pm (C + bD)[C^2 + 4bD^2 + 4bC\gamma_3(E_{F_h}, \eta_g) - 4bCD + 4b^2\gamma_3(E_{F_h}, \eta_g)D]^{-1/2}] \quad (1.249)$$

and

$$m_s^*(E_{F_h}, \eta_g) = \frac{\hbar^2}{2}[t_{11}\gamma_3'(E_{F_h}, \eta_g) - t_{41}t_5'(E_{F_h}, \eta_g)] \quad (1.250)$$

where, $b = \frac{\hbar^2}{2m_\parallel^*}$, $C = (\hbar^2 k_0/m_\parallel^*)^2$, $D = |V_G|$, $t_{11} = \frac{1}{a}$, $a = \frac{\hbar^2}{2m_\perp^*} + \bar{A}' b$, $t_{41} = \frac{\sqrt{g_3}}{2a^2}$, $g_3 = (4abc + 4a^2 c)$, $t_5^2(E_{F_h}, \eta_g) = [g_2 - 4aC\gamma_3(E_{F_h}, \eta_g)](g_3)^{-1}$, $g_2 = (4a^2b^2 + C^2 + 4aCD)$

The electron concentration can be expressed as

$$n_0 = \frac{g_v}{4\pi^2}\left[\bar{I}_{127}(E_{F_h},\eta_g) + \sum_{r=1}^{s} L(r)[\bar{I}_{127}(E_{F_h},\eta_g)]\right] \quad (1.251)$$

where,

$$\bar{I}_{127}(E_{F_h},\eta_g) = [M_{1HD}(E_{F_h},\eta_g)],$$

$$M_{1HD}(E_{F_h},\eta_g) = \Big[2(t_{11}\gamma_3(E_{F_h},\eta_g) + t_{21})\sqrt{t_{81} + t_{91}\gamma_3(E_{F_h},\eta_g)} + (t_{31}/3)\theta^3_{,-}(E_{F_h},\eta_g) + (t_{41}/2)\Big[\theta_{,-}(E_{F_h},\eta_g)\sqrt{\theta^2_{-}(E_{F_h},\eta_g) + t_5(E_{F_h},\eta_g)} - \sqrt{t_5(E_{F_h},\eta_g)}\Big] + (t_{41}t_5(E_{F_h},\eta_g)/2) \ln\left|\frac{\theta_{,-}(E_{F_h},\eta_g) + \sqrt{\theta^2_{-}(E_{F_h},\eta_g) + t_5(E_{F_h},\eta_g)}}{\sqrt{t_5(E_{F_h},\eta_g)}}\right|\Big]\Big],$$

$t_{21} = \frac{g_1}{2a^2}, g_1 = -(C + 2aD)$, $t_{81} = [t_{41}^4 + 4t_{41}^2 t_{21} t_{31} + (4t_{31}^2 t_{41}^2 g_2)(g_3)^{-1}]$, $t_{31} = \frac{b}{a}$, $t_{91} = [4t_{11}t_{31}t_{41}^2 + 8t_{11}t_{21}t_{31}^2 - (16t_{31}^2 t_{41}^2 aC)(g_3)^{-1}], \theta_{-}(E_{F_h},\eta_g) = (t_{31}\sqrt{2})^{-1}$ $[t_{61} + t_{71}\gamma_3(E_{F_h},\eta_g) - \sqrt{t_{81} + t_{91}\gamma_3(E_{F_h},\eta_g)}]$, $t_{61} = (t_{41}^2 + 2t_{21}t_{31})$ and $t_{71} = (2t_{11}t_{31})$

The $\bar{E}_{hd}$ in this case is given by the equation

$$\gamma_3(\bar{E}_{hd},\eta_g) = 0 \quad (1.252)$$

Therefore using (1.251), (1.252), and (1.31c) we can study the ER in this case.

The 2D dispersion relation in QW of HD GaP can be expressed following (1.248) as

$$k_s^2 = t_{11}\gamma_3(E,\eta_g) + t_{21} - t_{31}\left(\frac{\pi n_z}{d_z}\right)^2 - t_{41}\left[\left(\frac{\pi n_z}{d_z}\right)^2 + t_5^2(E,\eta_g)\right]^{1/2} \quad (1.253)$$

The EEM in this case can be written following (1.253) as

$$m^*(E_{F1HD},\eta_g,n_z) = \frac{\hbar^2}{2}\left[t_{11}\gamma_3'(E_{F1HD},\eta_g) - t_{41}t_5(E_{F1HD},\eta_g)t_5'(E_{F1HD},\eta_g)\left[\left(\frac{\pi n_z}{d_z}\right)^2 + t_5^2(E_{F1HD},\eta_g)\right]^{-1/2}\right] \quad (1.254)$$

The total DOS function assumes the form

$$N_{2DT}(E,\eta_g)=\frac{g_v}{2\pi}\sum_{n_z=1}^{n_{z\max}}[t_{11}\gamma'_3(E,\eta_g)-t_{41}t_5(E,\eta_g)t'_5(E,\eta_g)\left[\left(\frac{\pi n_z}{d_z}\right)^2+t_5^2(E,\eta_g)\right]^{-1/2}]H(E-E_{n_{z8THD}}) \quad (1.255)$$

where, $E_{n_{z8THD}}$ is given by the equation

$$t_{11}\gamma_3(E_{n_{z8THD}},\eta_g)+t_{21}-t_{31}\left(\frac{\pi n_z}{d_z}\right)^2-t_{41}\left[\left(\frac{\pi n_z}{d_z}\right)^2+t_5^2(E_{n_{z8THD}},\eta_g)\right]^{1/2}=0 \quad (1.256)$$

The surface electron concentration in QW of HD n-GaP can be written as

$$n_s=\frac{g_v}{\pi}\sum_{n_z=1}^{n_{z\max}}[t_{3HDGaP}(E_{F1HD},\eta_g,n_z)+t_{4HDGaP}(E_{F1HD},\eta_g,n_z)] \quad (1.257)$$

where, $t_{1HDGaP}(E_{F1HD},\eta_g,n_z)=t_{11}\gamma_3(E_{F1HD},\eta_g,n_z)\ +t_{21}-t_{31}\left(\frac{\pi n_z}{d_z}\right)^2-t_{41}$ $\left[\left(\frac{\pi n_z}{d_z}\right)^2+t_5^2(E_{F1HD},\ \eta_g,n_z)\right]^{1/2}$ and $t_{4HDGaP}(E_{F1HD},\eta_g,n_z)=\sum_{r=1}^{s}L(r)[t_{3HDGaP}$ $(E_{F1HD},\eta_g,n_z)]$

Thus using (1.257) and (1.38) we can study the ER in this case.

The 2D electron dispersion relation in size-quantized n-GaP in the absence of band tails assumes the form

$$E=ak_s^2+C(n_z\pi/d_z)^2+|V_G|-\left[Dk_s^2+|V_G|^2+D(n_z\pi/d_z)^2\right]^{1/2} \quad (1.258)$$

The sub-band energy $(E_{n_{z13}})$ are given by

$$E_{n_{z13}}=C(\pi n_z/d_z)^2+|V_G|-\left[|V_G|^2+D(\pi n_z/d_z)^2\right]^{1/2} \quad (1.259)$$

The (1.258) can be expressed as

$$k_s^2=t_{42}(E,n_z) \quad (1.260)$$

in which, $t_{42}(E,n_z)\equiv\left[\{2a(E-t_1)+D\}-\{[2a(E-t_1)+D]^2-4a^2[(E-t_1)^2-t_2]\}^{1/2}\right]$, $t_1\equiv|V_G|+C(\pi n_z/d_z)^2$ and $t_2\equiv|V_G|^2+D(\pi n_z/d_z)^2$.

The total DOS function is given by

$$N_{2DT}(E) = \frac{g_v}{4\pi a^2}\sum_{n_z=1}^{n_{z\max}}[t'_{42}(E,n_z)]H(E - E_{n_{z13}}) \tag{1.261a}$$

Using (1.260) the EEM can be expressed as

$$m^*(E_{F_s},n_z) = \frac{\hbar^2}{2}t'_{42}(E_{F_s},n_z) \tag{1.261b}$$

The electron statistics in QWs in n-GaP assumes the form

$$\begin{aligned} n_{2D} &= \left[\left(\frac{g_v}{4\pi a^2}\right)\sum_{n_z=1}^{n_{z\max}}[t_{42}(E_{Fs},n_z) + t_{43}(E_{Fs},n_z)]\right] \\ t_{43}(E_{Fs},n_z) &= \sum_{r=1}^{s} L(r)[t_{42}(E_{Fs},n_z)] \end{aligned} \tag{1.262}$$

where, $t_{43}(E_{Fs},n_z) = \sum_{r=1}^{s} L(r)[t_{42}(E_{Fs},n_z)]$

The ER in this case is given by

$$\frac{D}{\mu} = \left(\frac{1}{e}\right)\left[\sum_{n_z=1}^{n_{z\max}}[t'_{42}(E_{Fs},n_z) + t'_{43}(E_{Fs},n_z)]\right]^{-1}\left[\sum_{n_z=1}^{n_{z\max}}(t_{42}(E_{Fs},n_z) + t_{43}(E_{Fs},n_z))\right] \tag{1.263}$$

The EEMs in bulk specimens of n-GaP in the absence of band tails can be written as

$$m_s^*(E_F) = \frac{\hbar^2}{2}[t_{11} - t_{41}t'_5(E_F)] \tag{1.264}$$

and

$$m_z^*(E_F) = \frac{\hbar^2}{b}[1 - C[4bCE_F + 4b^2D^2 + C^2 - 4bCD]^{-1/2}] \tag{1.265}$$

where $t_5(E_F) = \left[\frac{g_2 - 4aCE_F}{g_3}\right]^{1/2}$

The electron concentration and the ER in this case assume the forms

$$n_0 = \frac{g_v}{4\pi^2}[M_1(E_F) + N_1(E_F)] \tag{1.266}$$

$$\frac{D}{\mu} = \frac{1}{e}[M_1(E_F) + N_1(E_F)][M'_1(E_F) + N'_1(E_F)]^{-1} \tag{1.267}$$

where $M_1(E_F) = [2(t_{11}E_F + t_{21})\sqrt{t_{91}E_F + t_{81}} + \frac{t_{31}}{3}\phi^3(E_F) + \frac{t_{41}}{2}[\phi(E_F)\sqrt{\phi^2(E_F) + t_5(E_F)} + \frac{t_{41}t_5(E_F)}{2}[\ln\left|\frac{\phi(E_F)+\sqrt{\phi^2(E_F)+t_5(E_F)}}{\sqrt{t_5(E_F)}}\right|]]$, $\phi(E_F) = (t_{31}\sqrt{2})^{-1}[t_{61} + E_F t_{71} - [t_{81} + t_{91} E_F]^{1/2}]$ and $N_1(E_F) = \sum_{r=1}^{s_0}[L(r)M_1(E_F)]$

1.2.8 The ER in QWs of HD Platinum Antimonide

The dispersion relation for the n-type $PtSb_2$ can be written as [195]

$$\left[E + \lambda_0\frac{a^2}{4}k^2 - lk_s^2\frac{a^2}{4}\right]\left[E + \delta_0 - v\frac{a^2}{4}k^2 - n'k_s^2\frac{a^2}{4}\right] = I\left(\frac{a^4}{16}\right)k^4 \tag{1.268}$$

The (1.268) assumes the form

$$\left[E + \omega_1 k_s^2 + \omega_2 k_z^2\right]\left[E + \delta_0 + \omega_3 k_s^2 - \omega_4 k_z^2\right] = I_1\left(k_s^2 + k_z^2\right)^2 \tag{1.269}$$

where $\omega_1 = [\lambda_0\frac{a^2}{4} + l\frac{a^2}{4}]$, $\omega_2 = \lambda_0\frac{a^2}{4}$, $\omega_3 = [n'\frac{a^2}{4} - \upsilon\frac{a^2}{4}]$, $\omega_4 = \upsilon\frac{a^2}{4}$, $I_1 = I(\frac{a^2}{4})^2$, $\lambda_0, l, \delta_0, v, n'$ and a are the band constants.

The carrier dispersion law in HD $PtSb_2$ can be written as

$$T_{11}k_s^4 - k_s^2[T_{21}(E,\eta_g) - T_{31}k_z^2] + [T_{41}k_z^4 - T_{51}(E,\eta_g)k_z^2 - T_{61}(E,\eta_g)] = 0 \tag{1.270}$$

where, $T_{11} = (I_1 - \omega_2\omega_3)$, $T_{21}(E,\eta_g) = [\omega_1\delta_0 + \omega_1\gamma_3(E,\eta_g) + \omega_3\gamma_3(E,\eta_g)]$, $T_{31} = [2I_1 + \omega_2\omega_4 - \omega_2\omega_3]$, $T_{41} = [2I_1 + \omega_2\omega_4]$, $T_{51}(E,\eta_g) = [\omega_2\gamma_0(E,\eta_g) - \omega_4\gamma_3\ (E,\eta_g) + \omega_2\gamma_3(E,\eta_g)]$, $T_{61}(E,\eta_g) = [\gamma_8(E,\eta_g) + \gamma_0(E,\eta_g)\gamma_3(E,\eta_g)]$ and $\gamma_8(E,\eta_g) = 2\theta_0\ (E,\eta_g)[1 + Erf(E/\eta_g)]^{-1}$

The EEMs are given by

$$m_s^*(E_{F_h},\eta_g) = \frac{\hbar^2}{2T_{11}}\left[T'_{21}(E_{F_h},\eta_g) + \frac{(T_{21}(E_{F_h},\eta_g)T'_{21}(E_{F_h},\eta_g) + 2T_{11}T'_{61}(E_{F_h},\eta_g))}{\sqrt{T_{21}^2(E_{F_h},\eta_g) + 4T_{11}T_{61}(E_{F_h},\eta_g)}}\right] \tag{1.271}$$

and

$$\begin{aligned} m_z^*(E_{F_h},\eta_g) = (\frac{\hbar^2}{2T_{41}})[T'_{51}(E_{F_h},\eta_g) + [T_{51}(E_{F_h},\eta_g)T'_{51}(E_{F_h},\eta_g) \\ + 2T_{41}T'_{61}(E_{F_h},\eta_g)][T_{51}^2(E_{F_h},\eta_g) + 4T_{41}T_{61}(E_{F_h},\eta_g)]^{-1/2}] \end{aligned} \tag{1.272}$$

The electron concentration assumes the form

$$n_0 = \frac{g_v}{3\pi^2}\left[\bar{I}_{128}(E_{F_h}, \eta_g) + \sum_{r=1}^{s} L(r)[\bar{I}_{128}(E_{F_h}, \eta_g)]\right] \tag{1.273a}$$

where, $\bar{I}_{128}(E_{F_h}, \eta_g) = [M_{6HD}(E_{F_h}, \eta_g)]$,

$$M_{6HD}(E_{F_h}, \eta_g) = [T_{91HD}(E_{F_h}, \eta_g)\rho_{2HD}(E_{F_h}, \eta_g) - T_{101}\frac{\rho_{2HD}^3(E_{F_h}, \eta_g)}{3} - T_{11}J_3(E_{F_h}, \eta_g)],$$

$$T_{91HD}(E_{F_h}, \eta_g) = \frac{T_{21}(E_{F_h}, \eta_g)}{2T_{11}},$$

$$\rho_{2HD}(E_{F_h}, \eta_g) = [(2T_{41})^{-1}[T_{51}(E_{F_h}, \eta_g) + \sqrt{T_{51}^2(E_{F_h}, \eta_g) + 4T_{41}T_{61}(E_{F_h}, \eta_g)}]]^{1/2}$$

$$T_{101} = [T_{31}/2T_{11}],$$

$$J_3(E_{F_h}, \eta_g) = \frac{\rho_{2HD}(E_{F_h}, \eta_g)}{3}[[A_{3HD}^2(E_{F_h}, \eta_g) + B_{3HD}^2(E_{F_h}, \eta_g)]E_0(\eta(E_{F_h}, \eta_g), t(E_{F_h}, \eta_g)) - [A_{3HD}^2(E_{F_h}, \eta_g) - B_{3HD}^2(E_{F_h}, \eta_g)]F_0(\eta(E_{F_h}, \eta_g), t(E_{F_h}, \eta_g))] + \frac{\rho_{2HD}(E_{F_h}, \eta_g)}{3}[(A_{3HD}^2(E_{F_h}, \eta_g) - \rho_{2HD}^2(E_{F_h}, \eta_g))(B_{3HD}^2(E_{F_h}, \eta_g) - \rho_{2HD}^2(E_{F_h}, \eta_g))]^{1/2},$$

$E_0(\eta(E_{F_h}, \eta_g), t(E_{F_h}, \eta_g))$ and $F_0(\eta(E_{F_h}, \eta_g), t(E_{F_h}, \eta_g))$ are the incomplete elliptic integrals of second and first respectively,

$$A_{3HD}^2(E_{F_h}, \eta_g) = \frac{1}{2}[T_{12}(E_{F_h}, \eta_g) + \sqrt{T_{12}^2(E_{F_h}, \eta_g) - 4T_{13}(E_{F_h}, \eta_g)}], T_{12}(E_{F_h}, \eta_g) = [T_7(E_{F_h}, \eta_g)/\bar{T}_{61}]$$

$$\bar{T}_{61} = [T_{31}^2 - 4T_{11}T_{41}],\ T_7(E_{F_h}, \eta_g) = [2T_{31}T_{21}(E_{F_h}, \eta_g) - 4T_{11}T_{51}(E_{F_h}, \eta_g)],$$

$$T_{13}(E_{F_h}, \eta_g) = (T_8(E_{F_h}, \eta_g)/\bar{T}_8),$$

$$T_8(E_{F_h}, \eta_g) = [T_{21}^2(E_{F_h}, \eta_g) + 4T_{11}T_{61}(E_{F_h}, \eta_g)],$$

$$B_{3HD}^2(E_{F_h}, \eta_g) = \frac{1}{2}[T_{12}(E_{F_h}, \eta_g) - \sqrt{T_{12}^2(E_{F_h}, \eta_g) - 4T_{13}(E_{F_h}, \eta_g)}],\ \bar{T}_{11} = [\sqrt{\bar{T}_{61}}/2T_{11}]$$

$$t(E_{F_h}, \eta_g) = [B_3(E_{F_h}, \eta_g)/A_3(E_{F_h}, \eta_g)]\eta(E_{F_h}\eta_g) = \sin^{-1}[\frac{\rho_2(E_{F_h}, \eta_g)}{B_3(E_{F_h}, \eta_g)}]$$

The $\bar{E}_{hd}$ in this case is given by the equation

$$T_{61}(\bar{E}_{hd}, \eta_g) = 0 \tag{1.273b}$$

Using (1.273a), (1.273b), and (1.38), we can study the ER in this case.
From (1.270) the dispersion relation in QWs of HD $PtSb_2$ can be expressed as

$$T_{11}k_s^4 - P_{1HD}(E,\eta_g,n_z)k_s^2 + P_{2HD}(E,\eta_g,n_z) = 0 \tag{1.274}$$

where,

$$P_{1HD}(E,\eta_g,n_z) = \left[T_{21}(E_{F_h},\eta_g) - T_{31}(\pi n_z/d_z)^2\right]$$
$$P_{2HD}(E,\eta_g,n_z) = \left[T_{41}(\pi n_z/d_z)^4 - T_{51}(E_{F_h},\eta_g)(\pi n_z/d_z)^2 - T_{61}(E_{F_h},\eta_g)\right]$$

(1.274) can be written as

$$k_s^2 = A_{60}(E,\eta_g,n_z) \tag{1.275}$$

where,

$$A_{60}(E,\eta_g,n_z) = \left[P_{1HD}(E,\eta_g,n_z) - \sqrt{P_{1HD}^2(E,\eta_g,n_z) - 4T_{11}P_{2HD}(E,\eta_g,n_z)}\right]$$

The EEM assumes the form

$$m^*(E_{F1HD},\eta_g,n_z) = \frac{\hbar^2}{2}A'_{60}(E_{F1HD},\eta_g,n_z) \tag{1.276}$$

The surface electron concentration is given by

$$n_0 = \frac{g_v}{2\pi}\sum_{n_z=1}^{n_{z\max}}\left[A_{60}(E_{F1HD},\eta_g,n_z) + B_{60}(E_{F1HD},\eta_g,n_z)\right] \tag{1.277}$$

where,

$$B_{60}(E_{F1HD},\eta_g,n_z) = \sum_{r=1}^{s_0} L(r)\left[A_{60}(E_{F1HD},\eta_g,n_z)\right]$$

From (1.269), we can write the expression of the 2D dispersion law in QWs of n-$PtSb_2$ in the absence of band tails as

$$k_s^2 = t_{44}(E,n_z) \tag{1.278}$$

where,

$$t_{44}(E,n_z) = [2A_9]^{-1}\left[-A_{10}(E,n_z) + \sqrt{A_{10}^2(E,n_z) + 4A_9A_{11}(E,n_z)}\right]$$
$$A_9 \equiv [I_1 + \omega_1\omega_3],\ A_{10}(E,n_z) \equiv \left[\omega_3 E + \omega_1\left\{E + \delta_0 - \omega_4\left(\frac{\pi n_z}{d_z}\right)^2\right\}\right.$$
$$\left. + \omega_2\omega_3\left(\frac{\pi n_z}{d_z}\right)^2 + 2I_1\left(\frac{\pi n_z}{d_z}\right)^4\right]$$

and

$$A_{11}(E,n_z) \equiv \left[E\left[E+\delta_0-\omega_4\left(\frac{\pi n_z}{d_z}\right)^2\right]+\omega_2\left(\frac{\pi n_z}{d_z}\right)^2\left[E+\delta_0-\omega_4\left(\frac{\pi n_z}{d_z}\right)^2\right]-I_1\left(\frac{\pi n_z}{d_z}\right)^4\right]$$

The area of k_s space can be expressed as

$$A(E,n_z) = \pi t_{44}(E,n_z) \tag{1.279}$$

The total DOS function assumes the form

$$N_{2DT}(E) = \frac{g_v}{2\pi}\sum_{n_z=1}^{n_{z\max}}[t'_{44}(E,n_z)]H(E-E_{n_{z14}}) \tag{1.280}$$

where the quantized levels $E_{n_{z14}}$ can be expressed through the equation

$$E_{n_{z14}} = (2)^{-1}\left[-\left[\omega_2\left(\frac{\pi n_z}{d_z}\right)^2+\delta_0-\omega_4\left(\frac{\pi n_z}{d_z}\right)^2\right]+\left\{\left[\omega_2\left(\frac{\pi n_z}{d_z}\right)^2+\delta_0-\omega_4\left(\frac{\pi n_z}{d_z}\right)^2\right]^2\right.\right.$$
$$\left.\left.+4\left[I_1\left(\frac{\pi n_z}{d_z}\right)^4+\omega_2\omega_4\left(\frac{\pi n_z}{d_z}\right)^4-\omega_2\delta_0\left(\frac{\pi n_z}{d_z}\right)^2\right]\right\}^{1/2}\right] \tag{1.281a}$$

Using (1.278), the EEM in this case can be written as

$$m^*(E_{F_s},n_z) = \frac{\hbar^2}{2}t'_{44}(E_{F_s},n_z) \tag{1.281b}$$

The electron statistics can be written as

$$n_{2D} = \frac{g_v}{2\pi}\sum_{n_z=1}^{n_{z\max}}[t_{44}(E_{Fs},n_z)+t_{45}(E_{Fs},n_z)] \tag{1.282}$$

where $t_{45}(E_{Fs},n_z) \equiv \sum_{r=1}^{S} L(r)[t_{44}(E_{Fs},n_z)]$

The ER in this case is given by

$$\frac{D}{\mu} = \frac{1}{e}\left[\sum_{n_z=1}^{n_{z\max}}[t'_{44}(E_{Fs},n_z)+t'_{45}(E_{Fs},n_z)]\right]^{-1}\cdot\left[\sum_{n_z=1}^{n_{z\max}}[t_{44}(E_{Fs},n_z)+t_{45}(E_{Fs},n_z)]\right] \tag{1.283}$$

1.2.9 The ER from QWs of HD Bismuth Telluride

The dispersion relation of the conduction electrons in Bi_2Te_3 can be written as [196–198]

$$E(1+\alpha E)=\bar{\omega}_1 k_x^2+\bar{\omega}_2 k_y^2+\bar{\omega}_3 k_z^2+2\bar{\omega}_4 k_z k_y \tag{1.284}$$

where $\overline{\omega}_1=\frac{\hbar^2}{2m_0}\overline{\alpha}_{11}, \overline{\omega}_2=\frac{\hbar^2}{2m_0}\overline{\alpha}_{22}, \overline{\omega}_3=\frac{\hbar^2}{2m_0}\overline{\alpha}_{33}, \overline{\omega}_4=\frac{\hbar^2}{2m_0}\overline{\alpha}_{23}$ in which $\overline{\alpha}_{11}, \overline{\alpha}_{22}, \overline{\alpha}_{33}$ and $\overline{\alpha}_{23}$ are system constants.

The dispersion relation in HD Bi_2Te_3 assumes the form

$$\gamma_2(E,\eta_g)=\bar{\omega}_1 k_x^2+\bar{\omega}_2 k_y^2+\bar{\omega}_3 k_z^2+2\bar{\omega}_4 k_z k_y \tag{1.285}$$

The EEMs can, respectively, be expressed as

$$m_x^*(E_{F_h},\eta_g)=\frac{\hbar^2}{2\bar{w}_1}\gamma_2'(E_{F_h},\eta_g) \tag{1.286}$$

$$m_y^*(E_{F_h},\eta_g)=\frac{\hbar^2}{2\bar{w}_2}\gamma_2'(E_{F_h},\eta_g) \tag{1.287}$$

$$m_z^*(E_{F_h},\eta_g)=\frac{\hbar^2}{2\bar{w}_3}\gamma_2'(E_{F_h},\eta_g) \tag{1.288}$$

The DOS function in this case is given by

$$N(E)=4\pi g_v\left(\frac{2m_0}{h^2}\right)^{3/2}\frac{\sqrt{\gamma_2(E,\eta_g)}\gamma_2'(E,\eta_g)}{\sqrt{\alpha_{11}\alpha_{22}\alpha_{33}-4\alpha_{11}\alpha_{23}^2}} \tag{1.289}$$

Thus combining (1.289) with the Fermi Dirac occupation probability factor, the electron concentration can be written as

$$n_0=\frac{g_v}{3\pi^2}\left(\frac{2m_0}{\hbar^2}\right)^{3/2}(\alpha_{11}\alpha_{22}\alpha_{33}-4\alpha_{11}\alpha_{23}^2)^{-1/2}[U_{1HD}(E_{F_h},\eta_g)+U_{2HD}(E_{F_h},\eta_g)] \tag{1.290a}$$

where,
$U_{1HD}(E_{F_h},\eta_g)=[\gamma_2(E_{F_h},\eta_g)]^{3/2}, U_{2HD}(E_{F_h},\eta_g)=\sum_{r=1}^{s}L(r)[U_{1HD}(E_{F_h},\eta_g)]$

The $\bar{E}_{hd}$ in this case is given by the equation

$$\gamma_2(\bar{E}_{hd},\eta_g)=0 \tag{1.290b}$$

Using (1.290a), (1.290b), and (1.38), we can study the ER in this case.

The dispersion relation in QWs of HD Bi_2Te_3 can be expressed as

$$\gamma_2(E, \eta_g) = \bar{\omega}_1 (\frac{\pi n_x}{d_x})^2 + \bar{\omega}_2 k_y^2 + \bar{\omega}_3 k_z^2 + 2\bar{\omega}_4 k_z k_y \quad (1.291)$$

The EEM can be expressed as

$$m^*(E_{F1HD}, \eta_g) = \frac{m_0}{\sqrt{\alpha_{11}\alpha_{33} - 4\alpha_{23}^2}} \gamma_2'(E_{F1HD}, \eta_g) \quad (1.292)$$

The surface electron concentration can be written as

$$n_{2D} = \frac{g_v}{2\pi} \left[\sum_{n_z=1}^{n_{z\max}} R_{60}(E_{F1HD}, \eta_g, n_z) + R_{61}(E_{F1HD}, \eta_g, n_z) \right] \quad (1.293)$$

$$R_{60}(E_{F1HD}, \eta_g, n_z) = \frac{1}{\sqrt{\alpha_{11}\alpha_{33} - 4\alpha_{23}^2}} [\frac{2m_0 \gamma_2(E_{F1HD}, \eta_g)}{\hbar^2} - \frac{2m_0}{\hbar^2} (\frac{\pi n_x}{d_x})^2 \bar{\alpha}_{11}]$$

and

$$R_{61}(E_{F1HD}, \eta_g, n_z) = \sum_{r=1}^{s} L(r)[R_{60}(E_{F1HD}, \eta_g, n_z)]$$

Using (1.293) and (1.38) we can study the ER in this case.

The 2D electron dispersion law in QWs of Bi_2Te_3 in the absence of band tails assumes the form

$$E(1 + \alpha E) = \overline{\omega}_1 (\frac{n_x \pi}{d_x})^2 + \overline{\omega}_2 k_y^2 + \overline{\omega}_3 k_z^2 + 2\overline{\omega}_4 k_z k_y \quad (1.294)$$

The area of the ellipse is given by

$$A_n(E, n_x) = \frac{\pi}{\sqrt{\bar{\alpha}_{22}\bar{\alpha}_{33} - 4\bar{\alpha}_{23}^2}} \left[\frac{2m_0 E(1 + \alpha E)}{\hbar^2} - \bar{\omega}_1 (\frac{n_x \pi}{d_x})^2 \right] \quad (1.295)$$

The total DOS function assumes the form

$$N_{2DT}(E) = \frac{g_v m_0}{\pi \hbar^2 \sqrt{\overline{\alpha}_{22}\overline{\alpha}_{33} - 4\overline{\alpha}_{23}^2}} \sum_{n_x=1}^{n_{x\max}} (1 + 2\alpha E) H(E - E_{n_{z15}}) \quad (1.296)$$

where, $(E_{n_{z15}})$ can be expressed through the equation

$$E_{n_{z15}}(1 + \alpha E_{n_{z15}}) = \overline{\omega}_1 (\frac{n_x \pi}{d_x})^2 \quad (1.297a)$$

The EEM in this case assumes the form as

$$m^*(E_{F_s}) = \frac{m_0(1+2\alpha(E_{F_s}))}{\sqrt{\bar{\alpha}_{22}\bar{\alpha}_{33} - 4\bar{\alpha}_{23}^2}} \tag{1.297b}$$

The electron concentration can be written as

$$n_{2D} = \frac{k_B T g_v}{\pi\hbar^2}\left(\frac{m_0}{\sqrt{\bar{\alpha}_{22}\bar{\alpha}_{33} - 4\bar{\alpha}_{23}^2}}\right)\sum_{n_x=1}^{n_{x\max}}[(1+2\alpha E_{n_{z15}})F_0(\eta_{n_{15}}) + 2\alpha k_B T F_1(\eta_{n_{15}})] \tag{1.298}$$

Using (1.298) the *ER* in this case is given by

$$\frac{D}{\mu} = \left[\sum_{n_z=1}^{n_{z\max}}\left[\left(\frac{k_B T}{e}\right)\left(1+2_{n_{z15}}\right)F_{-1}(\eta_{n_{15}}) + 2_B T F_0(\eta_{n_{15}})\right]\right]^{-1}$$
$$[\sum_{n_z=1}^{n_{z\max}}\left[\left(1+2\alpha E_{n_{z15}}\right)F_0(\eta_{n_{15}}) + 2\alpha k_B T F_1(\eta_{n_{15}})\right]] \tag{1.299}$$

where, $\eta_{n_{15}} = \frac{E_{F_s} - E_{n_{z15}}}{k_B T}$

1.2.10 The ER from QWs of HD Germanium

It is well known that the conduction electrons of n-Ge obey two different types of dispersion laws since band non-parabolicity has been included in two different ways as given in the literature [199–201].

(a) The energy spectrum of the conduction electrons in bulk specimens of n-Ge can be expressed in accordance with Cardona et al. [199], Gibson [200] as

$$E = -\frac{E_{g_0}}{2} + \frac{\hbar^2 k_z^2}{2m_\parallel^*} + \left[\frac{E_{g_0^2}}{4} + E_{g_0} k_s^2\left(\frac{\hbar^2}{2m_\perp^*}\right)\right]^{1/2} \tag{1.300}$$

where in this case $m_\parallel^*$ and $m_\perp^*$ are the longitudinal and transverse effective masses along <111> direction at the edge of the conduction band respectively

The (1.300) can be written as

$$\frac{\hbar^2 k_s^2}{2m_\perp^*} = E(1+\alpha E) + \alpha\left(\frac{\hbar^2 k_z^2}{2m_\parallel^*}\right)^2 - (1+2\alpha E)\left(\frac{\hbar^2 k_z^2}{2m_\parallel^*}\right) \quad (1.301)$$

The dispersion relation under the condition of heavy doping can be expressed from (1.301) as

$$\frac{\hbar^2 k_s^2}{2m_\perp^*} = \gamma_2(E,\eta_g) + \alpha\left(\frac{\hbar^2 k_z^2}{2m_\parallel^*}\right)^2 - (1+2\alpha\gamma_3(E,\eta_g))\frac{\hbar^2 k_z^2}{2m_\parallel^*} \quad (1.302)$$

The EEMs can be written as

$$m_s^*(E_{Fh},\eta_g) = m_\perp^*\gamma_2'(E_{Fh},\eta_g) \quad (1.303)$$

and

$$m_z^*(E_{Fh},\eta_g) = m_\parallel^*[\gamma_3'(E_{Fh},\eta_g) - \frac{[\gamma_3'(E_{Fh},\eta_g)[1+2\alpha\gamma_3(E_{Fh},\eta_g)] - \gamma_2'(E_{Fh},\eta_g)]}{\sqrt{[1+2\alpha\gamma_3(E_{Fh},\eta_g)]^2 - 4\alpha\gamma_2(E_{Fh},\eta_g)}}] \quad (1.304)$$

The electron concentration can be written as

$$n_0 = \frac{8\pi g_v m_\perp^* \sqrt{2m_\parallel^*}}{h^3}\left[\bar{I}_{129}(E_{F_h},\eta_g) + \sum_{r=1}^{s} L(r)[\bar{I}_{129}(E_{F_h},\eta_g)]\right] \quad (1.305a)$$

where,

$$\bar{I}_{129}(E_{F_h},\eta_g) = [M_{8HD}(E_{F_h},\eta_g)],$$

$$M_{8HD}(E_{F_h},\eta_g) = [\gamma_3(E_{F_h},\eta_g)]^{1/2}[\gamma_2(E_{F_h},\eta_g) + \frac{\alpha}{5}\gamma_3^2(E_{F_h},\eta_g)] - \frac{\gamma_3(E_{F_h},\eta_g)}{3}[1+2\alpha\gamma_3(E_{F_h},\eta_g)]]$$

The $\bar{E}_{hd}$ in this case is given by the equation

$$\gamma_2(\bar{E}_{hd},\eta_g) = 0 \quad (1.305b)$$

Thus by using (1.305a), (1.305b) and (1.38), we can study the ER in this case.

In the presence of size quantization, the dispersion law in QW of HD Ge can be written following (1.302) as

$$\frac{\hbar^2 k_s^2}{2m_\perp^*} = \gamma_2(E,\eta_g) + \alpha\left(\frac{\hbar^2(n_z\pi/d_z)^2}{2m_\parallel^*}\right)^2 - (1+2\alpha\gamma_3(E,\eta_g))\frac{\hbar^2(n_z\pi/d_z)^2}{2m_\parallel^*} \quad (1.306a)$$

The EEM assumes the form

$$m_s^*(E_{F1HD},\eta_g,n_z) = m_\perp^*[\gamma_2'(E_{F1HD},\eta_g) - \frac{\alpha\hbar^2}{m_\parallel^*}(\frac{n_z\pi}{d_z})^2\gamma_3'(E_{F_h},\eta_g)] \quad (1.306b)$$

The surface electron concentration per unit area is given by

$$n_{2D} = \frac{g_v m_\perp^*}{\pi\hbar^2}\sum_{n_z=1}^{n_{z\max}}[R_1(E_{F1HD},\eta_g,n_z) + S_1(E_{F1HD},\eta_g,n_z)] \quad (1.307)$$

where, $R_1(E_{F1HD},\eta_g,n_z) = [\gamma_2(E_{F1HD},\eta_g) + \alpha\left(\frac{\hbar^2(n_z\pi/d_z)^2}{2m_\parallel^*}\right)^2 - (1+2\alpha\gamma_3(E_{F1HD},\eta_g))$
$\frac{\hbar^2(n_z\pi/d_z)^2}{2m_\parallel^*}]$ and $S_1(E_{F1HD},\eta_g,n_z) = \sum_{r=1}^{S} L(r)[R_1(E_{F1HD},\eta_g,n_z)]$

Thus using (1.307) and (1.38) we can study the ER in this case.

In the presence of size quantization along k_z direction, the 2D dispersion relation of the conduction relations in QWs of n-Ge in the absence of band tails can be written by extending the method as given in [193] as

$$\frac{\hbar^2 k_x^2}{2m_1^*} + \frac{\hbar^2 k_y^2}{2m_2^*} = \gamma(E,n_z) \quad (1.308)$$

where, $m_1^* \equiv m_\perp^*$, $m_2^* = \frac{m_\perp^* + 2m_\parallel^*}{3}$, $\gamma(E,n_z) \equiv \left[E(1+\alpha E) - (1+2\alpha E)\frac{\hbar^2}{2m_3^*}\left(\frac{n_z\pi}{d_z}\right)^2 + \alpha\left[\frac{\hbar^2}{2m_3^*}\left(\frac{n_z\pi}{d_z}\right)^2\right]^2\right]$

and $m_3^* = \frac{3m_\parallel^* m_\perp^*}{2m_\parallel^* + m_\perp^*}$.

The area of ellipse of the 2D surface as given by (1.308) can be written as

$$A(E,n_z) = \frac{2\pi\sqrt{m_1^* m_2^*}}{\hbar^2}\gamma(E,n_z) \quad (1.309a)$$

The EEM in this case can be written as

$$m^*(E_{F_s},n_z) = (\sqrt{m_1^* m_2^*})[\gamma(E_{F_s},n_z)]' \quad (1.309b)$$

The DOS function per sub-band can be expressed as

$$N_{2D}(E) = \frac{4\sqrt{m_1^* m_2^*}}{\pi \hbar^2}\left[1 + 2\alpha E - 2\alpha\left(\frac{\hbar^2}{2m_3^*}\left(\frac{\pi n_z}{d_z}\right)^2\right)\right] \tag{1.310}$$

The total DOS function is given by

$$N_{2DT}(E) = \frac{4}{\pi \hbar^2}\sqrt{m_1^* m_2^*}\sum_{n_z=1}^{n_{z\max}}\left[1 + 2\alpha E - 2\alpha\left(\frac{\hbar^2}{2m_3^*}\left(\frac{\pi n_z}{d_z}\right)^2\right)\right] H(E - E_{n_{z16}}) \tag{1.311}$$

where, $E_{n_{z16}}$ is the positive root of the following equation

$$E_{n_{z16}}(1 + \alpha E_{n_{z16}}) - (1 + 2\alpha E_{n_{z16}})\left(\frac{\hbar^2}{2m_3^*}\left(\frac{\pi n_z}{d_z}\right)^2\right) + \alpha\left(\frac{\hbar^2}{2m_3^*}\left(\frac{\pi n_z}{d_z}\right)^2\right)^2 = 0 \tag{1.312}$$

Thus combining (1.311) with the Fermi Dirac occupation probability factor, the 2D electron statistics in this case can be written as

$$n_{2D} = \frac{4\sqrt{m_1^* m_2^*}k_B T}{\pi \hbar^2}\sum_{n_z=1}^{n_{z\max}}[(A_1(n_z) + 2\alpha E_{n_{z16}})F_0(E_{n_{z16}}) + 2\alpha k_B T F_1(E_{n_{z16}})] \tag{1.313}$$

where $A_1(n_z) \equiv \left[1 + 2\alpha(\hbar/2m_3^*)(\pi n_z/d_z)^2\right]$ and $\eta_{n_{z16}} \equiv \frac{1}{k_B T}\left[E_{Fs} - E_{n_{z16}}\right]$.

The ER in this case is given by

$$\frac{D}{\mu} = \frac{k_B T}{e}\left[\sum_{n_z=1}^{n_{z=\max}}[(A_1(n_z) + 2\alpha E_{n_{z16}})F_{-1}(E_{n_{z16}}) + 2\alpha k_B T F_0(E_{n_{z16}})]\right]^{-1} \left[\sum_{n_z=1}^{n_{z=\max}}[(A_1(n_z) + 2\alpha E_{n_{z16}})F_0(E_{n_{z16}}) + 2\alpha k_B T F_1(E_{n_{z16}})]\right] \tag{1.314}$$

The expressions of EEMs' in bulk specimens of Ge in the absence of band tails can be written following (1.301) as

$$m_z^*(E_F) = m_\parallel^* \tag{1.315}$$

$$m_s^*(E_F) = m_\perp^*(1 + 2\alpha E_F) \tag{1.316}$$

The DOS function for bulk specimens of Ge in the absence of band tails can be written following (1.301) as

$$N(E) = 4\pi g_v \left(\frac{2m_D^*}{h^2}\right)^{\frac{3}{2}} \left[E^{\frac{1}{2}} - \frac{5}{6}\alpha E^{\frac{3}{2}} + \frac{18\alpha}{5}\left(\frac{m_{11}^*}{\hbar^2}\right)^2 E^{\frac{7}{2}}\right]; \quad m_D = \left(m_\perp^{*2} \cdot m_\parallel^*\right)^{\frac{1}{3}} \tag{1.317}$$

Using (1.317), the electron concentration in bulk specimens of Ge can be written as

$$n_0 = N_{c1}\left[F_{\frac{1}{2}}(\eta) - \frac{5}{4}\alpha k_B T F_{\frac{3}{2}}(\eta) + \frac{189}{4}\alpha k_B T\left(\frac{m_{11}^* k_B T}{\hbar^2}\right)^2 F_{\frac{7}{2}}(\eta)\right]; \tag{1.318}$$

$$N_{c1} = 2g_v\left(\frac{2\pi m_D^* k_B T}{h^2}\right)^{\frac{3}{2}}$$

The use of (1.318) leads to the expression of ER in this case as

$$\frac{D}{\mu} = \frac{k_B T}{e}\left[\frac{[F_{\frac{1}{2}}(\eta) - \frac{5}{4}\alpha k_B T F_{\frac{3}{2}}(\eta) + \frac{189}{4}\alpha k_B T\left(\frac{m_{11}^* k_B T}{\hbar^2}\right)^2 F_{\frac{7}{2}}(\eta)]}{[F_{-\frac{1}{2}}(\eta) - \frac{5}{4}\alpha k_B T F_{\frac{1}{2}}(\eta) + \frac{189}{4}\alpha k_B T\left(\frac{m_{11}^* k_B T}{\hbar^2}\right)^2 F_{\frac{5}{2}}(\eta)]}\right] \tag{1.319}$$

(b) The dispersion relation of the conduction electron in bulk specimens of n-Ge can be expressed in accordance with the model of Wang and Ressler [201] can be written as

$$E = \frac{\hbar^2 k_z^2}{2m_\parallel^*} + \frac{\hbar^2 k_s^2}{2m_\perp^*} - \bar{\alpha}_4\left(\frac{\hbar^2 k_s^2}{2m_\perp^*}\right)^2 - \bar{\alpha}_5\left(\frac{\hbar^2 k_s^2}{2m_\perp^*}\right)\left(\frac{\hbar^2 k_z^2}{2m_\parallel^*}\right) - \bar{\alpha}_6\left(\frac{\hbar^2 k_z^2}{2m_\parallel^*}\right)^2 \tag{1.320}$$

where, $\bar{\alpha}_4 = \beta_4(\frac{2m_\perp^*}{\hbar^2})$, $\beta_4 = 1.4\beta_5$,

$$\beta_5 = \frac{\alpha\hbar^4}{4}[(m_\perp^*)^{-1} - (m_0)^{-1}]^2, \bar{\alpha}_5 = \bar{\alpha}_7(\frac{4m_\perp^* m_\parallel^*}{\hbar^4}), \bar{\alpha}_7 = 0.8\beta_5$$

$$\text{and } \bar{\alpha}_6 = (0.005\beta_5)(\frac{2m_\parallel^*}{\hbar^2})^2$$

The energy spectrum under the condition of heavy doping can be written as

$$\gamma_3(E, \eta_g) = \frac{\hbar^2 k_z^2}{2m_\parallel^*} + \frac{\hbar^2 k_s^2}{2m_\perp^*} - \bar{\alpha}_4\left(\frac{\hbar^2 k_s^2}{2m_\perp^*}\right)^2 - \bar{\alpha}_5\left(\frac{\hbar^2 k_s^2}{2m_\perp^*}\right)\left(\frac{\hbar^2 k_z^2}{2m_\parallel^*}\right) - \bar{\alpha}_6\left(\frac{\hbar^2 k_z^2}{2m_\parallel^*}\right)^2 \tag{1.321a}$$

The (1.321) can be expressed as

$$\frac{\hbar^2 k_s^2}{2m_\perp^*} = \bar{\alpha}_8 - \bar{\alpha}_9 k_z^2 - \bar{\alpha}_{10}\left[k_z^4 + \bar{\alpha}_{11}k_z^2 + \bar{\alpha}_{12}(E,\eta_g)\right]^{1/2} \quad (1.321b)$$

where

$$\bar{\alpha}_8 = \frac{1}{2\bar{\alpha}_4},\ \bar{\alpha}_9 = \frac{\bar{\alpha}_5}{2\bar{\alpha}_4}\left(\frac{\hbar^2}{2m_\parallel^*}\right),\ \bar{\alpha}_{10} = \frac{1}{2\bar{\alpha}_4}\left(\frac{\hbar^2}{2m_\parallel^*}\right)\sqrt{\bar{\alpha}_5^2 - 4\bar{\alpha}_4\bar{\alpha}_6},$$

$$\bar{\alpha}_{11} = \frac{2m_\parallel^*}{\hbar}\left[\frac{4\bar{\alpha}_4 - 2\bar{\alpha}_5}{\bar{\alpha}_5^2 - 4\bar{\alpha}_4\bar{\alpha}_6}\right] \quad and\ \bar{\alpha}_{12}(E,\eta_g) = \left(\frac{2m_\parallel^*}{\hbar^2}\right)^2\left[\frac{(1-4\bar{\alpha}_4\gamma_3(E,\eta_g))}{\bar{\alpha}_5^2 - 4\bar{\alpha}_4\bar{\alpha}_6}\right]$$

The EEMs' can be written as

$$m^*(E_{Fh},\eta_g) = \left[\frac{m_\parallel^*\gamma_3'(E_{Fh},\eta_g)}{\sqrt{1-4\bar{\alpha}_6\gamma_3(E_{Fh},\eta_g)}}\right] \quad (1.322)$$

$$m_\perp^*(E_{Fh},\eta_g) = \left[\frac{m_\perp^*\gamma_3'(E_{Fh},\eta_g)}{\sqrt{1-4\bar{\alpha}_4\gamma_3(E_{Fh},\eta_g)}}\right] \quad (1.323)$$

The electron concentration in HD Ge in accordance with the model of Wang and Ressler can be expressed as

$$n_0 = \frac{m_\perp^* g_v}{\pi^2\hbar^2}\left[I_3(E_{F_h},\eta_s) + I_4(E_{F_h},\eta_s)\right] \quad (1.324a)$$

where

$$I_3(E_{F_h},\eta_s) = \left[\bar{\alpha}_8\rho_{10}(E_{F_h},\eta_g) - \frac{\bar{\alpha}_9}{3}\rho_{10}^3(E_{F_h},\eta_s) - \bar{\alpha}_{10}J_{10}(E_{F_h},\eta_s)\right],$$

$$\rho_{10}(E_{F_h},\eta_g) = \frac{1}{\hbar}\left[\frac{m_\parallel^*}{\bar{\alpha}_6}\right]^{\frac{1}{2}}\left[1-\sqrt{1-4\bar{\alpha}_6\gamma_3(E_{F_h},\eta_g)}\right]^{\frac{1}{2}}$$

$$\begin{aligned} J_{10}(E_{F_h},\eta_g) = {} & \frac{\overline{A}_1(E_{F_h},\eta_g)}{3}[-E_0(\lambda(E_{F_h},\eta_g),q(E_{F_h},\eta_g))\cdot \\ & [\overline{A}_1^2(E_{F_h},\eta_g) + \overline{B}_1^2(E_{F_h},\eta_g)] + 2\overline{B}_1^2(E_{F_h},\eta_g)F_0(\lambda(E_{F_h},\eta_g),q(E_{F_h},\eta_g))] \\ & + \frac{\overline{A}_1(E_{F_h},\eta_g)}{3}[\rho_{10}^2(E_{F_h},\eta_g) + \overline{A}_1^2(E_{F_h},\eta_g) + 2\overline{B}_1^2(E_{F_h},\eta_g)] \\ & \left[\frac{\overline{A}_1^2(E_{F_h},\eta_g) + \rho_{10}^2(E_{F_h},\eta_g)}{\overline{B}_1^2(E_{F_h},\eta_g) + \rho_{10}^2(E_{F_h},\eta_g)}\right]^{\frac{1}{2}} \end{aligned}$$

$$\bar{A}_1^2(E_{F_h},\eta_g)=\frac{1}{2}[\bar{\alpha}_{11}+\sqrt{\bar{\alpha}_{11}^2-4\bar{\alpha}_{12}^2(E_{F_h},\eta_g)}],\ \bar{B}_1^2(E_{F_h},\eta_g)=\frac{1}{2}[\bar{\alpha}_{11}+\sqrt{\bar{\alpha}_{11}^2-4\bar{\alpha}_{12}^2(E_{F_h},\eta_g)}],$$

$$\lambda(E_{F_h},\eta_g)=\tan^{-1}[\frac{\rho_{10}(E_{F_h},\eta_g)}{\bar{B}_1(E_{F_h},\eta_g)}],q(E_{F_h},\eta_g)=\left[\frac{\bar{A}_1^2(E_{F_h},\eta_g)-\bar{B}_1^2(E_{F_h},\eta_g)}{\bar{A}_1^2(E_{F_h},\eta_g)}\right]$$

and $I_4(E_{F_h},\eta_g)=\sum_{r=1}^{s}L(r)[I_3(E_{F_h},\eta_g)]$

The $\bar{E}_{hd}$ in this case is given by the equation

$$\gamma_2(\bar{E}_{hd},\eta_g)=0 \tag{1.324b}$$

Thus using (1.324a), (1.324b), and (1.38) we can study the ER in this case.
The dispersion relation in QW of HD Ge can be written as

$$\frac{\hbar^2k_s^2}{2m_\perp^*}=\bar{\alpha}_8-\bar{\alpha}_9(\frac{n_z\pi}{d_z})^2-\bar{\alpha}_{10}\left[(\frac{n_z\pi}{d_z})^4+\bar{\alpha}_{11}(\frac{n_z\pi}{d_z})^2+\bar{\alpha}_{12}(E,\eta_g)\right]^{1/2}, \tag{1.325}$$

The (1.325) can be expressed as

$$\frac{\hbar^2k_s^2}{2m_\perp^*}=A_{75}(E,\eta_g,n_z) \tag{1.326}$$

where, $A_{75}(E,\eta_g,n_z)=[\bar{\alpha}_8-\bar{\alpha}_9(\frac{\pi n_z}{d_z})^2-\bar{\alpha}_{10}[(\frac{\pi n_z}{d_z})^4+\bar{\alpha}_{11}(\frac{\pi n_z}{d_z})^2+\bar{\alpha}_{12}(E,\eta_g)]^{1/2}]$

The EEM is given by

$$m_s^*(E_{F1HD},\eta_g,n_z)=m_\perp^*A'_{75}(E_{F1HD},\eta_g,n_z) \tag{1.327}$$

The electron concentration per unit area assumes the form

$$n_{2D}=\frac{m_\perp^*g_v}{\pi\hbar^2}\sum_{n_z=1}^{n_{z\max}}[A_{75}(E_{F1HD},\eta_g,n_z)+A_{76}(E_{F1HD},\eta_g,n_z)] \tag{1.328}$$

where, $A_{76}(E_{F1HD},\eta_g,n_z)=\sum_{r=1}^{s}L(r)[A_{75}(E_{F1HD},\eta_g,n_z)]$

Using (1.328) and (1.38) we can study the ER in this case

The 2D dispersion law in the absence of band tails can be expressed as

$$E=A_5(n_z)+A_6(n_z)\beta-\bar{\alpha}_4\beta^2 \tag{1.329}$$

where, $A_5(n_z)=\frac{\hbar^2}{2m_3^*}\left(\frac{\pi n_z}{d_z}\right)^2[1-\bar{\alpha}_6(\frac{\hbar^2}{2m_3^*})(\frac{\pi n_z}{d_z})^2]$, $A_6(n_z)=\left[1-\bar{\alpha}_5(\frac{\hbar^2}{2m_3^*})(\frac{\pi n_z}{d_z})^2\right]$ and

$$\beta \equiv \frac{\hbar^2 k_x^2}{2m_1^*} + \frac{\hbar^2 k_y^2}{2m_2^*}.$$

The (1.329) can be written as

$$\frac{\hbar^2 k_x^2}{2m_1^*} + \frac{\hbar^2 k_y^2}{2m_2^*} = I_1(E, n_z) \tag{1.330}$$

where, $I_1(E, n_z) = (2\bar{\alpha}_4)^{-1}[A_6(n_z) - [A_6^2(n_z) - 4\bar{\alpha}_4 E + 4\bar{\alpha}_4 A_5(n_z)]^{1/2}]$

From (1.330), the area of the 2D k_s-space is given by

$$A(E, n_z) = \frac{2\pi\sqrt{m_1^* m_2^*}}{\hbar^2} I_1(E, n_z) \tag{1.331a}$$

Using (1.331a) in this case can be expressed as

$$m^*(E_{F_s}, n_z) = (\sqrt{m_1^* m_2^*})[I_1(E_{F_s}, n_z)]' \tag{1.331b}$$

The DOS function per sub-band can be written as

$$N_{2D}(E) = \frac{4}{\pi} \frac{\sqrt{m_1^* m_2^*}}{\hbar^2} \{I_1(E, n_z)\}' \tag{1.332}$$

where $\{I_1(E, n_z)\}' \equiv \frac{\partial}{\partial E}[I_1(E, n_z)]$

The total DOS function assumes the form

$$N_{2D}(E) = \frac{4\sqrt{m_1^* m_2^*}}{\pi\hbar^2} \sum_{n_z=1}^{n_{z\max}} \{I_1(E, n_z)\}' H(E - E_{n_{z17}}) \tag{1.333}$$

where, the sub-band energy $(E_{n_{z17}})$ are given by

$$E_{n_{z17}} = \left(\frac{\hbar^2}{2m_3^*}\right)\left(\frac{\pi n_z}{d_z}\right)^2 \left[1 - \bar{\alpha}_6 (\frac{\hbar^2}{2m_3^*})(\frac{\pi n_z}{d_z})^2\right] \tag{1.334}$$

The electron statistics can be written as

$$n_{2D} = \frac{4\sqrt{m_1^* m_2^*}}{\pi\hbar^2} \sum_{n_z=1}^{n_{z\max}} [t_{46}(E_{Fs}, n_z) + t_{47}(E_{Fs}, n_z)] \tag{1.335}$$

where $t_{46}(E_{Fs}, n_z) \equiv I_1(E_{Fs}, n_z)$, $t_{47}(E_{Fs}, n_z) \equiv \sum_{r=1}^{s} L(r)(t_{46}(E_{Fs}, n_z))$

Using (1.335), the ER in this case is given by

$$\frac{D}{\mu}=\frac{1}{e}\left[\sum_{n_z=1}^{n_{z=\max}}[t'_{55}(E_{Fs},n_z)+t'_{56}(E_{Fs},n_z)]\right]^{-1}\left[\sum_{n_z=1}^{n_{z=\max}}[t_{55}(E_{Fs},n_z)+t_{56}(E_{Fs},n_z)]\right] \quad (1.336)$$

1.2.11 The ER from QWs of HD Gallium Antimonide

The dispersion relation of the conduction electrons in n-GaSb can be written as [202]

$$E=\frac{\hbar^2k^2}{2m_0}-\frac{\bar{E}'_{go}}{2}+\frac{\bar{E}'_{go}}{2}\left[1+\frac{2\hbar^2k^2}{\bar{E}'_{go}}\left(\frac{1}{m_c}-\frac{1}{m_0}\right)\right]^{\frac{1}{2}} \quad (1.337)$$

where $\bar{E}'_{go}=[E_{go}+\frac{5.10^{-5}T^2}{2(112+T)}]eV$

The (1.337) can be expressed as

$$\frac{\hbar^2k^2}{2m_c}=I_{36}(E) \quad (1.338)$$

where

$$I_{36}(E)=[E+\bar{E}'_{g0}-(m_c/m_0)(\bar{E}'_{g0}/2)-[(\bar{E}'_{g0}/2)^2 \\ +[((\bar{E}'_{g0})^2/2)(1-(m_c/m_0))]+[(\bar{E}'_{g0}/2)(1-(m_c/m_0))]^2+E\bar{E}'_{g0}(1-(m_c/m_0))]^{1/2}]$$

Under the condition of heavy doping (1.338) assumes the form

$$\frac{\hbar^2k^2}{2m_c}=I_{36}(E,\eta_g) \quad (1.339)$$

where,

$$I_{36}(E,\eta_g)=[\gamma_3(E,\eta_g)+E'_g-\frac{m_c}{m_0}\cdot\frac{E'_g}{2} \\ -[(\frac{E'_g}{2})^2+[\frac{E'_g}{2}(1-\frac{m_c}{m_0})]^2+(\frac{E'_g}{2})^2(1-\frac{m_c}{m_0})+\gamma_3(E,\eta_g)E'_g(1-\frac{m_c}{m_0})]^{1/2}]$$

The EEM can be written as

$$m^*(E_{F_h},\eta_g) = m_c\{I_{36}(E_{F_h},\eta_g)\}' \tag{1.340}$$

The DOS function in this case can be written as

$$N_{HD}(E,\eta_g) = \frac{g_v}{2\pi^2}\left(\frac{2m_c}{\hbar^2}\right)^{3/2}\sqrt{I_{36}(E,\eta_g)}\{I_{36}(E,\eta_g)\}' \tag{1.341}$$

Since, the original band model in this case is a no pole function, therefore, the HD counterpart will be totally real, and the complex band vanishes.

The electron concentration is given by

$$n_0 = \frac{g_v}{3\pi^2}\left(\frac{2m_c}{\hbar^2}\right)^{3/2}\{I_{36}(E_{F_h},\eta_g)\}^{3/2} \tag{1.342}$$

In this case, $\bar{E}_{hd}$ is given by

$$I_{36}(\bar{E}_{hd},\eta_g) = 0 \tag{1.343}$$

One can numerically compute the ER by using (1.342), (1.343) and (1.31c) and the allied definitions in this case.

For dimensional quantization along z-direction, the dispersion relation of the 2D electrons in QWs of HD GaSb can be written following (1.339) as

$$\frac{\hbar^2(\mathrm{n}_z\pi/\mathrm{d}_z)^2}{2m_c} + \frac{\hbar^2(k_s)^2}{2m_c} = I_{36}(E,\eta_g) \tag{1.344}$$

The expression of the $N_{2DT}(E)$ in this case can be written as

$$N_{2DT}(E) = \frac{m_c g_v}{\pi\hbar^2}\sum_{n_z=1}^{n_{z\max}} T'_{119D}(E,\eta_g,\mathrm{n}_z)H(\mathrm{E}-\mathrm{E}_{n_zD119}) \tag{1.345}$$

where, $T_{119D}(E,\eta_g,\mathrm{n}_z) = [I_{36}(E,\eta_g) - \hbar^2(\mathrm{n}_z\pi/\mathrm{d}_z)^2(2m_c)^{-1}]$,

The sub band energies E_{n_zD119} in this case given by

$$\left\{\hbar^2(\mathrm{n}_z\pi/\mathrm{d}_z)^2\right\}(2m_c)^{-1} = I_{36}(\mathrm{E}_{n_zD119},\eta_g) \tag{1.346}$$

The EEM in this case assumes the form

$$m^*(E_{F1HD},\eta_g,\mathrm{n}_z) = m_c[I'_{36}(E_{F1HD},\eta_g,\mathrm{n}_z)] \tag{1.347}$$

The 2-D electron statistics in this case can be written as

$$n_{2D} = \frac{m_c g_v}{\pi \hbar^2} \sum_{n_z=1}^{n_{z\max}} [T_{119D}(E_{F1HD}, \eta_g, \mathrm{n}_z) + T_{129D}(E_{F1HD}, \eta_g, \mathrm{n}_z)] \tag{1.348}$$

where, $T_{129D}(E_{F1HD}, \eta_g, \mathrm{n}_z) = \sum_{r=1}^{s} L(r)[T_{119D}(E_{F1HD}, \eta_g, \mathrm{n}_z)]$,

Therefore combining (1.348) and (1.38) we can get the ER in this case.

The total 2D DOS function in the absence of band tails in this case can be written as

$$N_{2DT}(E) = \left(\frac{m_c g_v}{\pi \hbar^2}\right) \sum_{n_z=1}^{n_{\max}} \{[I_{36}(E)]' H(E - E_{n_{z44}})\} \tag{1.349}$$

where, the sub-band energies $E_{n_{z3}}$ can be expressed as

$$I_{36}(E_{n_{z44}}) = \frac{\hbar^2}{2m_c} (\pi n_z / d_z)^2 \tag{1.350a}$$

The EEM in this case can be written as

$$m^*(E_{F_s}) = (m_c)[I_{36}(E_{F_s})]' \tag{1.350b}$$

The 2D carrier concentration assumes the form

$$n_{2D} = \left(\frac{m_c g_v}{\pi \hbar^2}\right) \sum_{n_z=1}^{n_{\max}} [\bar{T}_{55}(E_{Fs}, n_z) + \bar{T}_{56}(E_{Fs}, n_z)] \tag{1.351}$$

where

$\bar{T}_{55}(E_{Fs}, n_z) = [I_{36}(E_{Fs}) - \frac{\hbar^2}{2m_c}(\pi n_z/d_z)^2]$ and $\bar{T}_{56}(E_{Fs}, n_z) = \sum_{r=1}^{s} L(r)[\bar{T}_{55}(E_{Fs}, n_z)]$

Using (1.351), the ER in this case is given by

$$\frac{D}{\mu} = \frac{1}{|e|} \frac{\sum_{n_z=1}^{n_{z\max}} [\bar{T}_{55}(E_{Fs}, n_z) + \bar{T}_{56}(E_{Fs}, n_z)]}{\sum_{n_z=1}^{n_{z\max}} [\{\bar{T}_{55}(E_{Fs}, n_z)\}' + \{\bar{T}_{56}(E_{Fs}, n_z)\}']} \tag{1.352}$$

The expression of electron concentration for bulk specimens of GaSb (in the absence of band tails) can be expressed as

$$n_0 = \frac{g_v}{3\pi^2} \left(\frac{2m_c}{\hbar^2}\right)^{3/2} [\bar{M}_{A_{10}}(E_F) + \bar{N}_{A_{10}}(E_F)] \tag{1.353}$$

where, $\overline{M}_{A_{10}}(E_F) = [I_{36}(E_F)]^{3/2}$ and $\bar{N}_{A_{10}}(E_F) = \sum_{r=1}^{s} L(r)[\bar{M}_{A_{10}}(E_F)]$

The ER in this case can be expressed as

$$\frac{D}{\mu} = \frac{1}{e}[\bar{M}'_{A_{10}}(E_F) + \bar{N}'_{A_{10}}(E_F)]^{-1}[\bar{M}_{A_{10}}(E_F) + \bar{N}_{A_{10}}(E_F)] \quad (1.354)$$

Thus, we can summarize the whole mathematical background in the following way. In this chapter, we have investigated the 3D and 2D ERs in HD bulk and QWs of non-linear optical materials on the basis of a newly formulated electron dispersion law considering the anisotropies of the effective electron masses, the spin orbit splitting constants and the influence of crystal field splitting within the framework of **k.p** formalism. The results for 3D and 2D ERs for HD bulk and QWs of III-V, ternary and quaternary compounds in accordance with the three and two band models of Kane form a special case of our generalized analysis. We have also studied the ER in accordance with the models of Stillman et al. and Palik et al. respectively since these models find use to describe the electron energy spectrum of the aforesaid materials. The 3D and 2D ERs has also been derived for HD bulk and QWs of II-VI, IV-VI, stressed materials, Te, n-GaP, p-$PtSb_2$, Bi_2Te_3, n-Ge and n-GaSb compounds by using the models of Hopfield, Dimmock, Seiler, Bouat and Thuillier, Rees, Emtage, Kohler, Cardona, Wang et al. and Mathur et al. respectively on the basis of the appropriate carrier energy spectra. The well-known expressions of the ERs in the absence of band tails for wide gap materials have been obtained as special cases of our generalized analysis under certain limiting conditions. This indirect test not only exhibits the mathematical compatibility of our formulation but also shows the fact that our simple analysis is a more generalized one, since one can obtain the corresponding results for relatively wide gap materials having parabolic energy bands under certain limiting conditions from our present derivation.

1.3 Result and Discussions

Using the appropriate equations and taking the values of the energy band constants from Table 1.1, we have plotted, in Figs. 1.1 and 1.2, the normalized 2D ER at low temperatures, where the quantum effects become prominent for the QWs of HD n-Cd_3As_2 and n-$CdGeAs_2$ as functions of nano-thickness. The curves (a) and (b) corresponds to $\delta \neq 0$ and $\delta = 0$ respectively for the purpose of assessing the influence of crystal field splitting on the 2D ER in HD QWs of tetragonal and nonlinear optical materials. We have plotted the curve (c) in accordance with the HD three band model of Kane. The curve (d) and (e) have been in accordance with the HD two band model of Kane and that of the parabolic energy band models respectively.

The influence of quantum confinement is immediately apparent from all the curves of Figs. 1.1 and 1.2, since, the 2D ER in HD QWs depends strongly on the

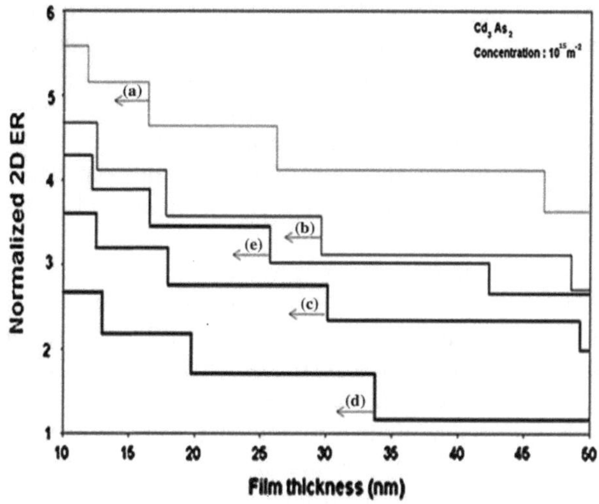

Fig. 1.1 Plot of the normalized 2D ER as a function of film thickness for the QWs of HD n-Cd_3As_2 in accordance with *a* the generalized HD band model with $\delta \neq 0$, *b* the generalized HD band model with $\delta = 0$, *c* the simplified HD three band model of Kane, *d* the HD two band model of Kane and *e* the HD parabolic energy bands

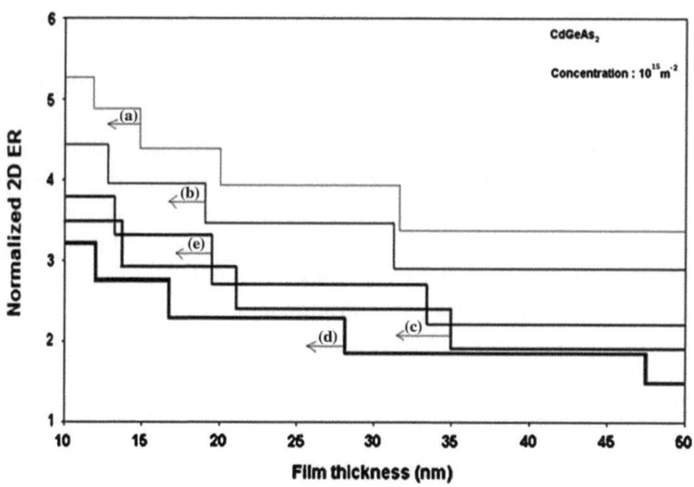

Fig. 1.2 Plot of the normalized 2D ER as a function of film thickness for the QWs of HD n-$CdGeAs_2$ for all cases of Fig. 1.1

nano-thickness, which is in direct contrast with the corresponding bulk specimens (graphs for bulk HD materials have also been drawn for the purpose of relative comparison) and exibits the signature of quantum confinement. It appears from the said Figs. that the 2D ER decreases with the increasing film thickness in a step like manner as considered here although the numerical values vary widely and determined by the constants of the energy spectra. The oscillatory dependence is due to the crossing over of the Fermi level by the size quantized levels. For each coincidence of a size quantized level with the Fermi level, there would be a discontinuity in the DOS function resulting in a peak of oscillations. With large values of film thickness, the height of the steps decreases and the ER decreases with increasing film thickness in non-oscillatory manner and exhibit monotonic decreasing dependence.

The height of step size and the rate of decrement are totally dependent on the band structure. The influence of crystal field splitting is immediately apparent by the comparing the curves (a) and (b) of Figs. 1.1 and 1.2. The crystal field splitting enhances the numerical values of the 2D ER in both the cases. The numerical values of the 2D ER in accordance with the three band model of Kane are different as compared with the corresponding two band model, which reflects that fact that the presence of the spin orbit splitting constant changes the magnitude of the 2D ER. It may be noted that the presence of the band non-parabolicity in accordance with the two-band model of Kane further changes the peaks of the oscillatory 2D ER for all cases of quantum confinements. The appearance of the humps of the respective curves is due to the redistribution of the electrons among the quantized energy levels when the quantum numbers corresponding to the highest occupied level changes from one fixed value to the others. With varying electron concentration, a change is reflected in the 2D ER through the redistribution of the electrons among the quantized levels. Although the 2D ER varies in various manners with all the variables in all the limiting cases as evident from all the curves of Figs. 1.1 and 1.2, the rates of variations are totally band-structure dependent.

In Figs. 1.3 and 1.4, we have plotted the HD 2D ER as a function of surface electron concentration per unit area for all cases of Figs. 1.1 and 1.2 respectively. It appears that the HD 2D ER increases with increasing carrier degeneracy and also reflects the signature of the 1D confinement through the step like dependence with the HD 2D electron statistics. This oscillatory dependence will be less and less prominent with increasing carrier concentration and ultimately, for bulk specimens of the same material, the ER will be found to increase continuously with increasing electron concentration in a non-oscillatory manner. We have plotted the normalized HD 2D ER as functions of nano-thickness for QWs of HD GaAs, InAs and InSb in Figs. 1.5, 1.6 and 1.7, where, the curve (a), (b) and (c) correspond to the HD three and the two band models of Kane together with the parabolic energy band respectively. The dependence of the 2D ER on the surface electron concentration per unit area for all the said cases is shown in Figs. 1.8, 1.9, and 1.10 respectively.

Using the same set of equations as for III-V materials, we have plotted the normalized 2D ER for QWs of HD $Hg_{1-x-}Cd_xTe$ and $In_{1-x}Ga_xAs_yP_{1-y}$ as a function of nano-thickness and 2D electron statistics as shown by Figs. 1.11, 1.12, 1.13 and 1.14 respectively in which, the curve (a), (b) and (c) corresponds to the HD three and the two band models of Kane together with the parabolic energy bands respectively. The numerical values of the HD 2D ER depend on the energy band

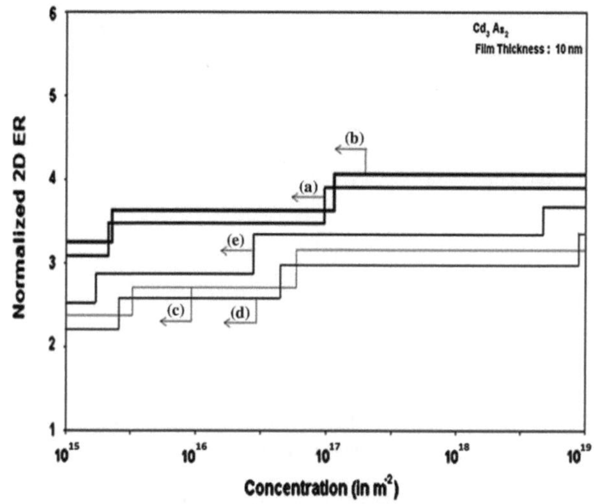

Fig. 1.3 Plot of the normalized HD 2D ER as a function of surface electron concentration per unit area for the QWs of HD n-Cd_3As_2 in accordance with *a* the HD generalized band model with $\delta \neq 0$, *b* the generalized HD band model with $\delta = 0$, *c* the simplified HD three band model of Kane, *d* the HD two band model of Kane and *e* the HD parabolic energy bands

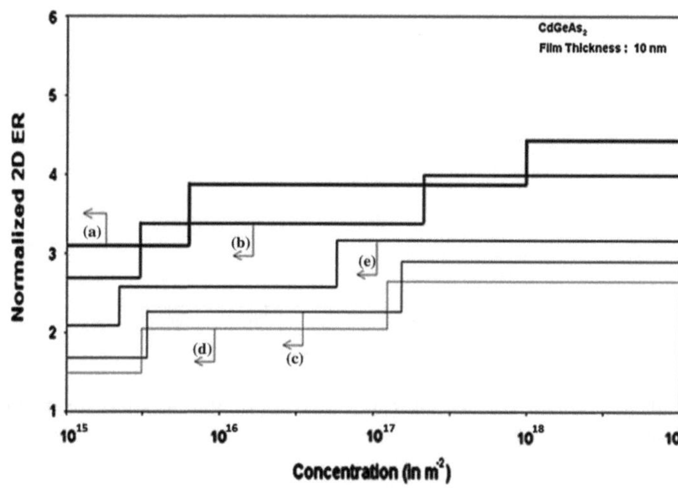

Fig. 1.4 Plot of the normalized HD 2D ER as a function of surface electron concentration per unit area for the QWs of HD n-$CdGeAs_2$ for all cases of Fig. 1.3

constants of different materials. The Figs. 1.15 and 1.16 exhibit the dependence of the 2D ER as a function of alloy composition for all the cases of the QWs of HD ternary and quaternary materials as considered above. The ER decreases with increasing alloy composition. We have investigated the normalized 2D ER, for

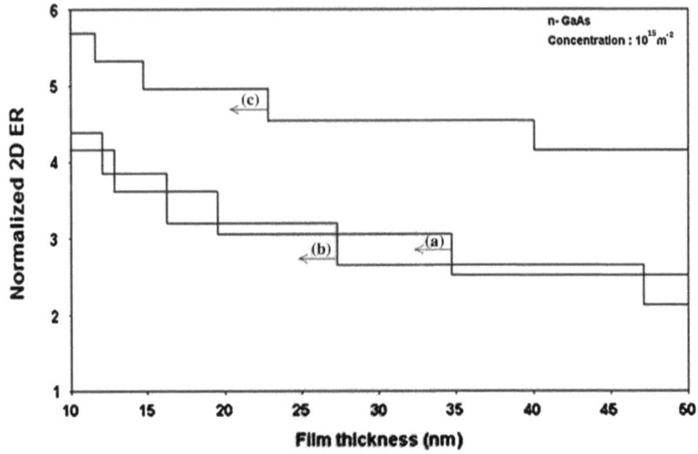

Fig. 1.5 Plot of the normalized HD 2D ER as a function of film thickness for the QWs of HD n-GaAs in accordance with *a* the simplified HD three band model of Kane, *b* the HD two band model of Kane and *c* the HD parabolic energy bands

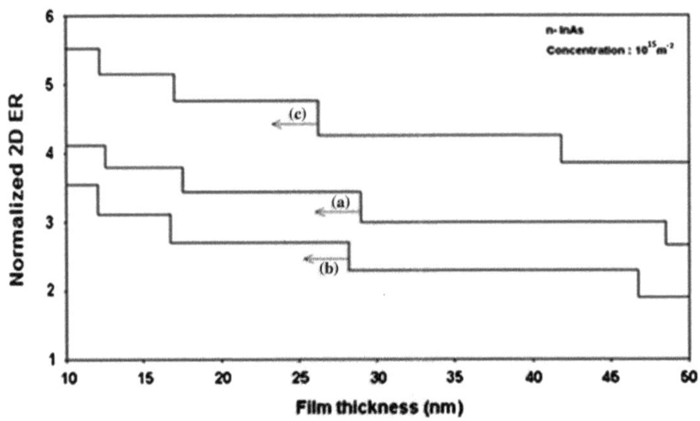

Fig. 1.6 Plot of the normalized HD 2D ER as a function of film thickness for the QWs of HD n-InAs for all cases of Fig. 1.5

QWs of HD p-CdS materials as functions of nano-thickness and surface electron concentration in Figs. 1.17 and 1.18 respectively, where, the curve (a) refers to $\bar{\lambda}_0 \neq 0$ and the curve (b) refers to $\bar{\lambda}_0 = 0$, which has been used for the purpose of assessing the influence of the splitting of the two-spin states by the spin orbit coupling and the crystalline field in this case.

It appears from Figs. 1.17 and 1.18, that the influence of the term $\bar{\lambda}_0$ is the reduction of the quantum jumps of the oscillatory 2D ER in QWs of HD II-VI

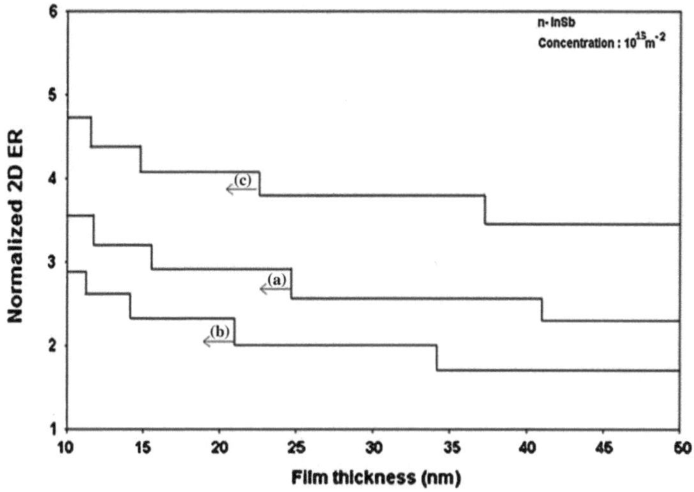

Fig. 1.7 Plot of the normalized HD 2D ER as a function of film thickness for the QWs of HD n-InAs for all cases of Fig. 1.5

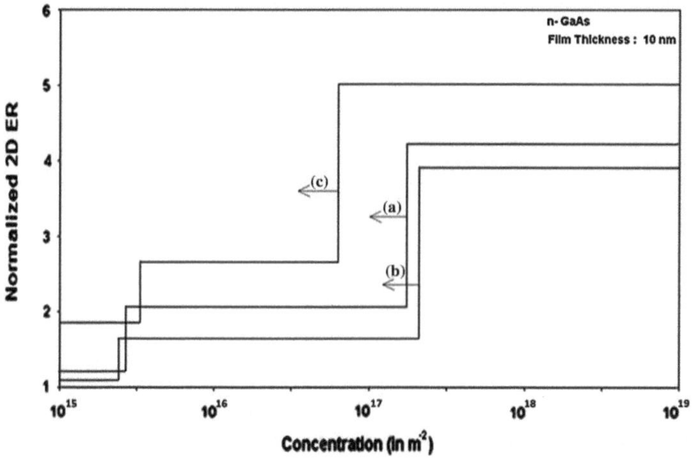

Fig. 1.8 Plot of the normalized HD 2D ER as a function of surface electron concentration per unit area for the QWs of HD GaAs in accordance with *a* the simplified HD three band model of Kane, *b* the HD two band model of Kane and *c* the HD parabolic energy bands

materials for both the variables. Using the appropriate equations we have plotted in Figs. 1.19 and 1.20, the normalized 2D ER for the QWs of HD Te, GaP, $PtSb_2$ and Bi_2Te_3 as functions of nano-thickness and surface electron concentration per unit area as shown by the curves (a), (b), (c) and (d) respectively. It appears from both the Figs. 1.19 and 1.20 that the numerical magnitudes of the 2D E R are due to the

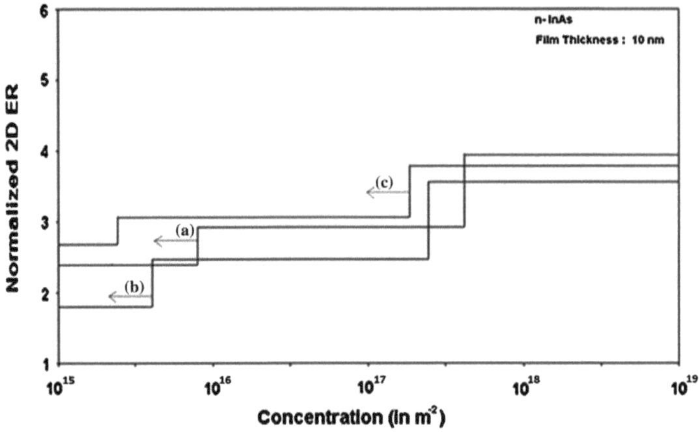

Fig. 1.9 Plot of the normalized HD 2D ER as a function of surface electron concentration per unit area for the QWs of HD n-InAs for all cases of Fig. 1.8

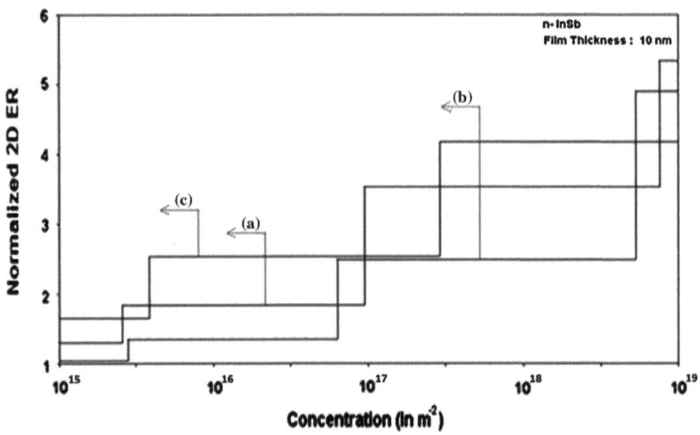

Fig. 1.10 Plot of the normalized HD 2D ER as a function of surface electron concentration per unit area for the QWs of HD n-InSb for all cases of Fig. 1.8

influences of the energy band constants of the respective material as considered here. The Figs. 1.21 and 1.22 exhibit the normalized 2D ER for the QWs of HD IV-VI materials as functions of nano-thickness and surface electron concentration respectively. The curves (a), (b) and (c) correspond to PbTe, n-PbSnTe and n-$Pb_{1-x}Sn_xSe$ respectively.

The influence of the energy band constants on the ER in both the cases is apparent for all the three different materials as considered here. The normalized 2D ER for QWs of HD stressed Kane type n- InSb has been plotted in Figs. 1.23 and 1.24 as functions of nano-thickness and surface electron concentration respectively

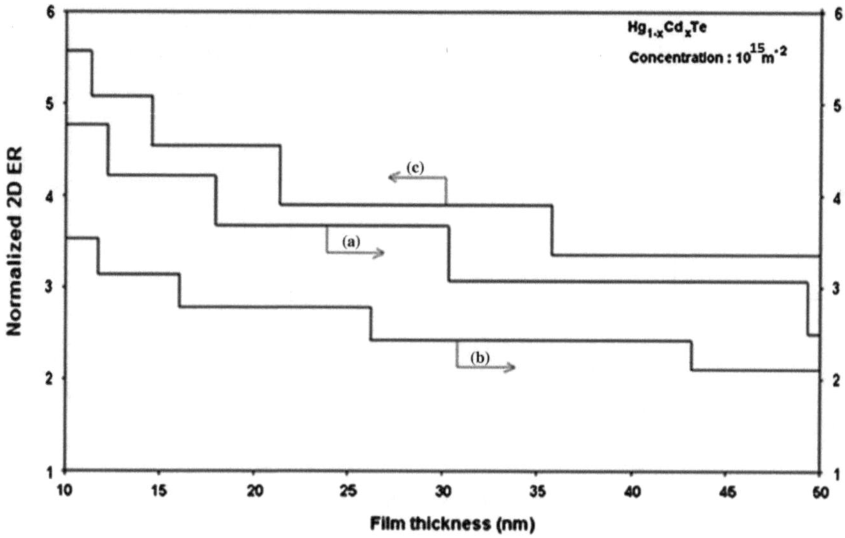

Fig. 1.11 Plot of the normalized HD 2D ER as a function of film thickness for the QWs of HD n-$Hg_{1-x}Cd_xTe$ in accordance with *a* the simplified HD three band model of Kane, *b* the HD two band model of Kane and *c* the HD parabolic energy bands

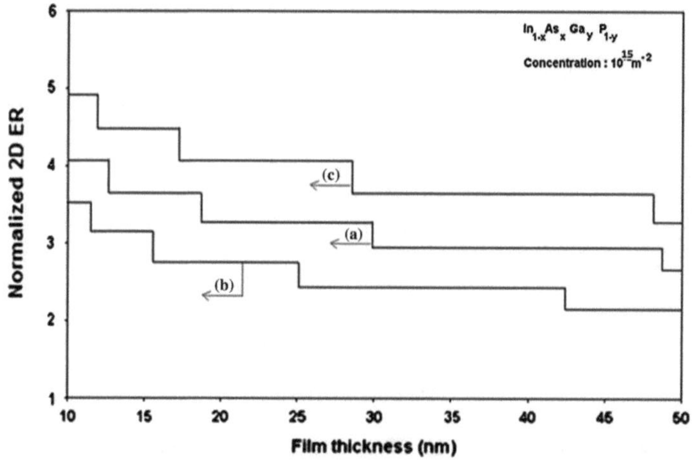

Fig. 1.12 Plot of the normalized HD 2D ER as a function of film thickness for the QWs of HD n-$In_{1-x}As_xGa_yP_{1-y}$ for all cases of Fig. 1.11

as shown in plot (a) in the presence of stress while the plot (b) exhibits the same in the absence of stress for the purpose of assessing the influence of stress on the 2D ER in QWs of HD of stressed n-InSb. In the presence of stress, the magnitude of the 2D ER is being increased as compared with the same under stress free

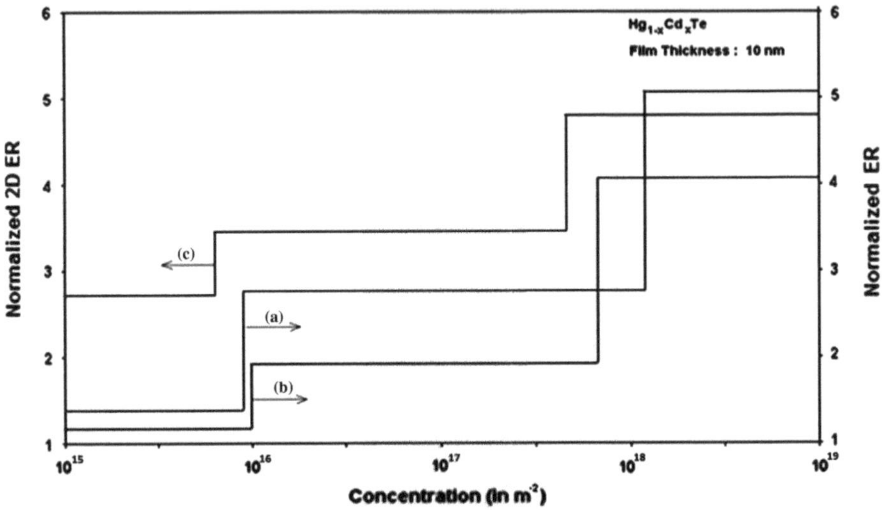

Fig. 1.13 Plot of the normalized HD 2D ER as a function of surface electron concentration for the QWs of HD n-$Hg_{1\text{-}x}Cd_xTe$ in accordance with *a* the simplified HD three band model of Kane, *b* the HD two band model of Kane and *c* the HD parabolic energy bands

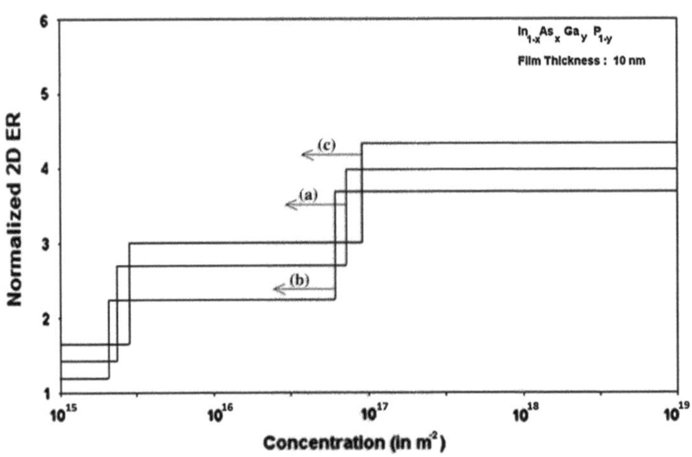

Fig. 1.14 Plot of the normalized HD 2D ER as a function of surface electron concentration for the QWs of HD n-$In_{1\text{-}x}As_xGa_yP_{1\text{-}y}$ for all cases of Fig. 1.13

condition. It may be noted that with the advent of modern experimental techniques, it is possible to fabricate quantum-confined structures with an almost defect-free surface. If the direction normal to the film was taken differently from that as assumed in this work, the expressions for the 2D ER in quasi two-dimensional structures would be different analytically, since the basic dispersion laws of many important materials are anisotropic.

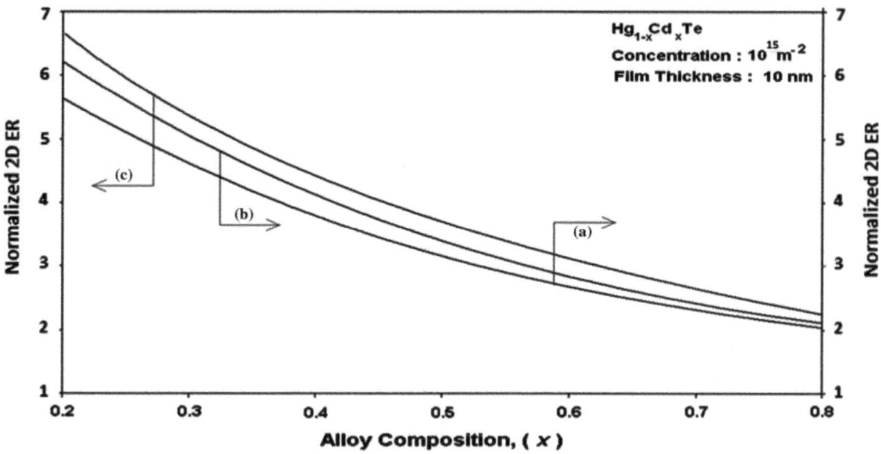

Fig. 1.15 Plot of the normalized HD 2D ER as a function of alloy composition (x) for the QWs of HD n-$Hg_{1-x}Cd_xTe$ in accordance with *a* the simplified HD three band model of Kane, *b* the HD two band model of Kane and *c* the HD parabolic energy bands

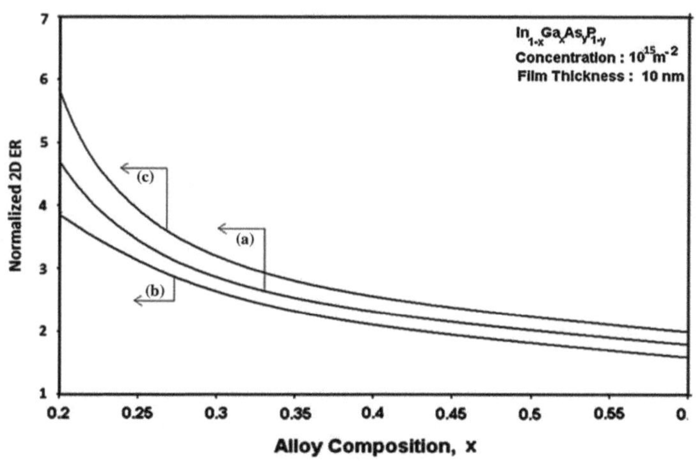

Fig. 1.16 Plot of the normalized HD 2D ER as a function of alloy composition (x) for the QWs of HD n-$In_{1-x}As_xGa_yP_{1-y}$ for all cases of Fig. 1.15

It may be noted that under certain limiting conditions, all the results for all the models as derived here get simplified to have transformed into the well-known expressions of 3D and 2D ERs. This indirect test not only exhibits the mathematical compatibility of the present formulation but also shows the fact that our simple analysis is a more generalized one, since one can obtain the corresponding results for relatively wide gap 2D materials having parabolic energy bands under certain limiting conditions from the present generalized analysis. Thus, the present

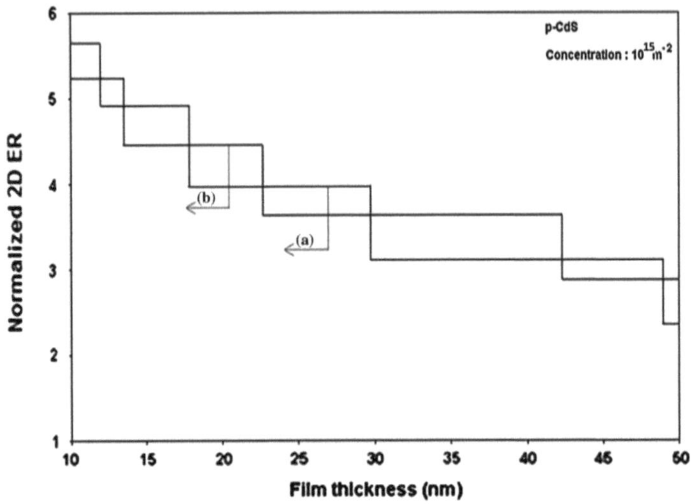

Fig. 1.17 Plot of the normalized HD 2D ER for as a function of film thickness for QWs of HD p- CdS in accordance with *a* Hopfield model with *b* Hopfield model with $\bar{\lambda}_0 = 0$

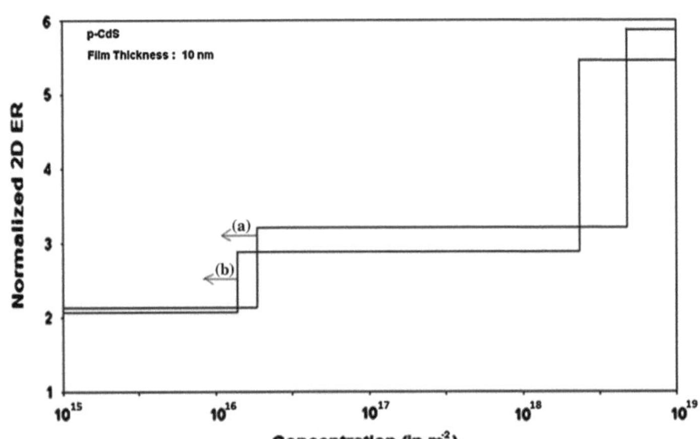

Fig. 1.18 Plot of the normalized HD 2D ER for as a function of surface electron concentration for QWs of HD p-CdS in accordance with *a* Hopfield model with $\bar{\lambda}_0 \neq 0$ *b* Hopfield model with $\bar{\lambda}_0 = 0$

investigations cover the study of 2D ER for QWs of HD nonlinear optical, III-V, ternaries, quaternaries, II-VI, IV-VI, stressed compounds, Te, GaP, $PtSb_2$, Bi_2Te_3, Ge and GaSb having different band structures.

One striking understanding as a collateral study as considered here, is that, the EEM becomes a function of the size quantum number, the Fermi energy and other

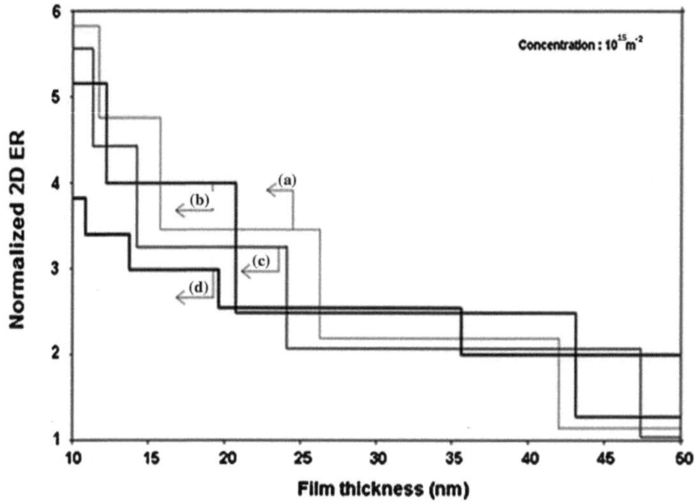

Fig. 1.19 Plot of the normalized HD 2D ER for QWs of HD *a* Te, *b* GaP, *c* PtSb2, and *d* Bi_2Te_3 as a function of film thickness

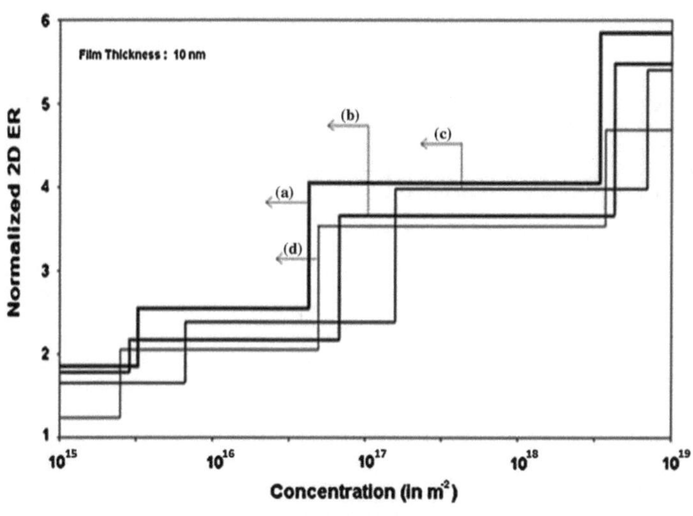

Fig. 1.20 Plot of the normalized HD 2D ER for QWs of HD *a* Te, *b* GaP, *c* $PtSb_2$, and *d* Bi_2Te_3 as a function of electron concentration per unit area

energy band constants depending on the respective HD 2D dispersion laws as formulated already in the respective theoretical background of this chapter together with the fact that the ***EEMs exists in the band gap, a phenomena which is impossible without the concept of band tailing***. It must be mentioned that a direct research application of the quantized materials is in the area of band structure. The

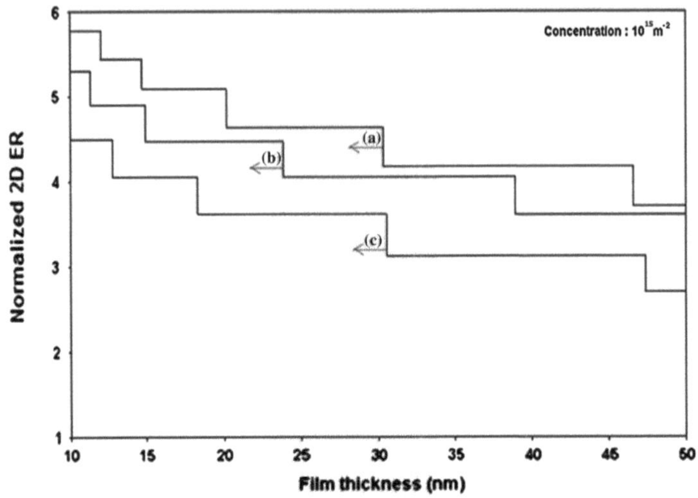

Fig. 1.21 Plot of the normalized HD 2D ER as a function of film thickness for QWs of HD *a* n-PbTe, *b* n-PbSnTe and *c* n-$Pb_{1-x}Sn_xSe$

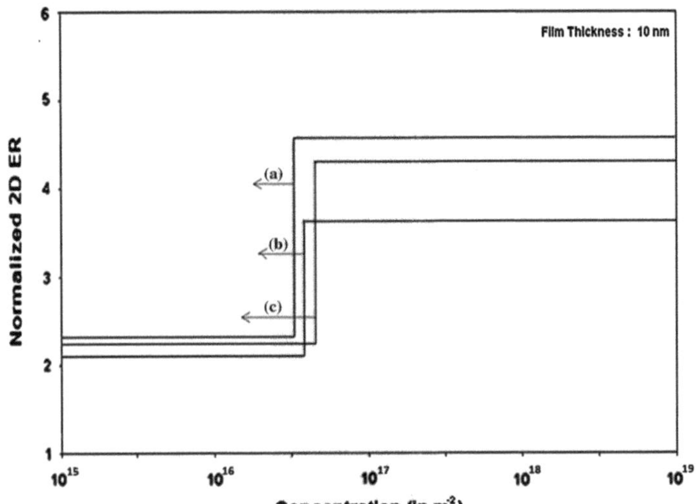

Fig. 1.22 Plot of the normalized HD 2D ER as a function of surface electron concentration for the QWs of HD *a* PbTe, *b* n-PbSnTe and *c* n-$Pb_{1-x}Sn_xSe$N-Pb1-xSnxSe

theoretical results as derived in this chapter can be used to determine the 2D diffusivity and the 3D diffusivity of the constituent HD bulk materials in the absence of quantum effects and this simplified formulation exhibits the basic qualitative features of 2D ER for different quantum confined materials.

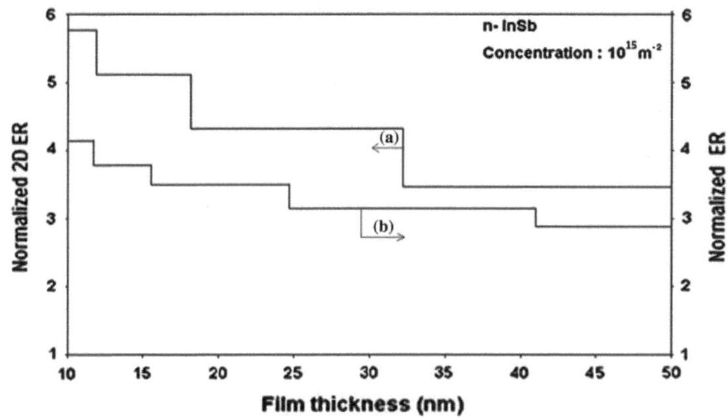

Fig. 1.23 Plot of the normalized HD 2D ER as a function of film thickness for the QWs of HD stressed n- InSb in which the curve *a* shows the 2D ER in the presence of stress while the curve *b* is applicable in the absence of the stress

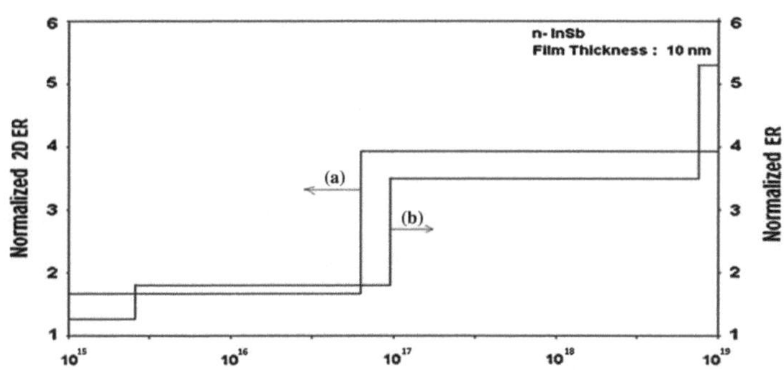

Fig. 1.24 Plot of the normalized HD 2D ER as a function of surface electron concentration for the QWs of HD stressed n-InSb in which the curve *a* shows the 2D ER in the presence of stress while the curve *b* is applicable in the absence of the stress

For the purpose of relative comparison, the 3D ER has been numerically computed for HD n-Cd_3As_2 as a function of electron concentration as shown in curve (a) of Fig. 1.25. The curve (b) corresponds to $\delta = 0$ and the curve (c) exhibits the dependence of the ER on n_0 in accordance with the HD three-band model of Kane, respectively. The plots (d) and (e) correspond to the HD two-band model of Kane and that of parabolic energy bands respectively. By comparing the curves (a) and (b) of Fig. 1.25, one can assess the influence of crystal field splitting of the ER in HD nonlinear optical compounds. The Fig. 1.26 represents all cases of Fig. 1.25 for heavily n-$CdGeAs_2$. It appears from Figs. 1.25 and 1.26 that, the ER in HD n-Cd_3As_2 and n-$CdGeAs_2$ increases with increasing carrier degeneracy as expected for degenerate materials without band tails.

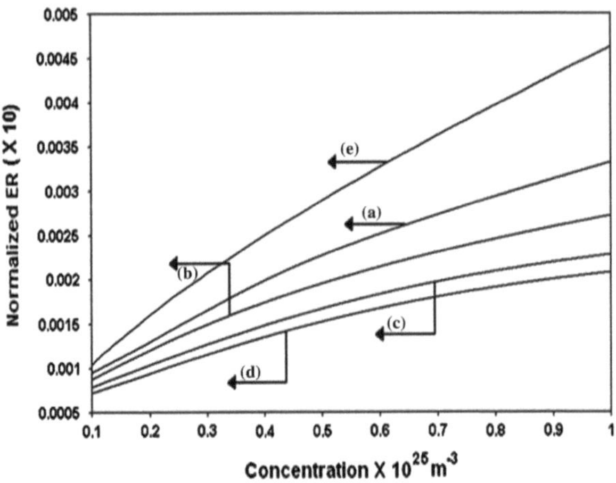

Fig. 1.25 The plot of the normalized ER in the HD n-Cd_3As_2 as function of electron concentration in accordance with *a* the generalized band model, *b* $\delta = 0$, *c* the three band model of Kane, *d* the two band model of Kane and *e* the parabolic energy bands

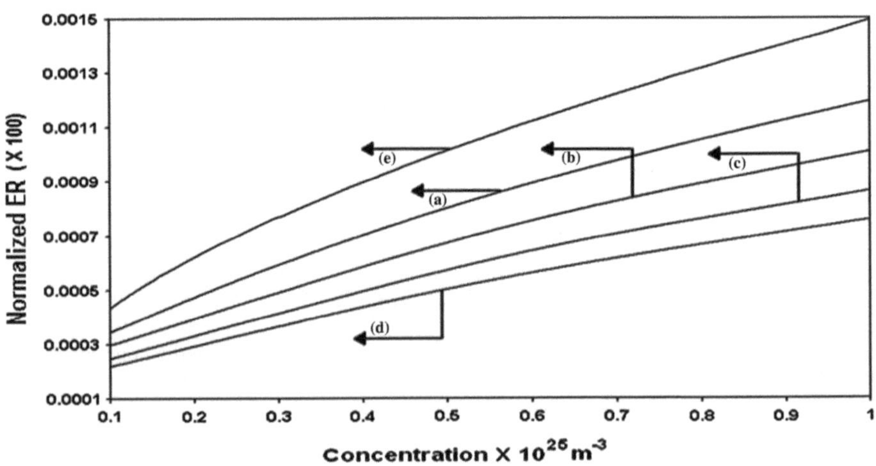

Fig. 1.26 The plot of the normalized ER in the HD n-$CdGeAs_2$ as function of electron concentration in accordance with *a* the generalized band model, *b* $\delta = 0$, *c* the three band model of Kane, *d* the two band model of Kane and *e* the parabolic energy bands

One can numerically evaluate the ER as a function of electron concentration in HD III-V compounds by using n-InAs, and n-InSb as examples as shown in Figs. 1.27 and 1.28 by curves (a), (b) and (c) respectively, in accordance with the HD three and two band models of Kane together with the model of parabolic energy bands . Using n-$Hg_{1-x}Cd_xTe$ as an example of HD ternary compounds, the ER has been numerically plotted for all the band models as a function of electron

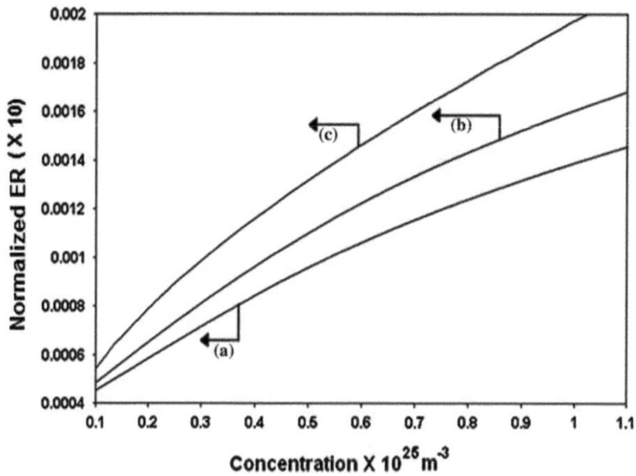

Fig. 1.27 The plot of the normalized ER in HD n-InAs as function of electron concentration in accordance with *a* the three band model of Kane, *b* the two band model of Kane and *c* the parabolic energy bands

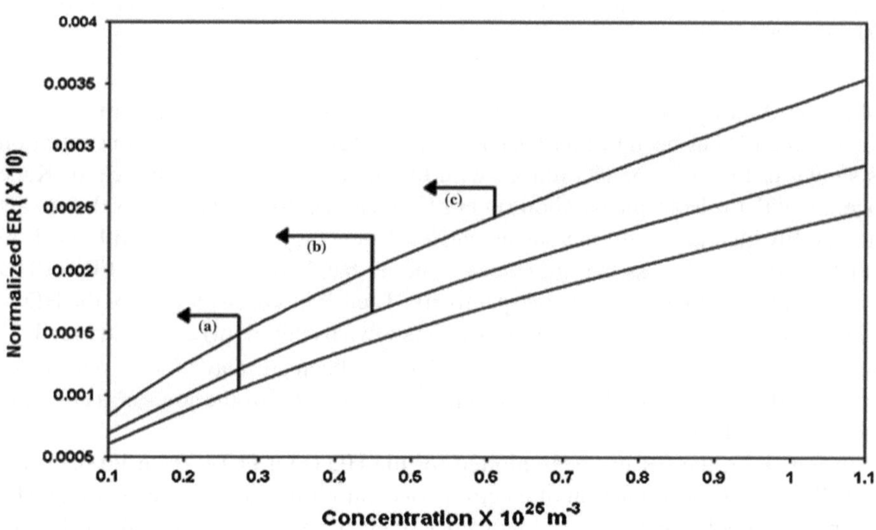

Fig. 1.28 The plot of the normalized ER in HD n-InSb as function of electron concentration in accordance with *a* the three band model of Kane, *b* the two band model of Kane and *c* the parabolic energy bands

concentration as shown in Fig. 1.29. It appears from the Fig. 1.29 that the ER in both cases of HD ternary compounds increases with increasing electron concentration as usual for the degenerate compounds without band tails. Taking

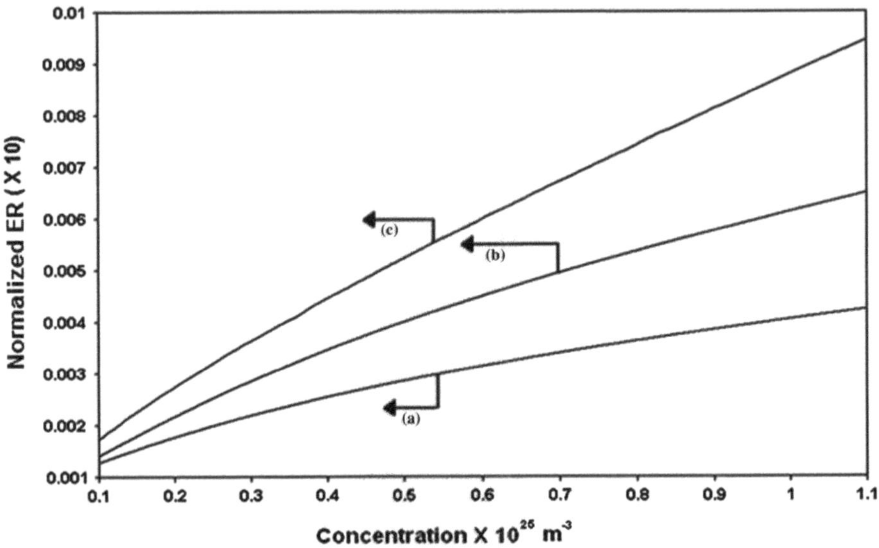

Fig. 1.29 The plot of the normalized ER in HD n-$Hg_{1-x}Cd_x$ Te as function of electron concentration in accordance with *a* the three band model of Kane, *b* the two band model of Kane and *c* the parabolic energy bands

n-$In_{1-x}Ga_xAs_yP_{1-y}$ lattice matched to InP as an example of HD quaternary compounds, the ER has been further been plotted as a function of electron concentration as shown in Fig. 1.30 in accordance with the three and two band models of Kane together with the isotropic parabolic energy band model for both the cases. It appears that the ER increases with increasing carrier degeneracy as usual. From Figs. 1.29 and 1.30, one can assess the influence of energy band constants on the ER for HD ternary and quaternary materials respectively. The ER has been plotted for the HD p-CdS, as a function of hole concentration p_0 as shown by curves (a) and (b) in Fig. 1.31 for which $\bar{\lambda}_0 \neq 0$ and $\bar{\lambda}_0 = 0$ respectively. This has been presented for the purpose of assessing the influence of the splitting of the two spin states by the spin-orbit coupling and the crystalline field on the ER.

In Fig. 1.32, the ER has been plotted for the HD (a) n-PbTe, (b) n-PbSnTe and (c) n-$Pb_{1-x}Sn_xSe$ as a function of electron concentration in accordance with the HD Dimmock model. For relatively low values of electron concentration, the values of the ER for the three materials exhibit convergence behavior where as for relatively large values of n_0, the numerical values differ widely from each other. Our present analysis is also valid for p-type IV-VI compounds with the proper change in the energy band constants.

In Fig. 1.33, the ER has been plotted for the HD stressed n-InSb as a function of electron concentration. For the purpose of assessing the influence of stress on the ER in bulk specimens of stressed HD n-InSb, the plot (a) exhibits the ER in the presence of the stress while the plot (b) shows the same in the absence of the stress.

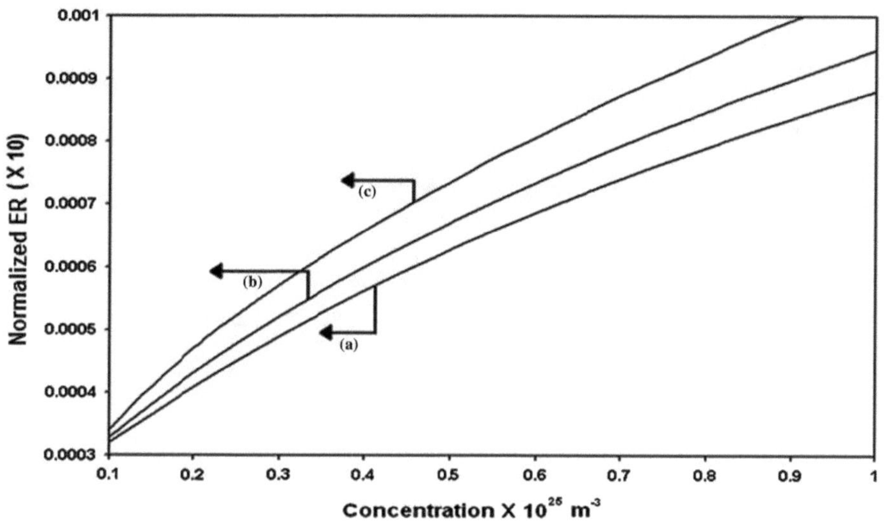

Fig. 1.30 The plot of the normalized ER in HD n-$In_{1-x}Ga_xAs_yP_{1-y}$ lattice matched to InP as function of electron concentration in accordance with *a* the three band model of Kane, *b* the two band model of Kane and *c* the parabolic energy bands

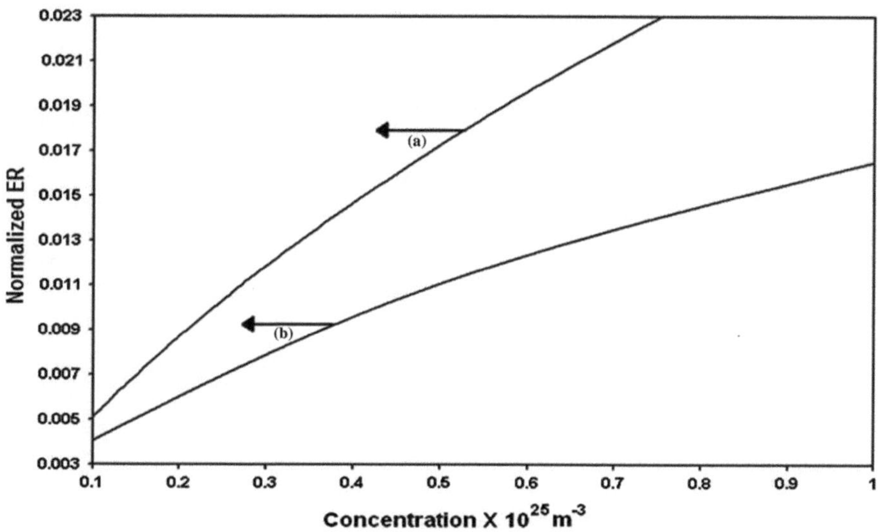

Fig. 1.31 The plot of the normalized ER in HD p-CdS as function of as function of hole concentration in accordance with *a* $\bar{\lambda}_0 \neq 0$ and *b* $\bar{\lambda}_0 = 0$

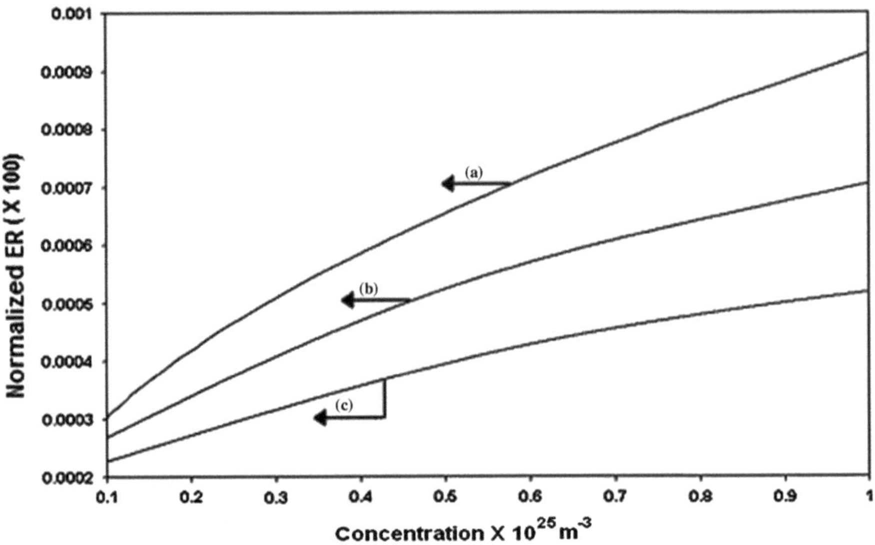

Fig. 1.32 The plot of the normalized ER in HD *a* n-PbTe, *b* PbSnTe and *c* $Pb_{1\text{-}x}Sn_xSe$ as function of electron concentration

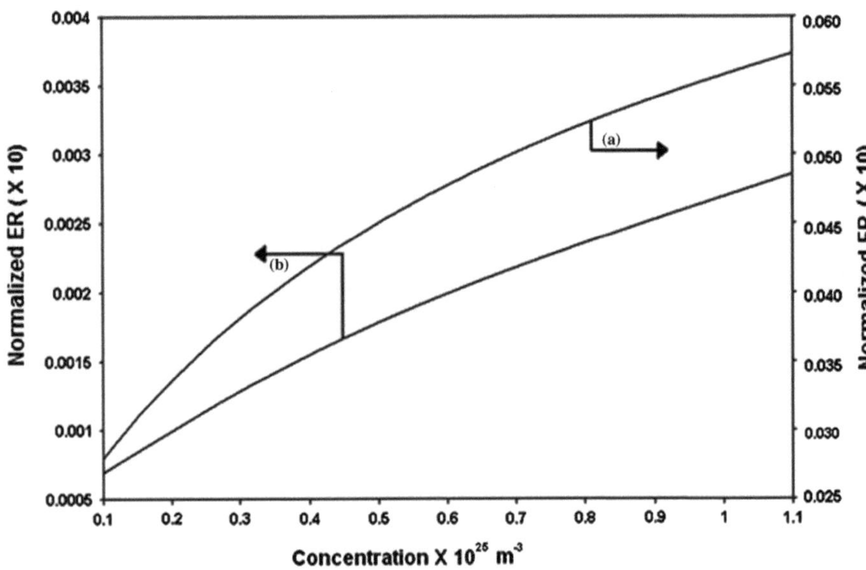

Fig. 1.33 The plot of the normalized ER in HD stressed n-InSb as a function of electron concentration in which the curve *a* is in the presence of stress and curve *b* is under absence of stress

In the presence of the stress, the magnitude of the ER is being increased as compared with the same under stress free condition for both the cases. One important concept of this chapter is the presence of poles in the finite complex plane in the dispersion relation of the materials in the absence of band tails creates the complex energy spectrum in the corresponding HD samples. Besides, from the DOS function in this case, it appears that a new forbidden zone has been created in addition to the normal band gap of the semiconductor. If the basic dispersion relation in the absence of band tails contains no poles in the finite complex plane, the corresponding HD energy band spectrum will be real, although it may be the complicated functions of exponential and error functions and deviate considerably from that in the absence of band tailing.

Another important point in this context is the existence of the effective mass within the forbidden zone, which is impossible without the formation of band tails. It is an amazing fact that the study of the carrier transport in HD quantized materials through proper formulation of the Boltzmann transport equation which needs in turn, the corresponding HD carrier energy spectra is still one of the open research problems.

It is already noted that with the advent of MBE and other experimental techniques, it is possible to fabricate quantum-confined structures with an almost defect-free surface. In formulating the generalized electron energy spectrum for non-linear optical materials, we have considered the crystal-field splitting parameter, the anisotropies in the momentum-matrix elements, and the spin-orbit splitting parameters, respectively. In the absence of the crystal field splitting parameter together with the assumptions of isotropic effective electron mass and isotropic spin orbit splitting, our basic relation as given by (1.2) converts into (1.48). The (1.48) is the well-known three-band Kane model and is valid for III-V compounds, in general. It should be used as such for studying the electronic properties of n-InAs where the spin-orbit splitting parameter (Δ) is of the order of band gap (E_g). For many important materials $\Delta \gg E_g$ and under this inequality, (1.48) assumes the form $E(1 + EE_g^{-1}) = \hbar^2 k^2/2m_c$ which is the well-known two-band Kane model. Also under the condition, $E_g \to \infty$, the above equation gets simplified to the well-known form of parabolic energy bands as $E = \hbar^2 k^2/2m_c$. It is important to note that under certain limiting conditions, all the results for all the models as derived here have transformed into the well-known expression of the 2D ER for size quantized materials having parabolic bands. We have not considered other types of compounds or external physical variables for numerical computations in order to keep the presentation brief. With different sets of energy band constants, we shall get different numerical values of the 2D ER though the nature of variations of the 2D ER as shown here would be similar for the other types of materials and the simplified analysis of this chapter exhibits the basic qualitative features of the 2D ER for such compounds.

By mapping, the discrete quantum state energies as a function of film thickness is a given crystal direction, information about the effective masses and the dispersion relations may be derived. We must note that the study of transport phenomena and the formulation of the electronic properties of HD nano-compounds are based on the dispersion relations in such materials. The theoretical results of our chapter can be used to determine the 2D ER and the constituent heavily-doped bulk materials in the absence of size effects It is worth remarking that this simplified formulation exhibits the basic qualitative features of 2D ER for nano-materials. The basic objective of this chapter is not solely to demonstrate the influence of quantum confinement on the 2D ER for QWs of non-parabolic materials but also to formulate the appropriate electron statistics in the most generalized form, since the transport and other phenomena in HD nano-materials having different band structures and the derivation of the expressions of many important electronic properties are based on the temperature-dependent electron statistics in such compounds.

Our method is not at all related to the DOS technique as used in the literature. From the E-k dispersion relation, we can obtain the DOS, but the DOS technique as used in the literature cannot provide the E-k dispersion relation. Therefore, our study is more fundamental than those of the existing literature because the Boltzmann transport equation, which controls the study of the charge transport properties of semiconductor devices, can be solved if and only if the E-k dispersion relation is known. We wish to note that we have not considered the many body effects in this simplified theoretical formalism due to the lack of availability in the literature of proper analytical techniques for including them for the generalized systems as considered in this chapter. Our simplified approach will be useful for the purpose of comparison when methods of tackling the formidable problem after inclusion of the many body effects for the present generalized systems appear. It is worth remarking in this context that from our simple theory under certain limiting conditions we get the well-known result of the DSL for wide gap materials having parabolic energy bands. The inclusion of the said effects would certainly increase the accuracy of the results, although the qualitative features of the 2D ER in QWs of HD materials discussed in this chapter would not change in the presence of the aforementioned effects. The influence of energy band models and the various band constants on the ER for different materials can also be studied from all the figures of this chapter.

The numerical results presented in this chapter would be different for other materials but the nature of variation would be unaltered. The theoretical results as given here would be useful in analyzing various other experimental data related to this phenomenon. Finally, we can write that the analysis as presented in this chapter can be used to investigate, the Burstein Moss shift, the carrier contribution to the elastic constants, the specific heat, screening length, activity coefficient, reflection coefficient, Hall coefficient, plasma frequency, various scattering mechanisms and other different transport coefficients of modern HD non-parabolic quantum confined HD devices operated under different external conditions having varying band structures.

1.4 Open Research Problems

The problems under these sections of this monograph are by far the most important part for the readers and few open research problems are presented from this chapter till end. The numerical values of the energy band constants for various semiconductors are given in Table 1.1 for the related computer simulations.

(R.1.1) Investigate the ER for the HD bulk semiconductors whose respective dispersion relations of the carriers in the absence of band tails and any externally applied field are given below:

(a) The electron dispersion law in n-GaP can be written as [203]

$$E = \frac{\hbar^2 k_z^2}{2m_{\parallel}^*} + \frac{\hbar^2 k_s^2}{2m_{\perp}^*} \mp \frac{\bar{\Delta}}{2} \pm \left[\left(\frac{\bar{\Delta}}{2}\right)^2 + P_1 k_z^2 + D_1 k_x^2 k_y^2\right]^{1/2} \quad \text{(R1.1)}$$

where, $\bar{\Delta} = 335\text{me}V$, $\text{P}_1 = 2 \times 10^{-10}$ eVm, $\text{D}_1 = \text{P}_1\text{a}_1$ and $\text{a}_1 = 5.4 \times 10^{-10}$ m.

(b) The dispersion relation for the conduction electrons for IV-VI semiconductors can also be described by the models of Cohen [204], McClure and Choi [205], Bangert et al. [206] and Foley et al. [207] respectively.

(i) In accordance with Cohen [204], the dispersion law of the carriers is given by

$$E(1 + \alpha E) = \frac{p_x^2}{2m_1} + \frac{p_z^2}{2m_3} - \frac{\alpha E p_y^2}{2m_2'} + \left(\frac{\alpha p_y^4}{4m_2 m_2'}\right) + \frac{p_y^2}{2m_2}(1 + \alpha E) \quad \text{(R1.2)}$$

where m_1, m_2 and m_3 are the effective carrier masses at the band-edge along x, y and z directions respectively and m_2' is the effective- mass tensor component at the top of the valence band (for electrons) or at the bottom of the conduction band (for holes).

(ii) The carrier energy spectra can be written, following McClure and Choi, [205] as

$$E(1 + \alpha E) = \frac{p_x^2}{2m_1} + \frac{p_y^2}{2m_2} + \frac{p_z^2}{2m_3} + \frac{p_y^2}{2m_2}\alpha E\left\{1 - \left(\frac{m_2}{m_2'}\right)\right\} + \frac{p_y^4 \alpha}{4m_2 m_2'} - \frac{\alpha p_x^2 p_y^2}{4m_1 m_2} - \frac{\alpha p_y^2 p_z^2}{4m_2 m_3} \quad \text{(R1.3)}$$

(iii) In accordance with Bangert and Kastner [206], the dispersion relation is given by

$$\Gamma(E) = F_1(E)k_s^2 + F_2(E)k_z^2 \tag{R1.4}$$

where, $\Gamma(E) \equiv 2E$, $F_1(E) \equiv \frac{R_1^2}{E+E_g} + \frac{S_1^2}{E+\Delta_c'} + \frac{Q_1^2}{E+E_g}$, $F_2(E) \equiv \frac{2C_5^2}{E+E_g} + \frac{(S_1+Q_1)^2}{E+\Delta_c''}$
$R_1^2 = 2.3 \times 10^{-19}(eVm)^2$, $C_5^2 = 0.83 \times 10^{-19}(eVm)^2$, $Q_1^2 = 1.3R_1^2$, $S_1^2 = 4.6R_1^2$, $\Delta_c' = 3.07eV$, $\Delta'' = 3028\,eV$ and $g_v = 4$. It may be noted that under the $S_1 = 0$, $Q_1 = 0$, $R_1^2 \equiv \frac{\hbar^2 E_g}{m_\perp^*}$, $C_5^2 \equiv \frac{\hbar^2 E_g}{2m_\parallel^*}$, (R1.4) assumes the form $E(1 + \alpha E) = \frac{\hbar^2 k_s^2}{2m_\perp^*} + \frac{\hbar^2 k_z^2}{2m_\parallel^*}$ which is the simplified Lax model.

(iv) The carrier energy spectrum of IV–VI semiconductors in accordance with Foley et al. [207] can be written as

$$E + \frac{E_g}{2} = E_-(k) + \left[\left[E_+(k) + \frac{E_g}{2}\right]^2 + P_\perp^2 k_s^2 + P_\parallel^2 k_z^2\right]^{1/2} \tag{R1.5}$$

where, $E_+(\mathrm{k}) = \frac{\hbar^2 k_s^2}{2m_\perp^+} + \frac{\hbar^2 k_z^2}{2m_\parallel^+}$, $E_-(\mathrm{k}) = \frac{\hbar^2 k_s^2}{2m_\perp^-} + \frac{\hbar^2 k_s^2}{2m_\parallel^-}$ represents the contribution from the interaction of the conduction and the valance band edge states with the more distant bands and the free electron term, $\frac{1}{m_\perp^\pm} = \frac{1}{2}\left[\frac{1}{m_{tc}} \pm \frac{1}{m_{tv}}\right]$, $\frac{1}{m_\parallel^\pm} = \frac{1}{2}\left[\frac{1}{m_{tc}} \pm \frac{1}{m_{1v}}\right]$,

For n-PbTe

$$\mathrm{P}_\perp = 4.61 \times 10^{-10}\,\mathrm{eVm},\ \mathrm{P}_\parallel = 4.61 \times 10^{-10}\,\mathrm{eVm},\ \frac{m_0}{m_{tv}} = 10.36,$$
$$\frac{m_0}{m_{tv}} = 0.75,\ \frac{m_0}{m_{tv}} = 11.36,\ \frac{m_0}{m_{tv}} = 1.20,\ \text{and}\ g_v = 4$$

(c) The hole energy spectrum of p-type zero-gap semiconductors (e.g. HgTe) is given by [208]

$$E = \frac{\hbar^2 k^2}{2m_v^*} + \frac{3e^2}{128\varepsilon_\infty}k - \left(\frac{2E_B}{\pi}\right)\ln\left|\frac{k}{k_0}\right| \tag{R1.6}$$

where m_v^* is the effective mass of the hole at the top of the valence band, $E_B = \frac{m_0 e^2}{2\hbar^2 \varepsilon_\infty^2}$ and $k_0 = \frac{m_0 e^2}{\hbar^2 \varepsilon_\infty}$.

(d) The conduction electrons of n-GaSb obey the following two dispersion relations:

1. In accordance with the model of Seiler et al. [209]

$$E = \left[-\frac{E_g}{2} + \frac{E_g}{2}[1 + \alpha_4 k^2]^{1/2} + \frac{\bar{\varsigma}_0 \hbar^2 k^2}{2m_o} + \frac{\bar{v}_0 f_1(k)\hbar^2}{2m_o} \pm \frac{\bar{\omega}_0 f_2(k)\hbar^2}{2m_o} \right] \tag{R1.7}$$

where $\alpha_4 \equiv 4P^2(E_g + \frac{2}{3}\Delta)\left[E_g^2(E_g + \Delta)\right]^{-1}$, P is the isotropic momentum matrix element, $f_1(k) \equiv k^{-2}\left[k_x^2 k_y^2 + k_y^2 k_z^2 + k_z^2 k_x^2\right]$ represents the warping of the Fermi surface, $f_2(k) \equiv \left[\{k^2 (k_x^2 k_y^2 + k_y^2 k_z^2 + k_z^2 k_x^2) - 9k_x^2 k_y^2 k_z^2\}^{1/2} k^{-1}\right]$ represents the inversion asymmetry splitting of the conduction bandand $\bar{\varsigma}_0, \bar{v}_0$ and $\bar{\omega}_0$ represent the constants of the electron spectrum in this case.

2. In accordance with the model of Zhang et al. [210]

$$E = \left[E_2^{(1)} + E_2^{(2)} K_{4,1}\right] k^2 + \left[E_4^{(1)} + E_4^{(2)} K_{4,1}\right] k^4 + \left[E_6^{(1)} + E_6^{(2)} K_{4,1} + E_6^{(3)} K_{6,1}\right] \tag{R1.8}$$

where $K_{4,1} \equiv \frac{5}{4}\sqrt{21}\left[\frac{k_x^4 + k_y^4 + k_z^4}{k^4} - \frac{3}{5}\right]$, $K_{6,1} \equiv \sqrt{\frac{639639}{32}}\left[\frac{k_x^2 k_y^2 k_z^2}{k^6} + \frac{1}{22}\left(\frac{k_x^4 + k_y^4 + k_z^4}{k^4} - \frac{3}{5}\right) - \frac{1}{105}\right]$, the coefficients are in eV, the values of k are $10\left(\frac{a}{2\pi}\right)$ times those of k in atomic units (a is the lattice constant), $E_2^{(1)} = 1.0239620$, $E_2^{(2)} = 0$, $E_4^{(1)} = -1.1320772$, $E_4^{(2)} = 0.05658$, $E_6^{(1)} = 1.1072073$, $E_6^{(2)} = -0.1134024$ and $E_6^{(3)} = -0.0072275$.

(e) In addition to the well-known band models of III-V semiconductors as discussed in this monograph, the conduction electrons of such compounds obey the following three dispersion relations:

1. In accordance with the model of Rossler [211]

$$E = \frac{\hbar^2 k^2}{2m_c} + \bar{\alpha}_{10} k^4 + \bar{\beta}_{10}[k_x^2 k_y^2 + k_y^2 k_z^2 + k_z^2 k_x^2] \pm \bar{\gamma}_{10}[k^2(k_x^2 k_y^2 + k_y^2 k_z^2 + k_z^2 k_x^2) - 9k_x^2 k_y^2 k_z^2]^{1/2} \tag{R1.9}$$

where, $\bar{\alpha}_{10} = \bar{\alpha}_{11} + \bar{\alpha}_{12} k$, $\bar{\beta}_{10} = \bar{\beta}_{11} + \bar{\beta}_{12} k$ and $\bar{\gamma}_{10} = \bar{\gamma}_{11} + \bar{\gamma}_{12} k$, in which, $\bar{\alpha}_{11} = -2{,}132 \times 10^{-40}\,\text{eVm}^4$, $\bar{\alpha}_{12} = 9{,}030 \times 10^{-50}\,\text{eVm}^5$, $\bar{\beta}_{11} = -2{,}493 \times 10^{-40}\,\text{eVm}^4$, $\bar{\beta}_{12} = 12{,}594 \times$

$10^{-50}\,\text{eVm}^5$, $\bar{\gamma}_{11} = 30 \times 10^{-30}\,\text{eVm}^3$ and $\bar{\gamma}_{12} = -154 \times 10^{-42}\ \text{eVm}^4$.

2. In accordance with Johnson and Dickey [212], the electron energy spectrum assumes the form

$E = -\frac{E_g}{2} + \frac{\hbar^2 k^2}{2}\left[\frac{1}{m_0} + \frac{1}{m_{\gamma b}}\right] + \frac{E_g}{2}\left[1 + 4\frac{\hbar^2 k^2}{2m_c'}\frac{\bar{f}_1(E)}{E_g}\right]^{\frac{1}{2}}$

where, $\frac{m_0}{m_c'} \equiv P^2\left[\frac{\left(E_g + \frac{2\Delta}{3}\right)}{E_g(E_g+\Delta)}\right]$, $\bar{f}_1(E) \equiv \frac{(E_g+\Delta)\left(E+E_g+\frac{2\Delta}{3}\right)}{\left(E_g+\frac{2\Delta}{3}\right)(E+E_g+\Delta)}$,

$m_c' = 0.139 m_0$ and $m_{\gamma b} = \left[\frac{1}{m_c'} - \frac{2}{m_0}\right]^{-1}$.

3. In accordance with Agafonov et al. [213], the electron energy spectrum can be written as

$$E = \frac{\bar{\eta} - E_g}{2}\left[1 - \frac{\hbar^2 k^2}{2\bar{\eta} m_c}\left\{\frac{D\sqrt{3} - 3\bar{B}}{2\left(\frac{\hbar^2}{2m_c}\right)}\right\}\left[\frac{k_x^4 + k_y^4 + k_z^4}{k^4}\right]\right] \tag{R1.10}$$

where, $\bar{\eta} \equiv \left(E_g^2 + \frac{8}{3}P^2 k^2\right)^{1/2}$, $\bar{B} \equiv -21\frac{\hbar^2}{2m_0}$ and $D \equiv -40\left(\frac{\hbar^2}{2m_0}\right)$.

(f) The dispersion relation of the carriers in n-type $Pb_{1\text{-}x}Ga_xTe$ with x = 0.01 can be written following Vassilev [214] as

$$[E - 0.606k_s^2 - 0.0722k_z^2][E + \bar{E}_g + 0.411k_s^2 + 0.0377k_z^2] = 0.23k_s^2 + 0.02k_z^2 \pm [0.06\bar{E}_g + 0.061k_s^2 + 0.0066k_z^2]k_s \tag{R1.11}$$

where, $\bar{E}_g (= 0.21\text{eV})$ is the energy gap for the transition point, the zero of the energy E is at the edge of the conduction band of the Γ point of the Brillouin zone and is measured positively upwards, k_x, k_y and k_z are in the units of $10^9\ \text{m}^{-1}$.

(g) The energy spectrum of the carriers in the two higher valance bands and the single lower valance band of Te can, respectively, be expressed as [215]

$$\bar{E} = A_{10}k_z^2 + B_{10}k_s^2 \pm \left[\Delta_{10}^2 + (\beta_{10}k_z)^2\right]^{1/2} \text{ and } \bar{E} = \Delta_{\parallel} + A_{10}k_z^2 + B_{10}k_s^2 \pm \beta_{10}k_z \tag{R1.12}$$

where, $\bar{E}$ is the energy of the hole as measured from the top of the valance and within it, $A_{10} = 3.77 \times 10^{-19}\,\text{eVm}^2$, $B_{10} = 3.57 \times$

$10^{-19}\,\text{eVm}^2$, $\Delta_{10} = 0.628\,\text{eV}$, $(\beta_{10})^2 = 6 \times 10^{-20}\,(\text{eVm})^2$ and $\Delta_{\|} = 1{,}004 \times 10^{-5}\,\text{eV}$ are the spectrum constants.

(h) The dispersion relation in graphite can be written following Brandt [216] as

$$E = \frac{1}{2}[E_2 + E_3] \pm \left[\frac{1}{4}(E_2 - E_3)^2 + \eta_2^2 k^2\right]^{1/2} \tag{R1.13}$$

where,
$E_2 \equiv \bar{\Delta} = -2\bar{\gamma}_1 \cos\phi_0 + 2\bar{\gamma}_5 \cos^2\phi_0$, $\phi_0 \equiv \frac{c_6 k_z}{2}$, $E_3 \equiv 2\bar{\gamma}_2 \cos^2\phi_0$ and $\eta_2 \equiv \left(\frac{\sqrt{3}}{2}\right) a_6(\bar{\gamma}_0 + 2\bar{\gamma}_4 \cos\phi_0)$ in which the band constants are $\bar{\Delta}$, $\bar{\gamma}_0$, $\bar{\gamma}_1$, $\bar{\gamma}_2$, $\bar{\gamma}_4$, $\bar{\gamma}_5$, a_6 and c_6 respectively.

(i) The dispersion relation of the conduction electrons in Antimony in accordance with Ketterson [217] can be written as

$$2m_0 E = \alpha_{11} p_x^2 + \alpha_{22} p_y^2 + \alpha_{33} p_z^2 + 2\alpha_{23} p_y p_z \tag{R1.14}$$

and

$$2m_0 E = a_1 p_x^2 + a_2 p_y^2 + a_3 p_z^2 + a_4 p_y p_z \pm a_5 p_x p_z \pm a_6 p_x p_y \tag{R1.15}$$

where, $a_1 = \frac{1}{4}(\alpha_{11} + 3\alpha_{22})$, $a_2 = \frac{1}{4}(\alpha_{22} + 3\alpha_{11})$, $a_3 = \alpha_{33}$, $a_4 = \alpha_{33}$, $a_5 = \sqrt{3}$ and $a_6 = \sqrt{3}(\alpha_{22} - \alpha_{11})$ in which α_{11}, α_{22}, α_{33} and α_{23} are the system constants.

(j) The dispersion relation of the holes in p-InSb can be written in accordance with Cunningham [218] as

$$\bar{E} = c_4(1 + \gamma_4 f_4)k^2 \pm \frac{1}{3}[2\sqrt{2}\sqrt{c_4}\sqrt{16 + 5\gamma_4}\sqrt{E_4} g_4 k] \tag{R1.16}$$

where, $c_4 \equiv \frac{\hbar^2}{2m_0} + \theta_4$, $\theta_4 \equiv 4.7\frac{\hbar^2}{2m_0}$, $\gamma_4 \equiv \frac{b_4}{c_4}$, $b_4 \equiv \frac{3}{2}b_5 + 2\theta_4$, $b_5 \equiv 2.4\frac{\hbar^2}{2m_0}$, $f_4 \equiv \frac{1}{4}[\sin^2 2\theta + \sin^4\theta \sin^2 2\phi]$, θ is measured from the positive z-axis, ϕ is measured from positive x-axis, $g_4 \equiv \sin\theta\left[\cos^2\theta + \frac{1}{4}\sin^4\theta \sin^2 2\phi\right]$ and $E_4 = 5 \times 10^{-4}\,\text{eV}$.

(k) The energy spectrum of the valance bands of CuCl in accordance with Yekimov et al. [219] can be written as

$$E_h = (\gamma_6 - 2\gamma_7)\frac{\hbar^2 k^2}{2m_0} \tag{R1.17}$$

and

$$E_{l,s} = (\gamma_6 + \gamma_7)\frac{\hbar^2 k^2}{2m_0} - \frac{\Delta_1}{2} \pm \left[\frac{\Delta_1^2}{2} + \gamma_7 \Delta_1 \frac{\hbar^2 k^2}{2m_0} + 9\left(\frac{\gamma_7 \hbar^2 k^2}{2m_0}\right)^2\right]^{1/2} \tag{R1.18}$$

where, $\gamma_6 = 0.53$, $\gamma_7 = 0.07$, $\Delta_1 = 70\,\text{meV}$.

(l) In the presence of stress, χ_6 along <001> and <111> directions, the energy spectra of the holes in semiconductors having diamond structure valance bands can be respectively expressed following Roman et al. [220] as

$$E = A_6 k^2 \pm \left[\bar{B}_7^2 k^4 + \delta_6^2 + B_7 \delta_6 (2k_z^2 - k_s^2)\right]^{1/2} \tag{R1.19}$$

and

$$E = A_6 k^2 \pm \left[\bar{B}_7^2 + \delta_7^2 + \frac{D_6}{\sqrt{3}} \delta_7 (2k_z^2 - k_s^2)\right]^{1/2} \tag{R1.20}$$

where, A_6, B_7, D_6 and C_6 are inverse mass band parameters in which $\delta_6 \equiv l_7(\bar{S}_{11} - \bar{S}_{12})\chi_6$, $\bar{S}_{ij}$ are the usual elastic compliance constants, $\bar{B}_7^2 \equiv \left(B_7^2 + \frac{c_6^2}{5}\right)$ and $\delta_7 \equiv \left(\frac{d_g S_{44}}{2\sqrt{3}}\right)\chi_6$. For gray tin, $d_8 = -4.1$ eV, $l_7 = -2.3$ eV, $A_6 = 19.2\frac{\hbar^2}{2m_0}$, $B_7 = 26.3\frac{\hbar^2}{2m_0}$, $D_6 = 31\frac{\hbar^2}{2m_0}$ and $c_6^2 = -1112\frac{\hbar^2}{2m_0}$.

(m) The dispersion relation of the carriers of cadmium and zinc diphosphides are given by [221]

$$E = \left[\beta_1 + \frac{\beta_2 \beta_3(k)}{8\beta_4}\right]k^2 \pm \left\{\left[\beta_4 \beta_3(k) x \left(\beta_5 - \frac{\beta_2 \beta_3(k)}{8\beta_4}\right)k^2\right] + 8\beta_4^2\left(1 - \frac{\beta_3^2(k)}{4}\right) - \beta_2\left(1 - \frac{\beta_3^2(k)}{4}\right)k^2\right\}^{1/2}$$

where β_1, β_2, β_4 and β_5 are system constants and $\beta_3(k) = \frac{k_x^2 + k_y^2 - 2k_z^2}{k^2}$

(R.1.2) Investigate the ER for bulk specimens of the heavily-doped semiconductors in the presences of Gaussian, exponential, Kane, Halperian, Lax and Bonch-Burevich types of band tails [37, 38] for all systems whose unperturbed carrier energy spectra are defined in R1.1.

(R.1.3) Investigate the ER for QWs of all the HD semiconductors as considered in R1.2.

(R.1.4) Investigate the ER for HD bulk specimens of the negative refractive index, organic, magnetic and other advanced optical materials in the presence of an arbitrarily oriented alternating electric field.

(R.1.5) Investigate the ER for the QWs of HD negative refractive index, organic, magnetic and other advanced optical materials in the presence of an arbitrarily oriented alternating electric field.

(R.1.6) Investigate the ER for the multiple QWs of HD materials whose unperturbed carrier energy spectra are defined in R1.1

(R.1.7) Investigate the ER for all the appropriate HD low dimensional systems of this chapter in the presence of finite potential wells.

(R.1.8) Investigate the ER for all the appropriate HD low dimensional systems of this chapter in the presence of parabolic potential wells.

(R.1.9) Investigate the ER for all the appropriate HD systems of this chapter forming quantum rings.

(R.1.10) Investigate the ER for all the above appropriate problems in the presence of elliptical Hill and quantum square rings.

(R.1.11) Investigate the ER for parabolic cylindrical HD low dimensional systems in the presence of an arbitrarily oriented alternating electric field for all the HD materials whose unperturbed carrier energy spectra are defined in R1.1.

(R.1.12) Investigate the ER for HD low dimensional systems of the negative refractive index and other advanced optical materials in the presence of an arbitrarily oriented alternating electric field and non-uniform light waves.

(R.1.13) Investigate the ER for triangular HD low dimensional systems of the negative refractive index, organic, magnetic and other advanced optical materials in the presence of an arbitrarily oriented alternating electric field in the presence of strain.

(R.1.14) Investigate the ER in HD quantum wires of non-parabolic semiconductors as discussed in this chapter.

(R.1.15) Investigate the ER for all the problems of (R1.14) in the presence of arbitrarily oriented magnetic field.

(R.1.16) Investigate the ER for all the problems of (R1.14) in the presence of alternating electric field.

(R.1.17) Investigate the ER for all the problems of (R1.14) in the presence of alternating magnetic field.

(R.1.18) Investigate the ER for all the problems of (R1.14) in the presence of crossed electric field and quantizing magnetic fields.

(R.1.19) Investigate the ER for all the problems of (R1.14) in the presence of crossed alternating electric field and alternating quantizing magnetic fields.

(R.1.20) Investigate the ER for HD quantum wires of the negative refractive index, organic and magnetic materials.

(R.1.21) Investigate the ER for HD quantum wires of the negative refractive index, organic and magnetic materials in the presence of alternating time dependent magnetic field.

(R.1.22) Investigate the ER for HD quantum wires of the negative refractive index, organic and magnetic materials in the presence of in the presence of crossed alternating electric field and alternating quantizing magnetic fields.

(R.1.23) (a) Investigate the ER for HD low dimensional systems of the negative refractive index, organic, magnetic and other advanced optical materials in the presence of an arbitrarily oriented alternating electric field considering many body effects.

(b) Investigate all the appropriate problems of this chapter for a Dirac electron.

(R.1.24) Investigate all the appropriate problems of this chapter by including the many body, image force, broadening and hot carrier effects respectively.

(R.1.25) Investigate all the appropriate problems of this chapter by removing all the mathematical approximations and establishing the respective appropriate uniqueness conditions.

References

1. P.M. Petroff, A.C. Gossard, W. Wiegmann, Appl. Phys. Lett. **45**, 620 (1984)
2. J.M. Gaines, P.M. Petroff, H. Kroemar, R.J. Simes, R.S. Geels, J.H. English, J. Vac. Sci. Technol., B **6**, 1378 (1988)
3. J. Cilbert, P.M. Petroff, G.J. Dolan, S.J. Pearton, A.C. Gossard, J.H. English, Appl. Phys. Lett. **49**, 1275 (1986)
4. T. Fujui, H. Saito, Appl. Phys. Lett. **50**, 824 (1987)
5. H. Sasaki, Jpn. J. Appl. Phys. **19**, 94 (1980)
6. P.M. Petroff, A.C. Gossard, R.A. Logan, W. Weigmann, Appl. Phys. Lett. **41**, 635 (1982)
7. H. Temkin, G.J. Dolan, M.B. Panish, S.N.G. Chu, Appl. Phys. Lett. **50**, 413 (1988)
8. I. Miller, A. Miller, A. Shahar, U. Koren, P.J. Corvini, Appl. Phys. Lett. **54**, 188 (1989)
9. L.L. Chang, H. Esaki, C.A. Chang, L. Esaki, Phys. Rev. Lett. **38**, 1489 (1977)
10. K. Hess, M.S. Shur, J.J. Drunnond, H. Morkoc, IEEE Trans. Electron. Devices **ED-30**, 07 (1983)
11. G. Bastard, *Wave Mechanics Applied to Semiconductor Hetero-Structures* (Halsted, New York, 1988). (Les Editions de Physique)
12. M.J. Kelly, *Low Dimensional Semiconductors: Materials, Physics, Technology, Devices* (Oxford University Press, Oxford, 1995)
13. C. Weisbuch, B. Vinter, *Quantum Semiconductor Structures* (Boston Academic Press, Boston, 1991)
14. N.T. Linch, Festkorperprobleme **23**, 27 (1985)
15. D.R. Sciferes, C. Lindstrom, R.D. Burnham, W. Streifer, T.L. Paoli, Electron. Lett. **19**, 170 (1983)
16. P.M. Solomon, Proc. IEEE **70**, 489 (1982)
17. T.E. Schlesinger, T. Kuech, Appl. Phys. Lett. **49**, 519 (1986)
18. D. Kasemet, C.S. Hong, N.B. Patel, P.D. Dapkus, Appl. Phys. Lett. **41**, 912 (1982)
19. K. Woodbridge, P. Blood, E.D. Pletcher, P.J. Hulyer, Appl. Phys. Lett. **45**, 16 (1984)
20. S. Tarucha, H.O. Okamoto, Appl. Phys. Lett. **45**, 16 (1984)
21. H. Heiblum, D.C. Thomas, C.M. Knoedler, M.I. Nathan, Appl. Phys. Lett. **47**, 1105 (1985)
22. O. Aina, M. Mattingly, F.Y. Juan, P.K. Bhattacharyya, Appl. Phys. Lett. **50**, 43 (1987)
23. I. Suemune, L.A. Coldren, IEEE J. Quant. Electron. **24**, 1178 (1988)
24. D.A.B. Miller, D.S. Chemla, T.C. Damen, J.H. Wood, A.C. Burrus, A.C. Gossard, W. Weigmann, IEEE J. Quant. Electron. **21**, 1462 (1985)
25. J.W. Rowe, J.L. Shay, Phys. Rev. B **3**, 451 (1973)
26. H. Kildal, Phys. Rev. B **10**, 5082 (1974)

27. J. Bodnar, in *Proceedings of the International Conference on Physics of Narrow-gap Semiconductors* (Polish Science Publishers, Warsaw, 1978)
28. G.P. Chuiko, N.N. Chuiko, Sov. Phys. Semicond. **15**, 739 (1981)
29. K.P. Ghatak, S.N. Biswas, Proc. SPIE **1484**, 149 (1991)
30. A. Rogalski, J. Alloys Comp. **371**, 53 (2004)
31. A. Baumgartner, A. Chaggar, A. Patanè, L. Eaves, M. Henini, Appl. Phys. Lett. **92**, 091121 (2008)
32. J. Devenson, R. Teissier, O. Cathabard, A.N. Baranov, Proc. SPIE **6909**, 69090U (2008)
33. B.S. Passmore, J. Wu, M.O. Manasreh, G.J. Salamo, Appl. Phys. Lett. **91**, 233508 (2007)
34. M. Mikhailova, N. Stoyanov, I. Andreev, B. Zhurtanov, S. Kizhaev, E. Kunitsyna, K. Salikhov, Y. Yakovlev, Proc. SPIE **6585**, 658526 (2007)
35. W. Kruppa, J.B. Boos, B.R. Bennett, N.A. Papanicolaou, D. Park, R. Bass, Electron. Lett. **42**, 688 (2006)
36. E.O. Kane, in *Semiconductors and Semimetals*, vol. 1, ed. by R.K. Willardson, A.C. Beer (Academic Press, New York, 1966), p. 75
37. A.N. Chakravarti, K.P. Ghatak, A. Dhar, K.K. Ghosh, S. Ghosh, Acta Phys. Polon A **60**, 151 (1981)
38. B.R. Nag, *Electron Transport in Compound Semiconductors* (Springer, Heidelberg, 1980)
39. G.E. Stillman, C.M. Wolfe, J.O. Dimmock, in *Semiconductors and Semimetals*, vol. **12**, ed. by R.K. Willardon, A.C. Beer (Academic Press, New York, 1977), p. 169
40. D.J. Newson, A. Karobe, Semicond. Sci. Tech. **3**, 786 (1988)
41. E.D. Palik, G.S. Picus, S. Teither, R.E. Wallis, Phys. Rev., 475 (1961)
42. P.Y. Lu, C.H. Wung, C.M. Williams, S.N.G. Chu, C.M. Stiles, Appl. Phys. Lett. **49**, 1372 (1986)
43. N.R. Taskar, I.B. Bhat, K.K. Prat, D. Terry, H. Ehasani, S.K. Ghandhi, J. Vac. Sci. Tech. **7A**, 281 (1989)
44. F. Koch, *Springer Series in Solid States Sciences* (Springer, Germany, 1984)
45. L.R. Tomasetta, H.D. Law, R.C. Eden, I. Reyhimy, K. Nakano, IEEE J. Quant. Electron. **14**, 800 (1978)
46. T. Yamato, K. Sakai, S. Akiba, Y. Suematsu, IEEE J. Quantum Electron. **14**, 95 (1978)
47. T.P. Pearsall, B.I. Miller, R.J. Capik, Appl. Phys. Lett. **28**, 499 (1976)
48. M.A. Washington, R.E. Nahory, M.A. Pollack, E.D. Beeke, Appl. Phys. Lett. **33**, 854 (1978)
49. M.I. Timmons, S.M. Bedair, R.J. Markunas, J.A. Hutchby, in *Proceedings of the 16th IEEE Photovoltaic Specialist Conference* (IEEE, San Diego, California 666, 1982)
50. J.A. Zapien, Y.K. Liu, Y.Y. Shan, H. Tang, C.S. Lee, S.T. Lee, Appl. Phys. Lett. **90**, 213114 (2007)
51. M. Park, Proc. SPIE **2524**, 142 (1995)
52. S.-G. Hur, E.T. Kim, J.H. Lee, G.H. Kim, S.G. Yoon, Electrochem. Solid-State Lett. **11**, H176 (2008)
53. H. Kroemer, Rev. Mod. Phys. **73**, 783 (2001)
54. T. Nguyen Duy, J. Meslage, G. Pichard, J. Crys. Growth **72**, 490 (1985)
55. T. Aramoto, F. Adurodija, Y. Nishiyama, T. Arita, A. Hanafusa, K. Omura, A. Morita, Solar Energy Mater. Solar Cells **75**, 211 (2003)
56. H.B. Barber, J. Electron. Mater. **25**, 1232 (1996)
57. S. Taniguchi, T. Hino, S. Itoh, K. Nakano, N. Nakayama, A. Ishibashi, M. Ikeda, Electron. Lett. **32**, 552 (1996)
58. J.J. Hopfield, J. Appl. Phys. **32**, 2277 (1961)
59. G.P. Agrawal, N.K. Dutta, *Semiconductor Lasers* (Van Nostrand Reinhold, New York, 1993)
60. S. Chatterjee, U. Pal, Opt. Eng. (Bellingham) **32**, 2923 (1993)
61. T.K. Chaudhuri, Int. J. Energy Res. **16**, 481 (1992)
62. J.H. Dughaish, Phys. B **322**, 205 (2002)

63. C. Wood, Rep. Prog. Phys. **51**, 459 (1988)
64. K.-F. Hsu, S. Loo, F. Guo, W. Chen, J.S. Dyck, C. Uher, T. Hogan, E.K. Polychroniadis, M.G. Kanatzidis, Science **303**, 818 (2004)
65. J. Androulakis, K.F. Hsu, R. Pcionek, H. Kong, C. Uher, J.J. D'Angelo, A. Downey, T. Hogan, M.G. Kanatzidis, Adv. Mater. **18**, 1170 (2006)
66. P.F.P. Poudeu, J. D'Angelo, A.D. Downey, J.L. Short, T.P. Hogan, M.G. Kanatzidis, Angew. Chem. Int. Ed. **45**, 3835 (2006)
67. P.F. Poudeu, J. D'Angelo, H. Kong, A. Downey, J.L. Short, R. Pcionek, T.P. Hogan, C. Uher, M.G. Kanatzidis, J. Am. Chem. Soc. **128**, 14347 (2006)
68. J.R. Sootsman, R.J. Pcionek, H. Kong, C. Uher, M.G. Kanatzidis, Chem. Mater. **18**, 4993 (2006)
69. A.J. Mountvala, G. Abowitz, J. Am. Ceram. Soc. **48**, 651 (1965)
70. E.I. Rogacheva, I.M. Krivulkin, O.N. Nashchekina, A.Y.U. Sipatov, V.A. Volobuev, M.S. Dresselhaus, Appl. Phys. Lett. **78**, 3238 (2001)
71. H.S. Lee, B. Cheong, T.S. Lee, K.S. Lee, W.M. Kim, J.W. Lee, S.H. Cho, J.Y. Huh, Appl. Phys. Lett. **85**, 2782 (2004)
72. K. Kishimoto, M. Tsukamoto, T. Koyanagi, J. Appl. Phys. **92**, 5331 (2002)
73. E.I. Rogacheva, O.N. Nashchekina, S.N. Grigorov, M.A. Us, M.S. Dresselhaus, S.B. Cronin, Nanotechnology **14**, 53 (2003)
74. E.I. Rogacheva, O.N. Nashchekina, A.V. Meriuts, S.G. Lyubchenko, M.S. Dresselhaus, G. Dresselhaus, Appl. Phys. Lett. **86**, 063103 (2005)
75. E.I. Rogacheva, S.N. Grigorov, O.N. Nashchekina, T.V. Tavrina, S.G. Lyubchenko, A.Y.U. Sipatov, V.V. Volobuev, A.G. Fedorov, M.S. Dresselhaus, Thin Solid Films **493**, 41 (2005)
76. X. Qiu, Y. Lou, A.C.S. Samia, A. Devadoss, J.D. Burgess, S. Dayal, C. Burda, Angew. Chem. Int. Ed. **44**, 5855 (2005)
77. C. Wang, G. Zhang, S. Fan, Y. Li, J. Phys. Chem. Solids **62**, 1957 (2001)
78. B. Poudel, W.Z. Wang, D.Z. Wang, J.Y. Huang, Z.F. Ren, J. Nanosci. Nanotechnol. **6**, 1050 (2006)
79. B. Zhang, J. He, T.M. Tritt, Appl. Phys. Lett. **88**, 043119 (2006)
80. W. Heiss, H. Groiss, E. KaQWmann, G. Hesser, M. Böberl, G. Springholz, F. Schäffler, K. Koike, H. Harada, M. Yano, Appl. Phys. Lett. **88**, 192109 (2006)
81. B.A. Akimov, V.A. Bogoyavlenskiy, L.I. Ryabova, V.N. Vasil'kov, Phys. Rev. B **61**, 16045 (2000)
82. Y.A. Ugai, A.M. Samoilov, M.K. Sharov, O.B. Yatsenko, B.A. Akimov, Inorg. Mater. **38**, 12 (2002)
83. Y.A. Ugai, A.M. Samoilov, S.A. Buchnev, YuV Synorov, M.K. Sharov, Inorg. Mater. **38**, 450 (2002)
84. A.M. Samoilov, S.A. Buchnev, YuV Synorov, B.L. Agapov, A.M. Khoviv, Inorg. Mater. **39**, 1132 (2003)
85. A.M. Samoilov, S.A. Buchnev, E.A. Dolgopolova, YuV Synorov, A.M. Khoviv, Inorg. Mater. **40**, 349 (2004)
86. H. Murakami, W. Hattori, R. Aoki, Phys. C **269**, 83 (1996)
87. H. Murakami, W. Hattori, Y. Mizomata, R. Aoki, Phys. C **273**, 41 (1996)
88. H. Murakami, R. Aoki, K. Sakai, Thin Solid Films **27**, 343 (1999)
89. B.A. Volkov, L.I. Ryabova, D.R. Khokhlov, Phys. Usp. **45**, 819 (2002). (and references therein)
90. F. Hüe, M. Hÿtch, H. Bender, F. Houdellier, A. Claverie, Phys. Rev. Lett. **100**, 156602 (2008)
91. S. Banerjee, K.A. Shore, C.J. Mitchell, J.L. Sly, M. Missous, IEE Proc. Circuits Devices Syst. **152**, 497 (2005)
92. M. Razeghi, A. Evans, S. Slivken, J.S. Yu, J.G. Zheng, V.P. Dravid, Proc. SPIE **5840**, 54 (2005)

93. R.A. Stradling, Semicond. Sci. Technol. **6**, C52 (1991)
94. P.K. Weimer, Proc. IEEE **52**, 608 (1964)
95. G. Ribakovs, A.A. Gundjian, IEEE J. Quant. Electron. **QE-14**, 42 (1978)
96. S.K. Dey, J. Vac. Sci. Technol. **10**, 227 (1973)
97. S.J. Lynch, Thin Solid Films **102**, 47 (1983)
98. V.V. Kudzin, V.S. Kulakov, D.R. Pape', S.V. Kulakov, V.V. Molotok, IEEE. Ultrason. Symp. **1**, 749 (1997)
99. F. Hatami, V. Lordi, J.S. Harris, H. Kostial, W.T. Masselink, J. Appl. Phys. **97**, 096106 (2005)
100. B.W. Wessels, J. Electrochem. Soc. **722**, 402 (1975)
101. D.W.L. Tolfree, J. Sci. Instrum. **41**, 788 (1964)
102. P.B. Hart, Proc. IEEE **61**, 880 (1973)
103. M.A. Hines, G.D. Scholes, Adv. Mater. **15**, 1844 (2003)
104. C.A. Wang, R.K. Huang, D.A. Shiau, M.K. Connors, P.G. Murphy, P.W. O'Brien, A.C. Anderson, D.M. DePoy, G. Nichols, M.N. Palmisiano, Appl. Phys. Lett. **83**, 1286 (2003)
105. C.W. Hitchcock, R.J. Gutmann, J.M. Borrego, I.B. Bhat, G.W. Charache, IEEE Trans. Electron Devices **46**, 2154 (1999)
106. H.J. Goldsmid, R.W. Douglas, Br. J. Appl. Phys. **5**, 386 (1954)
107. F.D. Rosi, B. Abeles, R.V. Jensen, J. Phys. Chem. Sol. **10**, 191 (1959)
108. T.M. Tritt (ed.), *Semiconductors and Semimetals*, vol. **69**, **70** and **71**: Recent Trends in Thermoelectric Materials Research I, II and III (Academic Press, New York, 2000)
109. D.M. Rowe (ed.), *CRC Handbook of Thermoelectrics* (CRC Press, Boca Raton, 1995)
110. D.M. Rowe, C.M. Bhandari, *Modern Thermoelectrics* (Reston Publishing Company, Virginia, 1983)
111. D.M. Rowe (ed.), *Thermoelectrics Handbook: Macro to Nano* (CRC Press, Boca Raton, 2006)
112. H. Choi, M. Chang, M. Jo, S.J. Jung, H. Hwang, Electrochem. Solid-State Lett. **11**, H154 (2008)
113. S. Cova, M. Ghioni, A. Lacaita, C. Samori, F. Zappa, Appl. Opt. **35**, 1956 (1996)
114. H.W.H. Lee, B.R. Taylor, S.M. Kauzlarich, Nonlinear Optics: Materials, Fundamentals, and Applications (Technical Digest, 2000), p. 12
115. E. Brundermann, U. Heugen, A. Bergner, R. Schiwon, G.W. Schwaab, S. Ebbinghaus, D.R. Chamberlin, E.E. Haller, M. Havenith, in *29th International Conference on Infrared and MillimeterWaves and 12th International Conference on Terahertz, Electronics*, vol. 283 (2004)
116. A.N. Baranov, T.I. Voronina, N.S. Zimogorova, L.M. Kauskaya, Y.P. Yakoviev, Sov. Phys. Semicond. **19**, 1676 (1985)
117. M. Yano, Y. Suzuki, T. Ishii, Y. Matsushima, M. Kimata, Jpn. J. Appl. Phys. **17**, 2091 (1978)
118. F.S. Yuang, Y.K. Su, N.Y. Li, Jpn. J. Appl. Phys. **30**, 207 (1991)
119. F.S. Yuang, Y.K. Su, N.Y. Li, K.J. Gan, J. Appl. Phys. **68**, 6383 (1990)
120. Y.K. Su, S.M. Chen, J. Appl. Phys. **73**, 8349 (1993)
121. S.K. Haywood, A.B. Henriques, N.J. Mason, R.J. Nicholas, P.J. Walker, Semicond. Sci. Technol. **3**, 315 (1988)
122. A.N. Chakravarti, K.P. Ghatak, K.K. Ghosh, S. Ghosh, H.M. Mukherjee, Phys. Stat. Sol. b **108**, 609 (1981)
123. K.P. Ghatak, M. Mondal, Z. fur Physik B **64**, 223 (1986)
124. P.K. Chakraborty, A. Sinha, S. Bhattacharya, K.P. Ghatak, Physica B **390**, 325 (2007)
125. K.P. Ghatak, G.C. Datta, K.P. Ghatak, Phys. Scr. **68**, 368 (2003)
126. P.K. Chakraborty, K.P. Ghatak, J. Phys. Chem. Solids **62**, 1061 (2001)
127. P.K. Chakraborty, K.P. Ghatak, Phys. Letts. A **288**, 335 (2001)
128. P.K. Chakraborty, K.P. Ghatak, Phys. D. Appl. Phys. **32**, 2438 (1999)

129. M. Abramowitz, I.A. Stegun, *Handbook of Mathematical Functions with Formulas, Graphs and Mathematical Tables* (Wiley, New York, 1964)
130. I.S. Gradshteyn, I.M. Ryzhik, *Tables of Integrals, Series and Products* (Academic, New York, 1965)
131. V. Heine, Proc. Phys. Soc. **81**, 300 (1963)
132. J.N. Schulman, Y.C. Chang, Phys. Rev. B **24**, 4445 (1981)
133. S. Adachi, J. Appl. Phys. **58**, R11 (1985)
134. K.P. Ghatak, J.P. Banerjee, D. Bhattacharya, Nanotechnology **7**, 110 (1996)
135. S. Bhattacharya, S .Choudhury, K.P Ghatak, Superlatt. Microstruct. **48**, 257 (2010)
136. K.P Ghatak, S Bhattacharya, S.Pahari, D. De, S.Ghosh, M.Mitra, Ann. Phys., **17**, 195 (2008)
137. S. Pahari, S. Bhattacharya, K.P Ghatak, J.Comput. Theor. Nanosci. (Invited Paper) **6**, 2088 (2009)
138. R.W. Cunningham, Phys. Rev. **167**, 761 (1968)
139. M. Krieehbaum, P. Kocevar, H. Pascher, G. Bauer, IEEE QE **24**, 1727 (1988)
140. M.S. Lundstrom, J. Guo, *Nanoscale Transistors, Device Physics, Modeling and Simulation* (Springer, USA, 2006)
141. R. Saito, G. Dresselhaus, M.S. Dresselhaus, *Physical Properties of Carbon Nanotubes* (Imperial College Press, London, 1998)
142. X. Yang, J. Ni, Phys. Rev. B **72**, 195426 (2005)
143. W. Mintmire, C.T. White, Phys. Rev. Letts. **81**, 2506 (1998)
144. D.G. Seiler, B.D. Bajaj, A.E. Stephens, Phys. Rev. B **16**, 2822 (1977)
145. A.V. Germaneko, G.M. Minkov, Phys. Stat. Sol. (b) **184**, 9 (1994)
146. G.L. Bir, G.E. Pikus, *Symmetry and Strain -Induced Effects in Semiconductors* (Nauka, Russia, 1972). (in Russian)
147. M. Mondal, K.P. Ghatak, Phys. Stat. Sol. (b) **135**, K21 (1986)
148. C.C. Wu, C.J. Lin, J. Low, Temp. Phys. **57**, 469 (1984)
149. G.P. Chuiko, Sov. Phys. Semi. **19**, 1381 (1985)
150. Y. Yamada, J. Phys. Soc. Jpn. **35**, 1600 (1973)
151. D.G. Seiler, W.M. Beeker, K.M. Roth, Phys. Rev. **1**, 764 (1970)
152. A.N. Chakravarti, K.P. Ghatak, K.K. Ghosh, S. Ghosh, H.M. Mukherjee, Phys. Stat. Sol. b **118**, 843 (1983)
153. S Bhattacharya, D. De, S.M.Adhikari, K.P. Ghatak, Superlatt. Microstruct. **51**, 203 (2012)
154. K.P. Ghatak, M. Mondal, Z.F. Naturforschung **41A**, 821 (1986)
155. A.N. Chakravarti, A.K. Choudhury, K.P. Ghatak, S. Ghosh, A. Dhar, Appl. Phys. **25**, 105 (1981)
156. P.K. Chakraborty, G.C. Datta, K.P. Ghatak, Phys. Scr. **68**, 368 (2003)
157. A.N. Chakravarti, K.P. Ghatak, A. Dhar, K.K. Ghosh, S. Ghosh, Appl. Phys. **A26**, (165) (1981)
158. K.P. Ghatak, S. Bhattacharya, S.K. Biswas, A. Dey, A.K. Dasgupta, Phys. Scrip **75**, 820 (2007)
159. K.P. Ghatak, M. Mondal, Z.F. Physik B. **B69**, 471 (1988)
160. A.N. Chakravarti, K.P.Ghatak, K.K.Ghosh, S. Ghosh, A. Dhar, Z.F, Physik B. **47**, 149 (1982)
161. H.A. Lyden, Phys. Rev. **135**, A514 (1964)
162. E.D Palik, G.B. Wright, in *Semiconductors and Semimetals*, vol. 3, ed. by R.K Willardson, A.C Beer (Academic Press, New York, 1967), p. 421
163. H.I. Zhang, Phys. Rev. **1B**, 3450 (1970)
164. M. Mondal, K.P.Ghatak, Phys. Letts. **131 A**, 529 (1988)
165. K.P. Ghatak, B. Mitra, Int. J. Electron. **72**, 541 (1992)
166. B. Mitra, A. Ghoshal, K.P. Ghatak, Nouvo Cimento D **12D**, 891 (1990)
167. K.P. Ghatak, S.N. Biswas, Nonlinear Opt. Quantum Opt. **4**, 347 (1993)
168. K.P. Ghatak, A. Ghoshal. B. Mitra., Nouvo Cimento **14D**, 903 (1992)
169. K.P. Ghatak, A. Ghoshal, B. Mitra, Nouvo Cimento **13D**, 867 (1991)

170. B. Mitra, K.P. Ghatak, Solid State Electron. **32**, 177 (1989)
171. M. Mondal, N. Chattapadhyay, K.P. Ghatak, J. Low Temp. Phys. **66**, 131 (1987)
172. P.N. Hai, W.M. Chen, I.A. Buyanova, H.P. Xin, C.W. Tu, Appl. Phys. Lett. **77**, 1843 (2000)
173. D.P. DiVincenzo, E.J. Mele, Phys. Rev. B **29**, 1685 (1984)
174. P. Perlin, E. Litwin-Staszewska, B. Suchanek, W. Knap, J. Camassel, T. Suski, R. Piotrzkowski, I. Grzegory, S. Porowski, E. Kaminska, J.C. Chervin, Appl. Phys. Lett. **68**, 1114 (1996)
175. G.E. Smith, Phys. Rev. Lett. **9**, 487 (1962)
176. D. Schneider, D. Rurup, A. Plichta, H.-U. Grubert, A. Schlachetzki, K. Hansen, Z.Phys. B **95**, 281 (1994)
177. F. Masia, G. Pettinari, A. Polimeni, M. Felici, A. Miriametro, M. Capizzi, A. Lindsay, S.B. Healy, E.P. O'Reilly, A. Cristofoli, G. Bais, M. Piccin, S. Rubini, F. Martelli, A. Franciosi, P.J. Klar, K. Volz, W. Stolz, Phys. Rev. B **73**, 073201 (2006)
178. V.K. Arora, H. Jeafarian, Phys. Rev. B. **13**, 4457 (1976)
179. S.E. Ostapov, V.V. Zhikharevich, V.G. Deibuk; Semicond. Phys., Quan. Electron. Optoelectron **9**, 29 (2006)
180. M.J. Aubin, L.G. Caron, J.P. Jay Gerin, Phys. Rev. B **15**, 3872 (1977)
181. S.L. Sewall, R.R. Cooney, P. Kambhampati, Appl. Phys. Lett. **94**, 243116 (2009)
182. K. Tanaka, N. Kotera; in *20th International Conference on Indium Phosphide and Related Materials*, 2008; 25–29 May 2008, p. 1–4, Versailles, France
183. M. Singh, P.R. Wallace, S.D. Jog, J. Erushanov, J. Phys. Chem. Solids **45**, 409 (1984)
184. W. Zawadzki, Adv. Phys. **23**, 435 (1974)
185. R.K. Pathria, *Statistical Mechanics*, 2nd edn. (Butterworth-Heinmann, Oxford, 1996)
186. K.P. Ghatak, B. Mitra, D.K. Basu, B. Nag, Nolinear Opt. **17**, 171 (1997)
187. B.R. Nag, A.N. Chakravarti, Solid State Electron. **18**, 109 (1975)
188. B.R. Nag, A.N. Chakravarti, Phys. Stal Sol. (a) **22**, K153 (1974)
189. J.O. Dimmock, in *The Physics of Semimetals and Narrowgap Semiconductors*, ed. by D.L. Carter, R.T. Bates (Pergamon Press, Oxford, 1971)
190. D.G. Seiler, B.D. Bajaj, A.E. Stephens, Phys. Rev. **B 16**, 2822 (1977)
191. A.V. Germaneko, G.M. Minkov, Phys. Stat. Sol. (b) **184**, 9 (1994)
192. G.L. Bir, G.E. Pikus, *Symmetry and Strain-Induced Effects in Semiconductors Nauka, Russia* (1972). (in Russian)
193. J. Bouat, J.C. Thuillier, Surf. Sci. **73**, 528 (1978)
194. G.J. Rees, Physics of Compounds, in *Proceedings of the 13th International Conference* ed. By F.G. Fumi (North Holland Company, New York, 1976), p. 1166
195. P.R. Entage, Phys. Rev. **138**, A246 (1965)
196. M. Stordeur, W. kuhnberger, Phys. Stat. Sol. **69**, 377 (1975)
197. D.R. Lovett, *Semimetals and Narrow-Bandgap Semiconductor* (Pion Limited, UK, 1977)
198. H. Kohler, Phys. Stat. Sol. **74**, 591 (1976)
199. M. Cardona, W. Paul, H. Brooks Helv, Acta Phys. **33**, 329 (1960)
200. A.F. Gibson in *Proceeding of International School of Physics*, *ENRICO FERMI, course XIII*, ed. By R.A. Smith (Academic Press, New York, 1963), p. 171
201. C.C. Wang, N.W. Ressler, Phys. Rev. **2**, 1970 (1827)
202. P.C. Mathur, S. Jain, Phys. Rev. **19**, 1359 (1979)
203. E.L. Ivchenko, G.E. Pikus Sov, Phys. Semicond. **13**, 579 (1979)
204. M. H. Cohen, Phys. Rev. **121**, 387 (1961)
205. J.W. McClure, K.H. Choi, Solid State Commun. **21**, 1977 (1015)
206. E. Bangert, P. Kastner, Phys. Stat. Sol (b) **61**, 503 (1974)
207. G.M.T. Foley, P.N. Langenberg, Phys. Rev. B **15B**, 4850 (1977)
208. V.I. Ivanov-Omskii, A.Sh. Mekhtisev, S.A. Rustambekova, E.N. Ukraintsev, Phys. Stat. Sol. (b) **119**, 159 (1983)
209. D.G. Seiler, W.M. Beeker, L.M. Roth, Phys. Rev. **1**, 764 (1970)
210. H.I. Zhang, Phys. Rev. B **1**, 3450 (1970)
211. U. Rossler, Solid State Commun. **49**, 943 (1984)

212. J. Johnson, D.H. Dickey, Phys. Rev. **1**, 2676 (1970)
213. V.G. Agafonov, P.M. Valov, B.S. Ryvkin, I.D. Yarashetskin, Sov. Phys. Semiconduct. **12**, 1182 (1978)
214. L.A. Vassilev, Phys. State sol (b), **121**, 203 (1984)
215. N.S. Averkiev, V.M. Asnin, A.A. Bakun, A.M. Danishevskii, E.L. Ivchenko, G.E. Pikus, A.A. Rogachev, Sov. Phys. Semicond. **18**, 379 (1984)
216. N.B. Brandt, V.N. Davydov, V.A. Kulbachinskii, O.M. Nikitina, Sov. Phys. Sol. Stat. **29**, 1987 (1014)
217. J.B. Ketterson, Phys. Rev. **129**, 18 (1963)
218. R.W. Cunningham, Phys. Rev. **167**, 761 (1968)
219. A.I. Yekimov, A.A. Onushchenko, A.G. Plyukhin, AI.L. Efros, J. Expt. Theor. Phys. **88**, 1490 (1985)
220. B.J. Roman, A.W. Ewald, Phys. Rev. **B5**, 3914 (1972)
221. G.P. Chuiko, Sov. Phys. Semiconduct. **19**(12), 1381 (1985)

Chapter 2
The ER in Doping Super Lattices of HD Non-parabolic Semiconductors

2.1 Introduction

The technological importance of super-lattices in general, and specifically doping super-lattices [1–51] has already been stated in the preface and also in the references of this chapter. In Sect. 2.2.1, of the theoretical background, the ER in doping superlattices of HD non-linear optical semiconductors has been investigated. The Sect. 2.2.2 contains the results for doping superlattices of HD III-V, ternary and quaternary semiconductors in accordance with the three and the two band models of Kane together with parabolic energy bands and they form the special cases of Sect. 2.2.1. Sections 2.2.3, 2.2.4 and 2.2.5 contain the study of the ER for doping superlattices of HD II-VI, IV-VI and stressed Kane type semiconductors respectively. Sections 2.3 and 2.4 contain the results and discussion and the open research problems for this chapter.

2.2 Theoretical Background

2.2.1 The ER in Doping Superlattices of HD Non-linear Optical Semiconductors

The dispersion relation of the conduction electrons in doping superlattices of HD nonlinear optical materials can be expressed by using (1.2) and following the method as given in [19–51] as

$$\frac{\left(n_i+\frac{1}{2}\right)}{\hbar T_{21}\left(E,\eta_g\right)}\omega_{8HD}\left(E,\eta_g\right)+\frac{\hbar^2 k_s^2}{2m_\perp^* T_{22}\left(E,\eta_g\right)}=1 \tag{2.1}$$

K. P. Ghatak and S. Bhattacharya, *Heavily-Doped 2D-Quantized Structures and the Einstein Relation*, Springer Tracts in Modern Physics 260,
DOI 10.1007/978-3-319-08380-3_2, © Springer International Publishing Switzerland 2015

where $\omega_{8HD}(E,\eta_g) \equiv Real\ part\ of \left(\frac{n_0|e|^2}{d_0\varepsilon_{sc}\left[m_{||}^{*}T'_{21}(E,\eta_g)\right]}\right)^{1/2}$, $n_i\,(=0,\,1,2\ldots)$ is the mini-band index for nipi structures and d_0 is the mini-band index for nipi structures and d_0 is the superlattice period.

The EEM in this case assumes the form

$$m^{*}(\mathrm{E}_{FnHD},\mathrm{n}_i\eta_g) = \text{Real part of}\left(\frac{\hbar^2}{2}\right)\mathrm{G}'_{21HD}(\mathrm{E}_{FnHD},\mathrm{n}_i\eta_g) \tag{2.2}$$

where, $\mathrm{G}_{21HD}(\mathrm{E},\eta_g,\mathrm{n}_i) = \frac{2m_{\perp}^{*}T_{22}(E,\eta_g)}{\hbar^2}\left[1-\frac{(n_i+\frac{1}{2})}{\hbar T_{21}(E,\eta_g)}\omega_{8HD}(E,\eta_g)\right]$ and $\overline{E}_{FnHD}$ is the Fermi energy in the present case as measured from the edge of the conduction band in vertically upward direction in the absence of any quantization.

From (2.2), we observe that the EEM is a function of the Fermi energy, nipi subband index, scattering potential and the other material constants which is the characteristic feature of doping superlattices of HD non-linear optical materials.

The subband energy (E_{1n_iHD}) can be written as

$$\frac{(n_i+\frac{1}{2})}{\hbar T_{21}(E_{1n_iHD},\eta_g)}\omega_{8HD}(E_{1n_iHD},\eta_g) = 1 \tag{2.3}$$

The DOS function for doping superlattices of HD non-linear optical materials can be expressed as

$$N_{nipiHD}(E,\eta_g) = \frac{g_v}{2\pi}\sum_{n_i=0}^{n_{imax}} G'_{21HD}(\mathrm{E},\eta_g,\mathrm{n}_i)H(E-E_{1n_iHD}) \tag{2.4}$$

The electron concentration, can be written as

$$n_{2DN} = \frac{g_v}{2\pi}\text{Real part of}\sum_{n_z=1}^{n_{zmax}}\left[G_{21HD}(\overline{E}_{FnHD},\eta_g,n_i)+G_{22HD}(\overline{E}_{FnHD},\eta_g,n_i)\right] \tag{2.5}$$

where, $G_{22HD}(\overline{E}_{FnHD},\eta_g,n_i) = \sum_{r=1}^{s} L(r)G_{21HD}(\overline{E}_{FnHD},\eta_g,\mathrm{n}_i)$

The ER of doping superlattices of HDS can, in general, be expressed as

$$\frac{D}{\mu} = \frac{n_{2DNO}}{|e|}\text{Real part of}\left[\frac{\partial n_{2DNO}}{\partial(\overline{E}_{FOHD}-E_{F_{10HD}})}\right]^{-1} \tag{2.6}$$

where, the aforementioned physical variables are applicable only for electric quantum limit. Thus by combining (2.5) and (2.6) we can study the ER in this case.

The dispersion relation of the conduction electrons in doping superlattices of nonlinear optical materials in the absence of band tails assumes the form

$$\psi_1(E) = \psi_2(E)k_s^2 + \psi_3(E)\left(n_i + \frac{1}{2}\right)\frac{2m_{\|}^*}{\hbar}\omega_8(E) \tag{2.7}$$

where $\omega_8(E) \equiv \left(\frac{n_0|e|^2}{d_0\varepsilon_{sc}[\theta_1(E)]}\right)^{1/2}$ and $\theta_1(E) \equiv \frac{\hbar}{2}\left\{\frac{\psi_3(E)[\psi_1(E)]' - \psi_1(E)[\psi_3(E)]'}{[\psi_3(E)]^2}\right\}$

The EEM in this case can be written as

$$m^*(E_{Fn}, n_i) = \left(\frac{\hbar^2}{2}\right)R_{81}(E, n_i)\Big|_{E=\overline{E}_{Fn}} \tag{2.8}$$

where,

$$R_{81}(E,n_i) \equiv [\psi_2(E)]^2\Bigg[\psi_2(E)\left\{[\psi_1(E)]' - \left(\frac{2m_{\|}^*}{\hbar}\right)[\psi_3(E)]'\left(n_i + \frac{1}{2}\right)[\omega_8(E)] - \left(\frac{2m_{\|}^*}{\hbar}\right)[\psi_3(E)]\left(n_i + \frac{1}{2}\right)[\omega_8(E)]'\right\}$$
$$- \left\{[\psi_1(E)] - \left(\frac{2m_{\|}^*}{\hbar}\right)[\psi_3(E)]\left(n_i + \frac{1}{2}\right)[\omega_8(E)]\right\}[\psi_2(E)]'\Bigg]$$

and $\overline{E}_{Fn}$ is the Fermi energy in the present case as measured from the edge of the conduction band in vertically upward direction in the absence of any quantization.

The subband energy (E_{1ni}) can be written as

$$\psi_1(E_{1ni}) = \psi_3(E_{1ni})\left(n_i + \frac{1}{2}\right)\frac{2m_{\|}^*}{\hbar}\omega_8(E_{1ni}) \tag{2.9}$$

The DOS function for doping superlattices of nonlinear optical materials can be expressed as

$$N_{nipi}(E) = \frac{g_v}{2\pi}\sum_{n_i=0}^{n_{imax}} R_{81}(E, n_i)H(E - E_{1ni}) \tag{2.10}$$

the electron concentration, can be written as

$$n_0 = \frac{g_v}{2\pi}\sum_{n_i=0}^{n_{imax}}\left[T_{81}(\overline{E}_{Fn}, n_i) + T_{82}(\overline{E}_{Fn}, n_i)\right] \tag{2.11}$$

where, $T_{81}(\overline{E}_{Fn}, n_i) \equiv \left[\psi_1(\overline{E}_{Fn}) - \psi_3(\overline{E}_{Fn})(n_i + \frac{1}{2})\frac{2m_{\|}^*}{\hbar}\omega_8(\overline{E}_{Fn})\right]\left[\psi_2(\overline{E}_{Fn})\right]^{-1}$ and

$$T_{82}(\overline{E}_{Fn}, n_i) \equiv \sum_{r=1}^{s} L(r)T_{81}(\overline{E}_{Fn}, n_i).$$

The electron concentration for the doping superlattices in quantum limit can be expressed as

$$\bar{n}_0 = \left[\frac{g_v}{2\pi}\right]\left[\psi_1(\overline{E}_{F0}) - \psi_3(\overline{E}_{F0})\frac{m_{||}^*}{\hbar}\omega_8(\overline{E}_{F0})\right]\left[\psi_2(\overline{E}_{F0})\right]^{-1} \tag{2.12}$$

where, $\overline{E}_{F0}$ is the Fermi energy in the present case in the quantum limit and

$$\omega_8(\overline{E}_{F0}) \equiv \left(\frac{\bar{n}_0|e|^2}{d_0\varepsilon_{sc}\left[\theta_1(\overline{E}_{F0})\right]}\right)^{1/2}.$$

The ER at the electric quantum limit for doping superlattices in the absence of band tails can be written as

$$\frac{D}{\mu} = \frac{\bar{n}_0}{e}\left[\frac{\partial \bar{n}_0}{\partial(\overline{E}_{F0} - E_{10})}\right]^{-1} \tag{2.13}$$

where, $\bar{n}_0$ is the electron concentration, $\overline{E}_{F0}$ is the Fermi energy and E_{10} is the sub-band energy at the electric quantum limit respectively.

In this case, E_{10} can be determined from the equation as given by

$$\psi_1(E_{10}) = \psi_3(E_{10})\frac{m_{||}^*}{\hbar}\omega_8(E_{10}) \tag{2.14}$$

Thus using (2.12), (2.13) and (2.14) we can study the ER in doping superlattices of non-linear optical materials in the absence of band tails.

2.2.2 The ER in Doping Superlattices of HD III-V, Ternary and Quaternary Semiconductors

(a) The electron energy spectrum in doping superlattices of HD III-V, ternary and quaternary materials can be expressed from (2.1) under the conditions $\Delta_{||} = \Delta_{\perp} = \Delta, \delta = 0$ and $m_{||}^* = m_{\perp}^* = m_c$, as

$$\frac{\hbar^2 k_s^2}{2m_c} = \left[T_{31}(\mathrm{E}, \eta_g) + \mathrm{i}T_{32}(\mathrm{E}, \eta_g) - \left(n_i + \frac{1}{2}\right)\hbar\omega_{9HD}(\mathrm{E}, \eta_g)\right] \tag{2.15}$$

where $\omega_{9HD}(E, \eta_g) \equiv \left(\dfrac{n_0|e|^2}{d_0\varepsilon_{sc}T'_{31}(E, \eta_g)m_c}\right)^{1/2}$

The EEM in this case assumes the form

$$m^*(\mathrm{E}_{FnHD}, n_i, \eta_g) = \text{Real Part of}\left(\frac{\hbar^2}{2}\right)G'_{23HD}(E_{FnHD}, \eta_g, n_i) \tag{2.16}$$

where $G_{23HD}(E_{FnHD}, \eta_g, n_i) = \frac{2m_c}{\hbar^2}[T_{31} \quad (\mathrm{E}_{FnHD}, \eta_g) + \mathrm{i}\mathrm{T}_{32}(\mathrm{E}_{FnHD}, \eta_g) - \left(n_i + \frac{1}{2}\right)$ $\hbar\omega_{9HD}(\mathrm{E}_{FnHD}, \eta_g)]$

The subband energy E_{2n_iHD} can be written as

$$[T_{31}(\mathrm{E}_{2n_iHD}, \eta_g) + \mathrm{i}\mathrm{T}_{32}(\mathrm{E}_{2n_iHD}, \eta_g) - \left(n_i + \frac{1}{2}\right)\hbar\omega_{9HD}(E_{2n_iHD}, \eta_g)] = 0 \tag{2.17}$$

The DOS function for doping superlattics of HD III-V, ternary and quaternary materials can be expressed as

$$N_{nipiHD}(E, \eta_g) = \frac{g_v m_c}{\pi\hbar^2}\sum_{n_i=0}^{n_{imax}} G'_{23HD}(E, \eta_g, n_i)H(E - E_{2n_iHD}) \tag{2.18}$$

The electron concentration, can be written as

$$n_{2DN} = \frac{g_v m_c}{\pi\hbar^2}\text{Real part of}\sum_{n_i=0}^{n_{imax}}[G_{23HD}(\overline{E}_{FnHD}, \eta_g, n_i) + G_{24HD}(\overline{E}_{FnHD}, \eta_g, n_i)] \tag{2.19}$$

where, $G_{24HD}(\overline{E}_{FnHD}, \eta_g, n_i) = \sum_{r=1}^{s} L(\mathrm{r})G_{23HD}(\overline{E}_{FnHD}, \eta_g, n_i)$

Using (2.6) and (2.19) at the electric quantum limit, we can study the ER in this case.

In the absence of band tails, the dispersion relation in this case assumes the form

$$I_{11}(E) = \left(n_i + \frac{1}{2}\right)\hbar\omega_{19}(E) + \frac{\hbar^2 k_s^2}{2m_c} \tag{2.20}$$

where $\omega_{19}(E) \equiv \left(\frac{n_0|e|^2}{d_0\varepsilon_{sc}I'_{11}(E)m_c}\right)$.

The EEM in this case can be written as

$$m^*(E_{Fn}, n_i) = m_c R_{82}(E, n_i)|_{E=E_{Fn}} \tag{2.21}$$

in which, $R_{82}(E, n_i) \equiv \left\{[I_{11}(E)]' - \left(n_i + \frac{1}{2}\right)\hbar[\omega_{19}(E)]'\right\}$.

From (2.21), we observe that the EEM in this case is a function of the Fermi energy, nipi subband index and the other material constants which is the characteristic feature of doping superlattices of III-V, ternary and quaternary compounds whose bulk dispersion relations is defined by the three band model of Kane.

The subband energies (E_{2ni}) can be written as

$$I_{11}(E_{2ni}) = \left(n_i + \frac{1}{2}\right)\hbar\omega_{19}(E_{2ni}) \tag{2.22}$$

The DOS function in this case can be expressed as

$$N_{nipi}(E) = \frac{m_c g_v}{\pi\hbar^2}\sum_{n_i=0}^{n_{imax}} R_{82}(E, n_i)H(E - E_{2ni}) \tag{2.23}$$

The use of (2.23) leads to the expression of the electron concentration as

$$n_0 = \frac{m_c g_v}{\pi\hbar^2}\sum_{n_i=0}^{n_{imax}}\left[T_{83}(\overline{E}_{Fn}, n_i) + T_{84}(\overline{E}_{Fn}, n_i)\right] \tag{2.24}$$

where $T_{83}(\overline{E}_{Fn}, n_i) \equiv \left[I_{11}(\overline{E}_{Fn}) - \left(n_i + \frac{1}{2}\right)\hbar\omega_{19}(\overline{E}_{Fn})\right]$ and $T_{84}(\overline{E}_{Fn}, n_i) \equiv \sum_{r=1}^{s} L(r)$ $T_{83}(\overline{E}_{Fn}, n_i)$.

Using (2.24), the electron concentration in the electric quantum limit for doping superlattices of III-V, ternary and quaternary materials can be written as

$$\overline{n}_0 = \left(\frac{m_c g_v}{\pi\hbar^2}\right)\left[I_{11}(E_{20}) - \{(1/2)\hbar\omega_{19}(E_{20})\}\right] \tag{2.25}$$

where, E_{20} is determined from the equation

$$I_{11}(E_{20}) = \frac{1}{2}\hbar\omega_{19}(E_{20}) \tag{2.26}$$

Using (2.13) and (2.26) we can study, the ER in this case.

(b) The electron energy spectrum in doping superlattics of HD III-V, ternary and quaternary materials whose energy band structures in the absence of band tails are described by the two band model of Kane can be expressed from (2.15) under the conditions $\Delta \gg E_g$ or $\Delta \ll E_g$, as

$$\frac{\hbar^2 k_s^2}{2m_c} = \left[\gamma_2(E, \eta_g) - \left(n_i + \frac{1}{2}\right)\hbar\omega_{10HD}(\mathrm{E}, \eta_g)\right] \tag{2.27}$$

where $\omega_{10HD}(E) \equiv \left(\frac{n_0|e|^2}{d_0\varepsilon_{sc}\gamma_2'(E, \eta_g)m_c}\right)^{1/2}$

The EEM in this case assumes the form

$$m^*(\mathrm{E}_{FnHD}, n_i, \eta_g) = \left(\frac{\hbar^2}{2}\right) G'_{25HD}(E_{FnHD}, \eta_g, n_i) \tag{2.28}$$

where $G_{25HD}(E_{FnHD}, \eta_g, n_i) = \frac{2m_c}{\hbar^2}\left[\gamma_2(E_{FnHD}, \eta_g) - \left(n_i + \frac{1}{2}\right)\hbar\omega_{10HD}(E_{FnHD}, \eta_g)\right]$

The subband energy E_{3n_iHD} can be written as

$$[\gamma_2(E_{3n_iHD}, \eta_g) - \left(n_i + \frac{1}{2}\right)\hbar\omega_{9HD}(E_{3n_iHD}, \eta_g)] = 0 \tag{2.29}$$

The DOS function in this case is given by

$$N_{nipiHD}(E, \eta_g) = \frac{g_v m_c}{\pi\hbar^2}\sum_{n_i=0}^{n_{i\max}} G'_{25HD}(E, \eta_g, n_i) H(E - E_{3n_iHD}) \tag{2.30}$$

The electron concentration, can be written as

$$n_{2DN} = \frac{g_v m_c}{\pi\hbar^2}\sum_{n_i=0}^{n_{i\max}} [G_{25HD}(\overline{\mathrm{E}}_{FnHD}, \eta_g, n_i) + G_{26HD}(\overline{\mathrm{E}}_{FnHD}, \eta_g, n_i)] \tag{2.31}$$

where, $G_{26HD}(\overline{\mathrm{E}}_{FnHD}, \eta_g, n_i) = \sum_{r=1}^{s} L(\mathrm{r}) G_{25HD}(\overline{\mathrm{E}}_{FnHD}, \eta_g, n_i)$

Using (2.31) and (2.6) at the electric quantum limit, we can study the ER in this case.

In the absence of band tails, the dispersion relation in this case assumes the form

$$E(1 + \alpha E) = \left(n_i + \frac{1}{2}\right)\hbar\omega_{20}(E) + \frac{\hbar^2 k_s^2}{2m_c} \tag{2.32}$$

where $\omega_{20}(E) \equiv \left(\frac{n_0|e|^2}{d_0\varepsilon_{sc}(1 + 2\alpha E)m_c}\right)^{1/2}$.

The EEM in this case can be written as

$$m^*(E_{Fn}, n_i) = m_c R_{182}(E, n_i)|_{E=E_{Fn}} \tag{2.33}$$

in which, $R_{182}(E, n_i) \equiv \left\{[1 + 2\alpha E] - \left(n_i + \frac{1}{2}\right)\hbar[\omega_{19}(E)]'\right\}$.

From (2.33), we observe that the EEM in this case is a function of the Fermi energy, nipi subband index and the other material constants which is the characteristic feature of doping superlattices of III-V, ternary and quaternary compounds whose bulk dispersion relations is defined by the three band model of Kane.

The subband energies (E_{3ni}) can be written as

$$E_{3ni}(1+\alpha E_{3ni}) = \left(n_i + \frac{1}{2}\right)\hbar\omega_{20}(E_{3ni}) \tag{2.34}$$

The DOS function in this case can be expressed as

$$N_{nipi}(E) = \frac{m_c g_v}{\pi\hbar^2}\sum_{n_i=0}^{n_{imax}} R_{182}(E, n_i)H(E - E_{3ni}) \tag{2.35}$$

The use of (2.35) leads to the expression of the electron concentration as

$$n_0 = \frac{m_c g_v}{\pi\hbar^2}\sum_{n_i=0}^{n_{imax}} \left[T_{85}(\overline{E}_{Fn}, n_i) + T_{86}(\overline{E}_{Fn}, n_i)\right] \tag{2.36}$$

where $T_{85}(\overline{E}_{Fn}, n_i) \equiv \left[\overline{E}_{Fn}(1+\alpha\overline{E}_{Fn}) - (n_i + \frac{1}{2})\hbar\omega_{20}(\overline{E}_{Fn})\right]$ and $T_{86}(\overline{E}_{Fn}, n_i) \equiv \sum_{r=1}^{s} L(r)T_{85}(\overline{E}_{Fn}, n_i)$.

Thus using (2.36) in the electric quantum limit, we can study the ER in this case.

(c) The electron energy spectrum in nipi structures of HD III-V, ternary and quaternary materials whose energy band structures in the absence of band tails are described by the parabolic energy bands can be expressed as

$$\frac{\hbar^2 k_s^2}{2m_c} = \left[\gamma_3(E, \eta_g) - \left(n_i + \frac{1}{2}\right)\hbar\omega_{11HD}(\mathrm{E}, \eta_g)\right] \tag{2.37}$$

where $\omega_{11HD}(E) \equiv \left(\frac{n_0|e|^2}{d_0\varepsilon_{sc}\gamma_3'(E,\eta_g)m_c}\right)^{1/2}$

The EEM in this case assumes the form

$$m^*\left(\mathrm{E}_{FnHD}, n_i, \eta_g\right) = \left(\frac{\hbar^2}{2}\right)G_{27HD}'\left(E_{FnHD}, \eta_g, n_i\right) \tag{2.38}$$

where $G_{27HD}(\mathrm{E}_{FnHD}, \eta_g, n_i) = \frac{2m_c}{\hbar^2} = \left[\gamma_3(\mathrm{E}_{FnHD}, \eta_g) - (n_i + \frac{1}{2})\hbar\omega_{11HD}(\mathrm{E}_{FnHD}, \eta_g)\right]$

The subband energy E_{4n_iHD} can be expressed as

$$\left[\gamma_3(E_{4n_iHD}, \eta_g) - \left(n_i + \frac{1}{2}\right)\hbar\omega_{11HD}(E_{4n_iHD}, \eta_g)\right] = 0 \tag{2.39}$$

The DOS function in this case is given by

$$N_{nipiHD}(E,\eta_g) = \frac{g_v m_c}{\pi\hbar^2}\sum_{n_i=0}^{n_{imax}} G'_{27HD}(E,\eta_g,n_i)H(E - E_{4n_iHD}) \tag{2.40}$$

The electron concentration, can be written as

$$n_{2DN} = \frac{g_v m_c}{\pi\hbar^2}\sum_{n_i=0}^{n_{imax}} [G_{27HD}(\overline{E}_{FnHD},\eta_g,n_i) + G_{28HD}(\overline{E}_{FnHD},\eta_g,n_i) \tag{2.41}$$

where, $G_{28HD}(\overline{E}_{FnHD},\eta_g,n_i) = \sum_{r=1}^{s} L(\mathrm{r})G_{27HD}(\overline{E}_{FnHD},\eta_g,n_i)$

Using (2.41) and (2.6) at the electric quantum limit, we can study the ER in this case.

In the absence of band tails, the dispersion relation in this case assumes the form

$$E = \left(n_i + \frac{1}{2}\right)\hbar\omega_{21} + \frac{\hbar^2 k_s^2}{2m_c} \tag{2.42}$$

where $\omega_{21} \equiv \left(\frac{n_0|e|^2}{d_0 \varepsilon_{sc} m_c}\right)^{1/2}$.

The EEM in this case can be written as

$$m^*(E_{Fn},n_i) = m_c \tag{2.43}$$

Thus the EEM in this case is a constant quantity.

The subband energies (E_{4ni}) can be written as

$$E_{4ni} = \left(n_i + \frac{1}{2}\right)\hbar\omega_{21} \tag{2.44}$$

The DOS function in this case can be expressed as

$$N_{nipi}(E) = \frac{m_c g_v}{\pi\hbar^2}\sum_{n_i=0}^{n_{imax}} H(E - E_{4ni}) \tag{2.45}$$

The use of (2.45) leads to the expression of the surface electron concentration as

$$n_0 = \frac{m_c g_v k_B T}{\pi\hbar^2}\sum_{n_i=0}^{n_{imax}} F_0(\eta_{4i}) \tag{2.46}$$

where $\eta_{4i} = (\overline{\mathrm{E}}_{Fn} - \mathrm{E}_{4n_i})/(\mathrm{k}_B\mathrm{T})$

Thus using (2.46) in the electric quantum limit, we can study the ER in this case.

2.2.3 The ER in Doping Superlattices of HD II-VI Semiconductors

The 2D electron dispersion law in doping superlattices of HD II-VI semiconductors can be expressed as

$$\gamma_3(\mathrm{E},\eta_g) = a_0' k_s^2 + \left(n_i + \frac{1}{2}\right)\hbar\omega_{20}(\mathrm{E},\eta_g) \pm \bar{\lambda}_0 k_s, \omega_{30}(\mathrm{E},\eta_g) \equiv \left(\frac{n_0|e|^2}{d_0\gamma_3'(\mathrm{E},\eta_g)\varepsilon_{sc}m_{\|}^*}\right)^{1/2} \quad (2.47)$$

The EEM in this case assumes the form as

$$m^*(\mathrm{E}_{FnHD}, n_i, \eta_g) = m_\perp^*\left\{1 - \bar{\lambda}_0\left[(\bar{\lambda}_0)^2 + 4a_0'\gamma_3(\mathrm{E}_{FnHD},\eta_g) - 4a_0'\left(n_i + \frac{1}{2}\right)\hbar\omega_{30}(\mathrm{E}_{FnHD},\eta_g)\right]^{1/2}\right\}\gamma_3'(\mathrm{E}_{FnHD},\eta_g) \quad (2.48)$$

The subband energy can be written as

$$\gamma_3(E_{6n_iHD},\eta_g) = \left(n_i + \frac{1}{2}\right)\hbar\omega_{30}(E_{6n_iHD},\eta_g) \quad (2.49)$$

The surface electron concentration per unit area in this case is given by

$$n_{2DN} = \frac{g_v}{4\pi a_0'^2}\sum_{n_i=0}^{n_{imax}}\left[G_{30HD}(\bar{\mathrm{E}}_{FnHD},\eta_g,n_i) + G_{31HD}(\bar{\mathrm{E}}_{FnHD},\eta_g,n_i)\right] \quad (2.50)$$

where

$$G_{30HD}(\bar{\mathrm{E}}_{FnHD},\eta_g,n_i) = \left[(\bar{\lambda}_0)^2 - 2a_0'\left\{\left(n_i + \frac{1}{2}\right)\hbar\omega_{30}(\bar{\mathrm{E}}_{FnHD},\eta_g,n_i) - \gamma_3(\bar{\mathrm{E}}_{FnHD},\eta_g,n_i)\right\}\right]$$

and $G_{31HD}(\bar{\mathrm{E}}_{FnHD},\eta_g,n_i) = \sum_{r=1}^{s} L(\mathrm{r})G_{30HD}(\bar{\mathrm{E}}_{FnHD},\eta_g,n_i)$

Using (2.50) and (2.6) at the electric quantum limit, we can study the ER in this case.

In the absence of band-tails, the carrier dispersion law in doping superlattices of II-VI compounds can be expressed as

$$E = a_0' k_s^2 + \left(n_i + \frac{1}{2}\right)\hbar\bar{\omega}_{10} \pm \bar{\lambda}_0 k_s, \quad \bar{\omega}_{10} = \left(\frac{n_0|e|^2}{d_0\varepsilon_{sc}m_{\|}^*}\right)^{\frac{1}{2}} \quad (2.51)$$

Using (2.51), the EEM in this case can be written as

$$m^*(E_{Fn}, n_i) = m_\perp^* \left\{ 1 - \bar{\lambda}_0 \left[\left(\bar{\lambda}_0\right)^2 + 4a_0' E_{Fn} - 4a_0' \left(n_i + \frac{1}{2} \right) \hbar\bar{\omega}_{10} \right]^{-1/2} \right\} \quad (2.52)$$

Thus, the EEM in this case is a function of the Fermi energy, the nipi subband index number and the energy spectrum constants due to the only presence of $\bar{\lambda}_0$.

The subband energies (E_{8ni}) assume the form as

$$E_{8ni} = \left(n_i + \frac{1}{2} \right) \hbar\bar{\omega}_{10} \quad (2.53)$$

The DOS function in this case can be expressed as

$$N_{nipi}(E) = \frac{m_\perp^* g_v}{\pi\hbar^2} \sum_{n_i=0}^{n_{imax}} \left[1 - \frac{a_{81}}{\sqrt{E + b_{81}(n_i)}} \right] H(E - E_{8ni}) \quad (2.54)$$

in which, $a_{81} \equiv \frac{\bar{\lambda}_0}{2\sqrt{a_0'}}$ and $b_{81}(n_i) \equiv \left[\frac{1}{4a_0'} \left[\left(\bar{\lambda}_0\right)^2 - 4a_0' \left(n_i + \frac{1}{2}\right) \hbar\bar{\omega}_{10} \right] \right]$.

The use of the (2.54) leads to the electron concentration as

$$n_0 = \frac{m_\perp^* g_v k_B T}{\pi\hbar^2} \sum_{n_i=0}^{n_{imax}} \left[F_0(\eta_{81}) - \left(\frac{a_{81}}{\sqrt{k_B T}} \left[2\left(\sqrt{\eta_{81} + c_{81}(n_i)} - \sqrt{c_{81}(n_i)} \right) \right] \right) + \sum_{r=1}^{s} 2(1 - 2^{1-2r})\zeta(2r) \frac{(-1)^{2r-1}(2r-1)!}{(\eta_{81} + c_{81}(n_i))^{2r}} \right] \quad (2.55)$$

where, $\eta_{81} \equiv \frac{E_{Fn} - E_{8ni}}{k_B T}$ and $c_{81}(n_i) \equiv \frac{b_{81}(n_i) + E_{8ni}}{k_B T}$.

Using (2.55) in the electric quantum limit we can study the ER in this case.

2.2.4 The ER in Doping Superlattices of HD IV-VI Semiconductors

The 2D electron dispersion law in this case is given by

$$k_s^2 = \delta_{15}(E, \eta_g, n_i) \quad (2.56)$$

where $\delta_{15}(E, \eta_g, n_i) = [2\delta_{12}(E, \eta_g)]^{-1}[-\delta_{13}(E, \eta_g, n_i) + \sqrt{\delta_{13}^2(E, \eta_g, n_i) - 4\delta_{12}(E, \eta_g)\delta_{14}(E, \eta_g, n_i)}]$,
$\delta_{12}(E, \eta_g) = \frac{\alpha\hbar^4 Z_0(E, \eta_g)}{4m_t^+ m_l^-}$, $\delta_{13}(E, \eta_g, n_i) = \hbar^2[\lambda_{71}(E, \eta_g)\delta_{11}(E, \eta_g, n_i) + \lambda_{12}(E, \eta_g)]$,
$\delta_{14}(E, \eta_g, n_i) = [\lambda_{73}(E, \eta_g)\delta_{11}^2(E, \eta_g, n_i) + \lambda_{74}(E, \eta_g)\delta_{11}^4(E, \eta_g, n_i) - \lambda_{74}(E, \eta_g)]$,
$\delta_{11}(E, \eta_g, n_i) = \frac{2}{\hbar} m_{HD}^*(0, \eta_g)\left(n_i + \frac{1}{2}\right) \left[\frac{e^2 n_0}{d_0 \varepsilon_{sc} m_{HD}^*(E, \eta_g)} \right]^{1/2}$

and $m^*_{HD}(E,\eta_g) = \dfrac{\hbar^2}{4\lambda^2_{76}(E,\eta_g)}[2\lambda_{74}(E,\eta_g)\{-\lambda'_{73}(E,\eta_g)$

$$+\frac{\lambda_{73}(E,\eta_g)\lambda'_{73}(E,\eta_g)+2\lambda'_{74}(E,\eta_g)\lambda_{75}(E,\eta_g)+2\lambda_{74}(E,\eta_g)\lambda'_{75}(E,\eta_g)}{\sqrt{\lambda^2_{73}(E,\eta_g)+4\lambda_{74}(E,\eta_g)\lambda_{75}(E,\eta_g)}}\}$$

$$-2\lambda'_{74}(E,\eta_g)\{-\lambda_{73}(E,\eta_g)+\sqrt{\lambda^2_{73}(E,\eta_g)+4\lambda_{74}(E,\eta_g)\lambda_{75}(E,\eta_g)}\}]$$

The EEM in this case assumes the form

$$m^*(\mathrm{E}_{FnHD},n_i,\eta_g) = \left(\frac{\hbar^2}{2}\right)\delta'_{15}(E_{FnHD},\eta_g,n_i) \tag{2.57}$$

The sub-band energy E_{9n_iHD} can be expressed as in this case as

$$0 = \delta_{15}(E_{9n_iHD},\eta_g,n_i) \tag{2.58}$$

The surface electron concentration in this case is given by

$$n_{2DN} = \frac{g_v}{2\pi}\sum_{n_i=0}^{n_{imax}}[G_{32HD}(\overline{\mathrm{E}}_{FnHD},\eta_g,n_i)+G_{33HD}(\overline{\mathrm{E}}_{FnHD},\eta_g,n_i)] \tag{2.59}$$

where

$$G_{32HD}(\overline{\mathrm{E}}_{FnHD},\eta_g,n_i) = \delta_{15}(\overline{\mathrm{E}}_{FnHD},\eta_g,n_i) \text{ and } G_{33HD}(\overline{\mathrm{E}}_{FnHD},\eta_g,n_i) = \sum_{r=1}^{s}L(\mathrm{r})G_{32HD}(\overline{\mathrm{E}}_{FnHD},\eta_g,n_i)$$

Using (2.59) and (2.6) at the electric quantum limit, we can study the ER in this case.

The carrier energy spectrum in doping superlattices of IV-VI compounds in the absence of band tails can be written as

$$k_s^2 = (\hbar^2 S_{19})^{-1}\left[-S_{20}(E,n_i)+\sqrt{S^2_{20}(E,n_i)+4S_{19}S_{21}(E,n_i)}\right] \tag{2.60}$$

in which, $S_{19}\equiv\left(\frac{\alpha}{m_t^+m_t^-}\right)$, $S_{20}(E,n_i)\equiv\left\{\frac{1}{m_t^+}-\left(\frac{\alpha E}{m_t^+}\right)+\frac{1+\alpha E}{m_t^-}+\frac{\alpha\hbar^2}{2m_l^+m_t^-}\left(n_i+\frac{1}{2}\right)T(E)+\frac{\alpha\hbar^2}{2m_l^-m_t^+}\left(n_i+\frac{1}{2}\right)T(E)\right\}$

$$T(E) \equiv \frac{2m^*(0)}{\hbar}\omega_{11}(E), m^*(0) \equiv \left(\frac{m_l^* m_l^-}{m_l^* + m_l^-}\right), \quad \omega_{11}(E) \equiv \left(\frac{n_0|e|^2}{d_0 \varepsilon_{sc} m^*(E)}\right)^{\frac{1}{2}},$$

$$m^*(E) \equiv \frac{1}{4t_1}\left[-(t_2(E))' + \frac{t_2(E)(t_2(E))' + 2t_1(1+2\alpha E)}{\sqrt{t_2^2(E) + 4Et_1(1+\alpha E)}}\right],$$

$$t_1 \equiv \left(\frac{\alpha}{4m_l^+ m_l^-}\right), \; t_2(E) \equiv \frac{1}{2}\left[\left(\frac{1}{m_l^*}\right) - \left(\frac{\alpha E}{m_l^+}\right) + \left(\frac{1+\alpha E}{m_l^-}\right)\right], \; (t_2(E))' \equiv \frac{\alpha}{2}\left(\frac{1}{m_l^-} - \left(\frac{1}{m_l^+}\right)\right)$$

and

$$S_{21}(E, n_i) \equiv \left[E(1+\alpha E) + \frac{\alpha E \hbar^2}{2m_l^+}\left(n_i + \frac{1}{2}\right)T(E) + \frac{\hbar^2}{2m_l^-}\left(n_i + \frac{1}{2}\right)T(E)(1+\alpha E) + \frac{\hbar^4}{4m_l^- m_l^+}\left(n_i + \frac{1}{2}\right)T(E) + \left(\frac{\hbar^2}{2m_l^*}\right)T(E)\left(n_i + \frac{1}{2}\right)\right].$$

Using (2.60) the EEM in this case can be written as

$$m^*(E_{Fn}, n_i) = R_{84}(E, n_i)|_{E=E_{Fn}} \tag{2.61}$$

where,

$$R_{84}(E, n_i) \equiv (2S_{19})^{-1}\left[-(S_{20}(E, n_i))' + \frac{S_{20}(E, n_i)[S_{20}(E, n_i)]' + 2S_{19}[S_{21}(E, n_i)]'}{\left[\{[S_{20}(E, n_i)]'\}^2 + 4S_{19}S_{21}(E, n_i)\right]^{1/2}}\right].$$

Thus, one can observe that the EEM in this case is a function of both the Fermi energy and the nipi subband index number together with the spectrum constants of the system due to the presence of band non-parabolicity.

The subband energies (E_{10ni}) can be written as

$$\left[E_{10ni} - \frac{\hbar^2}{2m_l^-}T(E_{10ni})\left(n_i + \frac{1}{2}\right)\right]\left[1 + \alpha E_{10ni} + +\alpha\frac{\hbar^2}{2m_l^+}T(E_{10ni})\left(n_i + \frac{1}{2}\right)\right] = \left[\frac{\hbar^2}{2m_l^*}T(E_{10ni})\left(n_i + \frac{1}{2}\right)\right] \tag{2.62}$$

The DOS function in this case assumes the form as

$$N_{nipi}(E) = \frac{g_v}{\pi\hbar^2}\sum_{n_i=0}^{n_{imax}} R_{84}(E, n_i)H(E - E_{10ni}) \tag{2.63}$$

The use of (2.63) leads to the expression of the electron concentration as

$$n_0 = \frac{g_v}{2\pi\hbar^2 S_{19}} \sum_{n_i=0}^{n_{i\max}} T_{85}(\overline{E}_{Fn}, n_i) + T_{86}(\overline{E}_{Fn}, n_i) \tag{2.64}$$

where, $T_{85}(\overline{E}_{Fn}, n_i) \equiv \left[-S_{20}(E_{Fn}, n_i) + \sqrt{[S_{20}(E_{Fn}, n_i)]^2 + 4S_{19}S_{21}(E_{Fn}, n_i)} \right]$ and $T_{86}(\overline{E}_{Fn}, n_i) \equiv \sum_{r=1}^{s} L(r) T_{85}(\overline{E}_{Fn}, n_i)$.

The electron concentration at the quantum limit can be defined through the equation

$$\overline{n}_0 = \frac{g_v}{2\pi\hbar^2 S_{19}} \left[-S_{20}(\overline{E}_{F0}, 0) + \sqrt{\left[S_{20}(\overline{E}_{F0}, 0)\right]^2 + 4S_{19}S_{21}(\overline{E}_{F0}, 0)} \right] \tag{2.65}$$

2.2.5 *The ER in Doping Superlattices of HD Kane Type Semiconductors*

The 2D dispersion relation in this case is given by

$$P_{11}(E, \eta_g)k_x^2 + Q_{11}(E, \eta_g)k_y^2 + S_{11}(E, \eta_g)\delta_{19}(E, \eta_g, n_i) = 1 \tag{2.66}$$

where

$$\delta_{19}(E, \eta_g, n_i) = \frac{2}{\hbar} m_{zz}^*(0, \eta_g)\left(n_i + \frac{1}{2}\right)\left[\frac{n_0 e^2}{d_0 \varepsilon_{sc} m_{zz}(E, \eta_g)}\right]^{1/2}$$

and the expression for $m_{zz}(E, \eta_g)$ has already been given in (1.209) of Chap. 1.

The EEM in this case assumes the form

$$m^*(E_{FnHD}, n_i, \eta_g) = \left(\frac{\hbar^2}{2}\right)\delta_{20}'(E_{FnHD}, \eta_g, n_i) \tag{2.67}$$

where $\delta_{20}(E_{FnHD}, \eta_g, n_i) = \dfrac{[1 - S_{11}(E_{FnHD}, \eta_g)\delta_{19}(E_{FnHD}, \eta_g, n_i)]}{\sqrt{P_{11}(E_{FnHD}, \eta_g)Q_{11}(E_{FnHD}, \eta_g)}}$

The sub-band energy E_{15n_iHD} can be expressed as in this case as

$$S_{11}(E_{15n_iHD}, \eta_g)\delta_{19}(E_{15n_iHD}, \eta_g, n_i) = 1 \tag{2.68}$$

The surface electron concentration in this case is given by

$$n_{2DN} = \frac{g_v}{2\pi} \sum_{n_i=0}^{n_{imax}} [G_{34HD}(\overline{E}_{FnHD}, \eta_g, n_i) + G_{35HD}(\overline{E}_{FnHD}, \eta_g, n_i)] \tag{2.69}$$

where $G_{34HD}(\overline{E}_{FnHD}, \eta_g, n_i) = \delta_{20}(\overline{E}_{FnHD}, \eta_g, n_i)$ and $G_{35HD}(\overline{E}_{FnHD}, \eta_g, n_i) = \sum_{r=1}^{s} L(r) G_{34HD}(\overline{E}_{FnHD}, \eta_g, n_i)$.

Using (2.69) and (2.6) at the electric quantum limit, we can study the ER in this case.

The electron dispersion law in the doping superlattices of stressed Kane type semiconductors can be written as

$$\frac{k_x^2}{[\bar{a}_0(E)]^2} + \frac{k_y^2}{[\bar{b}_0(E)]^2} + \frac{1}{[\bar{c}_0(E)]^2} \frac{2m_z^*(0)}{\hbar} \left(n_i + \frac{1}{2}\right) \omega_{12}(E) = 1 \tag{2.70}$$

where $\omega_{12}(E) \equiv \left(\frac{n_0 |e|^2}{d_0 \varepsilon_{sc} m_z^*(E)}\right)^{\frac{1}{2}}$ and $m_z^*(E) \equiv \hbar^2 \bar{c}_0(E) \frac{\partial}{\partial E} [\bar{c}_0(E)]$.

The use of (2.70) leads to the expression of the EEM as

$$m^*(E_{Fn}, n_i) = \left(\frac{\hbar^2}{2}\right) R_{85}(E, n_i)\Big|_{E=E_{Fn}} \tag{2.71}$$

where,

$$\begin{aligned} R_{85}(E, n_i) \equiv & \left[\left[(\bar{a}_0(E))' b_0(E) + (\bar{b}_0(E))' \bar{a}_0(E) \right] \left[1 - \frac{1}{[\bar{c}_0(E)]^2} \frac{2m_z^*(0)}{\hbar} \left(n_i + \frac{1}{2}\right) \omega_{12}(E) \right] \right. \\ & \left. - \left[\frac{\bar{a}_0(E) \bar{b}_0(E)}{[\bar{c}_0(E)]^2} \frac{2m_z^*(0)}{\hbar} \left(n_i + \frac{1}{2}\right) [\omega_{12}(E)]' \right] + \left[\frac{\bar{a}_0(E) \bar{b}_0(E) [\bar{c}_0(E)]'}{[\bar{c}_0(E)]^3} \frac{4m_z^*(0)}{\hbar} \left(n_i + \frac{1}{2}\right) [\omega_{12}(E)] \right] \right] \end{aligned} \tag{2.72}$$

Thus, the EEM is a function of the Fermi energy and the nipi subband index due to the presence of stress and band non-parabolicity only.

The subband energies (E_{25ni}) can be written as

$$\frac{1}{[\bar{c}_0(E_{25ni})]^2} \frac{2m_z^*(0)}{\hbar} \left(n_i + \frac{1}{2}\right) \omega_{12}(E_{25ni}) = 1 \tag{2.73}$$

The DOS function can be written as

$$N_{nipi}(E) = \frac{g_v}{\pi \hbar^2} \sum_{n_i=0}^{n_{imax}} R_{85}(E, n_i) H(E - E_{25ni}) \tag{2.74}$$

Thus, using (2.74), the electron concentration in doping superlattices of stressed compounds can be expressed as

$$n_0 = \frac{g_v}{2\pi} \sum_{n_i=0}^{n_{imax}} C_3(\overline{E}_{Fn}, n_i) + C_4(\overline{E}_{Fn}, n_i) \tag{2.75}$$

where $C_3(\overline{E}_{Fn}, n_i) \equiv \overline{a}_0(\overline{E}_{Fn})\overline{b}_0(E_{Fn})\left[1 - \frac{2m_z^*(0)}{\hbar}\left(n_i + \frac{1}{2}\right)\frac{\omega_{12}(\overline{E}_{Fn})}{\left(\overline{c}_0(\overline{E}_{Fn})\right)^2}\right]$ and $C_4(\overline{E}_{Fn}, n_i) \equiv \sum_{r=1}^{s} L(r) C_3(\overline{E}_{Fn}, n_i)$

The use of (2.75) leads to the expression of the electron statistics at the electric quantum limit and at low temperatures as

$$\overline{n}_0 = \frac{g_v}{2\pi}\overline{a}_0(\overline{E}_{F_0})\overline{b}_0(\overline{E}_{F_0})\left[1 - \frac{m_z^*(0)}{\left(\overline{c}_0(\overline{E}_{F_0})\right)^2\hbar}\omega_{12}(\overline{E}_{F_0})\right] \tag{2.76}$$

Using (2.76), we can study the ER in this case.

2.3 Result and Discussions

Using the appropriate equations together with the energy band constants as given in Table 1.1, the ER in the quantum limit has been plotted for the doping superlattices of HD tetragonal compounds (taking HD Cd_3As_2 as an example) as a function of electron concentration as shown in curve (a) of Fig. 2.1. The curve (b) corresponds to $\delta = 0$ and the curve (c) exhibits the dependence of the ER on n_0 in accordance with the HD three-band model of Kane, respectively. The plots (d) and (e) correspond to the HD two-band model of Kane and that of HD parabolic energy bands. By comparing the curves (a) and (b) of Fig. 2.1, one can assess the influence of crystal field splitting of the ER in doping superlattices of HD Cd_3As_2. Figure 2.2 represents all cases of Fig. 2.1 for doping superlattices of HD nonlinear optical materials taking HD $CdGeAs_2$ as an example. It appears from Figs. 2.1 and 2.2 that, the ER in doping superlattices of HD nonlinear optical materials increases with increasing carrier degeneracy as expected for degenerate materials.

Using the appropriate equations one can numerically evaluate the ER in the quantum limit as a function of electron concentration in doping superlattices of HD III-V compounds by using the HD InAs, and InSb as shown in Figs. 2.3 and 2.4 by curves (a), (b) and (c) respectively, in accordance with three and two band models of Kane together with the model of parabolic energy bands.

Taking doping superlattices of HD $Hg_{1-x}Cd_xTe$ as an example of HD ternary compounds, the ER has been plotted for both the structures as a function of electron concentration as shown in Fig. 2.5 for all cases of the Fig. 2.3. It appears

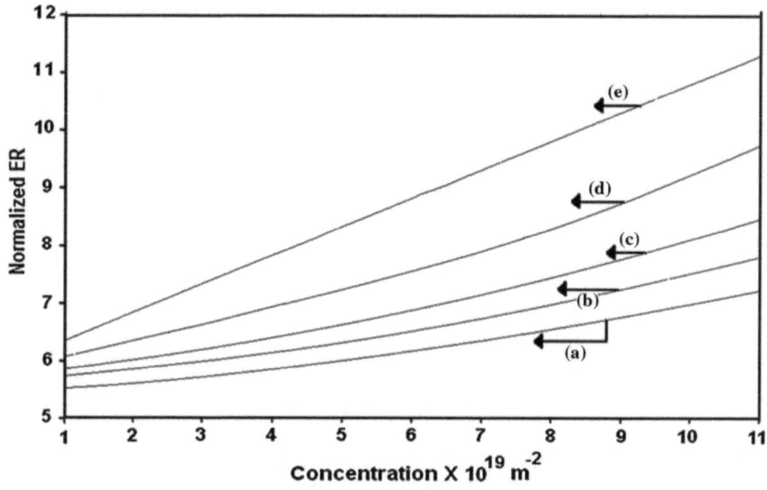

Fig. 2.1 The plot of the ER in the quantum limit for doping superlattices of HD Cd_3As_2 as a function of electron concentration in accordance with *a* the generalized band model, *b* $\delta = 0$, *c* the three band model of Kane, *d* the two band model of Kane and *e* the parabolic energy bands

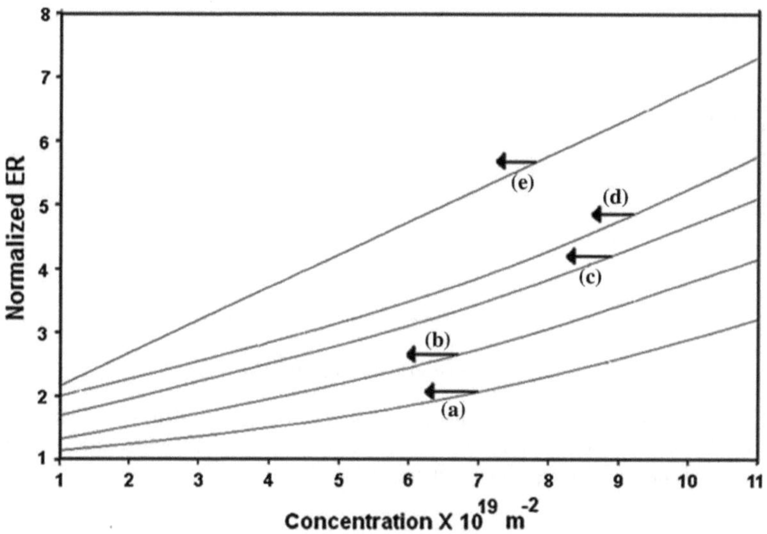

Fig. 2.2 The plot of the ER in the quantum limit for doping superlattices of HD $CdGeAs_2$ as a function of electron concentration in accordance with *a* the generalized band model, *b* $\delta = 0$, *c* the three band model of Kane, *d* the two band model of Kane and *e* the parabolic energy bands

from the Fig. 2.5 that the ER in the quantum limit in both cases of doping superlattices of ternary compounds increases with increasing electron concentration as usual for the degenerate compounds. Taking doping superlattices of HD

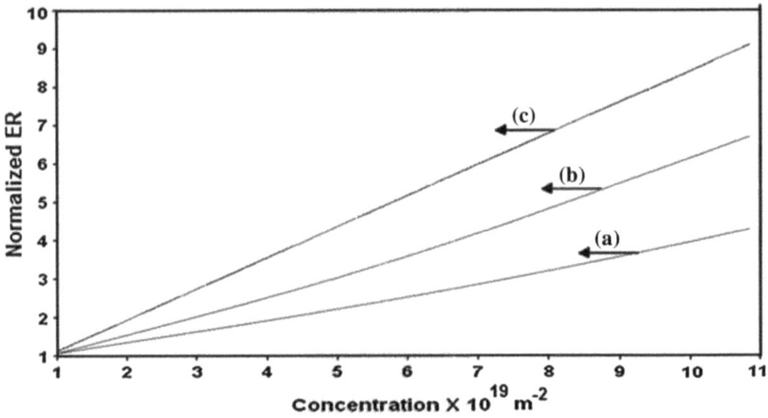

Fig. 2.3 The plot of the ER in the quantum limit for doping superlattices of HD InAs as a function of electron concentration in accordance with *a* the three band model of Kane, *b* the two band model of Kane and *c* the parabolic energy bands

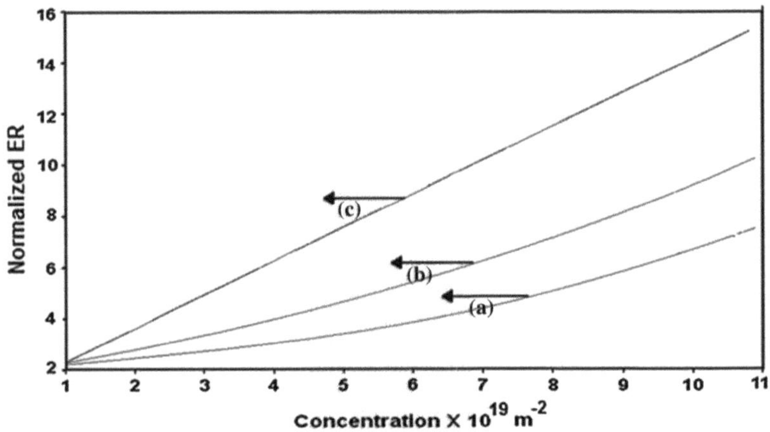

Fig. 2.4 The plot of the ER in the quantum limit for doping superlattices of HD InSb as a function of electron concentration in accordance with *a* the three band model of Kane, *b* the two band model of Kane and *c* the parabolic energy bands

$In_{1-x}Ga_xAs_yP_{1-y}$ lattice matched to HD InP as an example of quaternary compounds the ER in the quantum limit has been further been plotted as a function of electron concentration as shown in Fig. 2.6 in accordance with the three and two band models of Kane together with the isotropic parabolic energy band model for both the cases. It appears that the ER increases with increasing n_0 as usual. From Figs. 2.5 and 2.6, one can assess the influence of energy band constants on the ER for doping superlattices of ternary and quaternary materials respectively. Using the appropriate equations, the ER in the quantum limit has been plotted for the doping

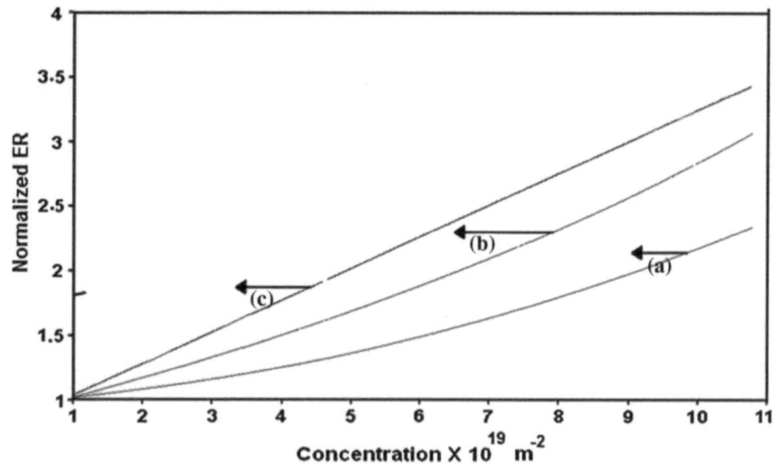

Fig. 2.5 The plot of the ER in the quantum limit for doping superlattices of HD $Hg_{1-x}Cd_xTe$ as a function of electron concentration in accordance with *a* the three band model of Kane, *b* the two band model of Kane and *c* the parabolic energy bands

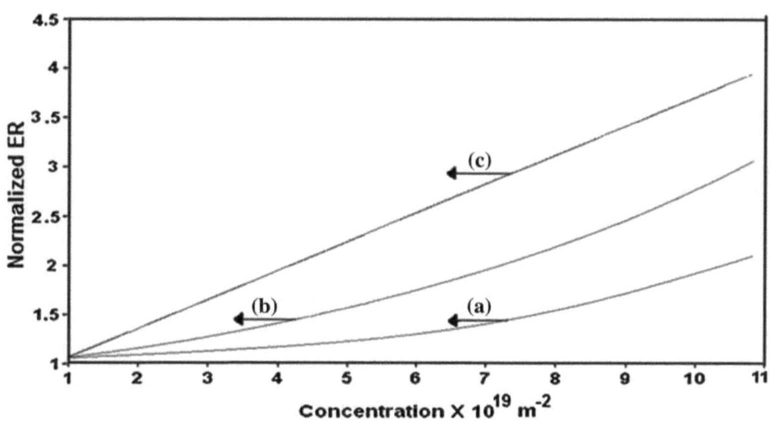

Fig. 2.6 The plot of the ER in the quantum limit for doping superlattices of HD $In_{1-x}Ga_xAs_yP_{1-y}$ lattice matched to InP as a function of electron concentration in accordance with *a* the three band model of Kane, *b* the two band model of Kane and *c* the parabolic energy bands

superlattices of CdS, as a function of carrier concentration as shown by curves (a) and (b) in Fig. 2.7 for both $\bar{\lambda}_0 \neq 0$ and $\bar{\lambda}_0 = 0$ respectively. This has been presented for the purpose of assessing the influence of the splitting of the two spin states by the spin-orbit coupling and the crystalline field on the ER for doping superlattices of II-VI materials. In Fig. 2.8, the ER in the quantum limit has been plotted for the HD doping superlattices of (a) PbTe, (b) PbSnTe and (c) $Pb_{1-x}Sn_xSe$

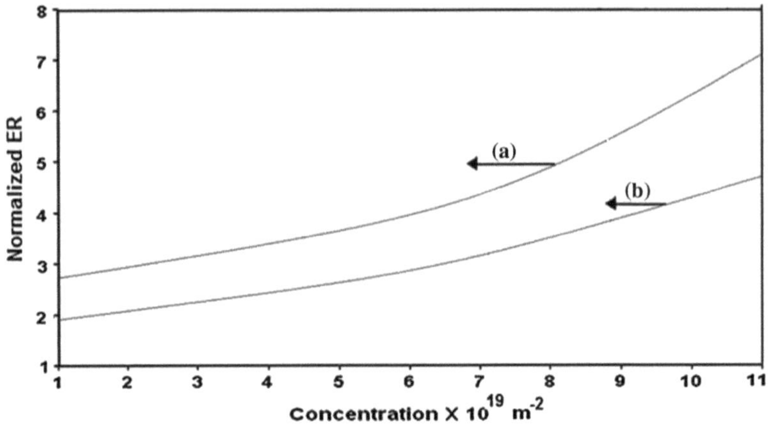

Fig. 2.7 The plot of the ER in the quantum limit for doping superlattices of HD CdS as a function of carrier concentration in accordance with *a* $\bar{\lambda}_0 \neq 0$ and *b* $\bar{\lambda}_0 = 0$

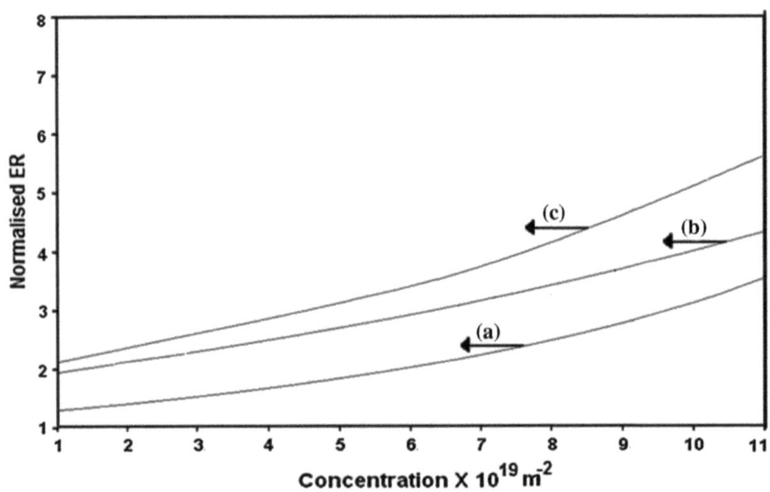

Fig. 2.8 The plot of the ER in the quantum limit as a function of electron concentration for the doping superlattices of HD *a* PbTe, *b* PbSnTe and *c* $Pb_{1-x}Sn_xSe$

as a function of electron concentration in accordance with the Dimmock model. For relatively low values of electron concentration, the values of the ER for the three materials exhibit convergence behavior whereas for relatively large values of n_0, the numerical values differ widely from each other in this case. In Fig. 2.9, the ER in the quantum limit has been plotted for the doping superlattices of stressed HD InSb as a function of electron concentration.

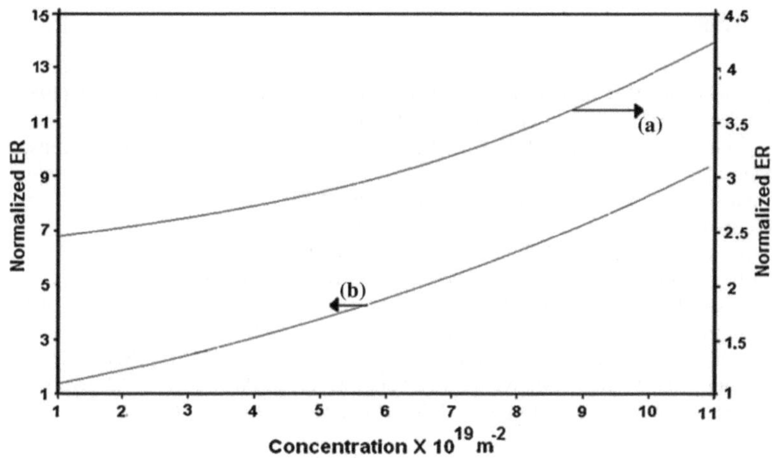

Fig. 2.9 The plot of the ER in the quantum limit as a function of electron concentration for the doping superlattices of stressed HD InSb in which the curve *a* is in the presence of stress and curve *b* is under absence of stress

The plot (a) of Fig. 2.9 exhibits the ER for the doping superlattices of stressed HD InSb in the presence of the stress while the plot (b) shows the same in the absence of the stress. In the presence of the stress, the magnitude of the ER is being increased as compared with the same under stress free condition.

2.4 Open Research Problems

R.2.1 Investigate the ER in the presence of an arbitrarily oriented non-quantizing magnetic field for nipi structures of HD nonlinear optical semiconductors by including the electron spin. Study all the special cases for HD III-V, ternary and quaternary materials in this context.

R.2.2 Investigate the ERs in nipi structures of HD IV-VI, II-VI and stressed Kane type compounds in the presence of an arbitrarily oriented non-quantizing magnetic field by including the electron spin.

R.2.3 Investigate the ER for nipi structures of all the materials as stated in R.2.1.

R.2.4 Investigate the ER for all the problems from R.2.1 to R.2.3 in the presence of an additional arbitrarily oriented electric field.

R.2.5 Investigate the ER for all the problems from R.2.1 to R.2.3 in the presence of arbitrarily oriented crossed electric and magnetic fields.

References

1. N.G. Anderson, W.D. Laidig, R.M. Kolbas, Y.C. Lo, J. Appl. Phys. **60**, 2361 (1986)
2. F. Capasso, Semiconductors Semimetals **22**, 2 (1985)
3. F. Capasso, K. Mohammed, A.Y. Cho, R. Hull, A.L. Hutchinson, Appl. Phys. Letts. **47**, 420 (1985)
4. F. Capasso, R.A. Kiehl, J. Appl. Phys. **58**, 1366 (1985)
5. K. Ploog, G.H. Doheler, Adv. Phys. **32**, 285 (1983)
6. F. Capasso, K. Mohammed, A.Y. Cho, Appl. Phys. Lett. **48**, 478 (1986)
7. R. Grill, C. Metzner, G.H. Döhler, Phys. Rev. B **63**, 235316 (2001)
8. A.R. Kost, M.H. Jupina, T.C. Hasenberg, E.M. Garmire, J. Appl. Phys. **99**, 023501 (2006)
9. A.G. Smirnov, D.V. Ushakov, V.K. Kononenko, Proc. SPIE **4706**, 70 (2002)
10. D.V. Ushakov, V.K. Kononenko, I.S. Manak, Proc. SPIE **4358**, 171 (2001)
11. J.Z. Wang, Z.G. Wang, Z.M. Wang, S.L. Feng, Z. Yang, Phys. Rev. B **62**, 6956 (2000)
12. A.R. Kost, L. West, T.C. Hasenberg, J.O. White, M. Matloubian, G.C. Valley, Appl. Phys. Lett. **63**, 3494 (1993)
13. S. Bastola, S.J. Chua, S.J. Xu, J. Appl. Phys. **83**, 1476 (1998)
14. Z.J. Yang, E.M. Garmire, D. Doctor, J. Appl. Phys. **82**, 3874 (1997)
15. G.H. Avetisyan, V.B. Kulikov, I.D. Zalevsky, P.V. Bulaev, Proc. SPIE **2694**, 216 (1996)
16. U. Pfeiffer, M. Kneissl, B. Knüpfer, N. Müller, P. Kiesel, G.H. Döhler, J.S. Smith, Appl. Phys. Lett. **68**, 1838 (1996)
17. H.L. Vaghjiani, E.A. Johnson, M.J. Kane, R. Grey, C.C. Phillips, J. Appl. Phys. **76**, 4407 (1994)
18. P. Kiesel, K.H. Gulden, A. Hoefler, M. Kneissl, B. Knuepfer, S.U. Dankowski, P. Riel, X.X. Wu, J.S. Smith, G.H. Doehler, Proc. SPIE **1985**, 278 (1993)
19. G.H. Doheler, Phys. Script. **24**, 430 (1981)
20. S. Mukherjee, S.N. Mitra, P.K. Bose, A.R. Ghatak, A. Neoigi, J.P. Banerjee, A. Sinha, M. Pal, S. Bhattacharya, K.P. Ghatak, J. Compu, Theor. Nanosc. **4**, 550 (2007)
21. N. Paitya, K.P. Ghatak, J. Adv. Phys. **1**, 161 (2012)
22. N. Paitya, S. Bhattacharya, D. De, K.P. Ghatak, Adv. Sci. Eng. Med. **4**, 96 (2012)
23. S. Bhattacharya, D. De, S.M. Adhikari, K.P. Ghatak, Superlatt. Microst. **51**, 203 (2012)
24. D. De, S. Bhattacharya, S.M. Adhikari, A. Kumar, P.K. Bose, K.P. Ghatak, Beilstein J. Nanotech. **2**, 339 (2012)
25. D. De, A. Kumar, S.M. Adhikari, S. Pahari, N. Islam, P. Banerjee, S.K. Biswas, S. Bhattacharya, K.P. Ghatak, Superlattices Microstruct. **47**, 377 (2010)
26. S. Pahari, S. Bhattacharya, S. Roy, A. Saha, D. De, K.P. Ghatak, Superlattices Microstruct. **46**, 760 (2009)
27. S. Pahari, S. Bhattacharya, K.P. Ghatak, J. Comput. Theor. Nanosci. **6**, 2088 (2009)
28. S.K. Biswas, A.R. Ghatak, A. Neogi, A. Sharma, S. Bhattacharya, K.P. Ghatak, Phys. E: Low-dimen. Sys. Nanostruct. **36**, 163 (2007)
29. L.J. Singh, S. Choudhury, D. Baruah, S.K. Biswas, S. Pahari, K.P. Ghatak, Phys. B: Condens. Matter **368**, 188 (2005)
30. S. Chowdhary, L.J. Singh, K.P. Ghatak, Phys. B: Condens. Matter **365**, 5 (2005)
31. L.J. Singh, S. Choudhary, A. Mallik, K.P. Ghatak, J. Comput. Theor. Nanosci **2**, 287 (2005)
32. K.P. Ghatak, J. Mukhopadhyay, J.P. Banerjee, SPIE Proc Ser **4746**, 1292 (2002)
33. K.P. Ghatak, S. Dutta, D.K. Basu, B. Nag, Il Nuovo Cimento D **20**, 227 (1998)
34. K.P. Ghatak, D.K. Basu, B. Nag, J. Phys. Chem. Solids **58**, 133 (1997)
35. K.P. Ghatak, B. De, Mater. Resour. Soc. Proc. **300**, 513 (1993)
36. K.P. Ghatak, B. Mitra, Il Nuovo Cimento D **15**, 97 (1993)
37. K.P. Ghatak, Int. Soc. Opt. Photon. Proc. Soc. Photo Opt. Instrum. Eng. **1626**, 115 (1992)
38. K.P. Ghatak, A. Ghoshal, Phys. Stat. Sol. (b) **170**, K27 (1992)

39. K.P. Ghatak, S. Bhattacharya, S.N. Biswas, Proc. Soc. Photo Opt. Instrum. Eng. **836**, 72 (1988)
40. K.P. Ghatak, A. Ghoshal, S.N. Biswas, M. Mondal, Proc. Soc. Photo Opt. Instrum. Eng. **1308**, 356 (1990)
41. K.P. Ghatak, G. Mazumder, Proc. Mater. Res. Soc. **484**, 679 (1998)
42. K.P. Ghatak, B. De, Defect Eng. Semi. Growth, Process Device Tech. Symp., Mat. Res. Soc. **262**, 911 (1992)
43. S.N. Biswas, K.P. Ghatak, Int. J. Electr. Theor. Exp. **70**, 125 (1991)
44. B. Mitra, K.P. Ghatak, Phys. Lett. A **146**, 357 (1990)
45. B. Mitra, K.P. Ghatak, Phys. Lett. A **142**, 401 (1989)
46. K.P. Ghatak, B. Mitra, A. Ghoshal, Phy. Stat. Sol. (b) **154**, K121 (1989)
47. B. Mitra, K.P. Ghatak, Phys. Stat. Sol. (b). **149**, K117 (1988)
48. K.P. Ghatak, S.N. Biswas, Proc. Soc. Photo Optical Instrum. Eng. **792**, 239 (1987)
49. S. Bhattacharyya, K.P. Ghatak, S. Biswas, OE/Fibers' 87. Int. Soc. Opt. Photon. **836**, 73 (1988)
50. M. Mondal, K.P. Ghatak, Czech. J. Phys. B. **36**, 1389 (1986)
51. K.P. Ghatak, A.N. Chakravarti, Phys. Stat. Sol. (b). **117**, 707 (1983)

Chapter 3
The ER in Accumulation and Inversion Layers of Non-parabolic Semiconductors

3.1 Introduction

It is well known that the electrons in bulk semiconductors in general, have three dimensional freedom of motion. When, these electrons are confined in a one dimensional potential well whose width is of the order of the carrier wavelength, the motion in that particular direction gets quantized while that along the other two directions remains as free. Thus, the energy spectrum appears in the shape of discrete levels for the one dimensional quantization, each of which has a continuum for the two dimensional free motion. The transport phenomena of such one dimensional confined carriers have recently studied [1–34] with great interest. For the metal-oxide-semiconductor (MOS) structures, the work functions of the metal and the semiconductor substrate are different and the application of an external voltage at the metal-gate causes the change in the charge density at the oxide semiconductor interface leading to a bending of the energy bands of the semiconductor near the surface. As a result, a one dimensional potential well is formed at the semiconductor interface. The spatial variation of the potential profile is so sharp that for considerable large values of the electric field, the width of the potential well becomes of the order of the de Broglie wavelength of the carriers. The Fermi energy, which is near the edge of the conduction band in the bulk, becomes nearer to the edge of the valance band at the surface creating accumulation layers. The energy levels of the carriers bound within the potential well get quantized and form electric subbands. Each of the subband corresponds to a quantized level in a plane perpendicular to the surface leading to a quasi two dimensional electron gas. Thus, the extreme band bending at low temperature allows us to observe the quantum effects at the surface. Though considerable work has already been done, nevertheless it appears from the literature that the ER in accumulation layers of non-parabolic semiconductors has yet to be investigated in details. For the purpose of comparison we shall also study the ER for inversion layers of non-parabolic compounds.

In what follows in Sect. 3.2.1, of the theoretical background, the ER in accumulation and Inversion layers of nonlinear optical semiconductors has been

K. P. Ghatak and S. Bhattacharya, *Heavily-Doped 2D-Quantized Structures and the Einstein Relation*, Springer Tracts in Modern Physics 260,
DOI 10.1007/978-3-319-08380-3_3, © Springer International Publishing Switzerland 2015

studied under weak electric field limit. The Sect. 3.2.2 contains the results for accumulation and Inversion layers of III–V, ternary and quaternary semiconductors for the weak electric field limit whose bulk electrons obey the three and the two band models of Kane together with parabolic energy bands and they form the special cases of Sect. 3.2.1. The Sect. 3.2.3 contains the study of the ER for accumulation and Inversion layers of II–VI semiconductors, which is valid for all values of electric field. The Sects. 3.2.4 and 3.2.5 contain the study of the ER in accumulation and Inversion layers of IV–VI and stressed semiconductors respectively. The Sect. 3.2.6 contains the study of the ER in accumulation and Inversion layers of Ge. The Sect. 3.3 contains the results and discussion of this chapter. The last Sect. 3.4 contains open research problems of this chapter.

3.2 Theoretical Background

3.2.1 The ER in Accumulation and Inversion Layers of Non-linear Optical Semiconductors

In the presence of a surface electric field F_s along z direction and perpendicular to the surface, (1.26) assumes the form

$$\frac{\hbar^2 k_z^2}{2m_{\parallel}^*} + \frac{\hbar^2 k_s^2}{2m_{\perp}^*}\frac{T_{21}(\mathrm{E}-|e|\mathrm{F_s z},\eta_g)}{T_{22}(\mathrm{E}-|e|\mathrm{F_s z},\eta_g)} = T_{21}(\mathrm{E}-|e|\mathrm{F_s z},\eta_g) \tag{3.1}$$

where, for this chapter, E represents the electron energy as measured from the edge of the conduction band at the surface in the vertically upward direction.

The quantization rule for 2D carriers in this case, is given by [5]

$$\int_0^{z_t} k_z dz = \frac{2}{3}(S_i)^{3/2} \tag{3.2}$$

where, z_t is the classical turning point and S_i is the zeros of the Airy function $(Ai\,(-S_i) = 0)$.

Using (3.1) and (3.2) leads to the dispersion relation of the 2D electrons in accumulation layers of HD non-linear optical materials under the condition of weak electric field limit as

$$\frac{\hbar^2 k_s^2}{2m_{\parallel}^*} = L_6(E, i, \eta_g) \tag{3.3a}$$

where $L_6(E,i,\eta_g) = \dfrac{T_{21}(E,\eta_g) - L_3(E,i,\eta_g)}{L_4(E,i,\eta_g)}$, $L_3(E,i,\eta_g) = S_i\left[T'_{21}(E,\eta_g)\right]^{2/3}\left[\dfrac{\hbar|e|F_s}{\sqrt{2m_{||}^*}}\right]^{2/3}$

and

$L_4(E,i,\eta_g) = \left[\frac{T_{21}(E,\eta_g)}{T_{22}(E,\eta_g)} + L_3(E,i,\eta_g)\frac{T_{21}(E,\eta_g)}{T'_{21}(E,\eta_g)T_{22}(E,\eta_g)} \cdot \frac{2}{3}\left\{\frac{T'_{21}(E,\eta_g)}{T_{21}(E,\eta_g)} - \frac{T'_{22}(E,\eta_g)}{T_{22}(E,\eta_g)}\right\}\right]$

The EEM in this case can be written as

$$m^*(E'_f, i, \eta_g) = m_{||}^* \quad \textit{Real part of } \left[L'_6(E'_f, i, \eta_g)\right] \tag{3.3b}$$

The sub-band energy E_i can be determined from the equation

$$0 = \textit{Real part of } L_6(E_i, i, \eta_g) \tag{3.3c}$$

The surface electron concentration in the regime of very low temperatures where the quantum effects become prominent can be written as

$$n_s = 2g_v \textit{ Real part of the } \sum_{i=0}^{i_{\max}}\left[\left[\frac{m_\perp^*}{2\pi\hbar^2}L_6\left(E'_f, i, \eta_g\right)\right] + \frac{1}{(2\pi)^3}\frac{2m_\perp^*\sqrt{2m_{||}^*}}{\hbar^3}t_i \left[T_{22}(E_{FB},\eta_g)\sqrt{T_{21}(E_{FB},\eta_g)}\right]\right] \tag{3.4}$$

where $t_i = \frac{E_{i\max}}{eF_s(1+i_{\max})}$, $E_{i\max}$ is the root of the Real part of the equation

$$T_{21}(E_{i\max},\eta_g) - L_3(E_{i\max}, i_{\max}, \eta_g) = 0 \tag{3.5}$$

$E'_f = eV_g - \frac{e^2 n_s d_{ox}}{\varepsilon_{ox}} + E_{FB}$, V_g is the gate voltage, n_s is the surface electron concentration, d_{ox} is the thickness of the oxide layer, ε_{ox} is the permittivity of the oxide layer, $F_s = \frac{en_s}{\varepsilon_{sc}}$, ε_{sc} is the semiconductor permittivity and E_{FB} should be determined from the equation

$$n_B = \frac{2g_v}{(2\pi)^3}\frac{2m_\perp^*\sqrt{2m_{||}^*}}{\hbar^3} \textit{ Real part of } \left[T_{22}(E_{FB},\eta_g)\sqrt{T_{21}(E_{FB},\eta_g)}\right] \tag{3.6}$$

where n_B is the bulk electron concentration.

The ER in this case is given by

$$\frac{D}{\mu} = \textit{Real part of } \left[\frac{n_s}{|e|}\left[\frac{\partial n_s}{\partial(E'_F - E')}\right]^{-1}\right] \tag{3.7}$$

where, E'_F is the Fermi energy and E' is the sub-band energy at the electric quantum limit respectively.

Thus using (3.4), (3.7) and the allied definitions in the electric quantum limit, we can study the ER in this case.

In what follows, we shall discuss the ER in accumulation layers of non-linear optical materials for the purpose of relative comparison. In the presence of a surface electric field F_s along z direction and perpendicular to the surface, the (1.2) assumes the form

$$\psi_1(E-|e|F_sz)=\psi_2(E-|e|F_sz)k_s^2+\psi_3(E-|e|F_sz)k_z^2 \tag{3.8}$$

where $\psi_1(E)=\gamma(E), \psi_2(E)=f_1(E)$ and $\psi_3(E)=f_2(E)$.

Using (3.2) and (3.8), under the weak electric field limit, one can write,

$$\int_0^{z_t}\sqrt{A_7(E)-|e|F_szD_7(E)}dz=\frac{2}{3}(S_i)^{3/2} \tag{3.9}$$

in which, $A_7(E)\equiv\left[\frac{\psi_1(E)-\psi_2(E)k_s^2}{\psi_3(E)}\right]D_7(E)\equiv[B_7(E)-A_7(E)C_7(E)]$,

$B_7(E)\equiv\left[\frac{(\psi_1(E))'-(\psi_2(E))'k_s^2}{\psi_3(E)}\right]$ and $C_7(E)\equiv\left[\frac{(\psi_3(E))'}{\psi_3(E)}\right]$.

Thus, the 2D electron dispersion law in inversion layers of nonlinear optical materials under the weak electric field limit can approximately be written as

$$\psi_1(E)=P_7(E,i)k_s^2+Q_7(E,i) \tag{3.10}$$

where,

$$P_7(E,i)\equiv\left[\psi_2(E)-\left(\frac{2t_2(E)}{3[t_1(E)]^{1/3}}\right)\psi_3(E)S_i(|e|F_s)^{2/3}\right],$$

$$t_2(E)\equiv\left[\frac{[\psi_2(E)]'}{\psi_3(E)}-\left(\frac{\psi_2(E)[\psi_3(E)]'}{[\psi_3(E)]^2}\right)\right],$$

$$t_1(E)\equiv\left[\frac{[\psi_1(E)]'}{\psi_3(E)}-\left(\frac{\psi_1(E)[\psi_3(E)]'}{[\psi_3(E)]^2}\right)\right]$$ and $Q_7(E,i)\equiv S_i\psi_3(E)[|e|F_st_1(E)]^{2/3}$.

The EEM in the x–y plane can be expressed as

$$m^*(E_{Fiw},i)=\left(\frac{\hbar^2}{2}\right)G_7(E,i)\Big|_{E=E_{Fiw}} \tag{3.11}$$

where, $G_7(E,i) \equiv [P_7(E,i)]^{-2}[P_7(E,i)\{(\psi_1(E))' - (Q_7(E,i))'\} - \{\psi_1(E) - (Q_7(E,i))\} (P_7(E,i))']$ and E_{Fiw} is the Fermi energy under the weak electric field limit as measured from the edge of the conduction band at the surface in the vertically upward direction. Thus, we observe that the EEM is the function of subband index, the Fermi energy and other band constants due to the combined influence of the crystal filed splitting constant and the anisotropic spin-orbit splitting constants respectively.

The subband energy $(E_{n_{iw1}})$ in this case can be obtained from (3.10) as

$$\psi_1(E_{n_{iw1}}) = Q_7(E_{n_{iw1}}, i) \tag{3.12}$$

The general expression of the 2D total DOS density-of-states function in this case can be written as

$$N_{2D_i}(E) = \frac{2g_v}{(2\pi)^2} \sum_{i=0}^{i_{\max}} \frac{\partial}{\partial E}[A(E,i)H(E - E_{n_i})] \tag{3.13}$$

where, $A(E,i)$ is the area of the constant energy 2D wave vector space for inversion layers and E_{n_i} is the corresponding subband energy.

Using (3.10) and (3.13), the total 2D DOS function under the weak electric field limit weak electric field limit can be expressed as

$$N_{2D_i}(E) = \frac{gv}{(2\pi)} \sum_{i=0}^{i_{\max}} [G_7(E,i)H(E - E_{n_{iw1}})] \tag{3.14}$$

Using (3.14) and the Fermi-Dirac occupation probability factor, the 2D surface electron concentration in inversion of tetragonal materials under the weak electric field limit (n_{2Dw}) can be written as

$$n_{2Dw} = g_v(2\pi)^{-1} \sum_{i=0}^{i_{\max}} [P_{7w}(E_{Fiw}, i) + Q_{7w}(E_{Fiw}, i)] \tag{3.15}$$

where, $P_{7w}(E_{Fiw}, i) \equiv [\psi_1(E_{Fiw}, i) - Q_7(E_{Fiw}, i)]\{P_7(E_{Fiw}, i)\}^{-1}$ and $Q_{7w}(E_{Fiw}, i) \equiv \sum_{r=1}^{s} \{L(r)[P_7(E_{Fiw}, i)]\}$ and

$$F_s \equiv \frac{|e|n_{2Dw}}{\varepsilon_{sc}}$$

Thus the surface electron concentration under the weak electric field quantum limit assumes the form

$$\overline{n}_{2Dw} = g_v(2\pi)^{-1}\left[P_{7w}(\overline{E}_{Fw}, 0)\right] \tag{3.16}$$

where, $\overline{E}_{Fw}$ is the Fermi energy under the weak electric field quantum limit as measured from the edge of the conduction band at the surface.

The ER in this case can be written as

$$\frac{D}{\mu} = \frac{\overline{n}_{2Dw}}{|e|}\left[\frac{\partial \overline{n}_{2Dw}}{\partial(E'_{Fw} - E_{n_{0w}})}\right]^{-1} \quad (3.17)$$

where $E_{n_{0w}}$ is the electric sub-band energy at the electric quantum limit by using (3.16), (3.17) and the allied definitions, we can study ER in inversion layers of non-linear optical materials at the electric quantum limit and extreme degeneracy.

3.2.2 The ER in Accumulation and Inversion Layers of III–V, Ternary and Quaternary Semiconductors

(a) Using the substitutions $\delta = 0, \Delta_{\|} = \Delta_{\perp} = \Delta$ and $m^*_{\|} = m^*_{\perp} = m_c$, (3.3a) under the condition of weak electric field limit, assumes the form

$$T_{90}(E, \eta_g) = \frac{\hbar^2 k_s^2}{2m_c} + S_i\left[\frac{\hbar|e|F_s[T_{90}(E, \eta_g)]'}{\sqrt{2m_c}}\right]^{2/3} \quad (3.18)$$

where, $T_{90}(E, \eta_g) = T_{31}(E, \eta_g) + iT_{32}(E, \eta_g)$.
(3.18) represents the dispersion relation of the 2D electrons in accumulation layers of HD III–V, ternary and quaternary materials under the weak electric field limit limit whose bulk electrons obey the HD three band model of Kane. Since the electron energy spectrum in accordance with the HD three-band model of Kane is complex in nature, the (3.18) will also be complex. The both complexities occur due to the presence of poles in the finite complex plane of the dispersion relation of the materials in the absence of band tails.

The EEM can be expressed as

$$m^*(E'_f, i, \eta_g) = m_c \textit{ Real part of } P'_{3HD}(E'_f, i, \eta_g) \quad (3.19)$$

where, $P_{3HD}(E'_f, i, \eta_g) = \left[T_{90}(E'_f, \eta_g) - S_i\left[\frac{\hbar|e|F_s[T_{90}(E'_f, \eta_g)]'}{\sqrt{2m_c}}\right]^{2/3}\right]$

Thus, one can observe that the EEM is a function of the sub-band index, surface electric field, the Fermi energy and the other spectrum constants due to the combined influence of E_g and Δ.

The sub-band energy E_{i1} is given by

$$0 = \textit{Real part of } \left[T_{90}(E_{i1}, \eta_g) - S_i[\hbar|e|F_s[T_{90}(E_{i1}, \eta_g)]'.(2m_c)^{-1/2}]^{2/3}\right] \quad (3.20)$$

The DOS function can be written as

$$N_{2D_i}(E) = \frac{m_c g_v}{\pi \hbar^2} \sum_{i=0}^{i_{\max}} \left[P_{3HD}(E, i, \eta_g) H(E - E_{i1}) \right] \quad (3.21)$$

Thus the DOS function is complex in nature.

The surface electron concentration is given by

$$n_s = g_v \; \textit{Real part of the} \sum_{i=0}^{i_{\max}} \left[\left[\frac{m_c}{\pi \hbar^2} \mathrm{P}_{3HD}(E_f', i, \eta_g) \right] + \frac{1}{3\pi^2} \left(\frac{2m_c}{\hbar^2} \right)^{3/2} t_i \left[\mathrm{T}_{90}(\mathrm{E}_{\mathrm{FB}}, \eta_g) \right]^{3/2} \right] \quad (3.22a)$$

where E_{FB} should be determined from the following equation

$$n_B = \frac{g_v}{3\pi^2} \left(\frac{2m_c}{\hbar^2} \right)^{3/2} \; \textit{Real part of} \; \left[\mathrm{T}_{90}(\mathrm{E}_{\mathrm{FB}}, \eta_g) \right]^{3/2} \quad (3.22b)$$

Thus using (3.7), (3.22b) and the allied definitions, we can study the ER in this case in the electric quantum limit.

Using the substitutions $\delta = 0$, $\Delta_{\|} = \Delta_{\perp} = \Delta$ and $m_{\|}^* = m_{\perp}^* = m_c$, (3.10) under the condition of weak electric field limit, assumes the form

$$I_{11}(E) = \frac{\hbar^2 k_s^2}{2m_c} + S_i \left[\frac{\hbar |e| F_s [I_{11}(E)]'}{\sqrt{2m_c}} \right]^{2/3} \quad (3.23)$$

(3.23) represents the dispersion relation of the 2D electrons in inversion layers of III–V, ternary and quaternary materials under the weak electric field limit whose bulk electrons obey the three band model of Kane.

The EEM can be expressed as

$$m^*(E_{Fiw}, i) = m_c [P_3(E, i)]|_{E = E_{Fiw}} \quad (3.24)$$

where, $P_3(E, i) \equiv \left\{ [I_{11}(E)]' - \left\{ \frac{2}{3} S_i \left[\frac{\hbar |e| F_s}{\sqrt{2m_c}} \right]^{2/3} \{[I_{11}(E)]'\}^{-1/3} [I_{11}(E)]'' \right\} \right\}$.

Thus, one can observe that the EEM is a function of the subband index, surface electric field, the Fermi energy and the other spectrum constants due to the combined influence of E_g and Δ.

The subband energy $(E_{n_{iw2}})$ in this case can be obtained from the (3.23) as

$$I_{11}(E_{n_{iw2}}) = S_i \left[\frac{\hbar |e| F_s [I_{11}(E_{n_{iw2}})]'}{\sqrt{2m_c}} \right]^{2/3} \quad (3.25)$$

Thus the 2D total DOS function in weak electric field limit can be expressed as

$$N_{2D_i}(E) = \frac{m_c g_v}{\pi \hbar^2} \sum_{i=0}^{i_{\max}} [P_3(E,i) H(E - E_{n_{iw2}})] \tag{3.26}$$

Using (3.26) and the occupation probability, the n_{2Dw} in the present case can be written as

$$n_{2Dw} = \frac{g_v m_c}{\pi \hbar^2} \sum_{i=0}^{i_{\max}} [P_{4w}(E_{Fiw}, i) + Q_{4w}(E_{Fiw}, i)] \tag{3.27}$$

where, $P_{4w}(E_{Fiw}, i) \equiv \left\{ I_{11}(E_{Fiw}) - S_i \left[\frac{\hbar e F_s [I_{11}(E_{Fiw})]'}{\sqrt{2m_c}}\right]^{2/3} \right\}$ and $Q_4(E_{Fiw}, i) \equiv \sum_{r=1}^{s} \{L(r)$ $[P_{4w}(E_{Fiw}, i)]\}$.

The surface electron concentration under the weak electric field quantum limit assumes the form

$$\bar{n}_{2Dw} = \frac{g_v m_c}{\pi \hbar^2} [P_{4w}(\bar{E}_{Fw}, 0)] \tag{3.28}$$

Using (3.17), (3.28) and the allied definitions, we can study the ER in this case.

(b) Using the constraints $\Delta \gg E_g$ or $\Delta \ll E_g$, the (3.18) under the low electric field limit assumes the form

$$\gamma_2(E, \eta_g) = \frac{\hbar^2 k_s^2}{2m_c} + S_i \left[\frac{\hbar |e| F_s [\gamma_2(E, \eta_g)]'}{\sqrt{2m_c}}\right]^{2/3} \tag{3.29}$$

The (3.29) represents the dispersion relation of the 2D electrons in accumulation layers of HD III–V, ternary and quaternary materials under the weak electric field limit whose bulk electrons obey the HD two band model of Kane.

The EEM can be expressed as

$$m^*(E'_f, i, \eta_g) = m_c P'_{3HD1}(E'_f, i, \eta_g) \tag{3.30}$$

where, $P_{3HD1}(E'_f, i, \eta_g) = \left[\gamma_2(E'_f, \eta_g) - S_i \left[\frac{\hbar |e| F_s [\gamma_2(E'_f, \eta_g)]'}{\sqrt{2m_c}}\right]^{2/3} \right]$.

Thus, one can observe that the EEM is a function of the sub-band index, surface electric field, the Fermi energy and the other spectrum constants due to the combined influence of E_g and Δ.

The sub-band energy E_{i1} is given by

$$0 = \left[\gamma_2(E_{i2}, \eta_g) - S_i\left[\hbar|e|F_s\left[\gamma_2(E_{i2}, \eta_g)\right]'.(2m_c)^{-1/2}\right]^{2/3}\right] \tag{3.31}$$

The DOS function can be written as

$$N_{2D_i}(E) = \frac{m_c g_v}{\pi\hbar^2}\sum_{i=0}^{i_{\max}}\left[P'_{3HD1}(E, i, \eta_g)H(E - E_{i2})\right] \tag{3.32}$$

Thus the DOS function is complex in nature.
The surface electron concentration is given by

$$n_S = g_v\sum_{i=0}^{i_{\max}}\left[\left[\frac{m_c}{\pi\hbar^2}\mathrm{P}_{3HD1}(E'_f, i, \eta_g)\right] + \frac{1}{3\pi^2}\left(\frac{2m_c}{\hbar^2}\right)^{3/2} t_i\left[\gamma_2(\mathrm{E_{FB}}, \eta_g)\right]^{3/2}\right] \tag{3.33a}$$

where $\mathrm{E_{FB}}$ should be determined from the following equation

$$n_B = \frac{g_v}{3\pi^2}\left(\frac{2m}{\hbar^2}\right)^{3/2}\left[\gamma_2(\mathrm{E_{FB}}, \eta_g)\right]^{3/2} \tag{3.33b}$$

Thus using (3.7), (3.33b) and the allied definitions, we can study the ER in this case in the electric quantum limit.

Using the constraints $\Delta \gg E_g$ or $\Delta \ll E_g$, the (3.23) under the low electric field limit assumes the form

$$E(1 + \alpha E) = \frac{\hbar^2 k_s^2}{2m_c} + S_i\left[\frac{\hbar|e|F_s(1 + 2\alpha E)}{\sqrt{2m_c}}\right]^{2/3} \tag{3.34}$$

For large values of i, $S_i \to \left[\frac{3\pi}{2}\left(i + \frac{3}{4}\right)\right]^{2/3}$ [5], and the (3.34) gets simplified as

$$E(1 + \alpha E) = \frac{\hbar^2 k_s^2}{2m_c} + \left[\frac{3\pi\hbar|e|F_s}{2}\left(i + \frac{3}{4}\right)\frac{(1 + 2\alpha E)}{\sqrt{2m_c}}\right]^{2/3} \tag{3.35}$$

The (3.35) was derived ***for the first time by*** Antcliffe et al. [3].
The EEM in this case is given by

$$m^*(E_{Fiw}, i) = m_c[P_6(E, i)]|_{E=E_{Fiw}} \tag{3.36}$$

where, $P_6(E, i) \equiv \left\{1 + 2\alpha E - \frac{4\alpha}{3}S_i\left[\frac{\hbar|e|F_s}{\sqrt{2m_c}}\right]^{2/3}\{1 + 2\alpha E\}^{-1/3}\right\}$.

Thus, one can observe that the EEM is a function of the subband index, surface electric field and the Fermi energy due to the presence of band non-parabolicity only.

The subband energies ($E_{n_{iw3}}$) are given by

$$E_{n_{iw3}}(1+\alpha E_{n_{iw3}}) = S_i\left[\frac{\hbar|e|F_s(1+2\alpha E_{n_{iw3}})}{\sqrt{2m_c}}\right]^{2/3} \quad (3.37)$$

The total 2D DOS function can be written as

$$N_{2D}(E) = \frac{m_c g_v}{\pi\hbar^2}\sum_{i=0}^{i_{\max}}\left\{\left[1+2\alpha E-\frac{4\alpha}{3}S_i\left[\frac{\hbar|e|F_s}{\sqrt{2m_c}}\right]^{2/3}(1+2\alpha E)^{-1/3}\right]H(E-E_{n_{iw3}})\right\} \quad (3.38)$$

Under the condition $\alpha E \ll 1$, the use of (3.38) and the Fermi-Dirac integral leads to the expression of n_{2Dw} as

$$n_{2Dw} = \left(\frac{g_v m_c k_B T}{\pi\hbar^2}\right)\sum_{i=0}^{i_{\max}}\{[1+D_i+2\alpha E_{n_{iw3}}]F_0(\eta_{iw})+2\alpha k_B T F_1(\eta_{iw})\} \quad (3.39)$$

where, $D_i \equiv \frac{4\alpha S_i}{3}\left(\frac{\hbar|e|F_s}{\sqrt{2m_c}}\right)^{2/3}$ and $\eta_{iw} \equiv \left[\frac{E_{Fiw}-E_{n_{iw3}}}{k_B T}\right]$.

For all values of αE_{Fiw}, the n_{2Dw} can be written by using (3.39) as

$$n_{2Dw} = \left(\frac{g_v m_c}{\pi\hbar^2}\right)\sum_{i=0}^{i_{\max}}[P_{5w}(E_{Fiw},i)+Q_{5w}(E_{Fiw},i)] \quad (3.40)$$

where, $P_{5w}(E_{Fiw},i) \equiv \left[E_{Fiw}(1+\alpha E_{Fiw}) - S_i\left[\frac{\hbar|e|F_s}{\sqrt{2m_c}}(1+2\alpha E_{Fiw})\right]^{2/3}\right]$
and $Q_{5w}(E_{Fiw},i) \equiv \sum_{r=1}^{s} L(r)P_{5w}(E_{Fiw},i)$.

The electron concentration under the weak electric field quantum limit assumes the form

$$\bar{n}_{2Dw} = \frac{g_v m_c}{\pi\hbar^2}\left\{\bar{E}_{Fw}(1+\alpha\bar{E}_{Fw}) - S_0\left(\frac{\hbar|e|F_s}{\sqrt{2m_c}}\right)^{2/3}(1+2\alpha\bar{E}_{Fw})^{2/3}\right\} \quad (3.41)$$

Using (3.17), (3.41) and the allied definitions, we can study the ER in this case.

(c) Using the constraints $\alpha \to 0$, the (3.29) under the low electric field limit assumes the form

$$\gamma_3(E,\eta_g) = \frac{\hbar^2 k_s^2}{2m_c} + S_i\left[\frac{\hbar|e|F_s[\gamma_3(E,\eta_g)]'}{\sqrt{2m_c}}\right]^{2/3} \quad (3.42)$$

The (3.42) represents the dispersion relation of the 2D electrons in accumulation layers of HD III–V, ternary and quaternary materials under the weak electric field limit whose bulk electrons obey the HD parabolic band model.

The EEM can be expressed as

$$m^*(E'_f, i, \eta_g) = m_c P'_{3HD2}(E'_f, i, \eta_g) \tag{3.43}$$

where, $P_{3HD2}\left(E'_f, i, \eta_g\right) = \left[\gamma_3(E'_f, \eta_g) - S_i\left[\frac{\hbar|e|F_s[\gamma_3(E'_f,\eta_g)]'}{\sqrt{2m_c}}\right]^{2/3}\right]$

Thus, one can observe that the EEM is a function of the sub-band index, surface electric field, the Fermi energy and the other spectrum constants due to the combined influence of E_g and Δ.

The sub-band energy E_{i1} is given by

$$0 = \left[\gamma_3(E_{i2}, \eta_g) - S_i\left[\hbar|e|F_s[\gamma_3(E_{i2}, \eta_g)]'\cdot(2m_c)^{-1/2}\right]^{2/3}\right] \tag{3.44}$$

The DOS function can be written as

$$N_{2D_i}(E) = \frac{m_c g_v}{\pi\hbar^2}\sum_{i=0}^{i_{\max}}\left[P'_{3HD2}(E, i, \eta_g)H(E - E_{i3})\right] \tag{3.45}$$

The surface electron concentration is given by

$$n_s = g_v\sum_{i=0}^{i_{\max}}\left[\left[\frac{m_c}{\pi\hbar^2}\mathrm{P}_{3HD2}(E'_f, i, \eta_g)\right] + \frac{1}{3\pi^2}\left(\frac{2m_c}{\hbar^2}\right)^{3/2} t_i\left[\gamma_3(\mathrm{E_{FB}}, \eta_g)\right]^{3/2}\right] \tag{3.46a}$$

where E_{FB} should be determined from the following equation

$$n_B = \frac{g_v}{3\pi^2}\left(\frac{2m_c}{\hbar^2}\right)^{3/2}\left[\gamma_3(\mathrm{E_{FB}}, \eta_g)\right]^{3/2} \tag{3.46b}$$

Thus using (3.7), (3.46a) and the allied definitions, we can study the ER in this case.

For $\alpha \to 0$, as for inversion layers, whose bulk electrons are defined by the parabolic energy bands, from (3.34), we can write,

$$E = \frac{\hbar^2 k_s^2}{2m_c} + S_i\left[\frac{\hbar|e|F_s}{\sqrt{2m_c}}\right]^{2/3} \tag{3.47}$$

The (3.47) is valid for all values of the surface electric field [1].

The electric subband energy $(E_{n_{i4}})$ assumes the form, from (3.47) as

$$E_{n_{i4}} = S_i \left[\frac{\hbar|e|F_s}{\sqrt{2m_c}}\right]^{2/3} \tag{3.48}$$

The total DOS function can be written using (3.48) as

$$N_{2D}(E) = \frac{m_c g_v}{\pi\hbar^2} \sum_{i=0}^{i_{\max}} H(E - E_{n_{i4}}) \tag{3.49}$$

The use of (3.49) leads to the expression of n_{2Di} as [1]

$$n_{2Di} = \frac{g_v m_c k_B T}{\pi\hbar^2} \sum_{i=0}^{i_{\max}} F_0(\eta_i) \tag{3.50}$$

where, $\eta_i \equiv (k_B T)^{-1}\left[E_{Fi} - S_i\left[\frac{\hbar|e|F_s}{\sqrt{2m_c}}\right]^{2/3}\right]$, E_{Fi} is the Fermi energy as measured from the edge of the conduction band at the surface.

Thus by using (3.50), (3.17) and the allied definitions at the electric quantum limit we can study the ER in this case.

3.2.3 The ER in Accumulation and Inversion Layers of II–VI Semiconductors

The use of (1.140) and (3.2) leads to the expression of the quantization integral as

$$\frac{\sqrt{2m_{\|}^*}}{\hbar} \int_0^{z_t} \left[\gamma_3(\mathrm{E}, \eta_g) - |e|F_s z\gamma_3'(\mathrm{E}, \eta_g) - a_0' k_s^2 \mp (\bar{\lambda}_0)k_s\right]^{1/2} dz = \frac{2}{3}(S_i)^{3/2} \tag{3.51}$$

where, $z_t \equiv \left(|e|F_s\gamma_3'(\mathrm{E}, \eta_g)\right)^{-1}\left[\gamma_3(\mathrm{E}, \eta_g) - a_0' k_s^2 \mp (\bar{\lambda}_0)k_s\right]$.

Therefore, the 2D electron dispersion law for accumulation layers of HD II–VI semiconductors can be expressed as

$$\gamma_3(\mathrm{E}, \eta_g) = a_0' k_s^2 \pm (\bar{\lambda}_0)k_s + S_i\left(\frac{\hbar|e|F_s\gamma_3'(\mathrm{E}, \eta_g)}{\sqrt{2m_{\|}^*}}\right)^{2/3} \tag{3.52}$$

The area of the 2D surface as enclosed by the (3.52) can be expressed as

$$A(E, \eta_g, i) = \frac{\pi}{a_0'^2} \Delta_{10}(E, \eta_g, i)$$

where

$$\Delta_{10}(E, \eta_g, i) = \left[(\bar{\lambda}_0)^2 - 2a_0' \left\{ -\gamma_3(E, \eta_g) + S_i \left(\frac{\hbar e F_s \gamma_3'(E, \eta_g)}{\sqrt{2m_{\|}^*}} \right)^{2/3} \right\} \right] \quad (3.53)$$

The EEM in this case assumed the form

$$m^*(E_f', \eta_g, i) = m_\perp^* \Delta_{10}'(E_f', \eta_g, i) \quad (3.54)$$

The sub-band energy E_{i2} can be written as

$$\gamma_3(E_{i2}, \eta_g) = S_i \left(\frac{\hbar |e| F_s \gamma_3'(E_{i2}, \eta_g)}{\sqrt{2m_{\|}^*}} \right)^{2/3} \quad (3.55)$$

The surface electron concentration can be written as

$$n_s = g_v \sum_{i=0}^{i_{\max}} \left[\left\{ \left(\frac{m_\perp^*}{\pi \hbar^2} \right) \left[\Delta_{10}(E_f', i, \eta_g) + \Delta_{11}(E_f', i, \eta_g) \right] \right\} + \frac{t_i}{2} \left(\frac{k_B T}{\pi b_0'} \right)^{\frac{3}{2}} \left(\frac{b_0'}{a_0'} \right) \left[F_{\frac{1}{2}}\left(\frac{E_{FB}}{k_B T}\right) + \frac{(\bar{\lambda}_0)^2}{2a_0' k_B T} F_{\frac{-1}{2}}\left(\frac{E_{FB}}{k_B T} \right) \right] \right] \quad (3.56)$$

The E_{FB} can be determined from the following equation

$$n_B = \frac{g_v}{2} \left(\frac{k_B T}{\pi b_0'} \right)^{\frac{3}{2}} \left(\frac{b_0'}{a_0'} \right) \left[F_{\frac{1}{2}}\left(\frac{E_{FB}}{k_B T} \right) + \frac{(\bar{\lambda}_0)^2}{2a_0' k_B T} F_{\frac{-1}{2}}\left(\frac{E_{FB}}{k_B T} \right) \right] \quad (3.57)$$

$$\Delta_{11}\left(E_f', i, \eta_g\right) = \sum_{r=1}^{s} L(r) \left[\Delta_{10}\left(E_f', i, \eta_g\right) \right]$$

Thus using (3.7), (3.56) and the allied definitions, we can study the ER in this case.

The use of (1.140) and (3.2) leads to the expression of the quantization integral in this case as

$$\frac{\sqrt{2m_{||}^{*}}}{\hbar}\int_{0}^{z_t}\left[E-|e|F_s z-a_0' k_s^2 \mp (\bar{\lambda}_0)k_s\right]^{1/2}dz=\frac{2}{3}(S_i)^{3/2} \tag{3.58}$$

where, $z_t \equiv (|e|F_s)^{-1}\left[E-a_0' k_s^2 \mp (\bar{\lambda}_0)k_s\right]$.

Therefore, the 2D electron dispersion law for inversion layers of II–VI semiconductors can be expressed for all values of F_s as

$$E=a_0' k_s^2 \pm (\bar{\lambda}_0)k_s + S_i\left(\frac{\hbar|e|F_s}{\sqrt{2m_{||}^{*}}}\right)^{2/3} \tag{3.59}$$

The area of the 2D surface as enclosed by the (3.59) can be expressed as

$$A(E,i)=\frac{\pi(m_{\perp}^{*})^2}{\hbar^4}\left[\left\{2(\bar{\lambda}_0)^2-\frac{2\hbar^2}{m_{\perp}^{*}}S_i\left(\frac{\hbar|e|F_s}{\sqrt{2m_{||}^{*}}}\right)^{2/3}+\frac{2\hbar^2 E}{m_{\perp}^{*}}\right\}\right.$$
$$\left.-2(\bar{\lambda}_0)\left[(\bar{\lambda}_0)^2-\frac{2\hbar^2}{m_{\perp}^{*}}S_i\left(\frac{\hbar|e|F_s}{\sqrt{2m_{||}^{*}}}\right)^{2/3}+\frac{2\hbar^2 E}{m_{\perp}^{*}}\right]^{1/2}\right] \tag{3.60}$$

The EEM is given by

$$m^{*}(E_{Fi},i)=m_{\perp}^{*}\left[1-\frac{\rho_{71}}{\sqrt{E_{Fi}+\rho_{72}}}\right] \tag{3.61}$$

where, E_{Fi} is the Fermi energy in this case, $\rho_{71}\equiv\frac{\bar{\lambda}_0}{2\sqrt{a_0'}}$ and $\rho_{72}\equiv\left[(\rho_{71})^2-\left(\frac{\hbar|e|F_s}{\sqrt{2m_{||}^{*}}}\right)^{2/3}\right]$.

Thus, the EEM depends on both the Fermi energy and the subband index due to the presence of the term $\bar{\lambda}_0$.

The subband energy $(E_{n_{i6}})$ can be written as

$$E_{n_{i6}}=S_i\left(\frac{\hbar|e|F_s}{\sqrt{2m_{||}^{*}}}\right)^{2/3} \tag{3.62}$$

The total 2D DOS function can be written as

$$N_{2D_i}(E) = \frac{m_\perp^* g_v}{\pi\hbar^2}\sum_{i=0}^{i_{\max}}\left\{\left[1-\frac{\rho_{71}}{\sqrt{E+\rho_{72}}}\right]H(E-E_{ni6})\right\} \tag{3.63}$$

The surface electron concentration assumes the form

$$n_{2Di}\frac{g_v m_\perp^* k_B T}{\pi\hbar^2}\left\{\sum_{i=0}^{i_{\max}}\left[F_0(\eta_i)-\left\{\frac{\bar{\lambda}_0 f_7(E_{Fi},i)}{2\sqrt{a_0' k_B T}}\right\}\right]\right\} \tag{3.64}$$

where, $\eta_i \equiv \left[\frac{E_{Fi}-E_{n_{i6}}}{k_B T}\right]$, $f_7(E_{Fi},i) \equiv \left[2\left[\sqrt{\eta_i+\delta_{72}}-\sqrt{\delta_{72}}\right]+\sum_{r=1}^{s}\ \{2(1-2^{1-2r})\right.$
$\left.\zeta(2r)\frac{(-1)^{2r-1}(2r-1)!}{(\eta_i+\delta_{72})^{2r}}\}\right]$
and $\delta_{72} \equiv \frac{(\bar{\lambda}_0)^2}{4a_0' k_B T}$.

The electron concentration in the electric field quantum limit can be obtained by

$$\bar{n}_{2Di} = \frac{g_v}{2a_0'}\left[\bar{E}_{F0}+\frac{(\bar{\lambda}_0)^2}{4a_0'}-\bar{E}_0-\left\{\frac{\bar{\lambda}_0}{2\sqrt{a_0'}}\left[\bar{E}_{F0}+\frac{(\bar{\lambda}_0)^2}{4a_0'}-\bar{E}_0\right]^{1/2}\right\}\right] \tag{3.65}$$

Using (3.65), (3.17) and the allied definitions at the electric quantum limit we can study the ER in inversion layers of II–VI semiconductors.

3.2.4 The ER in Accumulation and Inversion Layers of IV–VI Semiconductors

The dispersion relation of the conduction electrons in bulk specimens of IV–VI semiconductors in accordance with the model of Bangert and Kastner is given by

$$\begin{aligned} 2E = {} & k_s^2\left[\frac{(\bar{R})^2}{E_{g_0}(1+\alpha_1 E)}+\frac{(\bar{S})^2}{\Delta_c'(1+\alpha_2 E)}+\frac{(\bar{Q})^2}{\Delta_{g_c}''(1+\alpha_3 E)}\right] \\ & + k_z^2\left[\frac{(\bar{A})^2}{E_{g_0}(1+\alpha_1 E)}+\frac{(\bar{S}+\bar{Q})^2}{\Delta_c''(1+\alpha_3 E)}\right] \end{aligned} \tag{3.66}$$

where $(\bar{R})^2 = 2.3\times10^{-10}(ev_m)^2$, $(\bar{S})^2 = 4.6(\bar{R})^2$, $\alpha_1 = \frac{1}{E_{g0}}$, $\alpha_2 = \frac{1}{\Delta_c'}$, $\alpha_3 = \frac{1}{\Delta_c''}$
$\Delta_c'' = 3.28ev$, $\Delta_c' = 3.07ev$, $(\bar{Q})^2 = 1.3(\bar{R})^2$, $(\bar{A})^2 = 0.8\times10^{-4}(ev_m)^2$.

The electron energy spectrum in heavily doped IV–VI materials in accordance with this model can be expressed by using the methods as given in Chap. 1 as

$$2I(4) = k_s^2\left[\{c_1(\alpha_1, E, E_g) - iD_1(\alpha_1, E, E_g)\}\frac{(\overline{R})^2}{E_{g0}} + \{c_2(\alpha_2, E, E_g) - iD_2(\alpha_2, E, E)\}\frac{(\overline{S})^2}{\Delta_c'}\right.$$
$$\left.+\{c_3(\alpha_3, E, E_g) - iD_3(\alpha_3, E, E_g)\}\frac{(\overline{Q})^2}{\Delta_c''}\right] + k_z^2\left[\frac{2(\overline{A})^2}{E_{g0}}\{c_1(\alpha_1, E, E_g) - iD_1(\alpha_1, E, E_g)\}\right. \quad (3.67)$$
$$\left.+\frac{(\overline{S}+\overline{Q})^2}{\Delta_c''}\{c_3(\alpha_3, E, E_g) - iD_3(\alpha_3, E, E_g)\}\right]$$

where

$$\alpha_1 = \frac{1}{E_g},\ \alpha_2 = \frac{1}{\Delta_c'},\ \alpha_3 = \frac{1}{\Delta_c''},\ u_i = \frac{1+\alpha_i E}{\eta_g \alpha_i},$$
$$c_i(\alpha_i, E, \eta_g) = \left[\frac{2}{\alpha_i \eta_g \sqrt{\pi}}\right]\exp(-u_i^2) \times \left[\sum_{p=1}^{\infty}\{\exp(-p^2/4)(\sinh(pu_i))\}p^{-1}\right]$$
$$i = 1,2,3 \quad D_i(\alpha_i, E, \eta_g) = \left[\frac{\sqrt{\pi}}{\alpha_i \eta_g}\right]\exp(-u_i^2),$$

Therefore (3.67) can be written as,

$$F_1(E, \eta_g)k_s^2 + F_2(E, \eta_g)k_z^2 = 1 \quad (3.68)$$

where,

$$F_1(E, \eta_g) = [2\gamma_0(E, \eta_g)]^{-1}\left[\frac{(\overline{R})^2}{E_g}\{C_1(\alpha_1, E, E_g) - iD_1(\alpha_1, E, E_g)\} + \frac{(\overline{S})^2}{\Delta_c'}\{C_2(\alpha_2, E, E_g) - iD_2(\alpha_2, E, E_g)\}\right.$$
$$\left.+\frac{(\overline{Q})^2}{\Delta_c''}\{C_3(\alpha_3, E, E_g) - iD_3(\alpha_3, E, E_g)\}\right]$$

and

$$F_2(E, \eta_g) = [2\gamma_0(E, \eta_g)]^{-1}\left[\frac{2(\overline{A})^2}{E_g}\{C_1(\alpha_1, E, \eta_g) - iD_1(\alpha_1, E, \eta_g)\}\right.$$
$$\left.+\frac{(\overline{S}+\overline{Q})^2}{\Delta_c''}\{C_3(\alpha_3, E, \eta_g) - iD_3(\alpha_3, E, \eta_g)\}\right]$$

Since $F_1(E, \eta_g)$ and $F_2(E, \eta_g)$ are complex, the energy spectrum is also complex in the presence of Gaussian band tails.

The 2D electron dispersion relation in accumulation layers of IV–VI semiconductors can be written as

$$\theta_1(E, i, \eta_g)k_x^2 + \theta_2(E, i, \eta_g)k_y^2 = \theta_3(E, i, \eta_g) \quad (3.69)$$

where

$$\theta_1(E,i,\eta_g) = [\mathrm{F}_1(E,\eta_g) + S_i(\mathrm{eF}_s\mathrm{a}_1(E,\eta_g))^{2/3}F_2(E,\eta_g)]$$

$$\mathrm{a}_1(E,\eta_g) = \frac{1}{\mathrm{F}_2(E,\eta_g)}\left[\frac{F_2'(E,\eta_g)}{F_2(E,\eta_g)}F_1(E,\eta_g) - F_1'(E,\eta_g)\right]$$

$$\theta_2(E,i,\eta_g) = \left[\left[\mathrm{F}_1(E,\eta_g) + \frac{2\,\mathrm{a}_2(E,\eta_g)}{3\,\mathrm{a}_1(E,\eta_g)}(\mathrm{eF}_s\mathrm{a}_1(E,\eta_g))^{2/3}S_iF_1(E,\eta_g)\right]\right]$$

$$\mathrm{a}_2(E,\eta_g) = \frac{1}{\mathrm{F}_2(E,\eta_g)}\left[\frac{F_2'(E,\eta_g)}{F_2(E,\eta_g)}F_1(E,\eta_g) - F_1'(E,\eta_g)\right]$$

$$\theta_3(E,i,\eta_g) = \left[1 + \frac{2\,\mathrm{C}(E,\eta_g)}{3\,\mathrm{a}_1(E,\eta_g)}\mathrm{S}_i(\mathrm{eF}_s\mathrm{a}_1(E,\eta_g))^{2/3}F_2(E,\eta_g)\right] \quad \text{and } \mathrm{C}(\mathrm{E},\eta_g) = \left[\frac{\mathrm{F}_2'(E,\eta_g)}{\mathrm{F}_2^2(E,\eta_g)}\right]$$

The EEM can be expressed as

$$m^*(E_f',i,\eta_g) = \frac{\hbar^2}{2}\theta_4'(E_f',i,\eta_g) \tag{3.70}$$

where $\theta_4(E_f',i,\eta_g) = \dfrac{\theta_3(E_f',i,\eta_g)}{\sqrt{\theta_1(E_f',i,\eta_g)\theta_2(E_f',i,\eta_g)}}$.

The subband energy E_{i3} is given by

$$\theta_3(\mathrm{E}_{i3},i,\eta_g) = 0 \tag{3.71}$$

The 2D electron concentration in accumulation layer of IV–VI materials under the condition of extreme degeneracy and low electric field limit can be written as

$$n_s = g_v \,\textit{Real part of} \sum_{i=0}^{i_{max}}\left[\theta_4(E_f',i,\eta_g) + \frac{t_i}{3\pi^2}\left[F_1(E_{FB},\eta_g)\sqrt{F_2(E_{FB},\eta_g)}\right]^{-1}\right] \tag{3.72}$$

where E_{FB} can be determined from the equation

$$n_B = g_v \,\textit{Real part of}\left[\frac{1}{3\pi^2}\left[F_1(E_{FB},\eta_g)\sqrt{F_2(E_{FB},\eta_g)}\right]^{-1}\right] \tag{3.73}$$

Thus using (3.7), (3.72) and the allied definitions in the electric quantum limit, we can study the ER in this case.

The 2D electron dispersion relation of the inversion layers of IV–VI semiconductors in the low electric field limit can be written using (3.2) and (3.66) as

$$\mathrm{k}_s^2 = \beta_3(E,i) \tag{3.74}$$

where $\beta_3(E,i) = \dfrac{\beta_1(E,i)}{\beta_2(E,i)}$,

$$\beta_1(E,i) = 1 - \left[\frac{eF_sV_2'(E)}{V_2^2(E)}\right]^{\frac{2}{3}} S_iV_2(E), V_2(E) = \left[\frac{2(\bar{A})^2}{E_{g_0}(1+\alpha_1 E)} + \frac{(\bar{S}+\bar{Q})^2}{\Delta_c''(1+\alpha_3 E)}\right](2E)^{-1},$$

$$\beta_2(E,i) = \left[V_1(E) + \left[\frac{eF_sV_2'(E)}{V_2^2(E)}\right]^{\frac{2}{3}} S_iV_2(E)\frac{2}{3}\frac{V_2^2(E)}{V_2'(E)}\left[\frac{V_1(E)V_2'(E)}{V_2^2(E)} - \frac{V_1'(E)}{V_2(E)}\right]\right]$$

and

$$V_1(E) = \left[\frac{(\bar{R})^2}{E_{g_0}(1+\alpha_1 E)} + \frac{(\bar{S})^2}{\Delta_c'(1+\alpha_2 E)} + \frac{(\bar{Q}^2)}{\Delta_c''(1+\alpha_3 E)}\right](2E)^{-1}$$

The EEM can be expressed as

$$m^*(E_{Fi},i) = \frac{\hbar^2}{2}\beta_3'(E_{Fi},i) \tag{3.75}$$

The subband energy (E_{i4}) can be written as

$$0 = \beta_3(E_{i4},i) \tag{3.76}$$

The surface electron concentration under the condition of extreme degeneracy assumes the form

$$\bar{n}_{2Di} = \frac{g_v}{2\pi}\sum_{i=0}^{i_{max}} \beta_3(\bar{E}_{F0},i) \tag{3.77}$$

Using (3.17), (3.77) and the allied definitions we can study the ER in this case in the electric quantum limit.

3.2.5 The ER in Accumulation and Inversion Layers of Stressed III–V Semiconductors

The 2D electron dispersion relation in accumulation layers of stressed III–V semiconductors can be written as

$$\theta_{13}(E,i,\eta_g)k_x^2 + \theta_{23}(E,i,\eta_g)k_y^2 = \theta_{33}(E,i,\eta_g) \tag{3.78}$$

where

$$\theta_{13}(E,i,\eta_g) = \left[f_1(E,\eta_g) + \mathrm{S}_i(\mathrm{eF}_s\mathrm{a}_{13}(E,\eta_g))^{2/3}f_3(E,\eta_g)\right]$$

$$\mathrm{a}_{13}(E,\eta_g) = \frac{1}{f_3(E,\eta_g)}\left[\frac{f_3'(E,\eta_g)}{f_3(E,\eta_g)}f_1(E,\eta_g) - f_1'(E,\eta_g)\right]$$

$$\theta_{23}(E,i,\eta_g) = \left[\left[f_2(E,\eta_g) + \frac{2}{3}\frac{\mathrm{a}_{23}(E,\eta_g)}{\mathrm{a}_{13}(E,\eta_g)}(\mathrm{eF}_s\mathrm{a}_{13}(E,\eta_g))^{2/3}\mathrm{S}_if_2(E,\eta_g)\right]\right]$$

$$\mathrm{a}_{23}(E,\eta_g) = \frac{1}{f_3(E,\eta_g)}\left[\frac{f_3'(E,\eta_g)}{f_3(E,\eta_g)}f_2(E,\eta_g) - f_1'(E,\eta_g)\right]$$

$$\theta_{33}(E,i,\eta_g) = \left[1 + \frac{2}{3}\frac{\mathrm{C}_3(E,\eta_g)}{\mathrm{a}_{13}(E,\eta_g)}\mathrm{S}_i\left(\mathrm{eF}_s\mathrm{a}_{13}(E,\eta_g)\right)^{2/3}f_3(E,\eta_g)\right]$$

$$\mathrm{C}_3(E,\eta_g) = \left[\frac{f_3'(E,\eta_g)}{f_3^2(E,\eta_g)}\right],$$

and

$f_1(E,\eta_g)$, $f_2(E,\eta_g)$, $f_3(E,\eta_g), P_{11}(E,\eta_g), Q_{11}(E,\eta_g)$ *and* $S_{11}(E,\eta_g)$ are defined in (1.206) respectively.

The EEM can be expressed as

$$m^*(E_f',i,\eta_g) = \frac{\hbar^2}{2}\theta_{43}'(E_f',i,\eta_g) \tag{3.79}$$

where $\theta_{43}(E_f',i,\eta_g) = \frac{\theta_{33}(E_f',i,\eta_g)}{\sqrt{\theta_{13}(E_f',i,\eta_g)\theta_{23}(E_f',i,\eta_g)}}$.

The subband energy E_{i33} is given by

$$\theta_{33}(\mathrm{E}_{i33},i,\eta_g) = 0 \tag{3.80}$$

The 2D electron concentration in accumulation layers of stressed III–V materials under the condition of extreme degeneracy and low electric field limit can be written as

$$n_s = g_v \text{ Real part of } \sum_{i=0}^{i_{\max}}\left[\theta_{43}(E_f',i,\eta_g) + \frac{t_i}{3\pi^2}\left[f_1(E_{FB},\eta_g)f_2(E_{FB},\eta_g)f_3(E_{FB},\eta_g)\right]^{-1/2}\right] \tag{3.81}$$

The E_{FB} can be determined from the following equation

$$n_B = \frac{g_v}{3\pi^2}\left[f_1(E_{FB},\eta_g)f_2(E_{FB},\eta_g)f_3(E_{FB},\eta_g)\right]^{-1/2} \quad (3.82)$$

Thus using (3.7), (3.81) and the allied definitions in the electric quantum limit, we can study the ER in this case.

The (1.195) can be written as

$$(E-\alpha_1)k_x^2 + (E-\alpha_2)k_y^2 + (E-\alpha_3)k_z^2 = t_1E^3 - t_2E^2 + t_3E + t_4 \quad (3.83)$$

where

$$\alpha_1 \equiv \left[E_g - C_1\varepsilon - (\bar{a}_0 + C_1)\varepsilon + \frac{3}{2}\bar{b}_0\varepsilon_{xx} - \frac{\bar{b}_0}{2}\varepsilon + \left(\sqrt{3}/2\right)\varepsilon_{xy}\bar{d}_0\right],$$

$$\alpha_2 \equiv \left[E_g - C_1\varepsilon - (\bar{a}_0 + C_1)\varepsilon + \frac{3}{2}\bar{b}_0\varepsilon_{xx} - \frac{\bar{b}_0}{2}\varepsilon - \left(\sqrt{3}/2\right)\varepsilon_{xy}\bar{d}_0\right],$$

$$\alpha_3 \equiv \left[E_g - C_1\varepsilon - (\bar{a}_0 + C_1)\varepsilon + \frac{3}{2}\bar{b}_0\varepsilon_{zz} - \frac{\bar{b}_0}{2}\varepsilon\right],$$

$$t_1 \equiv \left(3/2B_2^2\right),$$

$$t_2 \equiv \left(1/2B_2^2\right)\left[6(E_g - C_1\varepsilon) + 3C_1\varepsilon\right],$$

$$t_3 \equiv \left(1/2B_2^2\right)\left[3(E_g - C_1\varepsilon)^2 + 6C_1\varepsilon(E_g - C_1\varepsilon) - 2C_2^2\varepsilon_{xy}^2\right] \quad \text{and}$$

$$t_4 \equiv \left(1/2B_2^2\right)\left[-3C_1\varepsilon(E_g - C_1\varepsilon)^2 + 2C_2^2\varepsilon_{xy}^2\right].$$

The use of (3.69) and (3.2) leads to the expression of the dispersion relation of the 2D electrons in inversion layers of stressed III–V materials under the low electric field limit as

$$[T_{57}(E,i)]k_x^2 + [T_{67}(E,i)]k_y^2 = T_{77}(E,i) \quad (3.84)$$

where,

$$T_{57}(E,i) = \left[E - \alpha_1 + \frac{2}{3}S_i\left(\frac{|e|^2}{\varepsilon_{sc}}\right)^{2/3}(n_{2Dw})^{2/3}L_{17}(E)\right],$$

$$L_{17}(E) = \left[\frac{(E-\alpha_1)}{(E-\alpha_3)^{2/3}[\overline{T}_{47}(E)]^{1/3}} - (E-\alpha_3)^{1/3}[\overline{T}_{47}(E)]^{-1/3}\right],$$

$$[\overline{T}_{47}(E)] = \left[\{\rho_5(E)\}' - \left(\frac{\rho_5(E)}{E-\alpha_3}\right)\right],$$

$$T_{67}(E,i) = \left[E - T_2 + \frac{2}{3}S_i\left(\frac{|e|^2}{\varepsilon_{sc}}\right)^{2/3}(n_{2Dw})^{2/3}L_{27}(E)\right],$$

$$L_{27}(E) = \left[\frac{(E-\alpha_2)}{(E-\alpha_3)^{2/3}[\overline{T}_{47}(E)]^{1/3}} - \left(\frac{(E-\alpha_3)^{1/3}}{[\overline{T}_{47}(E)]^{1/3}}\right)\right],$$

$$T_{77}(E,i) = \left[\rho_5(E) - S_i\left(\frac{|e|^2}{\varepsilon_{sc}}\right)^{2/3}(n_{2Dw})^{2/3}L_{37}(E)\right],$$

$$L_{37}(E) \equiv (E-\alpha_3)^{1/3}[\overline{T}_{47}(E)]^{2/3}$$

and

$$\rho_5(E) \equiv \left[t_1E^3 - t_2E^2 + t_3E + t_4\right]$$

The area of the 2D surface under the weak electric field limit can be written as

$$A(E,i) = \frac{\pi T_{77}(E,i)}{\sqrt{T_{57}(E,i)T_{67}(E,i)}} \tag{3.85}$$

The sub-band energies ($E_{n_{iw8}}$) in this case are defined by

$$T_{47}(E_{n_{iw8}}) = S_i\left(\frac{|e|^2}{\varepsilon_{sc}}\right)^{2/3}(n_{2Dw})^{2/3}L_{37}(E_{n_{iw8}}) \tag{3.86}$$

The expression of the EEM in this case can be written as

$$m^*(E_{Fiw},i) = \frac{\hbar^2}{2}L_{47}(E,i)|_{E=E_{Fiw}} \tag{3.87}$$

where,

$$L_{47}(E,i) \equiv \left[\frac{1}{T_{57}(E,i)T_{67}(E,i)}\right]\left[\{T_{77}(E,i)\}'[T_{57}(E,i)T_{67}(E,i)]^{1/2} - \left(\frac{T_{77}(E,i)}{2}\right)\left\{\{T_{57}(E,i)\}'\left[\frac{T_{67}(E,i)}{T_{57}(E,i)}\right]^{1/2} + \{T_{67}(E,i)\}'\left[\frac{T_{57}(E,i)}{T_{67}(E,i)}\right]^{1/2}\right\}\right].$$

The total 2D DOS function can be expressed as

$$N_{2D}(E) = \frac{g_v}{2\pi}\sum_{i=0}^{i_{max}}\{L_{47}(E,i)H(E - E_{n_{iw8}})\} \tag{3.88}$$

The surface electron concentration under the weak electric field quantum limit assumes the form

$$n_{2Dw} = \frac{g_v}{(2\pi)}\left\{\sum_{i=0}^{i_{max}}[P_{8w}(E_{Fwi},i) + Q_{8w}(E_{Fwi},i)]\right\} \tag{3.89}$$

where, $P_{8w}(E_{Fwi},i) \equiv \frac{T_{77}(E_{Fwi},i)}{\sqrt{T_{57}(E_{Fwi},i)T_{67}(E_{Fwi},i)}}$ and $Q_{8w}(E_{Fiw},i) \equiv \sum_{r=1}^{s} L(r)P_{8w}(E_{Fiw},i)$.

Using (3.75), the expression of the 2D surface electron concentration under the weak electric field quantum limit can be written as

$$\bar{n}_{2Dw} = \frac{g_v}{2\pi}\frac{T_{77}(\bar{E}_{Fw},0)}{\sqrt{T_{57}(\bar{E}_{Fw},0)T_{67}(\bar{E}_{Fw},0)}} \tag{3.90}$$

3.2.6 The ER in Accumulation and Inversion Layers of Ge

The 2D dispersion relation in accumulation layers of Ge can be written as

$$\frac{\hbar^2 k_x^2}{2m_1^*} + \frac{\hbar^2 k_y^2}{2m_2^*} = \gamma_{10}(E,i,\eta_g) \tag{3.91}$$

where

$$\gamma_{10}(E,i,\eta_g)=\left[\gamma_3(E,\eta_g)[1+\alpha\gamma_3(E,\eta_g)]-S_i\left[\frac{\hbar eFs\gamma_s'(E,\eta_g)}{\sqrt{2m_3^*}}\right]^{\frac{2}{3}}[1+2\alpha\gamma_3(E,\eta_g)]\right.$$
$$\left.+\alpha\left[S_i\left[\frac{\hbar eFs\gamma_3'(E,\eta_g)}{\sqrt{2m_3^*}}\right]^{\frac{2}{3}}\right]^2\right]$$

The EEM can be expressed as

$$m^*(E_f',i,\eta_g)=\sqrt{m_1^*m_2^*}[\gamma_{10}'(E,i,\eta_g)] \tag{3.92}$$

The band non-parabolicity and heavy doping makes the mass quantum number dependent.

The sub band energy E_{i14} can be written as

$$\gamma_{10}(E_{i14},i,\eta_g)=0 \tag{3.93}$$

The surface electron concentration in accumulation layers can be written as

$$n_s=g_v\sum_{i=0}^{i_{\max}}\left[\frac{\sqrt{m_1^*m_2^*}}{\pi\hbar^2}\left[\gamma_{10}(E_f',i,\eta_g)\right]\right.$$
$$\left.+\,t_i\frac{8\pi m_\perp^*\sqrt{m_{||}^*}}{h^3}[2\gamma_3(E_{FB},\eta_g)]^{\frac{3}{2}}\left[1+\frac{4\alpha}{5}\gamma_3(E_{FB},\eta_g)\right]\right] \tag{3.94}$$

where E_{FB} can be determined from the following equation

$$n_B=g_v\left[\frac{8\pi m_\perp^*\sqrt{m_{||}^*}}{h^3}[2\gamma_3(E_{FB},\eta_g)]^{\frac{3}{2}}\left[1+\frac{4\alpha}{5}\gamma_3(E_{FB},\eta_g)\right]\right] \tag{3.95}$$

Using (3.7), (3.94) and the allied definitions at the electric quantum limit and under the condition of extreme carrier degeneracy we can study the ER in this case.

The 2D electron dispersion law in inversion layers of Ge can be expressed as

$$\frac{\hbar^2k_x^2}{2m_1}+\frac{\hbar^2k_y^2}{2m_2}=\left[E(1+\alpha E)+\alpha E_{i20}^2-E_{i20}(1+2\alpha E)\right] \tag{3.96}$$

where, $E_{i20}=S_i\left(\frac{\hbar eF_s}{\sqrt{2m_3}}\right)^{2/3}$.

The area of 2D space is

$$A = \frac{2\pi\sqrt{m_1 m_2}}{\hbar^2}\left[E(1+\alpha E)+\alpha E_{i20}^2 - E_{i20}(1+2\alpha E)\right] \tag{3.97}$$

The EEM assumes the form

$$m^*(E_{Fiw}, i) = \sqrt{m_1 m_2}[1 + 2\alpha E_{Fiw} - E_{i20}2\alpha] \tag{3.98}$$

Thus the EEM is the function of both Fermi energy and quantum number due to band nonparabolicity.

The DOS function is given by

$$N_{2D}(E) = \frac{2g_v}{(2\pi)^2}\cdot\frac{2\pi\sqrt{m_1 m_2}}{\hbar}\sum_{i=0}^{i_{max}}[1+2\alpha E - 2\alpha E_{i20}]H(E - E_{i20}) \tag{3.99}$$

The surface electron concentration is given by

$$n_{Ds} = \frac{g_v k_B T}{\pi}\cdot\frac{\sqrt{m_1 m_2}}{\hbar^2}\sum_{i=0}^{i_{max}}[F_0(\eta_4) + 2\alpha k_B T F_1(\eta_4)] \tag{3.100}$$

where, $\eta_4 = (k_B T)^{-1}[E_{Fw} - E_{i20}]$

Using (3.17), (3.100) and the allied definitions under the conditions of extreme degeneracy and electric quantum limit, we can study the ER in this case.

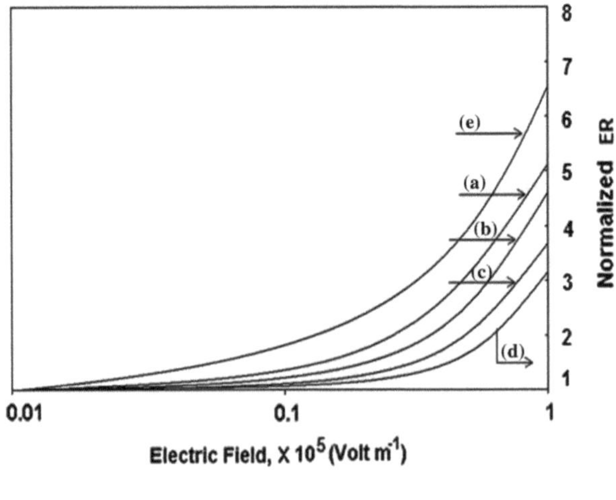

Fig. 3.1 The plot of the normalized ER in the accumulation layers of Cd_3As_2 under weak electric field quantum limit as a function of surface electric field in accordance with (*a*) the generalized band model, (*b*) $\delta = 0$, (*c*) the three band model of Kane, (*d*) the two band model of Kane and (*e*) the parabolic energy bands

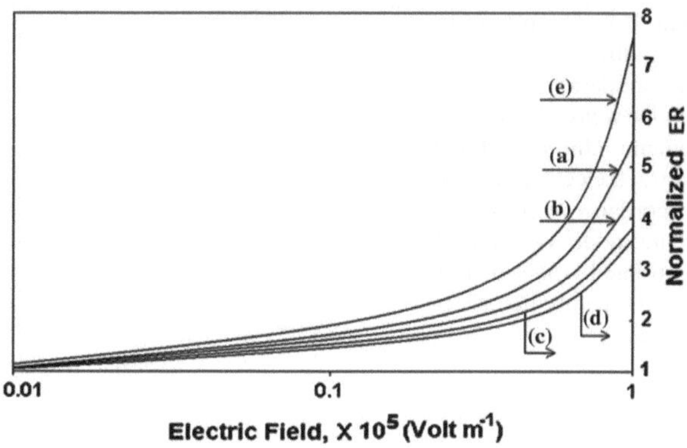

Fig. 3.2 The plot of the normalized ER in the accumulation layers of $CdGeAs_2$ under weak electric field quantum limit as a function of surface electric field in accordance with (*a*) the generalized band model, (*b*) $\delta = 0$, (*c*) the three band model of Kane, (*d*) the two band model of Kane and (*e*) the parabolic energy bands

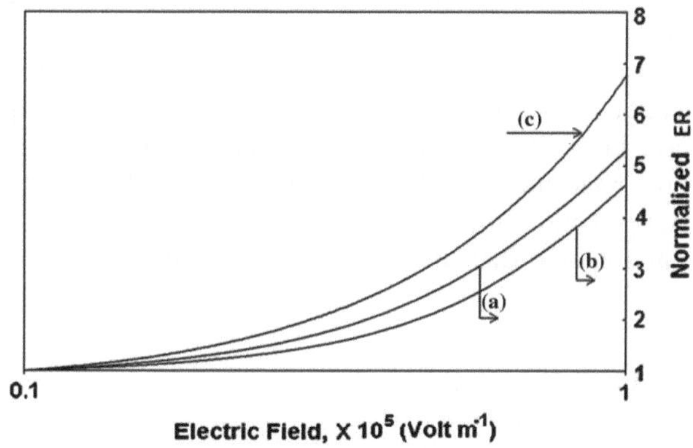

Fig. 3.3 The plot of the normalized ER in the accumulation layers of InAs under weak electric field quantum limit as a function of surface electric field in accordance with (*a*) the three band model of Kane, (*b*) the two band model of Kane and (*c*) the parabolic energy bands

3.3 Result and Discussions

Using the appropriate equations and the numerical values of the energy band constants as given in Table 1.1, the plot for the normalized ER for n-channel accumulation layers of Cd_3As_2 under weak electric field limit as a function of surface electric field has been shown in curve (a) of Fig. 3.1. The curve (b) corresponds to $\delta = 0$ and the curve (c) exhibits the dependence of the ER on the surface electric field in accordance with the three-band model of Kane, respectively.

The plots (d) and (e) correspond to the two-band model of Kane and that of parabolic energy bands respectively. By comparing the curves (a) and (b) of Fig. 3.1, one can assess the influence of crystal field splitting of the ER in accumulation layers of tetragonal compounds. It appears from Fig. 3.1 that, the ER increases with increasing surface electric field and the numerical values of the ER are totally band structure dependent. In Fig. 3.2 the ER for accumulation layers of $CdGeAs_2$ have been drawn as function of surface electric field for weak electric field limit respectively for all the cases of Fig. 3.1

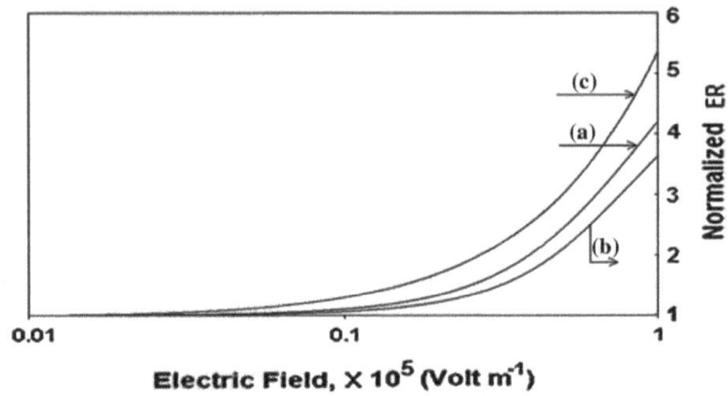

Fig. 3.4 The plot of the normalized ER in the accumulation layers of InSb under weak electric field quantum limit as a function of surface electric field in accordance with (*a*) the three band model of Kane, (*b*) the two band model of Kane and (*c*) the parabolic energy bands

The trend of variation of the ER for accumulation layers of $CdGeAs_2$ is more or less the same with different numerical magnitudes as compared with accumulation layers of Cd_3As_2 for both the limits.

Using the appropriate equations the Fig. 3.3 exhibits the normalized ER in accumulation layers of InAs for weak electric field quantum limit and the curves have been drawn as function of surface electric field in accordance with (a) the three band model of Kane, (b) the two band model of Kane and (c) the parabolic energy bands respectively. In Fig. 3.4 all cases of Fig. 3.3 have been drawn as function of surface electric field for accumulation layers of InSb for weak electric

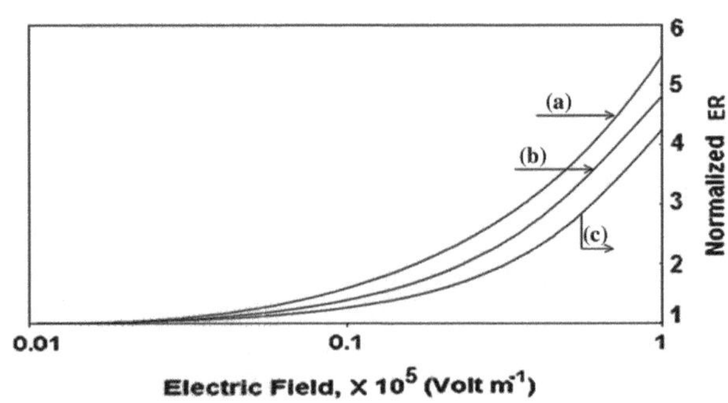

Fig. 3.5 The plot of the normalized ER in the accumulation layers of $Hg_{1-x}Cd_xTe$ under weak electric field quantum limit as a function of surface electric field in accordance with (*a*) the three band model of Kane, (*b*) the two band model of Kane and (*c*) the parabolic energy bands

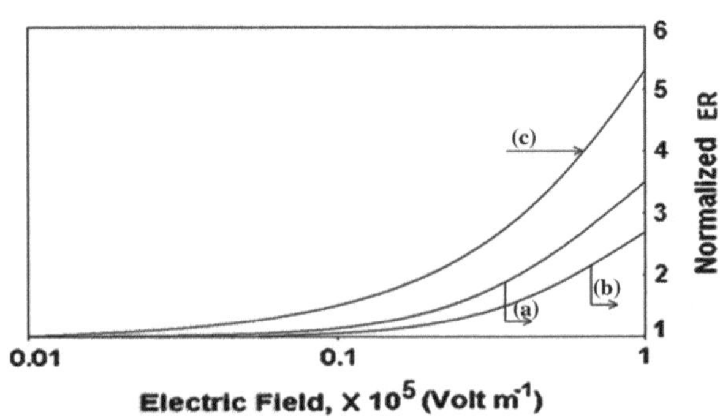

Fig. 3.6 The plot of the normalized ER in the accumulation layers of $In_{1-x}Ga_xAs_yP_{1-y}$ lattice matched to InP under weak electric field quantum limit as a function of surface electric field in accordance with (*a*) the three band model of Kane, (*b*) the two band model of Kane and (*c*) the parabolic energy bands

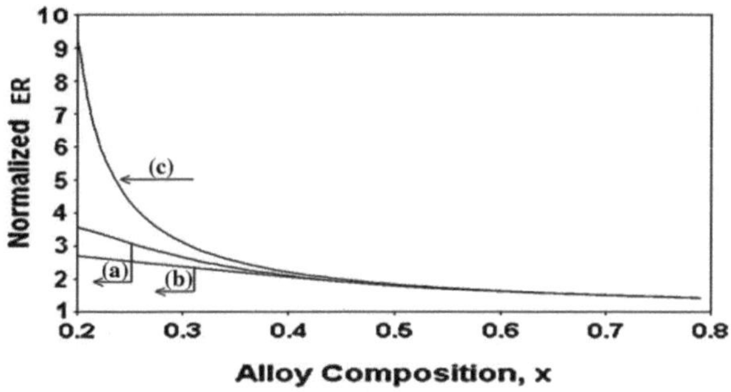

Fig. 3.7 The plot of the normalized ER in the accumulation layers of $Hg_{1-x}Cd_xTe$ under weak electric field quantum limit as a function of alloy composition in accordance with (*a*) the three band model of Kane, (*b*) the two band model of Kane and (*c*) the parabolic energy bands

field limit. In Fig. 3.5, the normalized ER in accumulation layers of $Hg_{1-x}Cd_xTe$ have been drawn for both the weak electric field limits as functions of surface electric field for all the cases of Fig. 3.4. In Fig. 3.6 the normalized ER in accumulation layers of $In_{1-x}Ga_xAs_yP_{1-y}$ lattice matched to InP have been drawn for low electric field limit as functions of surface electric field for all the cases of Fig. 3.5. It appears from Figs. 3.3, 3.4, 3.5 and 3.6 that the ER for accumulation layers of InAs, InSb, $Hg_{1-x}Cd_xTe$ and $In_{1-x}Ga_xAs_yP_{1-y}$ lattice matched to InP increases with increase in surface electric field with different numerical values and the influence of the energy band constants can also be assessed from the said Figs. In Figs. 3.7 and 3.8, the normalized ER for accumulation layers of $Hg_{1-x}Cd_xTe$ and $In_{1-x}Ga_xAs_yP_{1-y}$ lattice matched to InP have been drawn as a function of alloy composition under weak electric field limit in accordance with the three and two

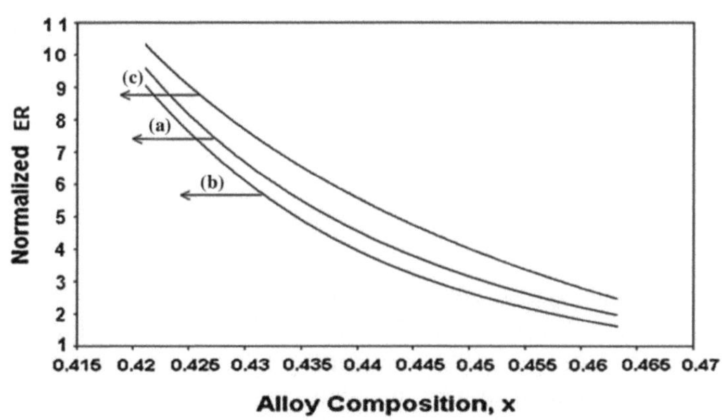

Fig. 3.8 The plot of the normalized ER in the accumulation layers of $In_{1-x}Ga_xAs_yP_{1-y}$ lattice matched to InP under weak electric field quantum limit as a function of alloy composition in accordance with (*a*) the three band model of Kane, (*b*) the two band model of Kane and (*c*) the parabolic energy bands

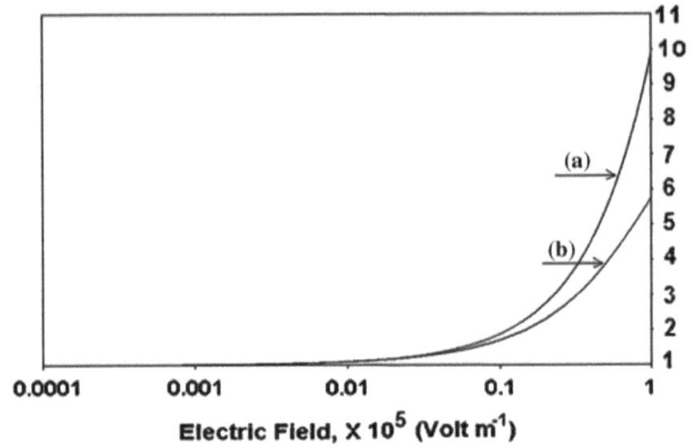

Fig. 3.9 The plot of the normalized ER in the p-channel accumulation layers of CdS as function of surface electric field in accordance with (*a*) $\bar{\lambda}_0 \neq 0$ and (*b*) $\bar{\lambda}_0 = 0$

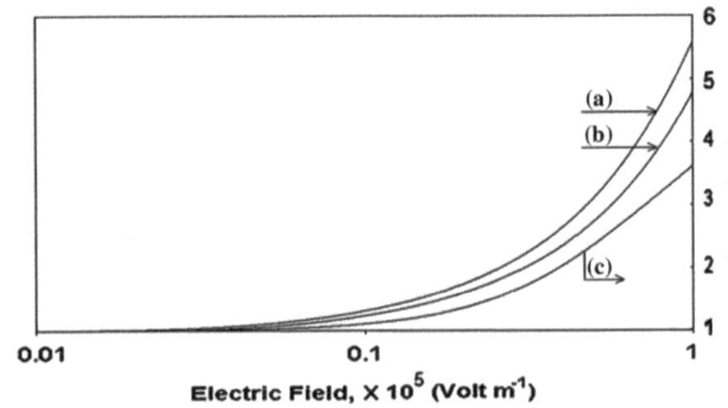

Fig. 3.10 The plot of the normalized ER in accumulation layers of (*a*) PbTe, (*b*) PbSnTe and (*c*) $Pb_{1-x}Sn_xSe$ under weak electric field quantum limit as a function of surface electric field

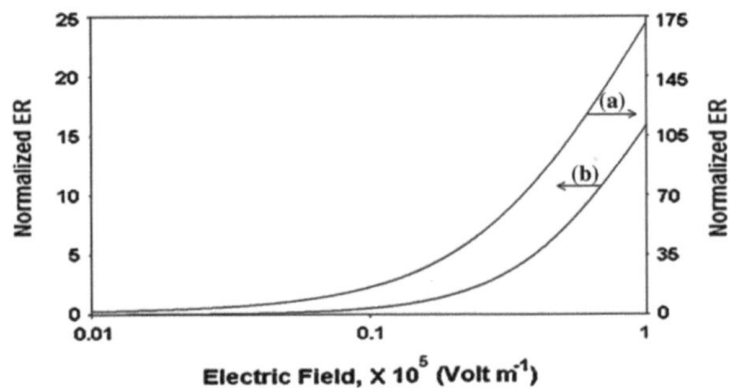

Fig. 3.11 The plot of the normalized ER in accumulation layers of stressed InSb under weak electric field quantum limit as a function of surface electric field in which the curve (*a*) is in the presence of stress and curve (*b*) is under absence of stress

band models of Kane together with parabolic energy bands respectively. It appears from Figs. 3.7 and 3.8 that the ER decreases with increasing alloy composition although the rate of decrease is determined by the respective energy band constants of the ternary and quaternary materials.

Using appropriate equations, the Fig. 3.9 exhibits the plot of the ER in p-channel accumulation layers of CdS as function of surface electric field in accordance with (a) $\bar{\lambda}_0 \neq 0$ and (b) $\bar{\lambda}_0 = 0$. The presence of the crystal field splitting enhances the numerical values of the ER for relatively large values of the surface electric field. Using appropriate equations, the Fig. 3.10 exhibits the plots of the normalized ER in the accumulation layers of PbTe, PbSnTe and $Pb_{1-x}Sn_xSe$ for weak electric field limits as a function of surface electric field respectively. It appears that the ER increases with increasing surface electric field with a diverging nature for relatively large values of the electric field. The numerical values of the ER are greatest for accumulation layers of PbTe and least for the corresponding $Pb_{1-x}Sn_xSe$.

Using the appropriate equations, Fig. 3.11 exhibits the ER in accumulation layers of stressed n-InSb for weak electric field limit as a function of surface electric field, in which the curve (a) is valid in the presence of stress whereas the curve (b) shows the stress free condition respectively. It appears from Fig. 3.11 that the ER exhibits the increasing dependence with increasing surface electric field and the stress enhances the value of the ER for relatively large value of the electric field.

In this chapter, we have investigated the ER in accumulation layers of tetragonal materials for under weak electric field limit on the basis of the generalized electron energy spectrum as given by (1.2). The results for the accumulation layers of III–V, ternary and quaternary materials whose bulk electrons obey the three and two band models of Kane together with parabolic energy bands form a special case of our generalized analysis. The ER for accumulation layers of II–VI has been studied on the basis of Hopfield model for all values of surface electric field. The ER has been investigated in accumulation layers of IV–VI and stressed materials on the basis of the model of Bangert and Kastner for weak electric field limit. For the purpose of relative comparison we have also discussed the ERs in the inversion layers of the aforementioned semiconductors.

It may be noted that if the direction of application of the surface electric field applied perpendicular to the surface be taken as either k_x or k_y and not as k_z as assumed in the present work, the ER would be different analytically for both the limits. Nevertheless, the arbitrary choice of the direction normal to the surface would not result in a change of the basic qualitative feature of the ER in accumulation layers of semiconductors. The approximation of the potential well at the surface by a triangular well introduces some errors, as for instance the omission of the free charge contribution to the potential. This kind of approach is reasonable if there are only few charge carriers in the accumulation layer, but is responsible for an overestimation of the splitting when the accumulation carrier density exceeds that of the depletion layer. It has been observed that the maximum error due to the triangular potential well is tolerable in the practical sense because for actual calculations, one need a self consistent solution which is a formidable problem, for the present generalized systems due to the non availability of the proper analytical techniques, without exhibiting a widely different qualitative behavior [1, 3]. The second assumption of electric quantum limit in the numerical calculation is valid in the range of low temperatures, where the quantum effects become prominent. The errors which are being introduced for these assumptions are found not to be serious enough at low temperatures [3, 4]. Thus, whenever the condition of the electric quantum limit has been applied, the temperature has been assumed to be low enough so that the assumption becomes well grounded because at low temperature, one can assume that nearly all electrons are at the lowest electric subband [3, 4]. We wish to note that the many body effects, the hot electron effects, the formation of band tails, arbitrary orientation of the direction of the electric quantization and the effects of surface of states have been neglected in our simplified theoretical formalism due to the lack of availability in the literature of the proper analytical techniques for including them for the generalized systems as considered in this paper. Our simplified approach will be useful for the purpose of comparison, when, the methods of tackling of the aforementioned formidable problems for the present generalized system appear. The inclusion of the said effects would certainly increase the accuracy of our results, and the qualitative features of the ER as discussed in this chapter would not change in the presence of the aforementioned influences.

3.4 Open Research Problems

R.3.1 Investigate the ER in the presence of an arbitrarily oriented electric quantization for accumulation layers of tetragonal semiconductors. Study all the special cases for III–V, ternary and quaternary materials in this context.

R.3.2 Investigate the ER in accumulation layers of IV–VI, II–VI and stressed Kane type compounds in the presence of an arbitrarily oriented quantizing electric field.

R.3.3 Investigate the ER in accumulation layers of all the materials as stated in R.2.1 of Chap. 2 in the presence of an arbitrarily oriented quantizing electric field.

R.3.4 Investigate the ER in the presence of an arbitrarily oriented non-quantizing magnetic field in accumulation layers of tetragonal semiconductors by including the electron spin. Study all the special cases for III–V, ternary and quaternary materials in this context.

R.3.5 Investigate the ER in accumulation layers of IV–VI, II–VI and stressed Kane type compounds in the presence of an arbitrarily oriented non-quantizing magnetic field by including the electron spin.

R.3.6 Investigate the ER in accumulation layers of all the materials as stated in R.2.1 of Chap. 2 in the presence of an arbitrarily oriented non-quantizing magnetic field by including electron spin.

R.3.7 Investigate the ER in accumulation layers for all the problems from R.3.1 to R.3.6 in the presence of an additional arbitrarily oriented electric field.

R.3.8 Investigate the ER in accumulation layers for all the problems from R.3.1 to R.3.3 in the presence of arbitrarily oriented crossed electric and magnetic fields.

R.3.9 Investigate the ER in accumulation layers for all the problems from R.3.1 to R.3.8 in the presence of surface states.

R.3.10 Investigate the ER in accumulation layers for all the problems from R.3.1 to R.3.8 in the presence of hot electron effects.

R.3.11 Investigate the ER in accumulation layers for all the problems from R.3.1 to R.3.6 by including the occupancy of the electrons in various electric subbands.

R.3.12 Investigate the problems from R.3.1 to R.3.11 for the appropriate p-channel accumulation layers.

References

1. T. Ando, H. Fowler, F. Stern, Rev. Mod. Phys. **54**, 437 (1982)
2. J.J. Quinn, P.J. Styles (eds.), *Electronic Properties of Quasi Two Dimensional Systems* (North Holland, Amsterdam, 1976)
3. G.A. Antcliffe, R.T. Bate, R.A. Reynolds, in *Proceedings of the International Conference, Physics of Semi-metals and Narrow-Gap semiconductors* (Pergamon Press, Oxford, 1971), p. 499 ed. by D.L. Carter, R.T. Bate
4. Z.A. Weinberg, Sol. Stat. Electron. **20**, 11 (1977)
5. G. Paasch, T. Fiedler, M. Kolar, I. Bartos, Phys. Stat. Sol. (b) **118**, 641 (1983)
6. S. Lamari, Phys. Rev. B **64**, 245340 (2001)
7. T. Matsuyama, R. Kürsten, C. Meißner, U. Merkt, Phys. Rev. B **61**, 15588 (2000)
8. P.V. Santos, M. Cardona, Phys. Rev. Lett. **72**, 432 (1994)
9. L. Bu, Y. Zhang, B.A. Mason, R.E. Doezema, J.A. Slinkman, Phys. Rev. B **45**, 11336 (1992)
10. P.D. Dresselhaus, C.M. Papavassiliou, R.G. Wheeler, R.N. Sacks, Phys. Rev. Lett. **68**, 106 (1992)
11. U. Kunze, Phys. Rev. B **41**, 1707 (1990)
12. E. Yamaguchi, Phys. Rev. B **32**, 5280 (1985)
13. Th Lindner, G. Paasch, J. Appl. Phys. **102**, 054514 (2007)
14. S. Lamari, J. Appl. Phys. **91**, 1698 (2002)
15. K.P. Ghatak, M. Mondal, J. Appl. Phys. **70**, 299 (1991)
16. K.P. Ghatak, S.N. Biswas, J. Vac. Sc. and Tech. **7B**, 104 (1989)
17. B. Mitra, K.P. Ghatak, Sol. State Electron. **32**, 177 (1989)
18. K.P. Ghatak, M. Mondal, J. Appl. Phys. **62**, 922 (1987)
19. M. Mondal, K.P. Ghatak, J. Magnet, Magnetic Mat. **62**, 115 (1986)
20. M. Mondal, K.P. Ghatak, Phys. Script. **31**, 613 (1985)
21. K.P. Ghatak, M. Mondal, Z. fur Physik B **64**, 223 (1986)
22. K.P. Ghatak, S.N. Biswas, Sol. State Electron. **37**, 1437 (1994)
23. D.R. Choudhury, A.K. Chowdhury, K.P. Ghatak, A.N. Chakravarti, Phys. Stat. Sol. (b) **98**, K141 (1980)
24. A.N. Chakravarti, A.K. Chowdhury, K.P. Ghatak, Phys. Stat. Sol. (a) **63**, K97 (1981)
25. M. Mondal, K.P. Ghatak, Acta Phys. Polon. A **67**, 983 (1985)
26. M. Mondal, K.P. Ghatak, Phys. Stat. Sol. (b) **128**, K21 (1985)
27. M. Mondal, K.P. Ghatak, Phys. Stat. Sol. (a) **93**, 377 (1986)
28. K.P. Ghatak, M. Mondal, Phys. Stat. Sol. (b) **135**, 819 (1986)
29. M. Mondal, K.P. Ghatak, Phys. Stat. Sol. (b) **139**, 185 (1987)

30. K.P. Ghatak, N. Chattopadhyay, S.N. Biswas, OE/Fibers' **87**, 203 (1987)
31. K.P. Ghatak, N. Chatterjee, M. Mondal, Phys. Stat. Sol. (b) **139**, K25 (1987)
32. K.P. Ghatak, M. Mondal, Phys. Stat. Sol. (b) **148**, 645 (1988)
33. K.P. Ghatak, A. Ghosal Phys. Stat. Sol. (b) **151**, K135 (1989)
34. K.P. Ghatak, N. Chattopadhyay, M. Mondal, Appl. Phys. A **48**, 365 (1989)

Chapter 4
Suggestion for Experimental Determinations of 2D and 3D ERs and Few Related Applications

4.1 Introduction

In this book we have discussed many aspects of ER based on the dispersion relations of different technologically important semiconductors. In this chapter we discuss regarding the suggestions for the experimental determinations of 3D and 2D ERs in Sect. 4.2. We shall also discuss few applications on the basis of the content of this monograph in this context in Sect. 4.3.

4.2 Suggestion for the Experimental Determinations of the 3D and 2D ERs for HDS Having Arbitrary Dispersion Laws

It is well-known that the thermoelectric power of the carriers in HDS in the presence of a classically large magnetic field is independent of scattering mechanisms and is determined only by their energy band spectra [1–9]. The magnitude of the thermoelectric power G can be written as [1–9]

$$G = \frac{1}{|e|Tn_0} \int_{-\infty}^{\infty} (E - E_{F_h}) R(E) \left[-\frac{\partial f_0}{\partial E} \right] dE \tag{4.1}$$

where $R(E)$ is the total number of states. The (4.1) can be written under the condition of carrier degeneracy [10–23] as

$$G = \left(\frac{\pi^2 k_B^2 T}{3|e|n_0}\right) \text{Real part of } \left(\frac{\partial n_0}{\partial (E_{F_h} - \bar{E}_{hd})}\right) \tag{4.2}$$

K. P. Ghatak and S. Bhattacharya, *Heavily-Doped 2D-Quantized Structures and the Einstein Relation*, Springer Tracts in Modern Physics 260,
DOI 10.1007/978-3-319-08380-3_4, © Springer International Publishing Switzerland 2015

The use of (1.31c) and (4.2) leads to the result

$$\frac{D}{\mu} = \left(\frac{\pi^2 k_B^2 T}{3|e|^2 G}\right) \quad (4.3)$$

Thus, the ER for degenerate materials can be determined by knowing the experimental values of G.

(a) The suggestion for the experimental determination of the ER for degenerate semiconductors having arbitrary dispersion laws as given by (4.3) does not contain any energy band constants. For a fixed temperature, the ER varies inversely as G. Only the experimental values of G for any material as a function of electron concentration will generate the experimental values of the ER for that range of n_0 for that system. Since G decreases with increasing n_0, from (4.3) one can infer that the ER will increase with increase in n_0. ***Besides, under magnetic quantization, crossed fields configuration, and super-lattices under quantizing magnetic field, the same*** (4.3) ***holds good***. This statement is the compatibility test so far as the suggestion for the experimental determination of ER for HDS and degenerate materials are concerned.

(b) ***For quantized 2D systems under the condition of electric quantum limit, it can be proved that same*** (4.3) ***is again true***. For quantum wires and heterostructures with small charge densities, the relation between D/μ and G is thus given by the (4.3) [24]. Thus (4.3) is independent of the dimensions of quantum confinement. We should note that the present analysis is not valid for totally k-space quantized systems such as quantum dots, magneto-inversion and accumulation layers, magneto size quantization, magneto nipis, quantum dot superlattices and quantum well superlattices under magnetic quantization. Under the said conditions, the electron motion is possible in the broadened levels. The experimental results of G for degenerate materials will provide an experimental check on the ER and a technique for probing the band structure of degenerate compounds having arbitrary dispersion laws.

Using $n - Cd_3As_2$ as an example of $A_{III}^2 B_{II}^5$ compounds for the purpose of numerical computations and using the (1.46a) and (1.47) together with the energy band constants at T = 4.2 K, as given in Table 1.1, the variation of the ER as a function of electron concentration has been shown in curve (a) of Fig. 4.1. The circular points exhibit the same dependence and have been obtained by using (4.3) and taking the experimental values of the thermoelectric power in $n - Cd_3As_2$ in the presence of a classically large magnetic field [25]. The curve (b) corresponds to $\delta = 0$. The curve (c) shows the dependence of the ER on n_0 in accordance with the three-band model of Kane using the energy band constants as $E_g = 0.095$ eV $m^* = (m_{\parallel}^* + m_{\perp}^*)/2$ and $\Delta = (\Delta_{\parallel} + \Delta_{\perp})/2$. The curves (d) and (e) correspond to two-band model of Kane and that of the parabolic energy bands. By comparing the curves (a) and (b) of Fig. 4.1, one can easily assess the influence of crystal field splitting on the ER in nonlinear optical compounds.

It appears from Fig. 4.1 that, the ER in nonlinear optical compounds increases with increasing carrier degeneracy as expected for degenerate semiconductors and

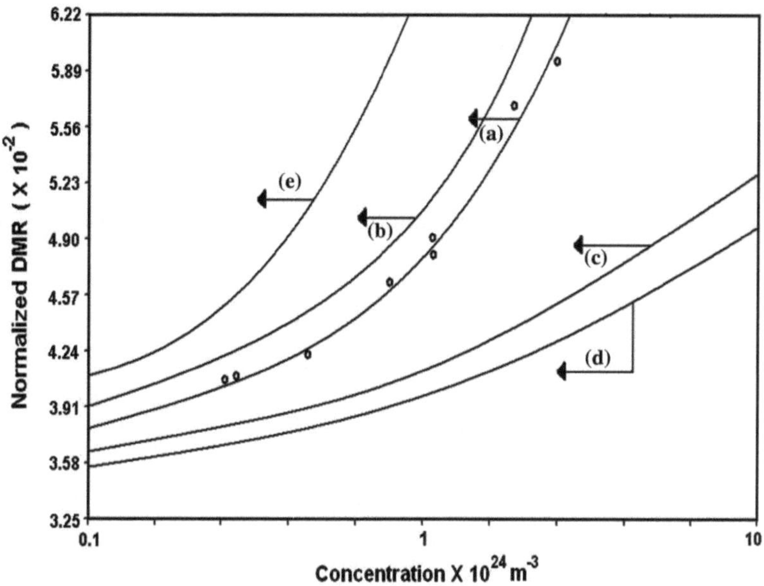

Fig. 4.1 The plot of the ER in the bulk specimens of n-Cd_3As_2 as a function of electron concentration in accordance with *a* the generalized band model; *b* $\delta = 0$; *c* the three band model of Kane; *d* the two band model of Kane and *e* the parabolic energy bands. The *dotted circular points* show the same dependence which have been obtained by using (4.3) and taking the experimental values of the thermoelectric power of the electrons in bulk n-Cd_3As_2 in the presence of a classically large magnetic field [25]

agrees well with the suggested experimental method of determining the same ratio for materials having arbitrary carrier energy spectra. It has been observed that the nonlinear optical crystal field affects the ER of the electrons quite significantly in this case. The dependence of the ER is directly determined by the band structure because of its immediate connection with the Fermi energy. The ER increases nonlinearly with the electron concentration in other limiting cases and the rates of increase are different from that in the generalized band model.

4.3 Different Related Applications

The content of this book finds six applications in the field of materials science and related disciplines in general.

1. **Carrier contribution to the elastic constants**: The knowledge of the carrier contribution to the elastic constants is important in studying the mechanical properties of the materials and has been investigated in the literature [26–35]. The electronic contribution to the second and third order elastic constants can be written as [26–35]

$$\Delta C_{44} = -\frac{G_0^2}{9}\frac{\partial n_0}{\partial(E_{F_h} - \bar{E}_{hd})}, \tag{4.4}$$

and

$$\Delta C_{456} = \frac{G_0^3}{27}\frac{\partial^2 n_0}{\partial(E_{F_h} - \bar{E}_{hd})^2}, \tag{4.5}$$

where G_0 is the deformation potential constant. Thus, using (4.2), (4.4) and (4.5), we can write

$$\Delta C_{44} = \left[-n_0 G_0^2 |e| G / \left(3\pi^2 k_B^2 T\right)\right] \tag{4.6}$$

and

$$\Delta C_{456} = \left(n_0 |e| G_0^3 G^2 / \left(3\pi^4 k_B^3 T\right)\right)\left(1 + \frac{n_0}{G}\frac{\partial G}{\partial n_0}\right). \tag{4.7}$$

Thus, again the experimental graph of G versus n_0 allows us to determine the electronic contribution to the elastic constants for materials having arbitrary spectra.

2. **Measurement of Band-gap in the presence of Light Waves**: With the advent of nano-photonics, there has been considerable interest in studying the optical processes in semiconductors and their nanostructures in the presence of intense light waves [36–44]. *It appears from the literature, that the investigations in the presence of external intense photo-excitation have been carried out on the assumption that the carrier energy spectra are invariant quantities under strong external light waves, which is not fundamentally true.* The physical properties of semiconductors in the presence of strong light waves which alter the basic dispersion relations have relatively been much less investigated in [45] as compared with the cases of other external fields needed for the characterization of the low dimensional semiconductors.

With the radical change in the dispersion relation, it is evident that the band gap will also change and in this section we study the normalized incremental band gap (ΔE_g) as functions of incident light intensity and the wave length respectively in the presence of strong light excitation.

Using (10.41–10.43), the ΔE_g has been plotted as a function of normalized I_0 (for a given wavelength and considering red light for which $\lambda = 660$ nm) at $T = 4.2$ K in Figs. 4.2 and 4.3 for n-$Hg_{1-x}Cd_xTe$ and n-$In_{1-x}Ga_xAs_yP_{1-y}$ lattice matched to InP in accordance with the perturbed three and two band models of Kane and that of perturbed parabolic energy bands respectively. In Figs. 4.4 and 4.5, the normalized incremental band gap has been plotted for the aforementioned optoelectronic compounds as a function of λ. It is worth remarking that the influence of an external photo-excitation is to change radically the original band structure of the material. Because of this change, the photon field causes to increase the band gap of

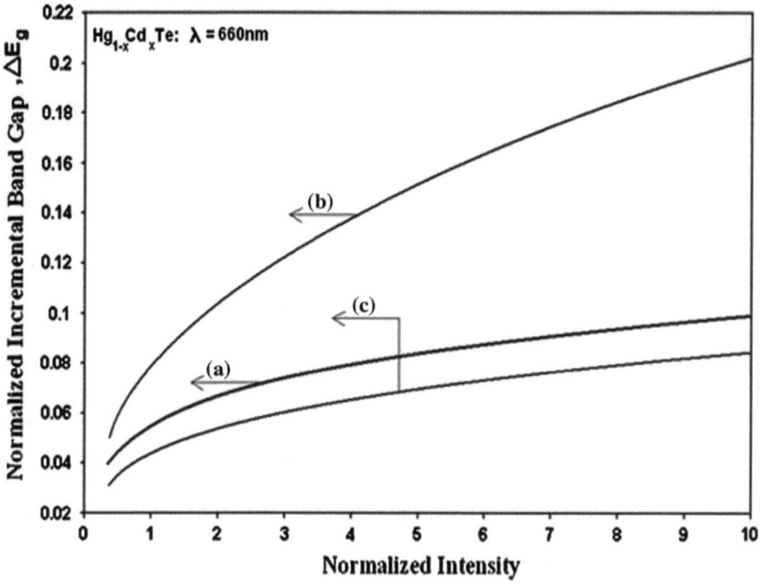

Fig. 4.2 Plots of the normalized incremental band gap (ΔE_g) for n-$Hg_{1-x}Cd_xTe$ as a function of normalized light intensity in which the *curves a* and *b* represent the perturbed three and two band models of Kane respectively. The *curve c* represents the same variation in n-$Hg_{1-x}Cd_xTe$ in accordance with the perturbed parabolic energy bands

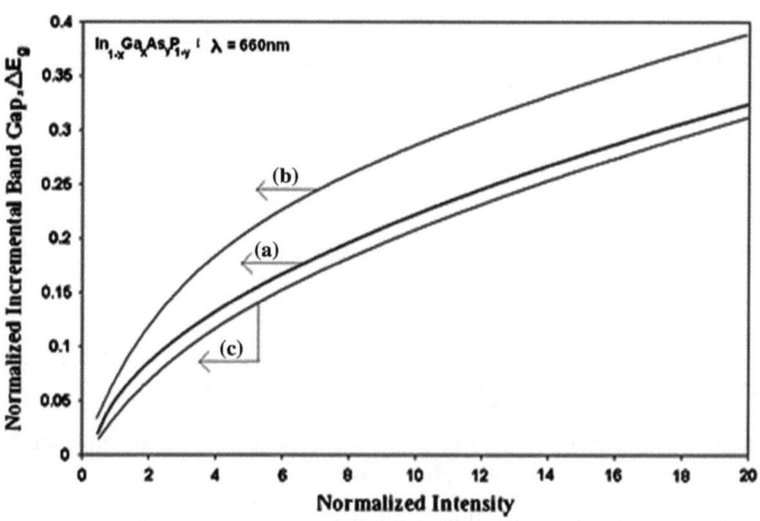

Fig. 4.3 Plots of the normalized incremental band gap (ΔE_g) for $In_{1-x}Ga_xAs_yP_{1-y}$ lattice matched to InP as a function of normalized light intensity for all cases of Fig. 4.2

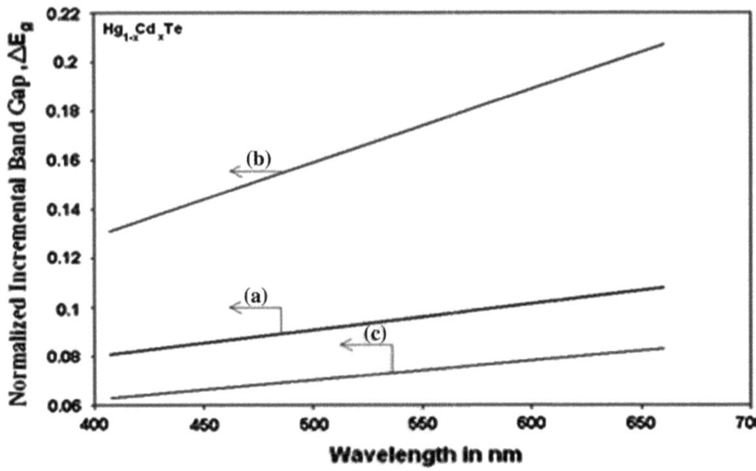

Fig. 4.4 Plots of the normalized incremental band gap (ΔE_g) for $Hg_{1\text{-}x}Cd_xTe$ as a function of wavelength for all cases of Fig. 4.2

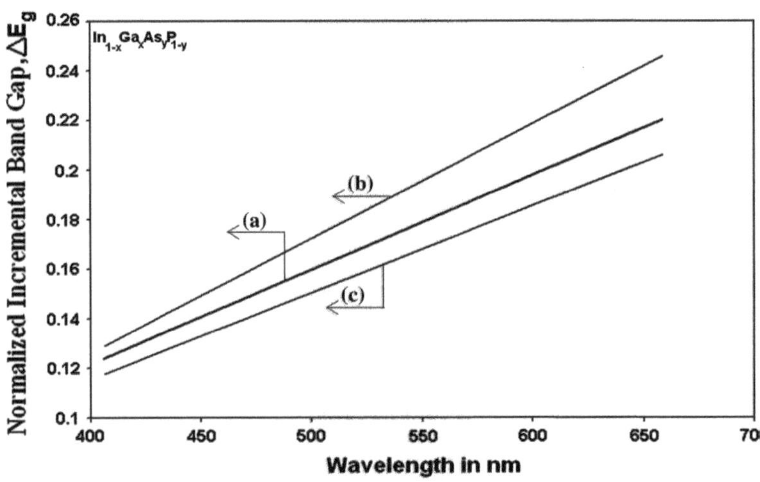

Fig. 4.5 Plots of the normalized incremental band gap (ΔE_g) for $In\text{-}_{1\text{-}x}Gs_xAs_yP_{1\text{-}y}$ lattice matched to InP as a function of wavelength for all cases of Fig. 4.2

semiconductors. We propose the following two experiments for the measurement of band gap of semiconductors under photo-excitation.

(a) A white light with colour filter is allowed to fall on a semiconductor and the optical absorption coefficient $(\bar{\alpha}_0)$ is being measured experimentally. For different colours of light, $(\bar{\alpha}_0)$ is measured and $(\bar{\alpha}_0)$ versus $\hbar\omega$ (the incident photon energy) is plotted and we extrapolate the curve such that $\bar{\alpha}_0 \to 0$ at a particular value $\hbar\omega_1$. This $\hbar\omega_1$ is the I_0 unperturbed band gap of the semiconductor. During this process, we vary the wavelength with fixed. From our present study, we have observed that the band gap of the semiconductor increases for various values of λ when I_0 is fixed (from Figs. 4.4 and 4.5). This implies that the band gap of the semiconductor measured (i.e. $\hbar\omega_1 = E_g$) is not the unperturbed band gap E_{g_0} but the perturbed band gap E_g; where $E_g = E_{g_0} + \Delta E$, ΔE_g is the increased band gap at $\hbar\omega_1$. Conventionally, we consider this E_g as the unperturbed band gap of the semiconductor and this particular concept needs modification. Furthermore, if we vary I_0 for a monochromatic light (when λ is fixed) the band gap of the semiconductor will also change consequently (Figs. 4.2 and 4.3). Consequently, the absorption coefficient will change with the intensity of light [45]. For the overall understanding, the detailed theoretical and experimental investigations are needed in this context for various materials having different band structures.

(b) The conventional idea for the measurement of the band gap of the semiconductors is the fact that the minimum photon energy hv (v is the frequency of the monochromatic light) should be equal to the band gap E_{g_0} (unperturbed) of the semiconductor, i.e.,

$$hv = E_{g_0} \tag{4.8}$$

In this case, λ is fixed for a given monochromatic light and the semiconductor is exposed to a light of wavelength λ. Also the intensity of the light is fixed. From Figs. 4.4 and 4.5, we observe that the band gap of the semiconductor is not $hv = E_{g_0}$ (for a minimum value of hv) but E_g, the perturbed band gap. Thus, we can rewrite the above equality as

$$hv = E_g \tag{4.9}$$

Furthermore, if we vary the intensity of light (Figs. 4.2 and 4.3) for the study of photoemission, the minimum photon energy should be

$$hv_1 = E_{g_1} \tag{4.10}$$

where E_{g_1} is the perturbed band gap of the semiconductor due to various intensity of light when v and v_1 are different.

Thus, we arrive at the following conclusions:

(a) Under different intensity of light, keeping λ fixed, the condition of band gap measurement is given by

$$hv_1 = E_{g_1} = E_{g_0} + \Delta E_{g_1} \tag{4.11}$$

(b) Under different colour of light, keeping the intensity fixed, the condition of band gap measurement assumes the form

$$h\nu = E_g = E_{g_0} + \Delta E_g \tag{4.12}$$

and not the conventional result as given by (4.8).

3. ***Diffusion Coefficient of the Minority Carriers***:
This particular coefficient in quantum confined lasers can be expressed as

$$D_i/D_0 = dE_{Fi}/dE_F \tag{4.13}$$

where D_i and D_0 are the diffusion coefficients of the minority carriers both in the presence and absence of quantum confinements and E_{Fi} and E_F are the Fermi energies in the respective cases. It appears then that, the formulation of the above ratio requires a relation between E_{Fi} and E_F, which, in turn, is determined by the appropriate carrier statistics. Thus, our present study plays an important role in determining the diffusion coefficients of the minority carriers of HD quantum-confined lasers with materials having arbitrary band structures. Therefore in the investigation of the optical excitation of the HD optoelectronic materials which lead to the study of the ambipolar diffusion coefficients the present results contribute significantly.

4. ***Nonlinear Optical Response***: The nonlinear response from the optical excitation of the free carriers is given by [46]

$$Z_0 = \frac{-e^2}{\omega^2\hbar^2}\int_0^\infty \left(k_x\frac{\partial k_x}{\partial E}\right)^{-1} f_0\, N(E)dE \tag{4.14}$$

where ω is the optical angular frequency, *N(E)* is the DOS function. From the various E-k relations of different HD materials under different physical conditions, we can formulate the expression of *N*(*E*) and from band structure we can derive the term $\left(k_x\frac{\partial k_x}{\partial E}\right)$ and thus by using the DOS function as formulated, we can study the Z_0 for all types of materials as considered in this monograph.

5. ***Third Order Nonlinear Optical Susceptibility***: This particular susceptibility can be written as [47]

$$\chi_{NP}(\omega_1,\omega_2,\omega_3) = \frac{n_0 e^4\langle\varepsilon^4\rangle}{24\,\omega_1\omega_2\omega_3(\omega_1+\omega_2+\omega_3)\hbar^4} \tag{4.15}$$

where $n_0\langle \varepsilon^4 \rangle = \int_0^\infty \frac{\partial^4 E}{\partial k_z^4} N(E) f_0 \, dE$ and the other notations are defined in [47]. The term $\left(\frac{\partial^4 E}{\partial k_z^4}\right)$ can be formulated by using the dispersion relations of different HD materials as given in appropriate sections of this monograph. Thus one can investigate the $\chi_{NP}(\omega_1, \omega_2, \omega_3)$ for all materials as considered in this monograph.

6. ***Generalized Raman Gain***: The generalized Raman gain in optoelectronic materials can be expressed as [48]

$$R_G = \bar{I}\left(\frac{16\pi^2 c^2}{\hbar\omega\rho g \omega_s^2 n_s n_p}\right)\left(\frac{\Gamma_\rho}{\Gamma}\right)\left(\left(\frac{e^2}{mc^2}\right)^2 m^2 R^2\right) \tag{4.16}$$

where, $\bar{I} = \sum_{n.t_z} [f_0(n, k_z \uparrow) - f_0(n, k_z \downarrow)]$, $f_0(n, k_z \uparrow)$ is the Fermi factor for spin-up Landau levels, $f_0\ (n, k_z \downarrow)$ is the Fermi factor for spin down Landau levels, n is the Landau quantum number and the other notations are defined in [48]. It appears then the formulation of R_G is determined by the appropriate derivation of the magneto-dispersion relations. By using the different appropriate formulas as formulated in various HD materials in different chapters of this monograph R_G can, in general, be investigated.

4.4 Open Research Problem

R.4.1 Investigate experimentally the diffusivity for all the HD systems as discussed in this monograph in the presence of arbitrarily oriented strain.

References

1. W. Zawadzki, Adv. Phys. **23**, 435 (1974)
2. B.M. Askerov, N.F. Gashimzade, M.M. Panakhov, Sov. Phys. Solid State **29**, 465 (1987)
3. S.P. Zelenin, A.S. Kondratev, A.I. Kuchma, Sov. Phys. Semicond. **16**, 355 (1982)
4. A.V. Germaneko, G.M. Minkov, Phys. Stat. Sol. (b) **184**, 9 (1994)
5. W. Zawadzki, In: Two-Dimensional Systems, Heterostructures and Superlattices edited by G. Bower, F. Kuches, H. Haiunich, Springer Ser. Solid-State Sci. (Springer, Berlin, Heidelberg) (1984) Vol. 53, p. 79
6. K.P. Ghatak, B. De, B. Nag, P.K. Chakraborty, Nonlin. Opt. Quant. Opt. **16**, 212 (1996)
7. K.P. Ghatak, M. Mitra, B. Goswami, B. Nag, Nonlin. Opt. **16**, 167 (1996)
8. K.P. Ghatak, B. Nag, Nanostruct. Mater. **2**, 14 (1995)
9. K.P. Ghatak, B. Mitra, Nouvo Cimento **15D**, 97 (1993)
10. S.K. Biswas, A.R. Ghatak, A. Neogi, A. Sharma, S. Bhattacharya, K.P. Ghatak, Physica E **36**, 163 (2007)
11. K.P. Ghatak, S. Bhattacharya, S. Bhowmik, R. Benedictus, S. Choudhury, J. Appl. Phys. **103**, 1 (2008)

12. M. Mondal, A. Ghoshal, K.P. Ghatak, Nouvo Cimento D **14D**, 63 (1992)
13. K.P. Ghatak, Nouvo Cimento D **13D**, 1321 (1992)
14. K.P. Ghatak, M. Mondal, J. Appl. Phys. **70**, 299 (1991)
15. K.P. Ghatak, S.N. Biswas, J. Low Temp. Phys. **78**, 219 (1990)
16. K.P. Ghatak, M. Mondal, J. Appl. Phys. **65**, 3480 (1989)
17. K.P. Ghatak, M. Mondal, J. Appl. Phys. **66**, 3056 (1989)
18. B. Mitra, K.P. Ghatak, Phys. Letts. **141A**, 81 (1989)
19. K.P. Ghatak, B. De, M. Modal, S.N. Biswas, Epitaxial heterostructures. MRS Symposium Proceedings Spring Meeting **198**, 327 (1990)
20. K.P. Ghatak, S.N. Biswas, Long wave length semiconductor devices. Materials and Processes MRS Symposium Proceedings, MRS Spring Meeting **216**, 465 (1990)
21. K.P. Ghatak, B. De, Modern perspective on thermoelectrics and related materials. MRS Symposium Proceedings, Spring Meeting **234**, 55 (1991)
22. Modern Perspective on Thermoelectrics and related materials, MRS Symposium Proceedings. Spring Meeting **234**, 59 (1991)
23. K. P. Ghatak, Proc. of SPIE, USA, Fiber Optic and Laser Sensors IX, **1584**, 435, (1992)
24. B.R. Nag, *Electron transport in compound semiconductors, springer series in solid state sciences*, vol. 11 (Springer, Germany, 1980)
25. E.A. Aruhanov, A.F. Knyazev, A.N. Nateprov, S.I. Radautsan, Sov. Phys. Semiconductors **15**, 828 (1981)
26. K. P. Ghatak, S. Bhattyacharya, *Thermoelectric power in nanostructured materials: Strong Magnetic Fields*, Springer Series in Materials Science, vol 137 (Springer, Germany, 2010)
27. R.W. Keyes, Solid State Phys. **20**, 37 (1967)
28. S. Bhattacharya, S. Chowdhury, K.P. Ghatak, J. Comput. Theory Nanosci. **3**, 423 (2006)
29. B. Nag, K.P. Ghatak, Nonlinear Opt. **19**, 1 (1998)
30. K.P. Ghatak, J.P. Banerjee, B. Goswami, B. Nag, Nonlinear Opt. Quant. Opt. **16**, 241 (1996)
31. K.P. Ghatak, Int. J. Electron. **71**, 239 (1991)
32. K.P. Ghatak, B. De, S.N. Biswas, M. Mondal, Mechanical behavior of materials and structures in microelectronics. MRS Symposium Proceedings, Spring Meeting **216**, 191 (1991)
33. K.P. Ghatak, B. De, MRS Symp. Proc. **226**, 191 (1991)
34. K.P. Ghatak, B. Nag, G. Majumdar, MRS Symp. Proc. **379**, 109 (1995)
35. D. Baruah, S. Choudhury, K.M. Singh, K.P. Ghatak, J. Phys. Conf. Series **61**, 80 (2007)
36. K.P. Ghatak, S. Bhattacharya, S. Pahari, D. De, S. Ghosh, M. Mitra, Ann. Phys. **17**, 5 (2008)
37. P.K. Bose, N. Paitya, S. Bhattacharya, D. De, S. Saha, K.M. Chatterjee, S. Pahari, K.P. Ghatak, Quantum Matter **1**, 89 (2012)
38. S. Bhattarchya, D. De, S. Ghosh, P. Banerjee, S. Saha, M. Mitra, B. Nag, M. Pal, S.K. Biswas, K.P. Ghatak, J. Comput. Theory Nanosci. **7**, 1066 (2010)
39. K.P. Ghatak, S. Bhattacharya, S. Pahari, S.N. Mitra, P.K. Bose, D. De, J. Phys. Chem. Solids **70**, 122 (2009)
40. S. Bhattacharya, N.C. Paul, D. De, K.P. Ghatak, Phys. B: Condens. Matter **403**, 4139 (2008)
41. K.P. Ghatak, S. Bhattacharya, D. De, R. Sarkar, S. Pahari, A. Dey, A.K. Dasgupta, J. Comput. Theory Nanosci. **5**, 1345 (2008)
42. K.P. Ghatak, S. Bhattacharya, S. Pahari, D. De, S. Ghosh, M. Mitra, Ann. Phys. **17**, 195 (2008)
43. K.P. Ghatak, S. Bhattacharya, J. Appl. Phys. **102**, 073704 (2007)
44. S. Mukherjee, D. De, D.J. Mukherjee, S. Bhattacharya, A. Sinha, K.P. Ghatak, Phys. B: Condens. Matter **393**, 347 (2007)
45. P.K. Chakraborty, S. Bhattacharya, K.P. Ghatak, J. Appl. Phys. **98**, 053517 (2005)
46. A.S. Filipchenko, I.G. Lang, D.N. Nasledov, S.T. Pavlov, L.N. Radaikine, Phys. Stat. Sol. (b) **66**, 417 (1974)
47. M. Wegener, *Extreme nonlinear optics* (Springer, Germany, 2005)
48. B.S. Wherreff, W. Wolland, C. R. Pidgeon, R.B. Dennis, S.D. Smith, in *Proceedings of the 12th International Conference of the Physics of the Semiconductors*, ed. by M.H. Pilkahn, R.G. Tenbner (Staffgard, 1978), p. 793

Chapter 5
Conclusion and Scope for Future Research

This monograph deals with the ER in various types of low dimensional HD materials. The external photo excitation, quantization and strong electric field alter profoundly the basic band structures, which, in turn, generate pinpointed knowledge regarding ER in various HDS and their nanostructures having different carrier energy spectra. The in-depth experimental investigations covering the whole spectrum of solid state and allied science in general, are extremely important to uncover the underlying physics and the related mathematics. The ER is basically the motion dependent phenomena and we have formulated the simplified expressions of ER for few HD quantized structures together with the fact that our investigations are based on the simplified ***k.p*** formalism of solid-state science without incorporating the advanced field theoretic techniques. In spite of such constraints, the role of band structure behind the curtain, which generates, in turn, new concepts are truly amazing and discussed throughout the text.

We present the set of few open research problems in this pin pointed topic of research of modern physics.

(R5.1) Investigate the ER in the presence of a quantizing magnetic field under exponential, Kane, Halperin, Lax and Bonch-Bruevich band tails [1] for all the problems of this monograph of all the whose unperturbed carrier energy spectra are defined in Chap. 1 by including spin and broadening effects.

(R5.2) Investigate all the appropriate problems after proper modifications introducing new theoretical formalisms for the problems as defined in (R5.1) for HD negative refractive index, macro molecular, nitride and organic materials.

(R5.3) Investigate all the appropriate problems of this monograph for all types of HD quantum confined p-InSb, p-CuCl and semiconductors having diamond structure valence bands whose dispersion relations of the carriers in bulk materials are given by Cunningham [2], Yekimov et al. [3] and Roman et al. [4] respectively.

K. P. Ghatak and S. Bhattacharya, *Heavily-Doped 2D-Quantized Structures and the Einstein Relation*, Springer Tracts in Modern Physics 260,
DOI 10.1007/978-3-319-08380-3_5, © Springer International Publishing Switzerland 2015

(R5.4) Investigate the influence of defect traps and surface states separately on the ER of the for all the appropriate problems of all the chapters after proper modifications.
(R5.5) Investigate the ER of the under the condition of non-equilibrium of the carrier states for all the appropriate problems of this monograph.
(R5.6) Investigate the ER for all the appropriate problems of this monograph for the corresponding HD p-type semiconductors and their nanostructures.
(R5.7) Investigate the ER for all the appropriate problems of this monograph for all types of HD semiconductors and their nanostructures under mixed conduction in the presence of strain.
(R5.8) Investigate the ER for all the appropriate problems of this monograph for all types of HD semiconductors and their nanostructures in the presence of hot electron effects.
(R5.9) Investigate the ER for all the appropriate problems of this monograph for all types of HD semiconductors and their nanostructures for nonlinear charge transport.
(R5.10) Investigate the ER for all the appropriate problems of this monograph for all types of HD semiconductors and their nanostructures in the presence of strain in an arbitrary direction.
(R5.11) Investigate all the appropriate problems of this monograph for strongly correlated electronic HD systems in the presence of strain.
(R5.12) Investigate all the appropriate problems of this chapter in the presence of arbitrarily oriented photon field and strain.
(R5.13) Investigate all the appropriate problems of this monograph for all types of HD nanotubes in the presence of strain.
(R5.14) Investigate all the appropriate problems of this monograph for various types of pentatellurides in the presence of strain.
(R5.15) Investigate all the appropriate problems of this monograph for HD Bi_2Te_3–Sb_2Te_3 super-lattices in the presence of strain.
(R5.16) Investigate the influence of temperature-dependent energy band constants for all the appropriate problems of this monograph.
(R5.17) Investigate the influence of the localization of carriers on the ER in HDS for all the appropriate problems of this monograph.
(R5.18) Investigate ER for HD p-type SiGe under different appropriate physical conditions as discussed in this monograph in the presence of strain.
(R5.19) Investigate ER for different metallic alloys under different appropriate physical conditions as discussed in this monograph in the presence of strain.
(R5.20) Investigate ER for different intermetallic compounds under different appropriate physical conditions as discussed in this monograph in the presence of strain.
(R5.21) Investigate ER for HD GaN under different appropriate physical conditions as discussed in this monograph in the presence of strain.

(R5.22) Investigate ER for different disordered HD conductors under different appropriate physical conditions as discussed in this monograph in the presence of strain.

(R5.23) Investigate ER for various semi metals under different appropriate physical conditions as discussed in this monograph in the presence of strain.

(R5.24) Investigate all the appropriate problems of this monograph for HD $Bi_2Te_{3-x}Se_x$ and $Bi_{2-x}Sb_xTe_3$ respectively in the presence of strain.

(R5.25) Investigate all the appropriate problems of this monograph for all types of skutterudites in the presence of strain.

(R5.26) Investigate all the appropriate problems of this monograph in the presence of crossed electric and quantizing magnetic fields.

(R5.27) Investigate all the appropriate problems of this monograph in the presence of crossed alternating electric and quantizing magnetic fields.

(R5.28) Investigate all the appropriate problems of this monograph in the presence of crossed electric and alternating quantizing magnetic fields.

(R5.29) Investigate all the appropriate problems of this monograph in the presence of alternating crossed electric and alternating quantizing magnetic fields.

(R5.30) Investigate all the appropriate problems of this monograph in the presence of arbitrarily oriented pulsed electric and quantizing magnetic fields.

(R5.31) Investigate all the appropriate problems of this monograph in the presence of arbitrarily oriented alternating electric and quantizing magnetic fields.

(R5.32) Investigate all the appropriate problems of this monograph in the presence of crossed in homogeneous electric and alternating quantizing magnetic fields.

(R5.33) Investigate all the appropriate problems of this monograph in the presence of arbitrarily oriented electric and alternating quantizing magnetic fields under strain.

(R5.34) Investigate all the appropriate problems of this monograph in the presence of arbitrarily oriented electric and alternating quantizing magnetic fields under light waves.

(R5.35) Investigate all the appropriate problems of this monograph in the presence of arbitrarily oriented pulsed electric and alternating quantizing magnetic fields under light waves.

(R5.36) Investigate all the appropriate problems of this monograph in the presence of arbitrarily oriented inhomogeneous electric and pulsed quantizing magnetic fields in the presence of strain and light waves.

(R5.37) ***(a) Investigate the ER for all types of of this monograph in the presence of many body effects strain and arbitrarily oriented respectively.***

(b) Investigate all the appropriate problems of this chapter for the Dirac electron.

(c) Investigate all the problems of this monograph by removing all the physical and mathematical approximations and establishing the respective appropriate uniqueness conditions.

The ER is the consequence of motion induced phenomena of solid state science and all the assumptions behind the said phenomena are also applicable to ER. ***The formulation of ER for all types of and their quantum confined counter parts considering the influence of all the bands created due to all types of quantizations after removing all the assumptions and establishing the respective appropriate uniqueness conditions is, in general, an extremely difficult problem***. **200 open research problems** have been presented in this monograph and we hope that the readers will not only solve them but also will generate new concepts, both theoretical and experimental. Incidentally, we can easily infer how little is presented and how much more is yet to be investigated in this exciting topic which is the signature of coexistence of new physics, advanced mathematics combined with the inner fire for performing creative researches in this context from the young scientists since like Kikoin [5] we firmly believe that "***A young scientist is no good if his teacher learns nothing from him and gives his teacher nothing to be proud of***". In the mean time our research interest has been shifted and we are leaving this particular beautiful topic with the hope that (R5.37) alone is sufficient to draw the attention of the researchers from diverse fields and our readers are surely in tune with the fact that "***Exposition, criticism, appreciation is the work for second-rate minds***" [6].

References

1. B.R. Nag, *Electron transport in compound semiconductors, springer series in solid state sciences*, vol. 11 (Springer, Germany, 1980)
2. R.W. Cunningham, Phys. Rev. **167**, 761 (1968)
3. A.I. Yekimov, A.A. Onushchenko, A.G. Plyukhin, Al L. Efros, J. Expt Theory Phys. **88**, 1490 (1985)
4. B.J. Roman, A.W. Ewald, Phys. Rev. B **5**, 3914 (1972)
5. I.K. Kikoin, *Science for everyone: encounters with physicists and physics* (Mir Publishers, Russia, 1989), p. 154
6. G.H. Hardy, *A mathematician's Apology* (Cambridge University Press, 1990, p. 61)

Chapter 6
Appendix A: The ER in HDS Under Magnetic Quantization

6.1 Introduction

It is well known that the band structure of semiconductors can be dramatically changed by applying the external fields. The effects of the quantizing magnetic field on the band structure of compound semiconductors are more striking and can be observed easily in experiments [1–67]. Under magnetic quantization, the motion of the electron parallel to the magnetic field remains unaltered while the area of the wave vector space perpendicular to the direction of the magnetic field gets quantized in accordance with the Landau's rule of area quantization in the wave-vector space [39–67]. The energy levels of the carriers in a magnetic field (with the component of the wave-vector parallel to the direction of magnetic field be equated with zero) are termed as the Landau levels and the quantized energies are known as the Landau sub-bands. It is important to note that the same conclusion may be arrived either by solving the single-particle time independent Schrödinger differential equation in the presence of a quantizing magnetic field or by using the operator method. The quantizing magnetic field tends to remove the degeneracy and increases the band gap. A semiconductor, placed in a magnetic field B, can absorb radiative energy with the frequency $(\omega_0 = (|e|B/m_c))$. This phenomenon is known as cyclotron or diamagnetic resonance. The effect of energy quantization is experimentally noticeable when the separation between any two consecutive Landau levels is greater than k_BT. A number of interesting transport phenomena originate from the change in the basic band structure of the semiconductor in the presence of quantizing magnetic field. These have been widely been investigated and also served as diagnostic tools for characterizing the different materials having various band structures [68–72]. The discreteness in the Landau levels leads to a whole crop of magneto-oscillatory phenomena, important among which are (i) Shubnikov-de Haas oscillations in magneto-resistance; (ii) De Haas-van Alphen oscillations in magnetic susceptibility; (iii) magneto-phonon oscillations in thermoelectric power, etc.

K. P. Ghatak and S. Bhattacharya, *Heavily-Doped 2D-Quantized Structures and the Einstein Relation*, Springer Tracts in Modern Physics 260,
DOI 10.1007/978-3-319-08380-3_6, © Springer International Publishing Switzerland 2015

In Sect. 6.2.1 of the theoretical background, the ER has been investigated in HD non linear optical semiconductors in the presence of a quantizing magnetic field. Section 6.2.2 contains the results for HD III-V, ternary and quaternary compounds in accordance with the three and the two band models of Kane. In the same section the ER in accordance with the models of Stillman et al. and Palik et al. have also been studied for the purpose of relative comparison. Section 6.2.3 contains the study of the ER for HD II-VI semiconductors under magnetic quantization. In Sect. 6.2.4, the ER in HD IV-VI materials has been discussed in accordance with the models of Cohen, Lax, Dimmock, Bangert and Kastner and Foley and Landenberg respectively. In Sect. 6.2.5, the magneto-ER for the stressed HD Kane type semiconductors has been investigated. In Sect. 6.2.6, the ER in HD Te has been studied under magnetic quantization. In Sect. 6.2.7, the magneto-ER in n-GaP has been studied. In Sect. 6.2.8, the ER in HD $PtSb_2$ has been explored under magnetic quantization. In Sect. 6.2.9, the magneto-ER in HD Bi_2Te_3 has been studied. In Sect. 6.2.10, the ER in HD Ge has been studied under magnetic quantization in accordance with the models of Cardona et al. and Wang and Ressler respectively. In Sects. 6.2.11 and 6.2.12, the magneto-ER in HD n-GaSb and II-V compounds has respectively been studied. In Sects. 6.2.13 the magneto ER in HD $Pb_{1-x}Ge_xTe$ has been discussed. The last Sect. 6.3 contains 52 open research problems for this chapter.

6.2 Theoretical Background

6.2.1 The ER in HD Nonlinear Optical Semiconductors Under Magnetic Quantization

The dispersion relation under magnetic quantization in non-linear optical materials can be written as

$$\gamma(E) = \frac{\hbar^2 k_s^2}{2m_\perp^*} f_3(E) + \frac{\hbar^2 k_z^2}{2m_\parallel^*} f_4(E) \pm \frac{eB\hbar E_g}{6}\left[\frac{(E_{g_0} + \Delta_\perp)}{(E_{g_0} + \frac{2}{3}\Delta_\perp)}\right]\left[E + E_{g_0} + \delta\frac{\Delta_\parallel^2 - \Delta_\perp^2}{3\Delta_\parallel}\right] \tag{6.1}$$

where

$$f_3(E) = \frac{E_g(E_g + \Delta_\perp)}{(E_g + \frac{2}{3}\Delta_\perp)}\left[(E + E_g)\left(E + E_g + \frac{2}{3}\Delta_\parallel\right) + \delta\left(E + E_g + \frac{1}{3}\Delta_\parallel\right) + \frac{1}{9}\left(\Delta_\parallel^2 - \Delta_\perp^2\right)\right]$$

and

$$f_4(E) = \frac{E_g(E_g + \Delta_\parallel)}{(E_g + \frac{2}{3}\Delta_\parallel)}[(E + E_g)\ (E + E_g + \frac{2}{3}\Delta_\parallel)$$

The (6.1) can be expressed as

$$\begin{aligned}
\frac{\hbar^2 k_z^2}{2m_\parallel^*} + (\frac{b_\parallel c_\perp}{b_\perp c_\parallel})(\frac{\hbar^2 k_s^2}{2m_\perp^*}) = & \left\{ \left[\frac{ab_\parallel}{c_\parallel} E^2 + \frac{(ac_\parallel + b_\parallel c_\parallel - ab_\parallel)}{c_\parallel^2} E \right.\right. \\
& \left. + \frac{1}{c_\parallel}(1 - \frac{a}{c_\parallel})(1 - \frac{b_\parallel}{c_\parallel}) - \frac{1}{c_\parallel}(1 - \frac{a}{c_\parallel})(1 - \frac{b_\parallel}{c_\parallel})\frac{1}{(c_\parallel E + 1)} \right] \\
& \left. + \frac{ab_\parallel}{c_\parallel}[\delta E + \frac{2}{9}(\Delta_\parallel^2 - \Delta_\perp^2)] - \frac{2}{9}\frac{ab_\parallel}{c_\parallel}\frac{(\Delta_\parallel^2 - \Delta_\perp^2)}{(c_\parallel E + 1)} \right\} \\
& - (\frac{\hbar^2 k_s^2}{2m_\perp^*}) \left\{ (\frac{b_\parallel c_\perp}{b_\perp c_\parallel})[(\frac{\delta}{2} + \frac{\Delta_\parallel^2 - \Delta_\perp^2}{6\Delta_\parallel})\frac{a}{(aE + 1)} \right. \\
& \left. + (\frac{\delta}{2} - \frac{\Delta_\parallel^2 - \Delta_\perp^2}{6\Delta_\parallel})\frac{c_\parallel}{(c_\parallel E + 1)}] \right\} \\
& \pm e_1 \left[\frac{\rho_1}{E + E_g} + \frac{\rho_2}{E + E_g + \frac{2}{3}\Delta_\parallel} \right]
\end{aligned} \quad (6.2)$$

where

$$e_1 = \left[\frac{eB\hbar(E_g + \frac{2}{3}\Delta_\parallel)}{6(E_g + \Delta_\parallel)} \frac{(E_g + \Delta_\perp)}{(E_g + \frac{2}{3}\Delta_\perp)} \right], \quad \rho_1 = \frac{(-E_g + G_1)}{(\frac{2}{3}\Delta_\parallel)},$$

$$G_1 = \left[E_{g_0} + \delta + \frac{\Delta_\parallel^2 - \Delta_\perp^2}{3\Delta_\parallel} \right]$$

and

$$\rho_2 = \frac{3}{2\Delta_\parallel}\left[\frac{\Delta_\parallel}{3} + \frac{\Delta_\perp^2}{3\Delta_\parallel} - \delta \right]$$

Therefore, the dispersion relation of the conduction electrons in heavily doped non-linear optical semiconductors in the presence of a quantizing magnetic field B can be written following the methods as developed in Chap. 1 as

$$\frac{\hbar^2 k_z^2}{2m_\parallel^*} = U_{1,\pm}(E, n, \eta_g) + iU_{2,\pm}(E, n, \eta_g) \quad (6.3a)$$

where

$$
\begin{aligned}
U_{1,\pm}(E,n,\eta_g) = \Bigg[& \frac{-eB}{m_\perp^*}(n+\frac{1}{2})(\frac{b_\| c_\perp}{c_\| b_\perp}) + [\frac{1+Erf(E/\eta_g)}{2}]^{-1}[\{[\frac{ab_\|}{c_\|}\theta_0(E,\eta_g) \\
& + \frac{ac_\| + b_\| c_\| - ab_\|}{c_\|^2}\gamma_0(E,\eta_g) + \frac{1}{c_\|}(1-\frac{a}{c_\|})(1-\frac{b_\|}{c_\|})[\frac{1+Erf(E/\eta_g)}{2}] \\
& - \frac{1}{c_\|}(1-\frac{a}{c_\|})(1-\frac{b_\|}{c_\|})c(\beta_1,E,\eta_g) + \frac{ab_\|}{c_\|}[\delta\gamma_0(E,\eta_g) \\
& + \frac{2}{9}(\Delta_\|^2 - \Delta_\perp^2)[\frac{1+Erf(E/\eta_g)}{2}]] - \frac{2}{9}\frac{ab_\|}{c_\|}(\Delta_\|^2 - \Delta_\perp^2)c(\beta_1,E,\eta_g)\} \\
& - \frac{\hbar eB}{m_\perp^*}(n+\frac{1}{2})\{(\frac{b_\| c_\perp}{c_\| b_\perp})[(\frac{\delta}{2} + \frac{\Delta_\|^2 - \Delta_\perp^2}{6\Delta_\|})ac(\beta_2,E,\eta_g) \\
& + (\frac{\delta}{2} - \frac{\Delta_\|^2 - \Delta_\perp^2}{6\Delta_\|})c_\| c(\beta_1,E,\eta_g)]\} \pm \frac{\rho_1 e_1}{E_g}c(\beta_2,E,\eta_g) \\
& \pm \frac{\rho_2 e_1}{(E_g + \frac{2}{3}\Delta_\|)}c(\beta_3,E,\eta_g)\Bigg]
\end{aligned}
$$

and

$$
\begin{aligned}
U_{2,\pm}(E,n,\eta_g) = & \left[\frac{1+Erf(E/\eta_g)}{2}\right]^{-1}\Bigg[\frac{1}{c_\|}(1-\frac{a}{c_\|})(1-\frac{b_\|}{c_\|})D(\beta_1,E,\eta_g) \\
& + \frac{2}{9}\frac{ab_\|}{c_\|}(\Delta_\|^2 - \Delta_\perp^2)D(\beta_1,E,\eta_g) \\
& + \frac{\hbar eB}{m_\perp^*}(n+\frac{1}{2})\{(\frac{b_\| c_\perp}{c_\| b_\perp})\Bigg[(\frac{\delta}{2} + \frac{\Delta_\|^2 - \Delta_\perp^2}{6\Delta_\|})aD(\beta_2,E,\eta_g) \\
& + (\frac{\delta}{2} - \frac{\Delta_\|^2 - \Delta_\perp^2}{6\Delta_\|})c_\| D(\beta_1,E,\eta_g)\Bigg]\} \\
& \mp \left[\frac{\rho_1 e_1}{E_g}D(\beta_2,E,\eta_g) \pm \frac{\rho_2 e_1}{(E_g + \frac{2}{3}\Delta_\|)}D(\beta_3,E,\eta_g)\right]\Bigg], \\
& \beta_1 = c_\|,\ \beta_2 = \frac{1}{E_g} = a,\ \beta_3 = \frac{1}{E_g + \frac{2}{3}\Delta_\|},
\end{aligned}
$$

$$
C(\beta_i,E,\eta_g) = \left[\frac{2}{\beta_i \eta_g \sqrt{\pi}}\right]\exp(-u_i^2) \times \left[\sum_{p=1}^{\infty}\{\frac{\exp(\frac{-p^2}{4})}{p}\}\sinh(pu_i)\right],\ u_i = \frac{1+\beta_i E}{\beta_i \eta_g}
$$

and

$$
D(\beta_i,E,\eta_g) = [\frac{\sqrt{\pi}}{\beta_i \eta_g}\exp(-u_i^2)]
$$

The EEM at the Fermi Level can be written from (6.3a) as

$$m^*_{\pm}(E_{FBHD}, n, \eta_g) = m^*_{\|} U'_{1,\pm}(E_{FBHD}, n, \eta_g) \tag{6.3b}$$

where E_{FBHD} is the Fermi energy in this case.

Therefore the double valued EEM in this case is a function of Fermi energy, magnetic field, quantum number and the scattering potential together with the fact that the EEM exists in the band gap which is the general characterstics of HD materials.

The complex density of states function under magnetic quantization is given by

$$\begin{aligned} N_B(E) &= N_{BR}(E) + iN_{BI}(E) \\ &= \frac{eB}{2\pi^2\hbar^2}\sqrt{2m^*_{11}} \sum_{n=0}^{n_{\max}} [\frac{x'}{2\sqrt{x}} + \frac{iy'}{2\sqrt{y}}] \end{aligned} \tag{6.4a}$$

where

$$x = \frac{\sqrt{(U_{1,\pm}(E,n,\eta_g))^2 + (U_{2,\pm}(E,n,\eta_g))^2} + (U_{1,\pm}(E,n,\eta_g))}{2},$$

$$y = \frac{\sqrt{(U_{1,\pm}(E,n,\eta_g))^2 + (U_{2,\pm}(E,n,\eta_g))^2} - (U_{1,\pm}(E,n,\eta_g))}{2}$$

and x' and y' are the differentiations of x and y with respect to energy E.

Therefore, from (6.4a) we can write

$$N_{BR}(E) = \frac{eB\sqrt{2m^*_{\|}}}{4\pi^2\hbar^2} \sum_{n=0}^{n_{\max}} \frac{x'}{\sqrt{x}} \tag{6.4b}$$

and

$$N_{BI}(E) = \frac{eB\sqrt{2m^*_{\|}}}{4\pi^2\hbar^2} \sum_{n=0}^{n_{\max}} \frac{y'}{\sqrt{y}} \tag{6.4c}$$

The electron concentration under the condition of extreme carrier degeneracy is given by

$$\begin{aligned} n_0 = \frac{g_v eB}{2\pi^2\hbar^2}\frac{\sqrt{2m^*_{\|}}}{\sqrt{2}} \sum_{n=0}^{n_{\max}} \Big[& \sqrt{(U_{1,\pm}(\mathrm{E}_{FBHD}, n, \eta_g))^2 + (U_{2,\pm}(\mathrm{E}_{FBHD}, n, \eta_g))^2} \\ & + (U_{1,\pm}(\mathrm{E}_{FBHD}, n, \eta_g))\Big]^{1/2} \end{aligned} \tag{6.5}$$

The ER in this case can be written as

$$\left(\frac{D}{\mu}\right)_{HDB} = Real\ part\ of\ \left[(n_0/e)\left(\frac{\partial n_0}{\partial(E_{FBHD} - E_{nHD})}\right)^{-1}\right] \tag{6.6}$$

where E_{nHD} is the complex Landau sub-band energy which can be obtained from (6.3a) by substituting $k_z = 0$ and $E = E_{nHD}$.

Thus using (6.6), (6.5) and (6.3a) we can study the ER in this case.

6.2.2 The ER in HD Kane Type III-V, Ternary and Quaternary Semiconductors Under Magnetic Quantization

(a) The electron energy spectrum in III-V semiconductors under magnetic quantizaion is given by

$$\frac{E(E+E_g)(E+E_g+\Delta)(E_g+\frac{2}{3}\Delta)}{E_g(E_g+\Delta)(E+E_g+\frac{2}{3}\Delta)} = (n+\frac{1}{2})\hbar\omega_0 + \frac{\hbar^2 k_z^2}{2m_c} \pm \frac{eB\hbar\Delta}{6m_c(E+E_g+\frac{2}{3}\Delta)}. \tag{6.7}$$

The (6.7) can be written as

$$\frac{ab}{c}E^2 + (\frac{ac+bc-ab}{c^2})E + \frac{1}{c}(1-\frac{a}{c})(1-\frac{b}{c}) - \frac{1}{c}(1-\frac{a}{c})(1-\frac{b}{c})\frac{1}{(1+cE)}$$
$$= (n+\frac{1}{2})\hbar\omega_0 + \frac{\hbar^2 k_z^2}{2m_c} \pm \frac{eB\hbar\Delta}{6m_c(1+cE)(E_g+\frac{2}{3}\Delta)}$$

where $a = \frac{1}{E_g}$, $b = \frac{1}{E_g+\Delta}$ and $c = \frac{1}{E_g+\frac{2}{3}\Delta}$

Therefore

$$\frac{ab}{c}I(5) + \left(\frac{ab+bc-ab}{c^2}\right)I(4) + \left[\frac{1}{c}(1-\frac{a}{c})(1-\frac{b}{c}) - (n+\frac{1}{2})\hbar\omega_0\right]I(1)$$
$$- g_{\pm}\left[G(C,E,\eta_g) - iH(C,E,\eta_g)\right] = \frac{\hbar^2 k_z^2}{2m_c}I(1) \tag{6.8}$$

where $g_{\pm} = \left[\frac{1}{c}(1-\frac{a}{c})(1-\frac{b}{c}) \pm \frac{eB\hbar\Delta}{6m_c(E_g+\frac{2}{3}\Delta)}\right]$,

$$G(C,E,\eta_g) = \left[\frac{2}{C\eta_g\sqrt{\pi}}\right]\exp(-u^2) \times \left[\sum_{p=1}^{\infty}\left\{\frac{\exp(\frac{-p^2}{4})}{p}\right\}\sinh(pu)\right], \quad u = \frac{1+CE}{C\eta_g},$$

$$H(C,E,\eta_g) = \left[\frac{\sqrt{\pi}}{C\eta_g}\exp(-u^2)\right]$$

Therefore

$$\frac{\hbar^2 k_z^2}{2m_c} = \left[\frac{ab}{c}\right]\theta_0(E,\eta_g)\left[\frac{1+Erf(E/\eta_g)}{2}\right]^{-1} + \left(\frac{ac+bc-ab}{c^2}\right)\gamma_0(E,\eta_g)\left[\frac{1+Erf(E/\eta_g)}{2}\right]^{-1} + \left[\frac{1}{c}(1-\frac{a}{c})(1-\frac{b}{c}) - (n+\frac{1}{2})\hbar\omega_0\right] - g_{\pm}\left[\frac{1+Erf(E/\eta_g)}{2}\right]^{-1}\left[G(C,E,\eta_g) - iH(C,E,\eta_g)\right] \tag{6.9}$$

Therefore the dispersion relation is given by

$$\frac{\hbar^2 k_z^2}{2m_c} = U_{3,\pm}(E,n,\eta_g) + iU_{4,\pm}(E,\eta_g) \tag{6.10a}$$

where

$$U_{3,\pm}(E,n,\eta_g) = \left[\frac{ab}{c}\theta_0(E,\eta_g) \times \left[\frac{1+Erf(E/\eta_g)}{2}\right]^{-1} + \left(\frac{ac+bc-ab}{c^2}\right)\gamma_0(E,\eta_g) \times \left[\frac{1+Erf(E/\eta_g)}{2}\right]^{-1} + \frac{1}{c}(1-\frac{a}{c})(1-\frac{b}{c}) - (n+\frac{1}{2})\hbar\omega_0 + g_{\pm}\left[\frac{1+Erf(E/\eta_g)}{2}\right]^{-1} G(C,E,\eta_g)\right]$$

and

$$U_{4,\pm}(E,\eta_g) = g_{\pm}\left[\frac{1+Erf(E/\eta_g)}{2}\right]^{-1} H(C,E,\eta_g)$$

The complex Landau energy E_{nHD1} in this case can be obtained by substituting $k_z = 0$ and $E = E_{nHD}$ in (6.10a).

The EEM at the Fermi Level can be written from (6.10a) as

$$m_{\pm}^*(E_{FBHD},n,\eta_g) = m_{\parallel}^* U'_{3,\pm}(E_{FBHD},n,\eta_g) \tag{6.10b}$$

Thus the EEM is a function of Fermi energy, Landau quantum number and scattering potential together with the fact it is double valued due to spin.

The complex density of states function under magnetic quantization is given by

$$N_B(E) = N_{BRI}(E) + iN_{BI1}(E) = \frac{eB}{2\pi^2\hbar^2}\sqrt{2m_c}\sum_{n=0}^{n_{\max}}\left[\frac{x_1'}{2\sqrt{x_1}} + \frac{iy_1'}{2\sqrt{y_1}}\right] \tag{6.11a}$$

where

$$x_1 = \frac{\sqrt{(U_{3,\pm}(E,n,\eta_g))^2 + (U_{4,\pm}(E,n,\eta_g))^2} + (U_{3,\pm}(E,n,\eta_g))}{2},$$

$$y_1 = \frac{\sqrt{(U_{3,\pm}(E,n,\eta_g))^2 + (U_{4,\pm}(E,n,\eta_g))^2} - (U_{3,\pm}(E,n,\eta_g))}{2}$$

and x_1' and y_1' are the differentiations of x and y with respect to energy E.

From (6.11a) we can write

$$N_{BR1}(E) = \frac{eB\sqrt{2m_c}}{4\pi^2\hbar^2}\sum_{n=0}^{n_{\max}}\frac{x_1'}{\sqrt{x_1}} \quad (6.11b)$$

and

$$N_{BI1}(E) = \frac{eB\sqrt{2m_c}}{4\pi^2\hbar^2}\sum_{n=0}^{n_{\max}}\frac{y_1'}{\sqrt{y_1}} \quad (6.11c)$$

The electron concentration under the condition of extreme degeneracy is given by

$$n_0 = \frac{g_v eB}{2\pi^2\hbar^2}\sqrt{m_c}\sum_{n=0}^{n_{\max}}\left[\sqrt{(U_{3,\pm}(\mathrm{E}_{FBHD},n,\eta_g))^2 + (U_{4,\pm}(\mathrm{E}_{FBHD},n,\eta_g))^2} + (U_{3,\pm}(\mathrm{E}_{FBHD},n,\eta_g))\right]^{1/2} \quad (6.12)$$

Thus using (6.6), (6.12) and the allied definitions we can study the ER in this case.

(b) Two band model of Kane

The magneto-dispersion law in this case is given by

$$\frac{\hbar^2k^2}{2m_c} = \gamma_2(E,\eta_g) - (n+\frac{1}{2})\hbar\omega_0 \mp \frac{1}{2}g^*\mu_0 B \quad (6.13a)$$

where g^* is the magnitude of the effective g factor at the edge of the conduction band and μ_0 is the Bohr magnetron.

The EEM at the Fermi Level can be written from (6.13a) as

$$m^*(E_{FBHD},\eta_g) = m_c\gamma_2'(E_{FBHD},\eta_g) \quad (6.13b)$$

Thus EEM is independent of quantum number.

The electron concentration under the condition of extreme degeneracy is given by

$$n_0 = \frac{g_v eB}{\pi^2 \hbar^2} \sqrt{m_c} \sum_{n=0}^{n_{\max}} (U_{5,\pm}(\mathrm{E}_{FBHD}, n, \eta_g))^{\frac{1}{2}} \tag{6.14}$$

where

$$U_{5,\pm}(\mathrm{E}_{FBHD}, n, \eta_g) = \gamma_2(\mathrm{E}_{FBHD}, \eta_g) - (n + \frac{1}{2})\hbar\omega_0 \mp \frac{1}{2} g^* \mu_0 B$$

Thus using (6.6), (6.14) and the allied definitions we can study the ER in this case.

(c) Parabolic Energy Bands

The magneto-dispersion law in this case is given by

$$\frac{\hbar^2 k^2}{2m_c} = \gamma_3(E, \eta_g) - (n + \frac{1}{2})\hbar\omega_0 \mp \frac{1}{2} g^* \mu_0 B \tag{6.15a}$$

The EEM at the Fermi Level can be written from (6.15a) as

$$m^*(E_{FBHD}, \eta_g) = m_c \gamma_3'(E_{FBHD}, \eta_g) \tag{6.15b}$$

Thus the EEM in heavily doped parabolic energy bands is a function of Fermi energy and scattering potential whereas in the absence of heavy doping the same mass is a constant quantity invariant of any variables.

The electron concentration under the condition of extreme degeneracy is given by

$$n_0 = \frac{g_v eB}{\pi^2 \hbar^2} \sqrt{m_c} \sum_{n=0}^{n_{\max}} \left(U_{6,\pm}(\mathrm{E}_{FBHD}, n, \eta_g) \right)^{\frac{1}{2}} \tag{6.16}$$

where, $U_{6,\pm}(\mathrm{E}_{FBHD}, n, \eta_g) = \gamma_3(\mathrm{E}_{FBHD}, \eta_g) - (n + \frac{1}{2})\hbar\omega_0 \mp \frac{1}{2} g^* \mu_0 B$

Thus using (6.6), (6.16) and the allied definitions we can study the ER in this case.

(d) The model of Stillman et al.

The (1.107) under the condition of band tailing assumes the form

$$k^2 = \frac{\left[\bar{t}_{11} - \sqrt{(\bar{t}_{11})^2 - 4\bar{t}_{12}\gamma_3(E, \eta_g)} \right]}{2\bar{t}_{12}} \tag{6.17}$$

Therefore the magneto dispersion law is given by

$$k_z^2 = U_7(E, n, \eta_g) \tag{6.18a}$$

where

$$U_7(E, n, \eta_g) = \left[\frac{\left[\bar{t}_{11} - \sqrt{(\bar{t}_{11})^2 - 4\bar{t}_{12}\gamma_3(E, \eta_g)} \right]}{2\bar{t}_{12}} - \frac{2eB}{\hbar}(n + \frac{1}{2}) \right]$$

The EEM at the Fermi Level can be written from (6.18a) as

$$m^*(E_{FBHD}, \eta_g) = \frac{\hbar^2}{2} U_7'(E_{FBHD}, n, \eta_g) \tag{6.18b}$$

The electron concentration under the condition of extreme degeneracy is given by

$$n_0 = \frac{g_v eB}{\pi^2 \hbar} \sum_{n=0}^{n_{\max}} \left(U_7(E_{FBHD}, n, \eta_g) \right)^{\frac{1}{2}} \tag{6.19}$$

Thus by using (6.6), (6.19) and the allied definitions we can study the ER in this case.

(e) The model of Palik et al.

To the fourth order in effective mass theory and taking into account the interactions of the conduction, light hole, heavy-hole and split-off hole bands, the electron energy spectrum in III-V semiconductors in the presence of a quantizing magnetic field $\vec{B}$ can be written as

$$\begin{aligned} E = J_{31} + (n + \frac{1}{2})\hbar\omega_0 + \frac{\hbar^2 k_z^2}{2m_c} \pm \frac{1}{4}(\frac{m_c}{m_0})\hbar\omega_0 g_0^* \pm k_{30}\alpha(n + \frac{1}{2})(\hbar\omega_0)^2 \\ \pm k_{31}\alpha\hbar\omega_0(\frac{\hbar^2 k^2}{2m_c}) + k_{32}\alpha\left[\hbar\omega_0\left(n + \frac{1}{2}\right) + \frac{\hbar^2 k_z^2}{2m_c}\right]^2 \end{aligned} \tag{6.20}$$

where

$$J_{31} = -\frac{1}{2}\alpha\hbar\omega_0\left[(1 - y_{11})/(2 + x_{11})^2\right] \cdot J_{32},$$

$$J_{32} = \left\{\left[\frac{1}{3}(1-x_{11})^2 - (2+x_{11}^2)\right](2+x_{11})\cdot y_{11} + \frac{1}{2}(1-x_{11}^2)(1+x_{11})(1+y_{11})\right\},$$

$$g_0^* = 2\left\{1 - \left[\frac{(1-x_{11})}{(2+x_{11})}\right]\left[\frac{(1-y_{11})}{y_{11}}\right]\right\},$$

$$k_{30} = (1-y_{11})(1-x_{11})\left\{\left[\left(2+\frac{3}{2}x_{11}+x_{11}^2\right)\cdot\frac{(1-y_{11})}{(2+x_{11})^2}\right] - \frac{2}{3}y_{11}\right\},$$

$$k_{31} = (1-y_{11})\left[\frac{(1-x_{11})}{(2+x_{11})}\right]\cdot\left\{\left[\left(2+\frac{3}{2}x_{11}+x_{11}^2\right)\cdot\frac{(1-y_{11})}{(2+x_{11})}\right] - \frac{2}{3}(1-x_{11})y_{11}\right\},$$

$$k_{32} = -\left[\left(1+\frac{1}{2}x_{11}^2\right)/\left(1+\frac{1}{2}x_{11}\right)\right](1-y_{11})^2,\ x_{11} = [1+(\frac{\Delta}{E_g})]^{-1} \text{ and } y_{11} = \frac{m_c}{m_0}$$

Under the condition of heavy doping, the (6.20) assumes the form

$$J_{34}k_z^4 + J_{35,\pm}(n)k_z^2 + J_{36,\pm}(n) - \gamma_3(\mathrm{E},\eta_g) = 0 \tag{6.21}$$

where

$$J_{34} = \alpha k_{32}\left(\hbar^2/2m_c\right)^2,\ J_{35,\pm}(n) = \left[\frac{\hbar^2}{2m_c} \pm \alpha k_{31}\hbar\omega_0\cdot\frac{\hbar^2}{2m_c} + \alpha k_{32}\hbar\omega_0\cdot\frac{\hbar^2}{2m_c}(n+\frac{1}{2})\right],$$

$$J_{36,\pm}(n) = \left[J_{31} \pm \frac{1}{4}(\frac{m_c}{m_0})\hbar\omega_0 g_0^* \pm k_{30}\alpha(\hbar\omega_0)^2(n+\frac{1}{2}) + k_{32}\alpha[(\hbar\omega_0)(n+\frac{1}{2})]^2\right]$$

The (6.21) can be written as

$$k_z^2 = A_{35HD,\pm}(E,n,\eta_g) \tag{6.22a}$$

where

$$A_{35HD,\pm}(E,n,\eta_g) = (2J_{34})^{-1}\left[-J_{35,\pm}(n) + \sqrt{(J_{35,\pm}(n))^2 - 4J_{34}\left[J_{36,\pm}(n) - \gamma_3(\mathrm{E},\eta_g)\right]}\right]$$

The EEM at the Fermi Level can be written from (6.22a) as

$$m_\pm^*(E_{FBHD},n,\eta_g) = \frac{\hbar^2}{2}A'_{35HD,\pm}(E_{FBHD},n,\eta_g) \tag{6.22b}$$

Thus, the EEM is a function of Fermi energy, Landau quantum number and the scattering potential.

The electron concentration is given by

$$n_0 = \frac{eBg_v}{2\pi^2 h} \sum_{n=0}^{n_{\max}} \left[Y_{34HD}(E_{FBHD}, n, \eta_g) + Z_{34HD}(E_{FBHD}, n, \eta_g) \right] \tag{6.23}$$

where $Y_{34HD}(E_{FBHD}, n, \eta_g) = \left[\sqrt{A_{35HD,+}(E_{FBHD}, n, \eta_g) + A_{35HD,-}(E_{FBHD}, n, \eta_g)} \right]$

and

$Z_{34HD}(E_{FBHD}, n, \eta_g) = \sum_{r=1}^{s_0} L_B(r) \left[Y_{34HD}(E_{FBHD}, n, \eta_g) \right]$.

Thus by using (6.6), (6.23) and the allied definitions we can study the ER in this case.

6.2.3 The ER in HD II-VI Semiconductors Under Magnetic Quantization

The magneto dispersion relation of the carriers in heavily doped II-VI semiconductors are given by

$$\gamma_3(E, \eta_g) = a'_0 \frac{2eB}{\hbar} \left(n + \frac{1}{2} \right) + b'_0 k_z^2 \pm \bar{\lambda}_0 \left[\frac{2eB}{\hbar} (n + \frac{1}{2}) \right]^{\frac{1}{2}} \tag{6.24}$$

The (6.24) can be written as

$$k_z^2 = U_{8\pm}(E, n, \eta_g) \tag{6.25a}$$

where $U_{8\pm}(E, n, \eta_g) = (b'_0)^{-1} [\gamma_3(E, \eta_g) - \frac{2eBa'_0}{\hbar} (n + \frac{1}{2}) \mp \bar{\lambda}_0 [\frac{2eB}{\hbar} (n + \frac{1}{2})]^{\frac{1}{2}}]$

The EEM at the Fermi Level can be written from (6.25a) as

$$m^*(E_{FBHD}, \eta_g) = \frac{\hbar^2}{2} U'_{8\pm}(E_{FBHD}, n, \eta_g) \tag{6.25b}$$

The electron concentration is given by

$$n_0 = \frac{eBg_v}{2\pi^2 \hbar} \sum_{n=0}^{n_{\max}} \left[Y_{35HD}(E_{FBHD}, n, \eta_g) + Z_{35HD}(E_{FBHD}, n, \eta_g) \right] \tag{6.26}$$

where

$$Y_{35HD}(E_{FBHD}, n, \eta_g) = \left[\sqrt{U_{8+}(E_{FBHD}, n, \eta_g)} + \sqrt{U_{8-}(E_{FBHD}, n, \eta_g)} \right]$$

and
$Z_{35HD}(E_{FBHD}, n, \eta_g) = \sum_{r=1}^{s_0} L_B(r)\left[Y_{35HD}(E_{FBHD}, n, \eta_g)\right].$

Thus by using (6.6), (6.26) and the allied definitions we can study the ER in this case.

6.2.4 The ER in HD IV-VI Semiconductors Under Magnetic Quantization

The electron energy spectrum in IV-VI semiconductors are defined by the models of Cohen, Lax, Dimmock and Bangert and Kastner respectively. The magneto ER in HD IV-VI semiconductors is discussed in accordance with the said model for the purpose of relative comparison.

(a) Cohen Model

In accordance with the Cohen model, the dispersion law of the carriers in IV-VI semiconductors is given by

$$E(1+\alpha E) = \frac{p_x^2}{2m_1} + \frac{p_z^2}{2m_3} - \frac{\alpha E p_y^2}{2m_2'} + \frac{p_y^2(1+\alpha E)}{2m_2} + \frac{\alpha p_y^4}{4m_2 m_2'} \tag{6.27}$$

where, $p_i = \hbar k_i$, $i = x, y, z$, m_1, m_2, and m_3 are the effective carrier masses at the band-edge along x, y and z directions respectively and m_2' is the effective-mass tensor component at the top of the valence band (for electrons) or at the bottom of the conduction band (for holes).

The magneto electron energy spectrum in IV-VI semiconductors in the presence of quantizing magnetic field$B$ along z-direction can be written as

$$E(1+\alpha E) = \left(n+\frac{1}{2}\right)\hbar\omega(E) \pm \frac{1}{2}g^*\mu_0 B + \frac{3}{8}\alpha\left(n^2+n+\frac{1}{2}\right)\hbar^2\omega^2(E) + \frac{\hbar^2 k_z^2}{2m_3} \tag{6.28a}$$

where, $\omega(E) \equiv \frac{|e|B}{\sqrt{m_1 m_2}}\left[1+\alpha E\left(1-\frac{m_2}{m_2'}\right)\right]^{1/2}$.

Therefore the magneto dispersion law in heavily doped IV-VI materials can be expressed as

$$\frac{\hbar^2 k_z^2}{2m_3} = U_{16,\pm}(E, n, \eta_g) \tag{6.28b}$$

where

$$U_{16,\pm}(E,n,\eta_g) = \left[\gamma_2(E,\eta_g) - (n+\frac{1}{2})\frac{\hbar eB}{\sqrt{m_1 m_2}} \mp \frac{1}{2}g^*\mu_0 B \right.$$
$$- \frac{3\alpha}{8}(n^2+n+\frac{1}{3})\left(\frac{\hbar eB}{\sqrt{m_1 m_2}}\right)^2$$
$$- \gamma_3(E,\eta_g)\left[\frac{\alpha}{2}(n+\frac{1}{2})\frac{\hbar eB}{\sqrt{m_1 m_2}}\left(1-\frac{m_2}{m_2'}\right)\right.$$
$$\left.\left. + \frac{3\alpha^2}{8}(n^2+n+\frac{1}{2})\left(\frac{\hbar eB}{\sqrt{m_1 m_2}}\right)^2\left(1-\frac{m_2}{m_2'}\right)\right]\right]$$

The EEM at the Fermi Level can be written from (6.28b) as

$$m^*_\pm(E_{FBHD},n,\eta_g) = m_3 U'_{16,\pm}(E_{FBHD},n,\eta_g) \tag{6.28c}$$

Thus, the EEM is a function of Fermi energy, Landau quantum number and the scattering potential.

The carrier statistics under the condition of extreme degeneracy in this case can be expressed as

$$n_0 = \frac{g_v eB}{\pi^2\hbar^2}\sqrt{m_3}\sum_{n=0}^{n_{\max}}\left(U_{16,\pm}(E_{FBHD},n,\eta_g)\right)^{\frac{1}{2}} \tag{6.29}$$

Thus by using (6.6), (6.29) and the allied definitions, we can study the ER in this case.

(b) Lax Model

In accordance with this model, the magneto dispersion relation assumes the form

$$E(1+\alpha E) = (n+\frac{1}{2})\hbar\omega_{03}(E) + \frac{\hbar^2 k_z^2}{2m_c} \pm \frac{1}{2}\mu_0 g^* B \tag{6.30}$$

where, $\omega_{03}(E) = \frac{eB}{\sqrt{m_1 m_2}}$.

The magneto dispersion relation in heavily doped IV-VI materials, can be written following (6.30) as

$$\gamma_2(E,\eta_g) = (n+\frac{1}{2})\hbar\omega_{03}(\mathrm{E}) + \frac{\hbar^2 k_z^2}{2m_3} \pm \frac{1}{2}g^*\mu_0 B \tag{6.31}$$

Equation (6.31) can be written as

$$\frac{\hbar^2 k_z^2}{2m_3} = U_{17,\pm}(E, n, \eta_g) \tag{6.32a}$$

where

$$U_{17,\pm}(E, n, \eta_g) = \gamma_2(E, \eta_g) - (n + \frac{1}{2})\hbar\omega_{03}(\mathrm{E}) \pm \frac{1}{2} g^* \mu_0 B$$

The EEM at the Fermi Level can be written from (6.32a) as

$$m^*(E_{FBHD}, \eta_g) = m_3 U'_{17,\pm}(E_{FBHD}, n, \eta_g) \tag{6.32b}$$

The electron concentration under the condition of extreme degeneracy can be written as

$$n_0 = \frac{g_v eB}{\pi^2 \hbar^2} \sqrt{m_3} \sum_{n=0}^{n_{\max}} (U_{17,\pm}(E_{FBHD}, n, \eta_g))^{\frac{1}{2}} \tag{6.33a}$$

Thus by using (6.6), (6.32a), (6.32b) and the allied definitions, we can study the ER in this case.

(c) Dimmock Model

The dispersion relation under magnetic quantization in HD IV-VI semiconductors can be expressed in accordance with Dimmock model as

$$\begin{aligned}
&\gamma_2(E, \eta_g) + \alpha\gamma_3(E, \eta_g)\frac{2eB}{\hbar}(n + \frac{1}{2})\frac{\hbar^2}{2}\left(\frac{1}{m_t^+} - \frac{1}{m_t^-}\right) + \alpha\gamma_3(E, \eta_g)x\frac{\hbar^2}{2}\left(\frac{1}{m_l^+} - \frac{1}{m_l^-}\right) \\
&= \frac{\hbar^2 k_s^2}{2m_t^*} + \frac{\hbar^2 k_z^2}{2m_l^*} + \frac{\hbar^2 k_s^2}{2m_t^-} + \frac{\hbar^2 k_z^2}{2m_l^-} + \alpha\left[\frac{\hbar^4 k_s^4}{4m_t^- m_t^+} + \frac{\hbar^4 k_s^2 k_z^2}{4m_l^- m_l^+} + \frac{\hbar^4 k_z^2 k_s^2}{4m_t^+ m_l^-} + \frac{\hbar^4 k_z^4}{4m_l^- m_l^+}\right] \\
&= \frac{2eB}{\hbar}(n + \frac{1}{2})\frac{\hbar^2}{2}\left(\frac{1}{m_t^*} - \frac{1}{m_t^-}\right) + x\frac{\hbar^2}{2}\left(\frac{1}{m_l^*} - \frac{1}{m_l^-}\right) \\
&\quad + \alpha\left[\frac{\hbar^4}{4m_t^- m_t^+}\left(\frac{2eB}{\hbar}(n + \frac{1}{2})\right)^2 + x\left[\frac{\hbar^4 eB}{2m_l^+ m_t^- \hbar} + \frac{\hbar^4 eB}{2m_t^+ m_l^- \hbar}\right](n + \frac{1}{2}) + \frac{\hbar^4}{4m_l^- m_l^+}x^2\right]
\end{aligned} \tag{6.33b}$$

where

$$x = k_z^2.$$

Therefore the magneto dispersion relation in heavily doped IV-VI materials, whose unperturbed carriers obey the Dimmock Model can be expressed as

$$k_z^2 = U_{17}(E, n, \eta_g) \tag{6.34}$$

where

$$U_{17}(E, n, \eta_g) = [2p_9]^{-1}\left[-q_9(E, n, \eta_g) + \left[q_9^2(E, n, \eta_g) + 4p_9 R_9(E, n, \eta_g)\right]^{\frac{1}{2}}\right] \tag{6.35}$$

$$p_9 = \frac{\alpha\hbar^4}{4m_l^- m_l^+},$$

$$q_9(\mathrm{E}, n, \eta_g) = \left[\frac{\hbar^2}{2}\left(\frac{1}{m_l^*} + \frac{1}{m_l^-}\right) + \frac{\alpha\hbar^3 eB}{2}\left(n + \frac{1}{2}\right)\left(\frac{1}{m_t^- m_l^+} + \frac{1}{m_l^- m_t^+}\right) - \alpha\gamma_3(\mathrm{E}, \eta_g)\left(\frac{1}{m_l^+} - \frac{1}{m_l^-}\right)\right]$$

and

$$R_9(\mathrm{E}, n, \eta_g) = \left[\gamma_2(\mathrm{E}, \eta_g) + \alpha \mathrm{eB}\gamma_3(\mathrm{E}, \eta_g)\left(n + \frac{1}{2}\right)\hbar\left(\frac{1}{m_t^+} + \frac{1}{m_t^-}\right)\right.$$
$$\left. - \hbar \mathrm{eB}\left(n + \frac{1}{2}\right)\left(\frac{1}{m_t^*} + \frac{1}{m_t^-}\right) - \frac{\alpha\hbar^2}{m_t^- m_t^+}\left[\mathrm{eB}\left(n + \frac{1}{2}\right)\right]^2\right] \tag{6.36a}$$

The EEM at the Fermi Level can be written from (6.34) as

$$m^*(E_{FBHD}, n, \eta_g) = \frac{\hbar^2}{2} U_{17}'(E_{FBHD}, n, \eta_g) \tag{6.36b}$$

Thus, the EEM is a function of Fermi energy, Landau quantum number and the scattering potential.

The electron concentration under the condition of extreme degeneracy can be written as

$$n_0 = \frac{g_v eB}{\pi^2\hbar}\sum_{n=0}^{n_{\max}}\left(U_{17}(\mathrm{E}_{FBHD}, n, \eta_g)\right)^{\frac{1}{2}} \tag{6.37}$$

Thus by using (6.6), (6.35) and the allied definitions, we can study the ER in this case.

(d) Model of Bangert and Kastner

In accordance with this model [73], the carrier energy spectrum in HD IV-VI semiconductors can be written following (3.68) as

$$\frac{k_s^2}{\rho_{11}^2(E, \eta_g)} + \frac{k_y^2}{\rho_{12}^2(E, \eta_g)} = 1 \tag{6.38}$$

where

$$\rho_{11}(E,\eta_g) = \frac{1}{\sqrt{S_1(E,\eta_g)}}, \quad \rho_{12}(E,\eta_g) = \frac{1}{\sqrt{S_2(E,\eta_g)}},$$

$$S_1(E,\eta_g) = \left[2\gamma_0(E,\eta_g)\right]^{-1}[\frac{(\overline{R})^2}{E_g}\left\{c_1(\alpha_1,E,E_g) - iD_1(\alpha_1,E,E_g)\right\}$$
$$+\frac{(\overline{S})^2}{\Delta_c'}\left\{c_2(\alpha_2,E,E_g) - iD_2(\alpha_2,E,E_g)\right\}$$
$$+\frac{(\overline{Q})^2}{\Delta_c''}\left\{c_3(\alpha_3,E,E_g) - iD_3(\alpha_3,E,E_g)\right\}\Big]$$

and

$$S_2(E,\eta_g) = \left[2\gamma_0(E,\eta_g)\right]^{-1}[\frac{2(\overline{A})^2}{E_g}\left\{c_1(\alpha_1,E,\eta_g) - iD_1(\alpha_1,E,\eta_g)\right\}$$
$$+\frac{(\overline{S}+\overline{Q})^2}{\Delta_c''}\left\{c_3(\alpha_3,E,\eta_g) - iD_3(\alpha_3,E,\eta_g)\right\}\Big]$$

Since $S_1(E,\eta_g)$ and $S_2(E,\eta_g)$ are complex, the energy spectrum is also complex in the presence of Gaussian band tails.

Therefore the magneto dispersion law in the presence of a quantizing magnetic field B which makes an angle θ with k_z axis can be written as

$$k_z^2 = U_{18}(E,n,\eta_g) \tag{6.39a}$$

where

$$U_{18}(E,n,\eta_g) = \left[\rho_{11}^2(E,\eta_g)\sin^2\theta + \rho_{12}^2(E,\eta_g)\cos^2\theta\right]$$
$$-\left[\frac{2eB}{\hbar}(n+\frac{1}{2})\left[\left(\rho_{11}^2(E,\eta_g)\rho_{12}(E,\eta_g)\right)^{-1}\{\rho_{11}^2(E,\eta_g)\sin^2\theta\right.\right.$$
$$+\rho_{12}^2(E,\eta_g)\cos^2\theta\}^{\frac{3}{2}}]]$$

The EEM at the Fermi Level can be written from (6.39a) as

$$m^*(E_{FBHD},n,\eta_g) = \frac{\hbar^2}{2}\,Real\,part\,of\,\left[U_{18}(E_{FBHD},n,\eta_g)\right]' \tag{6.39b}$$

Thus, the EEM is a function of Fermi energy, Landau quantum number and the scattering potential and the orientation of the applied quantizing magnetic field.

The electron concentration under the condition of extreme degeneracy can be written as

$$n_0 = \frac{g_v eB}{\pi^2 \hbar} \text{ Real Part of } \sum_{n=0}^{n_{\max}} (U_{18}(\mathrm{E}_{FBHD}, n, \eta_g))^{\frac{1}{2}} \tag{6.40}$$

Thus by using (6.6), (6.40) and the allied definitions, we can study the ER in this case.

(e) Model of Foley and Langenberg

The dispersion relation of the conduction electrons of IV-VI semiconductors in accordance with Foley et al. can be written as [74]

$$E + \frac{E_g}{2} = E_-(k) + \left[\left[E_+(k) + \frac{E_g}{2}\right]^2 + P_\perp^2 k_s^2 + P_{||}^2 k_s^2\right]^{\frac{1}{2}} \tag{6.41}$$

where $E_+(k) = \frac{\hbar^2 k_s^2}{2m_\perp^+} + \frac{\hbar^2 k_z^2}{2m_{||}^+}$, $E_-(k) = \frac{\hbar^2 k_s^2}{2m_\perp^-} + \frac{\hbar^2 k_z^2}{2m_{||}^-}$ represents the contribution from the interaction of the conduction and the valence band edge states with the more distant bands and the free electrons term, $\frac{1}{m_\perp^\pm} = \frac{1}{2}\left[\frac{1}{m_{tc}} \pm \frac{1}{m_{tv}}\right], \frac{1}{m_{||}^\pm} = \frac{1}{2}\left[\frac{1}{m_{lc}} \pm \frac{1}{m_{lv}}\right]$.

Following the methods as given in Chap. 1, the dispersion relation in heavily doped IV-VI materials in the present case is given by

$$\left[\left[\gamma_3(E,\eta_g) + \frac{E_{g0}}{2}\right] - \left[\frac{\hbar^2 k_s^2}{2m_\perp^-} + \frac{\hbar^2 k_z^2}{2m_{||}^-}\right]\right]^2 = \left[\frac{\hbar^2 k_s^2}{2m_\perp^+} + \frac{\hbar^2 k_z^2}{2m_{||}^+}\right]^2 + \frac{E_{g0}^2}{4} + E_{g0}\left[\frac{\hbar^2 k_s^2}{2m_\perp^+} + \frac{\hbar^2 k_z^2}{2m_{||}^+}\right] + P_{||}^2 k_z^2 + P_\perp^2 k_s^2 \tag{6.42}$$

Therefore the magneto-dispersion relation in heavily doped IV-VI materials can be written as

$$\gamma_3^2(E,\eta_g) + \frac{E_g^2}{4} + E_g \gamma_3(E,\eta_g) + \left[\frac{\hbar eB}{m_\perp^-}(n+\frac{1}{2}) + \frac{\hbar^2 x}{2m_{||}^-}\right]^2 - 2\left[\gamma_3(E,\eta_g) + \frac{E_g}{2}\right]\left[\frac{\hbar eB(n+\frac{1}{2})}{m_\perp^-} + \frac{\hbar^2 x}{2m_{||}^-}\right] = \left[\frac{\hbar eB(n+\frac{1}{2})}{m_\perp^+} + \frac{\hbar^2 x}{2m_{||}^+}\right]^2 + E_g\left[\frac{\hbar eB}{m_\perp^+}(n+\frac{1}{2}) + \frac{\hbar^2 x}{2m_{||}^+}\right] + P_{||}^2 x + P_\perp^2 \frac{2eB}{\hbar}(n+\frac{1}{2}) \tag{6.43}$$

where $k_z^2 = x$.

Therefore the magneto dispersion relation in IV-VI heavily doped materials, where unperturbed carriers follow the model of Foley et al. can be expressed as

$$k_z^2 = U_{19}(E, n, \eta_g) \tag{6.44a}$$

where

$$U_{19}(E, n, \eta_g) = [2p_{91}]^{-1}[-q_{91}(E, n, \eta_g) + \{q_{91}^2(E, n, \eta_g) + 4p_{91}R_{91}(E, n, \eta_g)\}^{\frac{1}{2}}]$$

$$p_{91} = \frac{\hbar^4}{4}\left[\frac{1}{(m_{\|}^+)^2} - \frac{1}{(m_{\|}^-)^2}\right],$$

$$q_{91}(E, n, \eta_g) = \left[\frac{\hbar^3 eB}{m_\perp^- m_{\|}^+}\left(n + \frac{1}{2}\right) + \mathrm{P}_{\|}^2 + \frac{\hbar^3 E_g}{2m_{\|}^+} - \frac{\hbar^3 eB(n + \frac{1}{2})}{m_\perp^- m_{\|}^-} + \frac{\hbar^2}{m_{\|}^-}\left(\gamma_3(E, \eta_g) + \frac{E_g}{2}\right)\right] \text{ and}$$

$$R_{91}(E, \eta_g, n) = [\gamma_3^2(E, \eta_g) + E_g\gamma_3(E, \eta_g) - \frac{2\hbar eB}{m_\perp^-}\left(\gamma_3(E, \eta_g) + \frac{E_g}{2}\right)\left(n + \frac{1}{2}\right)$$
$$- E_g\frac{\hbar eB}{m_\perp^+}\left(n + \frac{1}{2}\right) - P_\perp^2 \cdot \frac{2eB}{\hbar}\left(n + \frac{1}{2}\right)]$$

The EEM at the Fermi Level can be written from (6.44a) as

$$m^*(E_{FBHD}, n, \eta_g) = \frac{\hbar^2}{2} U'_{19}(E_{FBHD}, n, \eta_g) \tag{6.44b}$$

Thus, as noted already in this case also the EEM is a function of Fermi energy, Landau quantum number and the scattering potential.

The electron concentration under the condition of extreme degeneracy can be written as

$$n_0 = \frac{g_v eB}{\pi^2\hbar}\sum_{n=0}^{n_{\max}}\left(U_{19}(\mathrm{E}_{FBHD}, n, \eta_g)\right)^{\frac{1}{2}} \tag{6.45}$$

Thus by using (6.6), (6.45) and the allied definitions, we can study the ER in this case.

6.2.5 The ER in HD Stressed Kane Type Semiconductors Under Magnetic Quantization

The dispersion relation of the conduction electrons in heavily doped Kane type semiconductors can be written following (1.206) as

$$\frac{k_x^2}{a_{\|}^2(\mathrm{E}, \eta_g)} + \frac{k_y^2}{b_{\|}^2(\mathrm{E}, \eta_g)} + \frac{k_z^2}{c_{\|}^2(\mathrm{E}, \eta_g)} = 1$$

where

$$a_{||}(\mathrm{E},\eta_g)=\frac{1}{\sqrt{P_{||}(\mathrm{E},\eta_g)}},\ \mathrm{b}_{||}(\mathrm{E},\eta_g)=\frac{1}{\sqrt{Q_{||}(\mathrm{E},\eta_g)}}\ \text{and}\ c_{||}(\mathrm{E},\eta_g)=\frac{1}{\sqrt{S_{||}(\mathrm{E},\eta_g)}} \tag{6.46}$$

The electron energy spectrum in heavily doped Kane type semiconductors in the presence of an arbitrarily oriented quantizing magnetic field B which makes an angle $\overline{\alpha}_1, \overline{\beta}_1$ and $\overline{\gamma}_1$ with k_x, k_y and k_z axes respectively, can be written as

$$(k_z')^2 = U_{41}(\mathrm{E},n,\eta_g) \tag{6.47a}$$

where

$$U_{41}(\mathrm{E},n,\eta_g) = I_2(\mathrm{E},\eta_g)[1 - I_3(\mathrm{E},n,\eta_g)]$$

$$I_2(\mathrm{E},\eta_g) = [[a_{11}(\mathrm{E},\eta_g)]^2\cos^2\overline{\alpha}_1 + [b_{11}(\mathrm{E},\eta_g)]^2\cos^2\overline{\beta}_1 + [c_{11}(\mathrm{E},\eta_g)]^2\cos^2\overline{\gamma}_1]$$

and

$$I_3(\mathrm{E},n,\eta_g) = \frac{2eB}{\hbar}\left(n+\frac{1}{2}\right)[a_{11}(\mathrm{E},\eta_g)b_{11}(\mathrm{E},\eta_g)c_{11}(\mathrm{E},\eta_g)]^{-1}[I_2(\mathrm{E},\eta_g)]^{1/2}$$

The EEM at the Fermi Level can be written from (6.47a) as

$$m^*(E_{FBHD},n,\eta_g) = \frac{\hbar^2}{2}U_{41}'(E_{FBHD},n,\eta_g) \tag{6.47b}$$

From (6.47b) we observe that the EEM is a function of Fermi energy, Landau quantum number, the scattering potential and the orientation of the applied quantizing magnetic field.

The electron concentration under the condition of extreme degeneracy can be written as

$$n_0 = \frac{g_v eB}{\pi^2\hbar}\sum_{n=0}^{n_{\max}}\left(U_{41}(\mathrm{E}_{FBHD},n,\eta_g)\right)^{\frac{1}{2}} \tag{6.48}$$

Thus by using (6.6), (6.48) and the allied definitions, we can study the ER in this case.

6.2.6 The ER in HD Tellurium Under Magnetic Quantization

The magneto dispersion relation of the conduction electrons in HD Te can be expressed following (1.231) as

$$k_z^2 = U_{42\pm}(E, n, \eta_g) \tag{6.49a}$$

where

$$U_{42,\pm}(\mathrm{E}, n, \eta_g) = (2\psi_1^2)^{-1}\left[\left\{2\gamma_3(\mathrm{E}, \eta_g)\psi_1 + \psi_3^2 - 4\psi_1\psi_2\frac{eB}{\hbar}\left(n+\frac{1}{2}\right)\right\} - \left\{\psi_3^4 + 4\psi_1\psi_3^2\gamma_3(\mathrm{E}, \eta_g) + \frac{8eB}{\hbar}(n+\frac{1}{2})(\psi_1^2\psi_4^2 - \psi_1\psi_2\psi_3^2)\right\}^{-1/2}\right]$$

The EEM at the Fermi Level can be written from (6.49a) as

$$m_{\pm}^*(E_{FBHD}, n, \eta_g) = \frac{\hbar^2}{2}U'_{42\pm}(E_{FBHD}, n, \eta_g) \tag{6.49b}$$

Thus from (6.49b) we note that the EEM is a function of three variables namely Fermi energy, Landau quantum number and the scattering potential.

The electron concentration under the condition of extreme degeneracy can be written as

$$n_0 = \frac{g_v eB}{2\pi^2\hbar}\sum_{n=0}^{n_{\max}}\left(U_{42,\pm}(\mathrm{E}_{FBHD}, n, \eta_g)\right)^{\frac{1}{2}} \tag{6.49c}$$

Thus by using (6.6), (6.49) and the allied definitions, we can study the ER in this case.

6.2.7 The ER in HD Gallium Phosphide Under Magnetic Quantization

The magneto dispersion relation in HD GaP can be written following (1.248) as

$$k_z^2 = U_{43}(E, n, \eta_g) \tag{6.50a}$$

where

$$U_{43}(\mathrm{E},n,\eta_g)=(2b^2)^{-1}\left[\left\{2\gamma_3(\mathrm{E},\eta_g)b+c-2Db-4ab\frac{eB}{\hbar}\left(n+\frac{1}{2}\right)\right\}\right.$$
$$+\left\{[c^2+4bc\gamma_3(\mathrm{E},\eta_g)+4D^2b^2-4cDb]\right.$$
$$-\frac{8eB}{\hbar}(n+\frac{1}{2})(2ab^2D+4\gamma_3(\mathrm{E},\eta_g)b^2a$$
$$\left.\left.+abc-2b^2a\gamma_3(\mathrm{E},\eta_g)-b^2c\right\}^{1/2}\right]$$

The EEM at the Fermi Level can be expressed from (6.50a) as

$$m^*(E_{FBHD},n,\eta_g)=\frac{\hbar^2}{2}U'_{43}(E_{FBHD},n,\eta_g) \tag{6.50b}$$

Thus, from (6.50b) it appears that the EEM is the function of Fermi energy, Landau quantum number and the scattering potential.

The electron concentration under the condition of extreme degeneracy can be written as

$$n_0=\frac{g_veB}{\pi^2\hbar}\sum_{n=0}^{n_{\max}}\left(U_{43}(\mathrm{E}_{FBHD},n,\eta_g)\right)^{\frac{1}{2}} \tag{6.50c}$$

Thus by using (6.6), (6.50c) and the allied definitions, we can study the ER in this case.

6.2.8 The ER in HD Platinum Antimonide Under Magnetic Quantization

The magneto dispersion relation in HD $PtSb_2$ can be written following (1.270) as

$$k_z^2=U_{44}(\mathrm{E},n,\eta_g) \tag{6.51a}$$

where

$$U_{44}(\mathrm{E},n,\eta_g)=\frac{1}{2T_{41}}\left[T_{71}(\mathrm{E},n,\eta_g)+\sqrt{T_{71}^2(\mathrm{E},n,\eta_g)+4T_{41}T_{71}(\mathrm{E},n,\eta_g)}\right],$$
$$T_{71}(\mathrm{E},n,\eta_g)=\left[T_{51}(\mathrm{E},\eta_g)-T_{31}\frac{2eB}{\hbar}\left(n+\frac{1}{2}\right)\right]$$

and

$$T_{72}(\mathrm{E}, n, \eta_g) = \left[T_{61}(\mathrm{E}, \eta_g) + T_{21}(\mathrm{E}, \eta_g)\frac{2eB}{\hbar}\left(n + \frac{1}{2}\right) - T_{11}(\frac{2eB}{\hbar}\left(n + \frac{1}{2}\right))^2\right].$$

The EEM at the Fermi Level can be written from (6.51a) as

$$m^*(E_{FBHD}, n, \eta_g) = \frac{\hbar^2}{2} U'_{44}(E_{FBHD}, n, \eta_g) \tag{6.51b}$$

Thus, from the above equation we infer that the EEM is a function of Landau quantum number, the Fermi energy and the scattering potential.

The electron concentration under the condition of extreme degeneracy can be written as

$$n_0 = \frac{g_v eB}{\pi^2 \hbar} \sum_{n=0}^{n_{\max}} (U_{44}(\mathrm{E}_{FBHD}, n, \eta_g))^{\frac{1}{2}} \tag{6.52}$$

Thus by using (6.6), (6.52) and the allied definitions, we can study the ER in this case.

6.2.9 The ER in HD Bismuth Telluride Under Magnetic Quantization

The magneto dispersion relation in HD Bi_2Te_3 can be written following (1.285) as

$$k_x^2 = U_{45}(E, \eta_g, n) \tag{6.53a}$$

where

$$U_{45}(E, \eta_g, n) = \frac{\gamma_2(E, \eta_g) - (n + \frac{1}{2})\frac{e\hbar B}{M_{31}}}{\bar{\omega}_1} \quad \text{and} \quad M_{31} = \frac{m_0}{(\bar{\alpha}_{22}\bar{\alpha}_{33} - \frac{(\bar{\alpha}_{23})^2}{4})^{\frac{1}{2}}}$$

The EEM at the Fermi Level can be written from (6.53a) as

$$m^*(E_{FBHD}, \eta_g) = \frac{\hbar^2}{2} U'_{45}(E_{FBHD}, n, \eta_g) \tag{6.53b}$$

The electron concentration under the condition of extreme degeneracy can be written as

$$n_0 = \frac{g_v eB}{\pi^2 \hbar} \sum_{n=0}^{n_{\max}} \left(U_{45}(\mathrm{E}_{FBHD}, n, \eta_g) \right)^{\frac{1}{2}} \tag{6.54}$$

Thus by using (6.6), (6.54) and the allied definitions, we can study the ER in this case.

6.2.10 The ER in HD Germanium Under Magnetic Quantization

(a) Model of Cardona et al.

The magneto dispersion relation in HD Ge can be written following (1.300) as

$$k_x^2 = U_{46}(E, \eta_g, n) \tag{6.55a}$$

where

$$U_{46}(\mathrm{E}, n, \eta_g) = \frac{2m_{\parallel}^*}{\hbar^2} \left[\gamma_3(\mathrm{E}, \eta_g) + \frac{E_{go}}{2} - \left[\frac{E_{go}^2}{4} + \frac{E_{go}\hbar^2}{m_{\perp}^*} \frac{2eB}{\hbar} (n + \frac{1}{2}) \right] \right]^{1/2}$$

The EEM at the Fermi Level can be written from (6.55a) as

$$m^*(E_{FBHD}, n, \eta_g) = \frac{\hbar^2}{2} U'_{46}(E_{FBHD}, n, \eta_g) \tag{6.55b}$$

From (6.55b) it appears that the EEM is a function of Fermi energy and Landau quantum number due to band non-parabolicity.

The electron concentration under the condition of extreme degeneracy can be written as

$$n_0 = \frac{g_v eB}{\pi^2 \hbar} \sum_{n=0}^{n_{\max}} \left(U_{46}(\mathrm{E}_{FBHD}, n, \eta_g) \right)^{\frac{1}{2}} \tag{6.56}$$

Thus by using (6.6), (6.56) and the allied definitions, we can study the ER in this case.

(b) Model of Wang and Ressler

The magneto dispersion relation in HD Ge can be written following (1.321) as

$$k_z^2 = U_{47}(\mathrm{E}, n, \eta_g) \tag{6.57a}$$

where

$$U_{47}(\mathrm{E}, n, \eta_g) = \left(\frac{m_{\|}^*}{\hbar^2 \overline{\alpha}_6}\right)\left[1 - \overline{\alpha}_5(n + \frac{1}{2})\hbar\omega_\perp - \{\theta_7(n) - 4\overline{\alpha}_6\gamma_3(\mathrm{E}, \eta_g)\}^{1/2}\right],$$

$$\omega_\perp = \frac{eB}{m_\perp^*}$$

and

$$\theta_7(n) = \left[1 + (\overline{\alpha}_5)^2\{(n + \frac{1}{2})\hbar\omega_\perp\}^2 - 2\overline{\alpha}_5(n + \frac{1}{2})\hbar\omega_\perp + 4\overline{\alpha}_6(n + \frac{1}{2})\hbar\omega_\perp - 4\overline{\alpha}_6\overline{\alpha}_4\{(n + \frac{1}{2})\hbar\omega_\perp\}^2\right]$$

The EEM at the Fermi Level can be written from (6.57a) as

$$m^*(E_{FBHD}, n, \eta_g) = \frac{\hbar^2}{2} U_{47}'(E_{FBHD}, n, \eta_g) \tag{6.57b}$$

From (6.57b) we note that the mass is a function of Fermi energy and quantum number due to band non-parabolicity.

The electron concentration under the condition of extreme degeneracy can be written as

$$n_0 = \frac{g_v eB}{\pi^2 \hbar} \sum_{n=0}^{n_{\max}} \left(U_{47}(\mathrm{E}_{FBHD}, n, \eta_g)\right)^{\frac{1}{2}} \tag{6.58}$$

Thus by using (6.6), (6.58) and the allied definitions, we can study the ER in this case.

6.2.11 The ER in HD Gallium Antimonide Under Magnetic Quantization

The magneto dispersion relation in HD GaSb can be written following (1.338) as

$$k_z^2 = U_{48}(\mathrm{E}, n, \eta_g) \tag{6.59a}$$

where

$$U_{48}(\mathrm{E}, n, \eta_g) = \left[\frac{-2eB}{\hbar}\left(n+\frac{1}{2}\right) + (2\alpha_9^2)^{-1}\left[\left\{2\alpha_9\gamma_3^2(\mathrm{E}, \eta_g) + \alpha_9\overline{\mathrm{E}}'_{g0} + \frac{\alpha_{10}(\overline{\mathrm{E}}'_{g0})^2}{4}\right\}\right.\right.$$
$$\left.\left. - \left\{\alpha_9^2(\overline{\mathrm{E}}'_{g0})^2 + \frac{\alpha_{10}^2(\overline{\mathrm{E}}'_{g0})^4}{16} + \alpha_9\alpha_{10}\gamma_3(\mathrm{E}, \eta_g)(\overline{\mathrm{E}}'_{g0})^2 + \frac{\alpha_9\alpha_{10}(\overline{\mathrm{E}}'_{g0})^3}{2}\right\}^{1/2}\right]\right],$$

$$\alpha_9 = \frac{\hbar^2}{2m_0} \quad \text{and} \quad \alpha_{10} = \frac{2\hbar^2}{\overline{E}'_{go}}\left(\frac{1}{m_c} - \frac{1}{m_0}\right)$$

The EEM at the Fermi Level can be written from (6.59a) as

$$m^*(E_{FBHD}, \eta_g) = \frac{\hbar^2}{2} U'_{48}(E_{FBHD}, n, \eta_g) \tag{6.59b}$$

The electron concentration under extreme degeneracy can be written as

$$n_0 = \frac{g_v eB}{\pi^2\hbar}\sum_{n=0}^{n_{\max}}\left(U_{48}(\mathrm{E}_{FBHD}, n, \eta_g)\right)^{\frac{1}{2}} \tag{6.60}$$

Thus by using (6.6), (6.60) and the allied definitions, we can study the ER in this case.

6.2.12 The ER in HD II-V Materials Under Magnetic Quantization

The dispersion relation of the holes are given by [75–77]

$$E = \theta_1 k_x^2 + \theta_2 k_y^2 + \theta_3 k_z^2 + \delta_4 k_x \mp [\{\theta_5 k_x^2 + \theta_6 k_y^2 + \theta_7 k_z^2 + \delta_5 k_x\}^2 + G_3^2 k_y^2 + \Delta_3^2]^{\frac{1}{2}} \pm \Delta_3 \tag{6.61}$$

where, k_x, k_y and k_z are expressed in the units of $10^{10}\ \mathrm{m}^{-1}$,

$$\theta_1 = \frac{1}{2}(a_1 + b_1),\ \theta_2 = \frac{1}{2}(a_2 + b_2),\ \theta_3 = \frac{1}{2}(a_3 + b_3),\ \delta_4 = \frac{1}{2}(A + B),$$
$$\theta_5 = \frac{1}{2}(a_1 - b_1),\ \theta_6 = \frac{1}{2}(a_2 - b_2),\ \theta_7 = \frac{1}{2}(a_3 - b_3),\ \delta_5 = \frac{1}{2}(A - B),$$

$a_i\ (i = 1, 2, 3, 4), b_i, A, B, G_3$ and Δ_3 are system constants

The hole energy spectrum in HD II-V semiconductors can be expressed following the method of Chap. 1 as

$$\gamma_3(\mathrm{E},\eta_g) = \theta_1 k_x^2 + \theta_2 k_y^2 + \theta_3 k_z^2 + \delta_4 k_x \mp \left[\{\theta_5 k_x^2 + \theta_6 k_y^2 + \theta_7 k_z^2 + \delta_5 k_x\}^2 + G_3^2 k_y^2 + \Delta_3^2\right]^{\frac{1}{2}} \pm \Delta_3 \tag{6.62}$$

the magneto dispersion law in HD II-V semiconductors assumes the form

$$k_y^2 = U_{49,\pm}(\mathrm{E},n,\eta_g) \tag{6.63a}$$

where,

$$U_{49,\pm}(\mathrm{E},n,\eta_g) = \left[I_{35}\gamma_3(\mathrm{E},\eta_g) + I_{36,\pm}(n) \pm [\gamma_3^2(\mathrm{E},\eta_g) + \gamma_3(\mathrm{E},\eta_g) I_{38,\pm}(n) + I_{39,\pm}(n)\right]^{\frac{1}{2}},$$

$$I_{35} = \frac{\theta_2}{(\theta_2^2 - \theta_5^2)}, I_{36,\pm}(n) = \frac{I_{33,\pm}(n)}{2(\theta_2^2 - \theta_5^2)},$$

$$I_{38,\pm}(n) = (4\theta_5^2)^{-1}\left[4\theta_2 I_{33,\pm}(n) + 8\theta_2^2 I_{31,\pm}(n) - \theta_5^2 I_{31,\pm}(n)\right],$$

$$I_{39,\pm}(n) = (4\theta_5^2)^{-1}\left[I_{33,\pm}^2(n) + 4\theta_2^3 I_{34,\pm}(n) - 4\theta_5^3 I_{34,\pm}(n)\right],$$

$$I_{33,\pm}(n) = \left[G_3^2 + 2\theta_5 I_{32}(n) - 2\theta_2 I_{31,\pm}(n)\right],$$

$$I_{34,\pm}(n) = \left[I_{32,\pm}^2(n) + \Delta_3^2 - I_{31,\pm}(n)\right],$$

$$I_{31,\pm}(n) = \left[(n+\frac{1}{2})\hbar\omega_{31} - \frac{\delta_4^2}{4\theta_1} \pm \Delta_3\right],$$

$$I_{32}(n) = \left[(n+\frac{1}{2})\hbar\omega_{32} - \frac{\delta_5^2}{4\theta_5}\right],$$

$$\omega_{31} = \frac{eB}{\sqrt{M_{31}M_{32}}},\ \omega_{32} = \frac{eB}{\sqrt{M_{33}M_{34}}},\ M_{31} = \frac{\hbar^2}{2\theta_1},$$

$$M_{32} = \frac{\hbar^2}{2\theta_3},\ M_{33} = \frac{\hbar^2}{2\theta_5}\ \text{and}\ M_{34} = \frac{\hbar^2}{2\theta_7}.$$

The EEM at the Fermi Level can be written from (6.63a) as

$$m^*_{\pm}(E_{FBHD}, n, \eta_g) = \frac{\hbar^2}{2} U'_{49,\pm}(E_{FBHD}, n, \eta_g) \quad (6.63b)$$

From (6.63b) we note that the EEM is a function of Fermi energy, Landau quantum number and the scattering potential.

The electron concentration under extreme degeneracy can be written as

$$n_0 = \frac{g_v eB}{\pi^2 \hbar} \sum_{n=0}^{n_{max}} \left(U_{49,\pm}(\mathrm{E}_{FBHD}, n, \eta_g) \right)^{\frac{1}{2}} \quad (6.64)$$

Thus by using (6.6), (6.60) and the allied definitions, we can study the ER in this case.

6.2.13 The ER in HD Lead Germanium Telluride Under Magnetic Quantization

The dispersion relation of the carriers in n-type $Pb_{1-x}Ga_xTe$ with x = 0.01 can be written following Vassilev [78] as

$$\begin{aligned}&\left[E - 0.606k_s^2 - 0.0722k_z^2\right]\left[E + \overline{E}_g + 0.411k_s^2 + 0.0377k_z^2\right]\\ &\quad = 0.23k_s^2 + 0.02k_z^2 \pm \left[0.06\overline{E}_g + 0.061k_s^2 + 0.0066k_z^2\right]k_s\end{aligned} \quad (6.65)$$

where, $\overline{E}_g(= 0.21\, eV)$ is the energy gap for the transition point, the zero of the energy E is at the edge of the conduction band of the Γ point of the Brillouin zone and is measured positively upwards, k_x, k_y and k_z are in the units of 10^9 m^{-1}.

The magneto dispersion law in HD $Pb_{1-x}Ge_xTe$ can be expressed following the methods as given in Chap. 1 as

$$\begin{aligned}&\left[\frac{2\theta_0(E,\eta_g)}{1+Erf(E/\eta_g)}\right] + \gamma_3(E,\eta_g)\left[\overline{E_{g0}} - 0.345x - 0.390\frac{eB}{\hbar}(n+\frac{1}{2})\right]\\ &= \frac{0.46eB}{\hbar}(n+\frac{1}{2}) + 0.02x \pm \left[0.06\overline{E_{g0}} + 0.122\frac{eB}{\hbar}(n+\frac{1}{2}) + 0.0066x\right]\left(\frac{2eB}{\hbar}(n+\frac{1}{2})\right)^{\frac{1}{2}}\\ &+ \left[\overline{E_{g0}} + \frac{0.822eB}{\hbar}(n+\frac{1}{2}) + 0.377x\right]\left[\frac{1.212eB}{\hbar}(n+\frac{1}{2}) + 0.722x\right]\end{aligned} \quad (6.66)$$

Equation (6.66) assumes the form

$$k_z^2 = U_{50,\mp}(E, n, \eta_g) \quad (6.67a)$$

where

$$U_{50,\mp}(E, n, \eta_g) = (2p_{10})^{-1}\left[q_{10}(E, n, \eta_g) - \left[q_{10}^2(E, n, \eta_g) + 4p_{10}R_{10,\mp}(E, n, \eta_g)\right]^{\frac{1}{2}}\right]$$

$$p_{10} = (0.377 \times 0.722), q_{10}(E, n, \eta_g) = \Big[0.02 + 0.345\gamma_3(E, \eta_g)$$

$$\pm 0.0066\left(\frac{2eB}{\hbar}\left(n+\frac{1}{2}\right)\right)^{\frac{1}{2}} + 0.377 \times \frac{1.212eB}{\hbar}\left(n+\frac{1}{2}\right)$$

$$+ 0.722\left[\bar{E}_{go} + 0.822\frac{eB}{\hbar}(n+\frac{1}{2})\right]\Big] \text{ and}$$

$$R_{10,\mp}(E, n, \eta_g) = \Big[\frac{2\theta_0(E, \eta_g)}{1 + Eof(E/\eta_g)} + \gamma_3(E, \eta_g)\left[\bar{E}_{go} + 0.390\frac{eB}{\hbar}(n+\frac{1}{2})\right]$$

$$\mp \left(0.06\,\bar{E}_{go} + 0.122\frac{eB}{\hbar}\left(n+\frac{1}{2}\right)\right)\left(\frac{2eB}{\hbar}\left(n+\frac{1}{2}\right)\right)^{\frac{1}{2}}$$

$$- \left(\bar{E}_{go} + 0.822\frac{eB}{\hbar}\left(n+\frac{1}{2}\right)\right)\frac{1.212eB}{\hbar}\left(n+\frac{1}{2}\right) - \frac{0.46eB}{\hbar}\left(n+\frac{1}{2}\right)\Big].$$

The EEM at the Fermi Level can be written from (6.67a) as

$$m_{\mp}^*(E_{FBHD}, \eta_g) = \frac{\hbar^2}{2} U'_{50,\mp}(E_{FBHD}, n, \eta_g) \quad (6.67b)$$

Thus from (6.67b) we note that the EEM is a function of the Fermi energy, Landau quantum number and the scattering potential.

The electron concentration under extreme degeneracy can be written as

$$n_0 = \frac{g_v eB}{2\pi^2\hbar}\sum_{n=0}^{n_{\max}} (U_{50,\mp}(E_{FBHD}, n, \eta_g))^{\frac{1}{2}} \quad (6.68)$$

Thus by using (6.6), (6.68) and the allied definitions, we can study the ER in this case.

6.3 Open Research Problems

(R.6.1) Investigate the ER in the presence of an arbitrarily oriented quantizing magnetic field for all the materials as given in problems in R.1.1 of Chap. 1 in the presence of the Gaussian type band tails.

(R.6.2) Investigate the ER in the presence of an arbitrarily oriented quantizing magnetic field in HD nonlinear optical semiconductors by including broadening and the electron spin. Study all the special cases for HD III-V, ternary and quaternary materials in this context.

(R.6.3) Investigate the ERs for HD IV-VI, II-VI and stressed Kane type compounds in the presence of an arbitrarily oriented quantizing magnetic field by including broadening and electron spin.

(R.6.4) Investigate the ER for all the materials as stated in R.1.1 of Chap. 1 in the presence of an arbitrarily oriented quantizing magnetic field by including broadening and electron spin under the condition of heavily doping.

(R.6.5) Investigate the ER in the presence of an arbitrarily oriented quantizing magnetic field and crossed electric fields in HD nonlinear optical semiconductors by including broadening and the electron spin. Study all the special cases for HD III-V, ternary and quaternary materials in this context.

(R.6.6) Investigate the ERs for HD IV-VI, II-VI and stressed Kane type compounds in the presence of an arbitrarily oriented quantizing magnetic field and crossed electric field by including broadening and electron spin.

(R.6.7) Investigate the ER for all the materials as stated in R.1.1 of Chap. 1 in the presence of an arbitrarily oriented quantizing magnetic field and crossed electric field by including broadening and electron spin under the condition of heavy doping.

(R.6.8) Investigate the 2D ER in QWs of HD nonlinear optical, III-V, II-VI, IV-VI and stressed Kane type semiconductors.

(R.6.9) Investigate the 2D ER for HD QWs of all the materials as considered in problems R.1.1.

(R.6.10) Investigate the 2D ER in the presence of an arbitrarily oriented non-quantizing magnetic field for the QWs of HD nonlinear optical semiconductors by including the electron spin. Study all the special cases for III-V, ternary and quaternary materials in this context.

(R.6.11) Investigate the ERs in QWs of HD IV-VI, II-VI and stressed Kane type compounds in the presence of an arbitrarily oriented non-quantizing magnetic field by including the electron spin.

(R.6.12) Investigate the 2D ER for HD QWs of all the materials as stated in R.1.1 of Chap. 1 in the presence of an arbitrarily oriented magnetic field by including electron spin and broadening.

(R.6.13) Investigate the ER for all the problems of R.1.1 under an additional arbitrarily oriented electric field in the presence of heavy doping.

(R.6.14) Investigate the ER for all the problems of R.1.1 under the arbitrarily oriented crossed electric and magnetic fields in the presence of heavy doping.

(R.6.15) Investigate the 2D ER for all the problems in R.1.1 the presence of finite potential well under the conditions of formation of band tails and applied external parallel magnetic field.

(R.6.16) Investigate the 2D ER for all the problems in R.1.1 the presence of parabolic potential well under the conditions of formation of band tails and applied external parallel magnetic field.

(R.6.17) Investigate the 2D ER for all the problems in R.1.1 the presence of circular potential well under the conditions of formation of band tails and applied external parallel magnetic field.

(R.6.18) Investigate the 2D ER for accumulation layers of HD nonlinear optical, III-V, IV-VI, II-VI and stressed Kane type semiconductors in the presence of an arbitrary electric quantization.

(R.6.19) Investigate the 2D ER in accumulation layers of all the materials as stated in R.1.1 of Chap. 1 under the condition of heavy doping and in the presence of electric quantization along arbitrary direction.

(R.6.20) Investigate the 2D ER in the presence of an arbitrarily oriented electric quantization for accumulation layers of HD nonlinear optical semiconductors. Study all the special cases for III-V, ternary and quaternary materials in this context.

(R.6.21) Investigate the 2D ERs in accumulation layers of HD IV-VI, II-VI and stressed Kane type compounds in the presence of an arbitrarily oriented electric quantization.

(R.6.22) Investigate the 2D ER in accumulation layers of all the materials as stated in R.1.1 of Chap. 1 in the presence of an arbitrarily oriented quantizing electric field under the conditions of formation of band tails and applied external parallel magnetic field.

(R.6.23) Investigate the 2D ER in the presence of an arbitrarily oriented magnetic field in accumulation layers of HD nonlinear optical semiconductors by including the electron spin. Study all the special cases for HD III-V, ternary and quaternary materials in this context.

(R.6.24) Investigate the 2D ERs in accumulation layers of HD IV-VI, II-VI and stressed Kane type compounds in the presence of an arbitrarily oriented non-quantizing magnetic field by including the electron spin.

(R.6.25) Investigate the 2D ER in accumulation layers of all the materials as stated in R.1.1 of Chap. 1 in the presence of an arbitrarily oriented non-quantizing magnetic field by including electron spin and heavy doping.

(R.6.26) Investigate the 2D ER in accumulation layers for all the problems from R.A4.22 to R.A4.26 in the presence of an additional arbitrarily oriented electric field.

(R.6.27) Investigate the 2D ER in accumulation layers for all the problems from R.A4.22 to R.A4.26 in the presence of arbitrarily oriented crossed electric and magnetic fields.

(R.6.28) Investigate the 2D ER in accumulation layers for all the problems from R.A4.22 to R.A4.26 in the presence of surface states.

(R.6.29) Investigate the 2D ER in accumulation layers for all the problems from R.A4.22 to R.A4.26 in the presence of hot electron effects.

(R.6.30) Investigate the 2D ER in accumulation layers for all the problems from R.A4.22 to R.A4.26 by including the occupancy of the electrons in various electric subbands.

(R.6.31) Investigate the 2D ER in Doping superlattices of HD nonlinear optical, III-V, II-VI, IV-VI and stressed Kane type materials.

(R.6.32) Investigate the 2D ER in Doping superlattices of all types of materials as discussed in problem R.1.1 as given in Chap. 1 under the conditions of formation of band tails and applied external parallel magnetic field.

(R.6.33) Investigate the 2D ER in the presence of an arbitrarily oriented non-quantizing magnetic field for Doping superlattices of HD nonlinear optical semiconductors by including the electron spin. Study all the special cases for HD III-V, ternary and quaternary materials in this context.

(R.6.34) Investigate the 2D ERs in Doping superlattices of HD IV-VI, II-VI and stressed Kane type compounds in the presence of an arbitrarily oriented non-quantizing magnetic field by including the electron spin.

(R.6.35) Investigate the 2D ER for Doping superlattices of all the materials as stated in R.1.1 of Chap. 1 in the presence of an arbitrarily oriented non-quantizing magnetic field by including electron spin under the conditions of formation of band tails and applied external parallel magnetic field.

(R.6.36) Investigate the 2D ER for all the problems from R.A4.32 to R.A4.35 in the presence of an additional arbitrarily oriented non-quantizing electric field.

(R.6.37) Investigate the 2D ER for all the problems from R.A4.32 to R.A4.35 in the presence of arbitrarily oriented crossed electric and magnetic fields.

(R.6.38) Investigate all the problems from R.A4.1 to R.A4.37, in the presence of arbitrarily oriented light waves and magnetic quantization.

(R.6.39) Investigate all the problems from R.A4.1 upto R.A4.37 in the presence of exponential, Kane, Halperin and Lax and Bonch-Bruevich band tails [79].

(R.6.40) Investigate all the problems of this chapter by removing all the mathematical approximations and establishing the uniqueness conditions in each case.

(R.6.41) (a) Investigate the ER in all the bulk semiconductors as considered in this appendix in the presence of defects and magnetic quantization.

(b) Investigate the ER as defined in (R.A2.1) in the presence of an arbitrarily oriented quantizing magnetic field including broadening and the electron spin (applicable under magnetic quantization) for all the bulk semiconductors whose unperturbed carrier energy spectra are defined in Chap. 1.

(R.6.42) Investigate the ER as defined in (R.A2.1) in the presence of quantizing magnetic field under an arbitrarily oriented (a) non-uniform electric field and (b) alternating electric field respectively for all the semiconductors whose unperturbed carrier energy spectra are defined in Chap. 1 by including spin and broadening respectively.

(R.6.43) Investigate the ER as defined in (R.A2.1) under an arbitrarily oriented alternating quantizing magnetic field by including broadening and the electron spin for all the semiconductors whose unperturbed carrier energy spectra as defined in Chap. 1.

(R.6.44) Investigate the ER as defined in (R.A2.1) under an arbitrarily oriented alternating quantizing magnetic field and crossed alternating electric field by including broadening and the electron spin for all the semiconductors whose unperturbed carrier energy spectra as defined in Chap. 1.

(R.6.45) Investigate the ER as defined in (R.A2.1) under an arbitrarily oriented alternating quantizing magnetic field and crossed alternating non-uniform electric field by including broadening and the electron spin whose for all the semiconductors unperturbed carrier energy spectra as defined in Chap. 1.

(R.6.46) Investigate the ER as defined in (R.A2.1) in the presence and absence of an arbitrarily oriented alternating quantizing magnetic field under exponential, Kane, Halperin, Lax and Bonch-Bruevich band tails [69] for all the semiconductors whose unperturbed carrier energy spectra as defined in Chap. 1 by including spin and broadening (applicable under magnetic quantization).

(R.6.47) Investigate the ER as defined in (R.A2.1) in the presence of an arbitrarily oriented quantizing magnetic field for all the semiconductors as defined in (R.A2.6) under an arbitrarily oriented (a) non-uniform electric field and (b) alternating electric field respectively whose unperturbed carrier energy spectra as defined in Chap. 1.

(R.6.48) Investigate the ER as defined in (R.A2.1) under an arbitrarily oriented alternating quantizing magnetic field by including broadening and the electron spin for all semiconductors whose unperturbed carrier energy spectra as defined in Chap. 1.

(R.6.49) Investigate the ER as defined in (R.A2.1) under an arbitrarily oriented alternating quantizing magnetic field and crossed alternating electric field by including broadening and the electron spin for all the semiconductors whose unperturbed carrier energy spectra as defined in Chap. 1.

(R.A.50) Investigate all the appropriate problems of this section under magnetic quantization after proper modifications introducing new theoretical formalisms for functional, negative refractive index, macro molecular, organic and magnetic materials.

(R.A.51) Investigate all the appropriate problems of this section for HD p-InSb, p-CuCl and stressed semiconductors under magnetic quantization having diamond structure valence bands whose dispersion relations of the carriers in bulk semiconductors are given by Cunningham [79], Yekimov et al. [80] and Roman et al. [81] respectively.

(R.A.52) Investigate all the problems of this section by removing all the mathematical approximations and establishing the respective appropriate uniqueness conditions.

References

1. N. Miura, *Physics of Semiconductors in High Magnetic Fields, Series on Semiconductor Science and Technology* (Oxford University Press, USA, 2007)
2. K.H.J. Buschow, F.R. de Boer, *Physics of Magnetism and Magnetic Materials* (Springer, New York, 2003)
3. D. Sellmyer, R. Skomski (eds.), *Advanced Magnetic Nanostructures* (Springer, New York, 2005)
4. J.A.C. Bland, B. Heinrich (eds.), *Ultrathin Magnetic Structures III: Fundamentals of Nanomagnetism (Pt. 3)* (Springer, Germany, 2005)
5. B.K. Ridley, *Quantum Processes in Semiconductors*, 4th edn. (Oxford publications, Oxford, 1999)
6. J.H. Davies, *Physics of Low Dimensional Semiconductors* (Cambridge University Press, UK, 1998)
7. S. Blundell, *Magnetism in Condensed Matter, Oxford Master Series in Condensed Matter Physics* (Oxford University Press, USA, 2001)
8. C. Weisbuch, B. Vinter, *Quantum Semiconductor Structures: Fundamentals and Applications* (Academic Publishers, USA, 1991)
9. D. Ferry, *Semiconductor Transport* (CRC, USA, 2000)
10. M. Reed (ed.), *Semiconductors and Semimetals: Nanostructured Systems* (Academic Press, USA, 1992)
11. T. Dittrich, *Quantum Transport and Dissipation* (Wiley-VCH Verlag GmbH, Germany, 1998)
12. A.Y. Shik, *Quantum Wells: Physics and Electronics of Two dimensional Systems* (World Scientific, USA, 1997)
13. K.P. Ghatak, M. Mondal, Zietschrift fur Naturforschung A **41a**, 881 (1986)
14. K.P. Ghatak, M. Mondal, J. Appl. Phys. **62**, 922 (1987)
15. K.P. Ghatak, S.N. Biswas, Phys. Stat. Sol. (b) **140**, K107 (1987)
16. K.P. Ghatak, M. Mondal, J. Mag. Mag. Mat. **74**, 203 (1988)
17. K.P. Ghatak, M. Mondal, Phys. Stat. Sol. (b) **139**, 195 (1987)
18. K.P. Ghatak, M. Mondal, Phys. Stat. Sol. (b) **148**, 645 (1988)
19. K.P. Ghatak, B. Mitra, A. Ghoshal, Phys. Stat. Sol. (b) **154**, K121 (1989)
20. K.P. Ghatak, S.N. Biswas, J. Low Temp. Phys. **78**, 219 (1990)
21. K.P. Ghatak, M. Mondal, Phys. Stat. Sol. (b) **160**, 673 (1990)
22. K.P. Ghatak, B. Mitra, Phys. Lett. **A156**, 233 (1991)

23. K.P. Ghatak, A. Ghoshal, B. Mitra, Nouvo Cimento **D13D**, 867 (1991)
24. K.P. Ghatak, M. Mondal, Phys. Stat. Sol. (b) **148**, 645 (1989)
25. K.P. Ghatak, B. Mitra, Int. J. Elect. **70**, 345 (1991)
26. K.P. Ghatak, S.N. Biswas, J. Appl. Phys. **70**, 299 (1991)
27. K.P. Ghatak, A. Ghoshal, Phys. Stat. Sol. (b) **170**, K27 (1992)
28. K.P. Ghatak, Nouvo Cimento **D13D**, 1321 (1992)
29. K.P. Ghatak, B. Mitra, Int. J. Elect. **72**, 541 (1992)
30. K.P. Ghatak, S.N. Biswas, Nonlinear Opt. **4**, 347 (1993)
31. K.P. Ghatak, M. Mondal, Phys. Stat. Sol. (b) **175**, 113 (1993)
32. K.P. Ghatak, S.N. Biswas, Nonlinear Opt. **4**, 39 (1993)
33. K.P. Ghatak, B. Mitra, Nouvo Cimento **15D**, 97 (1993)
34. K.P. Ghatak, S.N. Biswas, Nanostruct. Mater. **2**, 91 (1993)
35. K.P. Ghatak, M. Mondal, Phys. Stat. Sol. (b) **185**, K5 (1994)
36. K.P. Ghatak, B. Goswami, M. Mitra, B. Nag, Nonlinear Opt. **16**, 9 (1996)
37. K.P. Ghatak, M. Mitra, B. Goswami, B. Nag, Nonlinear Opt. **16**, 167 (1996)
38. K.P. Ghatak, B. Nag, Nanostruct. Mater. **10**, 923 (1998)
39. D. Roy Choudhury, A.K. Choudhury, K.P. Ghatak, A.N. Chakravarti, Phys. Stat. Sol. (b) **98**, K141 (1980)
40. A.N. Chakravarti, K.P. Ghatak, A. Dhar, S. Ghosh, Phys. Stat. Sol. (b) **105**, K55 (1981)
41. A.N. Chakravarti, A.K. Choudhury, K.P. Ghatak, Phys. Stat. Sol. (a) **63**, K97 (1981)
42. A.N. Chakravarti, A.K. Choudhury, K.P. Ghatak, S. Ghosh, A. Dhar, Appl. Phys. **25**, 105 (1981)
43. A.N. Chakravarti, K.P. Ghatak, G.B. Rao, K.K. Ghosh, Phys. Stat. Sol. (b) **112**, 75 (1982)
44. A.N. Chakravarti, K.P. Ghatak, K.K. Ghosh, H.M. Mukherjee, Phys. Stat. Sol. (b) **116**, 17 (1983)
45. M. Mondal, K.P. Ghatak, Phys. Stat. Sol. (b) **133**, K143 (1984)
46. M. Mondal, K.P. Ghatak, Phys. Stat. Sol. (b) **126**, K47 (1984)
47. M. Mondal, K.P. Ghatak, Phys. Stat. Sol. (b) **126**, K41 (1984)
48. M. Mondal, K.P. Ghatak, Phys. Stat. Sol. (b) **129**, K745 (1985)
49. M. Mondal, K.P. Ghatak, Phys. Scr. **31**, 615 (1985)
50. M. Mondal, K.P. Ghatak, Phys. Stat. Sol. (b) **135**, 239 (1986)
51. M. Mondal, K.P. Ghatak, Phys. Stat. Sol. (b) **93**, 377 (1986)
52. M. Mondal, K.P. Ghatak, Phys. Stat. Sol. (b) **135**, K21 (1986)
53. M. Mondal, S. Bhattacharyya, K.P. Ghatak, Appl. Phys. **A42A**, 331 (1987)
54. S.N. Biswas, N. Chattopadhyay, K.P. Ghatak, Phys. Stat. Sol. (b) **141**, K47 (1987)
55. B. Mitra, K.P. Ghatak, Phys. Stat. Sol. (b) **149**, K117 (1988)
56. B. Mitra, A. Ghoshal, K.P. Ghatak, Phys. Stat. Sol. (b) **150**, K67 (1988)
57. M. Mondal, K.P. Ghatak, Phys. Stat. Sol. (b) **147**, K179 (1988)
58. M. Mondal, K.P. Ghatak, Phys. Stat. Sol. (b) **146**, K97 (1988)
59. B. Mitra, A. Ghoshal, K.P. Ghatak, Phys. Stat. Sol. (b) **153**, K209 (1989)
60. B. Mitra, K.P. Ghatak, Phys. Lett. **142A**, 401 (1989)
61. B. Mitra, A. Ghoshal, K.P. Ghatak, Phys. Stat. Sol. (b) **154**, K147 (1989)
62. B. Mitra, K.P. Ghatak, Sol. State Elect. **32**, 515 (1989)
63. B. Mitra, A. Ghoshal, K.P. Ghatak, Phys. Stat. Sol. (b) **155**, K23 (1989)
64. B. Mitra, K.P. Ghatak, Phys. Letts. **135A**, 397 (1989)
65. B. Mitra, K.P. Ghatak, Phys. Letts. **A146A**, 357 (1990)
66. B. Mitra, K.P. Ghatak, Phys. Stat. Sol. (b) **164**, K13 (1991)
67. S.N. Biswas, K.P. Ghatak, Int. J. Elect. **70**, 125 (1991)
68. P.R. Wallace, Phys. Stat. Sol. (b) **92**, 49 (1979)
69. B.R. Nag, *Electron Transport in Compound Semiconductors, Springer Series in Solid-State Sciences*, vol. 11 (Springer, Germany, 1980)
70. K.P. Ghatak, S. Bhattacharya, D. De, *Einstein Relation in Compound Semiconductors and Their Nanostructures, Springer Series in Materials Science*, vol. 116 (Springer, Germany, 2009)

71. C.C. Wu, C.J. Lin, J. Low Temp. Phys. **57**, 469 (1984)
72. M.H. Chen, C.C. Wu, C.J. Lin, J. Low Temp. Phys. **55**, 127 (1984)
73. E. Bangert, P. Kastner, Phys. Stat. Sol (b) **61**, 503 (1974)
74. G.M.T. Foley, P.N. Langenberg, Phys. Rev. B **15B**, 4850 (1977)
75. M. Singh, P.R. Wallace, S.D. Jog, E. Arushanov, J. Phys. Chem. Solids **45**, 409 (1984)
76. Y. Yamada, Phys. Soc. Japan **35**, 1600 (1973)
77. Y. Yamada, Phys. Soc. Japan **37**, 606 (1974)
78. L.A. Vassilev, Phys. Stat. Sol. (b) **121**, 203 (1984)
79. R.W. Cunningham, Phys. Rev. **167**, 761 (1968)
80. A.I. Yekimov, A.A. Onushchenko, A.G. Plyukhin, A.L. Efros, J. Expt, Theor. Phys. **88**, 1490 (1985)
81. B.J. Roman, A.W. Ewald, Phys. Rev. **B5**, 3914 (1972)

Chapter 7
Appendix B: The ER in Superlattices of HD Non-parabolic Semiconductors Under Magnetic Quantization

7.1 Introduction

In recent years, modern fabrication techniques have generated altogether a new dimension in the arena of quantum effect devices through the experimental realization of an important artificial structure known as semiconductor superlattice (SL) by growing two similar but different semiconducting compounds in alternate layers with finite thicknesses [1–33]. The materials forming the alternate layers have the same kind of band structure but different energy gaps. The concept of SL was developed for the first time by Keldysh [34] and was successfully fabricated by Esaki and Tsu [35–38]. The SLs are being extensively used in thermal sensors [39, 40], quantum cascade lasers [41–43], photodetectors [44, 45], light emitting diodes [46–49], multiplication [50], frequency multiplication [51], photocathodes [52], thin film transistor [53, 54], solar cells [55, 56], infrared imaging [57], thermal imaging [58], infrared sensing [59, 60] and also in other microelectronic devices.

The most extensively studied III-V SL is the one consisting of alternate layers of GaAs and $Ga_{1-x}Al_xAs$ owing to the relative easiness of fabrication. The GaAs and $Ga_{1-x}Al_xAs$ layers form the quantum wells and the potential barriers respectively. The III-V SL's are attractive for the realization of high speed electronic and optoelectronic devices [61]. In addition to SLs with usual structure, other types of SLs such as II-VI [62], IV-VI [63] and HgTe/CdTe [64] SL's have also been investigated in the literature. The IV-VI SLs exhibit quite different properties as compared to the III-V SL due to the specific band structure of the constituent materials [65]. The epitaxial growth of II-VI SL is a relatively recent development and the primary motivation for studying the mentioned SLs made of materials with the large band gap is in their potential for optoelectronic operation in the blue [65]. HgTe/CdTe SL's have raised a great deal of attention since 1979, when as a promising new materials for long wavelength infrared detectors and other

K. P. Ghatak and S. Bhattacharya, *Heavily-Doped 2D-Quantized Structures and the Einstein Relation*, Springer Tracts in Modern Physics 260,
DOI 10.1007/978-3-319-08380-3_7, © Springer International Publishing Switzerland 2015

electro-optical applications [66]. Interest in Hg-based SL's has been further increased as new properties with potential device applications were revealed [66, 67]. These features arise from the unique zero band gap material HgTe [68] and the direct band gap semiconductor CdTe which can be described by the three band mode of Kane [69]. The combination of the aforementioned materials with specified dispersion relation makes HgTe/CdTe SL very attractive, especially because of the tailoring of the material properties for various applications by varying the energy band constants of the SLs.

We note that all the aforementioned SLs have been proposed with the assumption that the interfaces between the layers are sharply defined, of zero thickness, i.e., devoid of any interface effects. The SL potential distribution may be then considered as a one dimensional array of rectangular potential wells. The aforementioned advanced experimental techniques may produce SLs with physical interfaces between the two materials crystallographically abrupt; adjoining their interface will change at least on an atomic scale. As the potential form changes from a well (barrier) to a barrier (well), an intermediate potential region exists for the electrons [70]. The influence of finite thickness of the interfaces on the electron dispersion law is very important, since; the electron energy spectrum governs the electron transport in SLs. In addition to it, for effective mass SLs, the electronic subbands appear continually in real space [71].

In this chapter, we shall study the ER under magnetic quantization in III-V, II-VI, IV-VI, HgTe/CdTe and strained layer, HDSLs with graded interfaces in Sects. 7.2.1–7.2.5 respectively. From Sects. 7.2.6–7.2.10, we shall investigate the same in III-V, II-VI, IV-VI, HgTe/CdTe and strained layer, HD effective mass SLs The last Sect. 7.3 contains open research problems.

7.2 Theoretical Background

7.2.1 The ER in HD III-V Superlattices with Graded Interfaces Under Magnetic Quantization

The electron dispersion law in bulk specimens of the heavily doped constituent materials of HD III-V SLs whose unperturbed energy band structures are defined by three-band model of Kane can be expressed as

$$\frac{\hbar^2 k^2}{2m_{cj}^*} = T_{1j}\left(E, \Delta_j, E_{gj}, \eta_{gj}\right) + iT_{2j}\left(E, \Delta_j, E_{gj}, \eta_{gj}\right) \tag{7.1}$$

where

$$j = 1,2,\ T_{ij}(E,\Delta_j,E_{gj},\eta_{gj})$$
$$= (2/(1+Erf(E/\eta_{gj}))[(\alpha_j b_j/c_j).\theta_0(E,\eta_{gj}) + [(\alpha_j c_j + b_j c_j - \alpha_j b_j)/c_j^2]$$
$$\times \gamma_0(E,\eta_{gj}) + \Big[(1/c_j)(1-(\alpha_j/c_j))(1-(b_j/c_j))\frac{1}{2}[1+Erf(E/\eta_{gj})]$$
$$-(1/c_j)(1-(\alpha_j/c_j))(1-(b_j/c_j))(2/(c_j\eta_{gj}\sqrt{\pi})]$$
$$\times \exp\left(-u_j^2\right)\left[\sum_{p=1}^{\infty}(\exp(-p^2/4)/p)\sinh(pu_j)\right]\Big],$$
$$b_j \equiv (E_{gj}+\Delta_j)^{-1},\ c_j \equiv \left(E_{gj}+\frac{2}{3}\Delta_j\right)^{-1},$$
$$u_j \equiv \frac{1+c_jE}{c_j\eta_{gj}}$$

and

$$T_{2j}(E,\Delta_j,E_{gj},\eta_{gj}) \equiv \left(\frac{2}{1+Erf(E/\eta_{gj})}\right)\frac{1}{c_j}\left(1-\frac{\alpha_j}{c_j}\right)\left(1-\frac{b_j}{c_j}\right)\frac{\sqrt{\pi}}{c_j\eta_{gj}}\exp\left(-u_j^2\right).$$

Therefore, the dispersion law of the electrons of heavily doped III-V SLs with graded interfaces can be expressed as

$$k_z^2 = G_8 + iH_8 \tag{7.2}$$

where

$$G_8 = \left[\frac{C_7^2 - D_7^2}{L_0^2} - k_s^2\right],$$
$$C_7 = \cos^{-1}(\overline{\omega_7}),\ \overline{\omega_7} = (2)^{\frac{-1}{2}}\left[(1-G_7^2-H_7^2) - \sqrt{(1-G_7^2-H_7^2)^2 + 4G_7^2}\right]^{\frac{1}{2}}$$
$$G_7 = [G_1 + (\rho_5 G_2/2) - (\rho_6 H_2/2) + (\Delta_0/2)\{\rho_6 H_2 - \rho_8 H_3 + \rho_9 H_4 - \rho_{10} H_4$$
$$+ \rho_{11} H_5 - \rho_{12} H_5 + (1/12)(\rho_{12} G_6 - \rho_{14} H_6)\}],$$
$$G_1 = [(\cos(h_1))(\cosh(h_2))(\cosh(g_1))(\cos(g_2))$$
$$+ (\sin(h_1))(\sinh(h_2))(\sinh(g_1))(\sin(g_2))],$$
$$h_1 = e_1(b_0 - \Delta_0),\ e_1 = 2^{\frac{-1}{2}}\left(\sqrt{t_1^2 + t_2^2} + t_1\right)^{\frac{1}{2}},$$
$$t_1 = [(2m_{c1}^*/\hbar^2).T_{11}(E,E_{g1},\Delta_1,\eta_{g1}) - k_s^2],$$
$$t_2 = [(2m_{c1}^*/\hbar^2)T_{21}(E,E_{g1},\Delta_1,\eta_{g1})],$$
$$h_2 = e_2(b_0 - \Delta_0),\ e_2 = 2^{\frac{-1}{2}}\left(\sqrt{t_1^2 + t_2^2} - t_1\right)^{\frac{1}{2}},$$

$$g_1 = d_1(a_0 - \Delta_0),\ d_1 = 2^{\frac{-1}{2}}\left(\sqrt{x_1^2 + y_1^2} + x_1\right)^{\frac{1}{2}},$$
$$x_1 = [-(2m_{c2}^*/\hbar^2).T_{11}(E - V_0, E_{g2}, \Delta_2, \eta_{g2}) + k_s^2],$$
$$y_1 = [(2m_{c2}^*/\hbar^2)T_{22}(E - V_0, E_{g2}, \Delta_2, \eta_{g2})],$$
$$g_2 = d_2(a_0 - \Delta_0), d_2 = 2^{\frac{-1}{2}}\left(\sqrt{x_1^2 + y_1^2} - x_1\right)^{\frac{1}{2}},$$
$$\rho_5 = (\rho_3^2 + \rho_4^2)^{-1}[\rho_1\rho_3 - \rho_2\rho_4],$$
$$\rho_1 = [d_1^2 + e_2^2 - d_2^2 - e_1^2], \rho_3 = [d_1e_1 + d_2e_2],$$
$$\rho_2 = 2[d_1d_2 + e_1e_2], \rho_4 = [d_1e_2 - e_1d_2],$$
$$G_2 = [(\sin(h_1))(\cosh(h_2))(\sinh(g_1))(\cos(g_2)) + (\cos(h_1))(\sinh(h_2))(\cosh(g_1))(\sin(g_2))],$$
$$\rho_6 = (\rho_3^2 + \rho_4^2)^{-1}[\rho_1\rho_4 + \rho_2\rho_3],$$

$$H_2 = [(\sin(h_1))(\cosh(h_2))(\sin(g_2))(\cosh(g_1)) - (\cos(h_1))(\sinh(h_2))(\sinh(g_1))(\cos(g_2))],$$
$$\rho_7 = [(e_1^2 + e_2^2)^{-1}[e_1(d_1^2 - d_2^2) - 2d_1d_2e_2] - 3e_1],$$
$$G_3 = [(\sin(h_1))(\cosh(h_2))(\cosh(g_1))(\cos(g_2)) + (\cos(h_1))(\sinh(h_2))(\sinh(g_1))(\sin(g_2))],$$
$$\rho_8 = [(e_1^2 + e_2^2)^{-1}[e_2(d_1^2 - d_2^2) + 2d_1d_2e_1] + 3e_2],$$
$$H_3 = [(\sin(h_1))(\cosh(h_2))(\sin(g_2))(\sinh(g_1)) - (\cos(h_1))(\sinh(h_2))(\cosh(g_1))(\cos(g_2))],$$
$$\rho_9 = [(d_1^2 + d_2^2)^{-1}[d_1(e_2^2 - e_1^2) + 2e_2d_2e_1] + 3d_1],$$
$$G_4 = [(\cos(h_1))(\cosh(h_2))(\cos(g_2))(\sinh(g_1)) - (\sin(h_1))(\sinh(h_2))(\cosh(g_1))(\sin(g_2))],$$
$$\rho_{10} = [-(d_1^2 + d_2^2)^{-1}[d_2(-e_2^2 + e_1^2) + 2e_2d_2e_1] + 3d_2],$$
$$H_4 = [(\cos(h_1))(\cosh(h_2))(\cosh(g_1))(\sin(g_2)) + (\sin(h_1))(\sinh(h_2))(\sinh(g_1))(\cos(g_2))],$$
$$\rho_{11} = 2[d_1^2 + e_2^2 - d_2^2 - e_1^2],$$
$$G_5 = [(\cos(h_1))(\cosh(h_2))(\cos(g_2))(\cosh(g_1)) - (\sin(h_1))(\sinh(h_2))(\sinh(g_1))(\sin(g_2))],$$
$$\rho_{12} = 4[d_1d_2 + e_1e_2],$$
$$H_5 = [(\cos(h_1))(\cosh(h_2))(\sinh(g_1))(\sin(g_2)) + (\sin(h_1))(\sinh(h_2))(\cosh(g_1))(\cos(g_2))],$$
$$\rho_{13} = [\{5(d_1e_1^3 - 3e_1e_2^2d_1) + 5d_2(e_1^3 - 3e_1^2e_2)\}(d_1^2 + d_2^2)^{-1} + (e_1^2 + e_2^2)^{-1}\{5(e_1d_1^3 - 3d_2e_1^2d_1) + 5(d_2^3e_2 - 3d_1^2d_2e_2)\} - 34(d_1e_1 + d_2e_2)],$$

$$G_6 = [(\sin(h_1))(\cosh(h_2))(\sinh(g_1))(\cos(g_2)) + (\cos(h_1))(\sinh(h_2))(\cosh(g_1))(\sin(g_2))],$$

$$\rho_{14} = [\{5(d_1e_2^3 - 3e_2e_1^2d_1) + 5d_2(-e_1^3 + 3e_2^2e_1)\}(d_1^2 + d_2^2)^{-1} + (e_1^2 + e_2^2)^{-1}\{5(-e_1d_2^3 + 3d_1^2d_2e_1) + 5(-d_1^3e_2 + 3d_2^2d_1e_2)\} + 34(d_1e_2 - d_2e_1)],$$

$$H_6 = [(\sin(h_1))(\cosh(h_2))(\cosh(g_1))(\sin(g_2)) - (\cos(h_1))(\sinh(h_2))(\sinh(g_1))(\cos(g_2))],$$

$$H_7 = [H_1 + (\rho_5H_2/2) + (\rho_6G_2/2) + (\Delta_0/2)\{\rho_8G_3 + \rho_7H_3 + \rho_{10}G_4 + \rho_9H_4 + \rho_{12}G_5 + \rho_{11}H_5 + (1/12)(\rho_{14}G_6 + \rho_{13}H_6)\}],$$

$$H_1 = [(\sin(h_1))(\sinh(h_2))(\cosh(g_1))(\cos(g_2)) + (\cos(h_1))(\cosh(h_2))(\sinh(g_1))(\sin(g_2))],$$

$$D_7 = \sinh^{-1}(\overline{\omega_7}),\ H_8 = (2C_7D_7/L_0^2)$$

The simplified dispersion relation of heavily doped III-V superlattices with graded interfaces under magnetic quantization can be expressed as

$$k_z^2 = G_{8E,n} + iH_{8E,n} \tag{7.3a}$$

where

$$G_{8E,n} = \left[\frac{C_{7E,n}^2 - D_{7E,n}^2}{L_0^2} - \left\{\frac{2eB}{\hbar}\left(n + \frac{1}{2}\right)\right\}\right],\ C_{7E,n} = \cos^{-1}(\overline{\omega_{7E,n}}),$$

$$\overline{\omega_{7E,n}} = (2)^{\frac{-1}{2}}\left[(1 - G_{7E,n}^2 - H_{7E,n}^2) - \sqrt{(1 - G_{7E,n}^2 - H_{7E,n}^2)^2 + 4G_{7E,n}^2}\right]^{\frac{1}{2}}$$

$$G_{7E,n} = [G_{1E,n} + (\rho_{5E,n}G_{2E,n}/2) - (\rho_{6E,n}H_{2E,n}/2) + (\Delta_0/2)\{\rho_{6E,n}H_{2E,n} - \rho_{8E,n}H_{3E,n} + \rho_{9E,n}H_{4E,n} - \rho_{10E,n}H_{4E,n} + \rho_{11E,n}H_{5E,n} - \rho_{12E,n}H_{5E,n} + (1/12)(\rho_{12E,n}G_{6E,n} - \rho_{14E,n}H_{6E,n})\}],$$

$$G_{1E,n} = [(\cos(h_{1E,n}))(\cosh(h_{2E,n}))(\cosh(g_{1E,n}))(\cos(g_{2E,n})) + (\sin(h_{1E,n}))(\sinh(h_{2E,n}))(\sinh(g_{1E,n}))(\sin(g_{2E,n}))],$$

$$h_{1E,n} = e_{1E,n}(b_0 - \Delta_0),\ e_{1E,n} = 2^{\frac{-1}{2}}\left(\sqrt{t_{1E,n}^2 + t_2^2} + t_{1E,n}\right)^{\frac{1}{2}},$$

$$t_{1E,n} = \left[(2m_{c1}^*/\hbar^2).T_{11}(E, E_{g1}, \Delta_1, \eta_{g1}) - \left\{\frac{2eB}{\hbar}\left(n + \frac{1}{2}\right)\right\}\right],$$

$$t_2 = \left[(2m_{c1}^*/\hbar^2)T_{21}(E, E_{g1}, \Delta_1, \eta_{g1})\right],$$

$$h_{2E,n} = e_{2E,n}(b_0 - \Delta_0),\ e_{2E,n} = 2^{\frac{-1}{2}}\left(\sqrt{t_{1E,n}^2 + t_2^2} - t_{1E,n}\right)^{\frac{1}{2}},$$

$$
\begin{aligned}
g_{1E,n} &= d_{1E,n}(a_0 - \Delta_0),\ d_{1E,n} = 2^{\frac{-1}{2}}\left(\sqrt{x_{1E,n}^2 + y_1^2} + x_{1E,n}\right)^{\frac{1}{2}}, x_{1E,n} \\
&= \left[-(2m_{c2}^*/\hbar^2).T_{11}(E - V_0, E_{g2}, \Delta_2, \eta_{g2}) + \left\{\frac{2eB}{\hbar}\left(n + \frac{1}{2}\right)\right\}\right], y_1 \\
&= [(2m_{c2}^*/\hbar^2)T_{22}(E - V_0, E_{g2}, \Delta_2, \eta_{g2})], g_{2E,n} = d_{2E,n}(a_0 - \Delta_0), d_{2E,n} \\
&= 2^{\frac{-1}{2}}\left(\sqrt{x_{1E,n}^2 + y_1^2} - x_{1E,n}\right)^{\frac{1}{2}}, \rho_{5E,n} \\
&= (\rho_{3E,n}^2 + \rho_{4E,n}^2)^{-1}[\rho_{1E,n}\rho_{3E,n} - \rho_{2E,n}\rho_{4E,n}], \rho_{1E,n} \\
&= [d_{1E,n}^2 + e_{2E,n}^2 - d_{2E,n}^2 - e_{1E,n}^2], \rho_{3E,n} = [d_{1E,n}e_{1E,n} + d_{2E,n}e_{2E,n}], \rho_{2E,n} \\
&= 2[d_{1E,n}d_{2E,n} + e_{1E,n}e_{2E,n}], \rho_{4E,n} = [d_{1E,n}e_{2E,n} - e_{1E,n}d_{2E,n}], G_{2E,n} \\
&= [(\sin(h_{1E,n}))(\cosh(h_{2E,n}))(\sinh(g_{1E,n}))(\cos(g_{2E,n})) \\
&\quad + (\cos(h_{1E,n}))(\sinh(h_{2E,n}))(\cosh(g_{1E,n}))(\sin(g_{2E,n}))], \rho_{6E,n} \\
&= (\rho_{3E,n}^2 + \rho_{4E,n}^2)^{-1}[\rho_{1E,n}\rho_{4E,n} + \rho_{2E,n}\rho_{3E,n}],
\end{aligned}
$$

$$
\begin{aligned}
H_{2E,n} &= [(\sin(h_{1E,n}))(\cosh(h_{2E,n}))(\sin(g_{2E,n}))(\cosh(g_{1E,n})) \\
&\quad - (\cos(h_{1E,n}))(\sinh(h_{2E,n}))(\sinh(g_{1E,n}))(\cos(g_{2E,n}))], \\
\rho_{7E,n} &= [(e_{1E,n}^2 + e_{2E,n}^2)^{-1}[e_{1E,n}(d_{1E,n}^2 - d_{2E,n}^2) \\
&\quad - 2d_{1E,n}d_{2E,n}e_{2E,n}] - 3e_{1E,n}], \\
G_{3E,n} &= [(\sin(h_{1E,n}))(\cosh(h_{2E,n}))(\cosh(g_{1E,n}))(\cos(g_{2E,n})) \\
&\quad + (\cos(h_{1E,n}))(\sinh(h_{2E,n}))(\sinh(g_{1E,n}))(\sin(g_{2E,n}))], \\
\rho_{8E,n} &= [(e_{1E,n}^2 + e_{2E,n}^2)^{-1}[e_{2E,n}(d_{1E,n}^2 - d_{2E,n}^2) + 2d_{1E,n}d_{2E,n}e_{1E,n}] + 3e_{2E,n}], \\
H_{3E,n} &= [(\sin(h_{1E,n}))(\cosh(h_{2E,n}))(\sin(g_{2E,n}))(\sinh(g_{1E,n})) \\
&\quad - (\cos(h_{1E,n}))(\sinh(h_{2E,n}))(\cosh(g_{1E,n}))(\cos(g_{2E,n}))], \\
\rho_{9E,n} &= [(d_{1E,n}^2 + d_{2E,n}^2)^{-1}[d_{1E,n}(e_{2E,n}^2 - e_{1E,n}^2) + 2e_{2E,n}d_{2E,n}e_{1E,n}] + 3d_{1E,n}], \\
G_{4E,n} &= [(\cos(h_{1E,n}))(\cosh(h_{2E,n}))(\cos(g_{2E,n}))(\sinh(g_{1E,n})) \\
&\quad - (\sin(h_{1E,n}))(\sinh(h_{2E,n}))(\cosh(g_{1E,n}))(\sin(g_{2E,n}))], \\
\rho_{10E,n} &= [-(d_{1E,n}^2 + d_{2E,n}^2)^{-1}[d_{2E,n}(-e_{2E,n}^2 + e_{1E,n}^2) + 2e_{2E,n}d_{2E,n}e_{1E,n}] + 3d_{2E,n}], \\
H_{4E,n} &= [(\cos(h_{1E,n}))(\cosh(h_{2E,n}))(\cosh(g_{1E,n}))(\sin(g_{2E,n})) \\
&\quad + (\sin(h_{1E,n}))(\sinh(h_{2E,n}))(\sinh(g_{1E,n}))(\cos(g_{2E,n}))], \\
\rho_{11E,n} &= 2[d_{1E,n}^2 + e_{2E,n}^2 - d_{2E,n}^2 - e_{1E,n}^2], \\
G_{5E,n} &= [(\cos(h_{1E,n}))(\cosh(h_{2E,n}))(\cos(g_{2E,n}))(\cosh(g_{1E,n})) \\
&\quad - (\sin(h_{1E,n}))(\sinh(h_{2E,n}))(\sinh(g_{1E,n}))(\sin(g_{2E,n}))], \\
\rho_{12E,n} &= 4[d_{1E,n}d_{2E,n} + e_{1E,n}e_{2E,n}], \\
H_{5E,n} &= [(cos(h_{1E,n}))(\cosh(h_{2E,n}))(\sinh(g_{1E,n}))(\sin(g_{2E,n})) \\
&\quad + (\sin(h_{1E,n}))(\sinh(h_{2E,n}))(\cosh(g_1))(\cos(g_2))],
\end{aligned}
$$

$$\rho_{13E,n} = [\{5(d_{1E,n}e_{1E,n}^3 - 3e_{1E,n}e_{2E,n}^2 d_{1E,n}) + 5d_{2E,n}(e_{1E,n}^3 - 3e_{1E,n}^2 e_{2E,n})\}(d_{1E,n}^2 + d_{2E,n}^2)^{-1} + (e_{1E,n}^2 + e_{2E,n}^2)^{-1}\{5(e_{1E,n}d_{1E,n}^3 - 3d_{2E,n}e_{1E,n}^2 d_{1E,n}) + 5(d_{2E,n}^3 e_{2E,n} - 3d_{1E,n}^2 d_{2E,n}e_{2E,n})\} - 34(d_{1E,n}e_{1E,n} + d_{2E,n}e_{2E,n})],$$

$$G_{6E,n} = [(\sin(h_{1E,n}))(\cosh(h_{2E,n}))(\sinh(g_{1E,n}))(\cos(g_{2E,n})) + (\cos(h_{1E,n}))(\sinh(h_{2E,n}))(\cosh(g_1))(\sin(g_2))],$$

$$\rho_{14E,n} = [\{5(d_{1E,n}e_{2E,n}^3 - 3e_{2E,n}e_{1E,n}^2 d_{1E,n}) + 5d_{2E,n}(-e_{1E,n}^3 + 3e_{2E,n}^2 e_{1E,n})\}(d_{1E,n}^2 + d_{2E,n}^2)^{-1} + (e_{1E,n}^2 + e_{2E,n}^2)^{-1}\{5(-e_{1E,n}d_{2E,n}^3 + 3d_{1E,n}^2 d_{2E,n}e_{1E,n}) + 5(-d_{1E,n}^3 e_{2E,n} + 3d_{2E,n}^2 d_{1E,n}e_{2E,n})\} + 34(d_{1E,n}e_{2E,n} - d_{2E,n}e_{1E,n})],$$

$$H_{6E,n} = [(\sin(h_{1E,n}))(\cosh(h_{2E,n}))(\cosh(g_{1E,n}))(\sin(g_{2E,n})) - (\cos(h_{1E,n}))(\sinh(h_{2E,n}))(\sinh(g_{1E,n}))(\cos(g_{2E,n}))],$$

$$H_{7E,n} = [H_{1E,n} + (\rho_{5E,n}H_{2E,n}/2) + (\rho_{6E,n}G_{2E,n}/2) + (\Delta_0/2)\{\rho_{8E,n}G_{3E,n} + \rho_{7E,n}H_{3E,n} + \rho_{10E,n}G_{4E,n} + \rho_{9E,n}H_{4E,n} + \rho_{12E,n}G_{5E,n} + \rho_{11E,n}H_{5E,n} + (1/12)(\rho_{14E,n}G_{6E,n} + \rho_{13E,n}H_{6E,n})\}],$$

$$H_{1E,n} = [(\sin(h_{1E,n}))(\sinh(h_{2E,n}))(\cosh(g_{1E,n}))(\cos(g_{2E,n})) + (\cos(h_{1E,n}))(\cosh(h_{2E,n}))(\sinh(g_{1E,n}))(\sin(g_{2E,n}))],$$

$$D_{7E,n} = \sinh^{-1}(\overline{\omega_{7E,n}}),\ \mathrm{H}_{8E,n} = (2C_{7E,n}D_{7E,n}/L_0^2)$$

The EEM can be written as

$$m^*(E_{F1}, n, \eta_g, B) = \frac{\hbar^2}{2} \mathit{Real\ part\ of}\ [G_{8E_{F1},n}]' \tag{7.3b}$$

where E_{F1} is the Fermi energy in this case.

Thus, from (7.3b) we can infer that the EEM in this case is a function of Fermi energy, Landau quantum number, magnetic field and scattering potential in rather complicated way.

The electron concentration is given by

$$n_0 = \frac{g_v eB}{\pi^2\hbar}\sum_{n=0}^{n_{\max}} [\phi_{1C}(E_{F1}, n) + \phi_{2C}(E_{F1}, n)] \tag{7.4}$$

where

$$\phi_{1C}(E_{F1},n) = \left[\left(G_{8E_{F1},n} + \sqrt{G^2_{8E_{F1},n} - H_{8E_{F1},n}}\right)/2\right]^{1/2},$$

$$\phi_{2C}(E_{F1},n) = \sum_{r=1}^{s} \theta_{2r,1}[\phi_{1C}(E_{F1},n)],\ \theta_{2r,i} = 2(k_BT)^{2r}\left(1-2^{1-2r}\right)\zeta(2r)\frac{\partial^{2r}}{\partial E_{Fi}^{2r}}$$

and $i = 1, 2, 3, \ldots$

The ER in this case can be written as

$$\frac{D}{\mu} = \frac{n_0}{|e|}\ Real\,part\,of\ \left[\frac{\partial n_0}{\partial(E_{Fi} - e_{ii})}\right]^{-1} \tag{7.5}$$

where E_{Fi} is the Fermi energy in this case and e_{ii} is the corresponding Landau level energy.

Thus using (7.3a), (7.3b), (7.4) and (7.5) we can study the ER in this case.

7.2.2 *The ER in HD II-VI Superlattices with Graded Interfaces Under Magnetic Quantization*

The electron energy spectra of the heavily doped constituent materials of II-VI SLs are given by

$$\gamma_3(E,\eta_{g1}) = \frac{\hbar^2 k_s^2}{2m^*_{\perp,1}} + \frac{\hbar^2 k_z^2}{2m^*_{\parallel,1}} \pm C_0 k_s \tag{7.6}$$

and

$$\frac{\hbar^2 k^2}{2m^*_{c2}} = T_{12}(E,\Delta_2,E_{g2},\eta_{g2}) + iT_{22}(E,\Delta_2,E_{g2},\eta_{g2}) \tag{7.7}$$

where $m^*_{\perp,1}$ and $m^*_{\parallel,1}$ are the transverse and longitudinal effective electron masses respectively at the edge of the conduction band for the first material.

Thus, the energy-wave vector dispersion relation of the conduction electrons in heavily doped II-VI SLs with graded interfaces can be expressed as

$$k_z^2 = G_{19} + iH_{19} \tag{7.8}$$

where

$$G_{19} = \left[\frac{C_{18}^2 - D_{18}^2}{L_0^2} - k_s^2\right],$$

$$C_{18} = \cos^{-1}(\omega_{18}), \omega_{18} = (2)^{\frac{-1}{2}}\left[(1 - G_{18}^2 - H_{18}^2) - \sqrt{(1 - G_{18}^2 - H_{18}^2)^2 + 4G_{18}^2}\right]^{\frac{1}{2}},$$

$$G_{18} = \frac{1}{2}[G_{11} + G_{12} + \Delta_0(G_{13} + G_{14}) + \Delta_0(G_{15} + G_{16})],$$

$$G_{11} = 2(\cos(g_1))(\cos(g_2))(\cos\gamma_{11}(E, k_s))$$

$$\gamma_{11}(E, k_s) = k_{21}(E, k_s)(b_0 - \Delta_0),$$

$$k_{21}(E, k_s) = \{[\gamma_3(E, \eta_{g1}) - \frac{\hbar^2 k_s^2}{2m_{\perp,1}^*} \pm C_0 k_s]\frac{2m_{\parallel,1}^*}{\hbar^2}\}^{1/2},$$

$$G_{12} = ([\Omega_1(E, k_s)(\sinh g_1)(\cos g_2) - \Omega_2(E, k_s)(\sin g_2)(\cosh g_1)](\sin\gamma_{11}(E, k_s)))$$

$$\Omega_1(E, k_s) = \left[\frac{d_1}{k_{21}(E, k_s)} - \frac{k_{21}(E, k_s)d_1}{d_1^2 + d_2^2}\right]$$

and

$$\Omega_2(E, k_s) = \left[\frac{d_2}{k_{21}(E, k_s)} + \frac{k_{21}(E, k_s)d_2}{d_1^2 + d_2^2}\right]$$

$$G_{13} = ([\Omega_3(E, k_s)(\cosh g_1)(\cos g_2) - \Omega_4(E, k_s)(\sinh g_1)(\sin g_2)](\sin\gamma_{11}(E, k_s)))$$

$$\Omega_3(E, k_s) = \left[\frac{d_1^2 - d_2^2}{k_{21}(E, k_s)} - 3k_{21}(E, k_s)\right],$$

$$\Omega_4(E, k_s) = \left[\frac{2d_1 d_2}{k_{21}(E, k_s)}\right]$$

$$G_{14} = ([\Omega_5(E, k_s)(\sinh g_1)(\cos g_2) - \Omega_6(E, k_s)(\sin g_1)(\cosh g_2)](\cos\gamma_{11}(E, k_s))).$$

$$\Omega_5(E, k_s) = \left[3d_1 - \frac{d_1}{d_1^2 + d_2^2}k_{21}^2(E, k_s)\right],$$

$$\Omega_6(E, k_s) = \left[3d_2 + \frac{d_2}{d_1^2 + d_2^2}k_{21}^2(E, k_s)\right]$$

$$G_{15} = ([\Omega_9(E, k_s)(\cosh g_1)(\cos g_2) - \Omega_{10}(E, k_s)(\sinh g_1)(\sin g_2)](\cos\gamma_{11}(E, k_s)))$$

$$\Omega_9(E, k_s) = [2d_1^2 - 2d_2^2 - k_{21}^2(E, k_s)],$$

$$\Omega_{10}(E, k_s) = [2d_1 d_2]$$

$$G_{16} = ([\Omega_7(E, k_s)(\sinh g_1)(\cos g_2) - \Omega_8(E, k_s)(\sin g_1)(\cosh g_2)](\sin\gamma_{11}(E, k_s)/12)),$$

$$\Omega_7(E, k_s) = \left[\frac{5d_1}{d_1^2 + d_2^2}k_{21}^3(E, k_s) + \frac{5(d_1^3 - 3d_2^2 d_1)}{k_{21}(E, k_s)} - 34k_{21}(E, k_s)d_1\right],$$

$$\Omega_8(E, k_s) = \left[\frac{5d_2}{d_1^2 + d_2^2}k_{21}^3(E, k_s) + \frac{5(d_2^3 - 3d_2^2 d_1)}{k_{21}(E, k_s)} + 34k_{21}(E, k_s)d_2\right]$$

$$H_{18} = \frac{1}{2}[H_{11} + H_{12} + \Delta_0(H_{13} + H_{14}) + \Delta_0(H_{15} + H_{16})],$$

$$H_{11} = 2(\sinh g_1 \sin g_2 \cos\gamma_{11}(E, k_s)),$$

$$H_{12} = ([\Omega_2(E, k_s)(\sinh g_1)(\cos g_2) + \Omega_1(E, k_s)(\sin g_2)(\cosh g_1)](\sin\gamma_{11}(E, k_s))),$$

$$H_{13} = ([\Omega_4(E, k_s)(\cosh g_1)(\cos g_2) + \Omega_3(E, k_s)(\sinh g_1)(\sin g_2)](\sin\gamma_{11}(E, k_s))),$$

$$H_{14} = ([\Omega_6(E,k_s)(\sinh g_1)(\cos g_2) + \Omega_5(E,k_s)(\sin g_1)(\cosh g_2)](\cos\gamma_{11}(E,k_s))),$$
$$H_{15} = ([\Omega_{10}(E,k_s)(\cosh g_1)(\cos g_2) + \Omega_9(E,k_s)(\sinh g_1)(\sin g_2)](\cos\gamma_{11}(E,k_s))),$$
$$H_{16} = ([\Omega_8(E,k_s)(\sinh g_1)(\cos g_2) + \Omega_7(E,k_s)(\sin g_1)(\cosh g_2)](\sin\gamma_{11}(E,k_s)/12)),$$
$$H_{19} = \left[\frac{2C_{18}D_{18}}{L_0^2}\right]$$

and

$$D_{18} = \sinh^{-1}(\omega_{18})$$

The simplified dispersion relation in heavily doped II-VI superllatices with graded interfaces under magnetic quantization can be expressed as

$$k_z^2 = G_{19E,n} + iH_{19E,n} \tag{7.9a}$$

where

$$\mathrm{G}_{19E,n} = \left[\frac{C_{18E,n}^2 - D_{18E,n}^2}{L_0^2} - \left(\frac{2eB}{\hbar}\left(n+\frac{1}{2}\right)\right)\right],$$
$$C_{18E,n} = \cos^{-1}(\omega_{18E,n}),\ \omega_{18E,n} = (2)^{\frac{-1}{2}}[(1 - G_{18E,n}^2 - H_{18E,n}^2) - \sqrt{(1 - G_{18E,n}^2 - H_{18E,n}^2)^2 + 4G_{180D}^2}]^{\frac{1}{2}},$$
$$G_{18E,n} = \frac{1}{2}[G_{11E,n} + G_{12E,n} + \Delta_0(G_{13E,n} + G_{14E,n}) + \Delta_0(G_{15E,n} + G_{16E,n})],$$
$$G_{11E,n} = 2(\cos(g_{1E,n}))(\cos(g_{2E,n}))(\cos\gamma_{11}(E,n)),$$
$$\gamma_{11}(E,n) = k_{21}(E,n)(b_0 - \Delta_0)$$
$$k_{21}(E,n) = \left\{\left[\gamma_3(E,\eta_{g1}) - \frac{\hbar^2}{2m_{\perp,1}^*}\left\{\frac{2eB}{\hbar}\left(n+\frac{1}{2}\right)\right\} \pm C_0\left\{\frac{2eB}{\hbar}\left(n+\frac{1}{2}\right)\right\}^{1/2}\right]\frac{2m_{\parallel,1}^*}{\hbar^2}\right\}^{1/2},$$
$$G_{12E,n} = ([\Omega_1(E,n)(\sinh g_{1E,n})(\cos g_{2E,n}) - \Omega_2(E,n)(\sin g_{2E,n})(\cosh g_{1E,n})](\sin\gamma_{11}(E,n)))$$
$$\Omega_1(E,n) = \left[\frac{d_{1E,n}}{k_{21}(E,n)} - \frac{k_{21}(E,n)d_{1E,n}}{d_{1E,n}^2 + d_{2E,n}^2}\right],$$
$$\Omega_2(E,n) = \left[\frac{d_{2E,n}}{k_{21}(E,n)} + \frac{k_{21}(E,n)d_{2E,n}}{d_{1E,n}^2 + d_{2E,n}^2}\right],$$
$$G_{13E,n} = ([\Omega_3(E,n)(\cosh g_{1E,n})(\cos g_{2E,n}) - \Omega_4(E,n)(\sinh g_{1E,n})(\sin g_{2E,n})](\sin\gamma_{11}(E,n)))$$

$$\Omega_3(E,n)=\left[\frac{d_{1E,n}^2-d_{2E,n}^2}{k_{21}(E,n)}-3k_{21}(E,n)\right],$$

$$\Omega_4(E,n)=\left[\frac{2d_{1E,n}d_{2E,n}}{k_{21}(E,n)}\right]$$

$$G_{14E,n}=([\Omega_5(E,n)(\sinh g_{1E,n})(\cos g_{2E,n})-\Omega_6(E,n)(\sin g_{1E,n})(\cosh g_{2E,n})](\cos\gamma_{11}(E,n))).$$

$$\Omega_5(E,n)=\left[3d_{1E,n}-\frac{d_{1E,n}}{d_{1E,n}^2+d_{2E,n}^2}k_{21}^2(E,n)\right],$$

$$\Omega_6(E,n)=\left[3d_{2E,n}+\frac{d_{2E,n}}{d_{1E,n}^2+d_{2E,n}^2}k_{21}^2(E,n)\right]$$

$$G_{15E,n}=([\Omega_9(E,n)(\cosh g_{1E,n})(\cos g_{2E,n})-\Omega_{10}(E,n)(\sinh g_{1E,n})(\sin g_{2E,n})](\cos\gamma_{11}(E,n)))$$

$$\Omega_9(E,n)=[2d_{1E,n}^2-2d_{2E,n}^2-k_{21}^2(E,n)],$$

$$\Omega_{10}(E,n)=[2d_{1E,n}d_{2E,n}]$$

$$G_{16E,n}=([\Omega_7(E,n)(\sinh g_{1E,n})(\cos g_{2E,n})-\Omega_8(E,n)(\sin g_{1E,n})(\cosh g_{2E,n})](\sin\gamma_{11}(E,n)/12)),$$

$$\Omega_7(E,n)=\left[\frac{5d_{1E,n}}{d_{1E,n}^2+d_{2E,n}^2}k_{21}^3(E,n)+\frac{5(d_{1E,n}^3-3d_{2E,n}^2d_{1E,n})}{k_{21}(E,n)}-34k_{21}(E,n)d_{1E,n}\right],$$

$$\Omega_8(E,n)=\left[\frac{5d_{2E,n}}{d_{1E,n}^2+d_{2E,n}^2}k_{21}^3(E,n)+\frac{5(d_{2E,n}^3-3d_{2E,n}^2d_{1E,n})}{k_{21}(E,n)}+34k_{21}(E,n)d_{2E,n}\right]$$

$$H_{18E,n}=\frac{1}{2}\left[H_{11E,n}+H_{12E,n}+\Delta_0(H_{13E,n}+H_{14E,n})+\Delta_0(H_{15E,n}+H_{16E,n})\right],$$

$$H_{11E,n}=2(\sinh g_{1E,n})(\sin g_{2E,n})(\cos\gamma_{11}(E,n))),$$

$$H_{12E,n}=([\Omega_2(E,n)(\sinh g_{1E,n})(\cos g_{2E,n})+\Omega_1(E,n)(\sin g_{2E,n})(\cosh g_{1E,n})](\sin\gamma_{11}(E,n))),$$

$$H_{13E,n}=([\Omega_4(E,n)(\cosh g_{1E,n})(\cos g_{2E,n})+\Omega_3(E,n)(\sinh g_{1E,n})(\sin g_{2E,n})](\sin\gamma_{11}(E,n))),$$

$$H_{14E,n}=([\Omega_6(E,n)(\sinh g_{1E,n})(\cos g_{2E,n})+\Omega_5(E,n)(\sin g_{1E,n})(\cosh g_{2E,n})](\cos\gamma_{11}(E,n))),$$

$$H_{15E,n}=([\Omega_{10}(E,n)(\cosh g_{1E,n})(\cos g_{2E,n})+\Omega_9(E,n)(\sinh g_{1E,n})(\sin g_{2E,n})](\cos\gamma_{11}(E,n))),$$

$$H_{16E,n}=([\Omega_8(E,n)(\sinh g_{1E,n})(\cos g_{2E,n})+\Omega_7(E,n)(\sin g_{1E,n})(\cosh g_{2E,n})](\sin\gamma_{11}(E,n)/2)),$$

$$\mathrm{H}_{19E,n}=\left[\frac{2C_{18E,n}D_{18E,n}}{L_0^2}\right]\quad\text{and}\quad D_{18E,n}=\sinh^{-1}(\omega_{18E,n})$$

The EEM can be written as

$$m^*(E_{F2},n,\eta_g,B)=\frac{\hbar^2}{2}\mathit{Real\ part\ of}\ [G_{19E_{F2},n}]' \tag{7.9b}$$

where E_{F2} is the Fermi energy in this case

Thus from (7.9b) it appears that the EEM is a function of Fermi energy, Landau quantum number, scattering potential and magnetic field.

The electron concentration is given by

$$n_0 = \frac{g_v eB}{\pi^2 \hbar} \sum_{n=0}^{n_{\max}} [\phi_{3C}(E_{F2}, n) + \phi_{4C}(E_{F2}, n)] \tag{7.10}$$

where

$$\phi_{3C}(E_{F2}, n) = \left[\left(G_{19E_{F2},n} + \sqrt{G_{19E_{F2},n}^2 - H_{19E_{F2},n}}\right)/2\right]^{1/2}$$

and

$$\phi_{4C}(E_{F2}, n) = \sum_{r=1}^{s} \theta_{2r,2}[\phi_{3C}(E_{F2}, n)]$$

Thus using (7.5), (7.9a), (7.10) and the allied definitions we can study the ER in this case.

7.2.3 The ER in HD IV-VI Superlattices with Graded Interfaces Under Magnetic Quantization

The **E-k** dispersion relation of the conduction electrons of the heavily doped constituent materials of the IV-VI SLs can be expressed as

$$k_z^2 = [2\bar{p}_{9,i}]^{-1}[-\bar{q}_{9,i}(E, k_s, \eta_{gi}) + [[\bar{q}_{9,i}(E, k_s, \eta_{gi})]^2 + 4\bar{p}_{9,i}\bar{R}_{9,i}(E, k_s, \eta_{gi})]^{\frac{1}{2}}] \tag{7.11}$$

where

$$\bar{p}_{9,i} = (\alpha_i \hbar^4)/(4m_{l,i}^- m_{li}^+), \quad \mathrm{i} = 1, 2, \; \bar{q}_{9,i}(E, k_s, \eta_{gi}) = [(\hbar^2/2)((1/m_{li}^*) + (1/m_{li}^-)) + \alpha_i(\hbar^4/4)k_s^2((1/m_{li}^+ m_{ti}^-) + (1/m_{ti}^+ m_{li}^-)) - \alpha_i \gamma_3(E, \eta_{gi})((1/m_{li}^+) - (1/m_{li}^-)]$$

and

$$\bar{R}_{9,i}(E, k_s, \eta_{gi}) = [\gamma_2(E, \eta_{gi}) + \gamma_3(E, \eta_{gi})[(\hbar^2/2)\alpha_i k_s^2((1/m_{ti}^*) - (1/m_{ti}^-))] - [(\hbar^2/2)k_s^2((1/m_{ti}^*) + (1/m_{ti}^-))] - \alpha_i(\hbar^6/4)k_s^4((1/m_{ti}^+ m_{ti}^-))]$$

The electron dispersion law in heavily doped IV-VI SLs with graded interfaces can be expressed as

$$\cos(L_o k) = \frac{1}{2}\Phi_2(E, k_s) \quad (7.12)$$

where

$$\begin{aligned}
\Phi_2(E,k_s) \equiv{}& [2\cosh\{\beta_2(E,k_s)\}\cos\{\gamma_2(E,k_s)\} + \varepsilon_2(E,k_s)\sinh\{\beta_2(E,k_s)\}\sin\{\gamma_{22}(E,k_s)\} \\
&+ \Delta_0\left[\left(\frac{\{K_{112}(E,k_s)\}^2}{K_{212}(E,k_s)} - 3K_{212}(E,k_s)\right)\cosh\{\beta_2(E,k_s)\}\sin\{\gamma_{22}(E,k_s)\}\right. \\
&+ \left.\left(3K_{112}(E,k_s) - \frac{\{K_{212}(E,k_s)\}^2}{K_{112}(E,k_s)}\right)\sinh\{\beta_2(E,k_s)\}\cos\{\gamma_{22}(E,k_s)\}\right] \\
&+ \Delta_0\left[2\left(\{K_{112}(E,k_s)\}^2 - \{K_{212}(E,k_s)\}^2\right)\cosh\{\beta_2(E,k_s)\}\cos\{\gamma_{22}(E,k_s)\}\right. \\
&+ \frac{1}{12}\left[\frac{5\{K_{112}(E,k_s)\}^3}{K_{212}(E,k_s)} + \frac{5\{K_{212}(E,k_s)\}^3}{K_{112}(E,k_s)} - 34K_{212}(E,k_s)K_{112}(E,k_s)\right] \\
&\left.\sinh\{\beta_2(E,k_s)\}\sin\{\gamma_{22}(E,k_s)\}\right]],
\end{aligned}$$

$$\begin{aligned}
\beta_2(E,k_s) &\equiv K_{112}(E,k_s)[a_0 - \Delta_0], \\
k_{112}^2(E,k_s) &= [2\bar{p}_{9,2}]^{-1}[-\bar{q}_{9,2}(E - V_0, k_s, \eta_{g2}) \\
&\quad - \left[[\bar{q}_{9,2}(E - V_0, k_s, \eta_{g2})]^2 + 4\bar{p}_{9,2}\bar{R}_{9,2}(E - V_0, k_s, \eta_{g2})\right]^{\frac{1}{2}}\right], \\
\gamma_{22}(E,k_s) &= K_{212}(E,k_s)[b_0 - \Delta_0], \\
k_{212}^2(E,k_s) &= [2\bar{p}_{9,1}]^{-1}[-\bar{q}_{9,1}(E, k_s, \eta_{g1}) \\
&\quad + \left[[\bar{q}_{9,1}(E, k_s, \eta_{g1})]^2 + 4\bar{p}_{9,1}\bar{R}_{9,1}(E, k_s, \eta_{g1})\right]^{\frac{1}{2}}\right]
\end{aligned}$$

and

$$\varepsilon_2(E,k_s) \equiv \left[\frac{K_{112}(E,k_s)}{K_{212}(E,k_s)} - \frac{K_{212}(E,k_s)}{K_{112}(E,k_s)}\right].$$

The simplified dispersion relation in heavily doped IV-VI superllatices with graded interfaces under magnetic quantization can be expressed as

$$k_z^2 = \frac{1}{L_0^2}\left[\cos^{-1}\left\{\frac{1}{2}\Phi_2(E,n)\right\}\right]^2 - \frac{2eB}{\hbar}\left(n + \frac{1}{2}\right) \quad (7.13a)$$

where

$$\Phi_2(E,n) \equiv [2\cosh\{\beta_2(E,n)\}\cos\{\gamma_2(E,n)\} + \varepsilon_2(E,n)\sinh\{\beta_2(E,n)\}\sin\{\gamma_{22}(E,n)\}$$
$$+ \Delta_0[((\{K_{112}(E,n)\}^2/K_{212}(E,n)) - 3K_{212}(E,n))\cosh\{\beta_2(E,n)\}\sin\{\gamma_{22}(E,n)\}$$
$$+ \left(3K_{112}(E,n) - \frac{\{K_{212}(E,n)\}^2}{K_{112}(E,n)}\right)\sinh\{\beta_2(E,n)\}\cos\{\gamma_{22}(E,n)\}]$$
$$+ \Delta_0\left[2\left(\{K_{112}(E,n)\}^2 - \{K_{212}(E,n)\}^2\right)\cosh\{\beta_2(E,n)\}\cos\{\gamma_{22}(E,n)\}\right.$$
$$+ \frac{1}{12}\left[\frac{5\{K_{112}(E,n)\}^3}{K_{212}(E,n)} + \frac{5\{K_{212}(E,n)\}^3}{K_{112}(E,n)} - 34K_{212}(E,n)K_{112}(E,n)\right]$$
$$\sinh\{\beta_2(E,n)\}\sin\{\gamma_{22}(E,n)\}]],$$

$$\beta_2(E,n) \equiv K_{112}(E,n)[a_0 - \Delta_0],$$

$$k_{112}^2(E,n) = [2\bar{p}_{9,2n}]^{-1}[-\bar{q}_{9,2n}(E - V_0, \eta_{g2})$$
$$- \left[[\bar{q}_{9,2n}(E - V_0,\eta_{g2})]^2 + 4\bar{p}_{9,2n}\bar{R}_{9,2n}(E - V_0,\eta_{g2})]^{\frac{1}{2}}\right],$$

$$\bar{q}_{9,2n}(E - V_0,\eta_{g2}) = [(\hbar^2/2)((1/m_{t2}^*) + (1/m_{l2}^-)) + \alpha_2(\hbar^4/4)\frac{2eB}{\hbar}\left(n+\frac{1}{2}\right)((1/m_{l2}^+ m_{t2}^-)$$
$$+ (1/m_{t2}^+ m_{l2}^-)) - \alpha_2\gamma_3(E - V_0,\eta_{g2})((1/m_{l2}^+) - (1/m_{l2}^-)],$$

$$\bar{R}_{9,2n}(E,\eta_{g2}) = [\gamma_2(E - V_0,\eta_{g2}) + \gamma_3(E - V_0,\eta_{g2})[(\hbar^2/2)\alpha_2\frac{2eB}{\hbar}\left(n+\frac{1}{2}\right)((1/m_{t2}^*)$$
$$- (1/m_{t2}^-))] - [(\hbar^2/2)k_{s0}^2((1/m_{t2}^*) + (1/m_{t2}^-))]$$
$$- \alpha_2(\hbar^6/4)\left[\frac{2eB}{\hbar}\left(n+\frac{1}{2}\right)\right]^2((1/m_{t2}^+ m_{t2}^-))],$$

$$\gamma_2(E,n) = K_{212}(E,n)[b_0 - \Delta_0], k_{212}^2(E,n) = [2\bar{p}_{9,1n}]^{-1}[-\bar{q}_{9,1n}(E,\eta_{g1})$$
$$+ [[\bar{q}_{9,1n}(E,\eta_{g1})]^2 + 4\bar{p}_{9,1n}\bar{R}_{9,1n}(E,\eta_{g1})]^{\frac{1}{2}}]$$

$$\bar{q}_{9,1n}(E,\eta_{g1}) = [(\hbar^2/2)((1/m_{t1}^*) + (1/m_{l1}^-)) + \alpha_1(\hbar^4/4)\frac{2eB}{\hbar}\left(n+\frac{1}{2}\right)((1/m_{l1}^+ m_{t1}^-)$$
$$+ (1/m_{t1}^+ m_{l1}^-)) - \alpha_1\gamma_3(E,\eta_{g1})((1/m_{l1}^+) - (1/m_{l1}^-))],$$

$$\bar{R}_{9,1}(E,\eta_{g1}) = [\gamma_2(E,\eta_{g1}) + \gamma_3(E,\eta_{g1})[(\hbar^2/2)\alpha_1(2eB/\hbar)\left(n+\frac{1}{2}\right)((1/m_{t1}^*)$$
$$- (1/m_{t1}^-))] - [(\hbar^2/2)k_{s0}^2((1/m_{t1}^*) + (1/m_{t1}^-))]$$
$$- \alpha_1(\hbar^6/4)\left((2eB/\hbar)\left(n+\frac{1}{2}\right)\right)^2((1/m_{t1}^+ m_{t1}^-))]$$

and

$$\varepsilon_2(E,n) \equiv \left[\frac{K_{112}(E,n)}{K_{212}(E,n)} - \frac{K_{212}(E,n)}{K_{112}(E,n)}\right].$$

and

$$\varepsilon_2(E,n) \equiv \left[\frac{K_{112}(E,n)}{K_{212}(E,n)} - \frac{K_{212}(E,n)}{K_{112}(E,n)}\right].$$

The EEM can be written as

$$m^*(E_{F3},n,\eta_g,B) = \hbar^2 \, Real \; part \; of \; [\phi_{5C}(E_{F3},n)\phi'_{5C}(E_{F3},n)] \tag{7.13b}$$

where E_{F3} is the Fermi energy in this case and

$$\phi_{5C}(E_{F3},n) = \left[\frac{1}{L_0^2}\left[\cos^{-1}\left\{\frac{1}{2}\Phi_2(E_{F3},n)\right\}\right]^2 - \frac{2eB}{\hbar}\left(n+\frac{1}{2}\right)\right]^{1/2}$$

From (7.13b) we observe that the EEM in this case is a function of Fermi energy, Landau quantum number, scattering potential and magnetic field.

The electron concentration is given by

$$n_0 = \frac{g_v eB}{\pi^2\hbar}\sum_{n=0}^{n_{\max}}[\phi_{5C}(E_{F3},n) + \phi_{6C}(E_{F3},n)] \tag{7.14}$$

where

$$\phi_{6C}(E_{F3},n) = \sum_{r=1}^{s}\theta_{2r,3}[\phi_{6C}(E_{F3},n)]$$

Thus using (7.5), (7.13a), (7.14) and the allied definitions we can study the ER in this case.

7.2.4 The ER in HD HgTe/CdTe Superlattices with Graded Interfaces Under Magnetic Quantization

The electron energy spectra of the constituent materials of HgTe/CdTe SLs are given by

$$k^2 = \left[\frac{B_{01}^2 + 4A_1E - B_{01}\sqrt{B_{01}^2 + 4A_1E}}{2A_1^2}\right] \tag{7.15}$$

and

$$\frac{\hbar^2 k^2}{2m_{c2}^*} = T_{12}(E, \Delta_2, E_{g2}, \eta_{g2}) + iT_{22}(E, \Delta_2, E_{g2}, \eta_{g2}) \tag{7.16}$$

where $B_{01} = \left(3|e|^2 \big/ 128\varepsilon_{sc1}\right)$, $A_1 = (\hbar^2/2m_{c1}^*) \cdot \varepsilon_{sc1}$ is the semiconductor permittivity of the first material.

The energy-wave vector dispersion relation of the conduction electrons in heavily doped HgTe/CdTe SLs with graded interfaces can be expressed as

$$k_z^2 = G_{192} + iH_{192} \tag{7.17}$$

where

$$G_{192} = [((C_{182}^2 - D_{182}^2)/L_0^2) - k_s^2],$$

$$C_{182} = \cos^{-1}(\omega_{182}), \omega_{182} = (2)^{\frac{-1}{2}}\left[(1 - G_{182}^2 - H_{182}^2) - \sqrt{(1 - G_{182}^2 - H_{182}^2)^2 + 4G_{182}^2}\right]^{\frac{1}{2}},$$

$$G_{182} = \frac{1}{2}[G_{112} + G_{122} + \Delta_0(G_{132} + G_{142}) + \Delta_0(G_{152} + G_{162})],$$

$$G_{112} = 2(\cos(g_{12}))(\cos(g_{22}))(\cos\gamma_8(E, k_s))$$

$$\gamma_8(E, k_s) = k_8(E, k_s)(b_0 - \Delta_0), \quad k_8(E, k_s) = \left[\frac{B_{01}^2 + 4A_1E - B_{01}\sqrt{B_{01}^2 + 4A_1E}}{2A_1^2} - k_s^2\right]^{1/2},$$

$$G_{122} = ([\Omega_{12}(E, k_s)(\sinh g_{12})(\cos g_{22}) - \Omega_{22}(E, k_s)(\sin g_{22})(\cosh g_{12})](\sin\gamma_8(E, k_s)))$$

$$\Omega_{12}(E, k_s) = \left[\frac{d_{12}}{k_8(E, k_s)} - \frac{k_8(E, k_s)d_{12}}{d_{12}^2 + d_{22}^2}\right], \quad \Omega_{22}(E, k_s) = \left[\frac{d_{22}}{k_8(E, k_s)} + \frac{k_8(E, k_s)d_{22}}{d_{12}^2 + d_{22}^2}\right],$$

$$G_{132} = ([\Omega_{32}(E, k_s)(\cosh g_{12})(\cos g_{22}) - \Omega_{42}(E, k_s)(\sinh g_{12})(\sin g_{22})](\sin\gamma_8(E, k_s))),$$

$$\Omega_{32}(E, k_s) = \left[\frac{d_{12}^2 - d_{22}^2}{k_8(E, k_s)} - 3k_8(E, k_s)\right], \quad \Omega_{42}(E, k_s) = \left[\frac{2d_{12}d_{22}}{k_8(E, k_s)}\right],$$

$$G_{142} = ([\Omega_{52}(E, k_s)(\sinh g_{12})(\cos g_{22}) - \Omega_{62}(E, k_s)(\sin g_{12})(\cosh g_{22})](\cos\gamma_8(E, k_s))),$$

$$\Omega_{52}(E, k_s) = \left[3d_{12} - \frac{d_{12}}{d_{12}^2 + d_{22}^2}k_8^2(E, k_s)\right], \quad \Omega_{62}(E, k_s) = \left[3d_{22} + \frac{d_{22}}{d_{12}^2 + d_{22}^2}k_8^2(E, k_s)\right],$$

$$G_{152} = ([\Omega_{92}(E, k_s)(\cosh g_{12})(\cos g_{22}) - \Omega_{102}(E, k_s)(\sinh g_{12})(\sin g_{22})](\cos\gamma_8(E, k_s))),$$

$$\Omega_{92}(E, k_s) = [2d_{12}^2 - 2d_{22}^2 - k_8^2(E, k_s)], \quad \Omega_{102}(E, k_s) = [2d_{12}d_{22}],$$

$$G_{162} = ([\Omega_{72}(E, k_s)(\sinh g_{12})(\cos g_{22}) - \Omega_{82}(E, k_s)(\sin g_{12})(\cosh g_{22})](\sin\gamma_8(E, k_s)/12)),$$

$$\Omega_{72}(E, k_s) = \left[\frac{5d_{12}}{d_{12}^2 + d_{22}^2}k_8^3(E, k_s) + \frac{5(d_{12}^3 - 3d_{22}^2d_{12})}{k_8(E, k_s)} - 34k_8(E, k_s)d_{12}\right],$$

$$\Omega_{82}(E, k_s) = \left[\frac{5d_{22}}{d_{12}^2 + d_{22}^2}k_8^3(E, k_s) + \frac{5(d_{22}^3 - 3d_{22}^2d_{12})}{k_8(E, k_s)} + 34k_8(E, k_s)d_{22}\right]$$

$$H_{182} = \frac{1}{2}[H_{112} + H_{122} + \Delta_0(H_{132} + H_{142}) + \Delta_0(H_{152} + H_{162})],$$
$$H_{112} = 2(\sinh g_{12} \sin g_{22} \cos \gamma_8(E, k_s)),$$
$$H_{122} = ([\Omega_{22}(E, k_s)(\sinh g_{12})(\cos g_{22}) + \Omega_{12}(E, k_s)(\sin g_{22})(\cosh g_{12})](\sin \gamma_8(E, k_s))),$$
$$H_{132} = ([\Omega_{42}(E, k_s)(\cosh g_{12})(\cos g_{22}) + \Omega_{32}(E, k_s)(\sinh g_{12})(\sin g_{22})](\sin \gamma_8(E, k_s))),$$
$$H_{142} = ([\Omega_{62}(E, k_s)(\sinh g_{12})(\cos g_{22}) + \Omega_{52}(E, k_s)(\sin g_{12})(\cosh g_{22})](\cos \gamma_8(E, k_s))),$$
$$H_{152} = ([\Omega_{102}(E, k_s)(\cosh g_{12})(\cos g_{22}) + \Omega_{92}(E, k_s)(\sinh g_{12})(\sin g_{22})](\cos \gamma_8(E, k_s))),$$
$$H_{162} = ([\Omega_{82}(E, k_s)(\sinh g_{12})(\cos g_{22}) + \Omega_{72}(E, k_s)(\sin g_{12})(\cosh g_{22})](\sin \gamma_8(E, k_s)/12)),$$
$$\mathrm{H}_{192} = [((2C_{182}D_{182})/L_0^2)] \text{ and } D_{182} = \sinh^{-1}(\omega_{182})$$

The simplified dispersion relation in heavily doped HgTe/CdTe superlattices with graded interfaces under magnetic quantization can be expressed as

$$(k_z)^2 = G_{192E,n} + iH_{192E,n} \tag{7.18a}$$

where

$$G_{192E,n} = \left[\frac{C_{182E,n}^2 - D_{182E,n}^2}{L_0^2} - (2eB/\hbar)(n + (1/2))\right],$$
$$C_{1820D} = \cos^{-1}(\omega_{182E,n}),$$
$$\omega_{182E,n} = (2)^{\frac{-1}{2}}\left[(1 - G_{182E,n}^2 - H_{182E,n}^2) - \sqrt{(1 - G_{182E,n}^2 - H_{182E,n}^2)^2 + 4G_{182E,n}^2}\right]^{\frac{1}{2}},$$
$$G_{182E,n} = \frac{1}{2}[G_{112E,n} + G_{122E,n} + \Delta_0(G_{132E,n} + G_{142E,n}) + \Delta_0(G_{152E,n} + G_{162E,n})],$$
$$G_{112E,n} = 2(\cos(g_{12}))(\cos(g_{22}))(\cos \gamma_8(E, n)), \quad \gamma_8(E, n) = k_8(E, n)(b_0 - \Delta_0)$$
$$k_8(E, n) = \left[\frac{B_{01}^2 + 4A_1E - B_{01}\sqrt{B_{01}^2 + 4A_1E}}{2A_1^2} - (2eB/\hbar)(n + (1/2))\right]^{1/2},$$
$$G_{120D} = ([\Omega_{12}(E, n)(\sinh g_{12E,n})(\cos g_{22E,n}) - \Omega_{22}(E, n)(\sin g_{22E,n})(\cosh g_{12E,n})](\sin \gamma_8(E, n))),$$

$$\Omega_{12}(E,n)=\left[\frac{d_{12E,n}}{k_8(E,n)}-\frac{k_8(E,n)d_{12E,n}}{d_{12E,n}^2+d_{22E,n}^2}\right],$$

$$\Omega_{22}(E,n)=\left[\frac{d_{22E,n}}{k_8(E,n)}+\frac{k_8(E,n)d_{22E,n}}{d_{12E,n}^2+d_{22E,n}^2}\right],$$

$$\begin{aligned}G_{1320D}&=([\Omega_{32}(E,n)(\cosh g_{12E,n})(\cos g_{22E,n})\\&\quad-\Omega_{42}(E,n)(\sinh g_{12E,n})(\sin g_{22E,n})](\sin\gamma_8(E,n))),\end{aligned}$$

$$\Omega_{32}(E,n)=\left[\frac{d_{12E,n}^2-d_{2E,n}^2}{k_8(E,n)}-3k_8(E,n)\right],$$

$$\Omega_{42}(E,n)=\left[\frac{2d_{12E,n}d_{22E,n}}{k_8(E,n)}\right],$$

$$\begin{aligned}G_{1420D}&=([\Omega_{52}(E,n)(\sinh g_{12E,n})(\cos g_{22E,n})\\&\quad-\Omega_{62}(E,n)(\sin g_{12E,n})(\cosh g_{22E,n})](\cos\gamma_8(E,n))),\end{aligned}$$

$$\Omega_{52}(E,n)=\left[3d_{12E,n}-\frac{d_{12E,n}}{d_{12E,n}^2+d_{22E,n}^2}k_8^2(E,n)\right],$$

$$\Omega_{62}(E,n)=\left[3d_{22E,n}+\frac{d_{22E,n}}{d_{12E,n}^2+d_{22E,n}^2}k_8^2(E,n)\right],$$

$$\begin{aligned}G_{1520D}&=([\Omega_{92}(E,n)(\cosh g_{12E,n})(\cos g_{22E,n})\\&\quad-\Omega_{102}(E,n)(\sinh g_{12E,n})(\sin g_{22E,n})](\cos\gamma_8(E,n))),\end{aligned}$$

$$\Omega_{92}(E,n)=[2d_{12E,n}^2-2d_{22E,n}^2-k_8^2(E,n)],$$

$$\Omega_{102}(E,n)=[2d_{12E,n}d_{22E,n}],$$

$$\begin{aligned}G_{162E,n}&=([\Omega_{72}(E,n)(\sinh g_{12E,n})(\cos g_{22E,n})\\&\quad-\Omega_{82}(E,n)(\sin g_{12E,n})(\cosh g_{22E,n})](\sin\gamma_{80D}(E,n)/12)),\end{aligned}$$

$$\Omega_{72}(E,n)=\left[\frac{5d_{12E,n}}{d_{12E,n}^2+d_{22E,n}^2}k_8^3(E,n)+\frac{5(d_{12E,n}^3-3d_{22E,n}^2d_{12E,n})}{k_8(E,n)}-34k_8(E,n)d_{12E,n}\right],$$

$$\Omega_{82}(E,n)=\left[\frac{5d_{22E,n}}{d_{12E,n}^2+d_{22E,n}^2}k_8^3(E,n)+\frac{5(d_{22E,n}^3-3d_{22E,n}^2d_{12E,n})}{k_8(E,n)}+34k_8(E,n)d_{22E,n}\right],$$

$$H_{182E,n}=\frac{1}{2}[H_{112E,n}+H_{122E,n}+\Delta_0(H_{132E,n}+H_{142E,n})+\Delta_0(H_{152E,n}+H_{162E,n})],$$

$$H_{112E,n}=2(\sinh g_{12E,n})(\sin g_{22E,n})(\cos\gamma_8(E,n))),$$

$$H_{1220D}=([\Omega_{22}(E,n)(\sinh g_{12E,n})(\cos g_{22E,n})+\Omega_{12}(E,n)(\sin g_{22E,n})(\cosh g_{12E,n})](\sin\gamma_8(E,n))),$$

$$H_{132E,n}=([\Omega_{42}(E,n)(\cosh g_{12E,n})(\cos g_{22E,n})+\Omega_{32}(E,n)(\sinh g_{12E,n})(\sin g_{22E,n})](\sin\gamma_8(E,n))),$$

$$H_{142E,n}=([\Omega_{62}(E,n)(\sinh g_{12E,n})(\cos g_{22E,n})+\Omega_{52}(E,n)(\sin g_{12E,n})(\cosh g_{22E,n})](\cos\gamma_8(E,n))),$$

$$H_{1520D}=([\Omega_{102}(E,n)(\cosh g_{12E,n})(\cos g_{22E,n})+\Omega_{92}(E,n)(\sinh g_{12E,n})(\sin g_{22E,n})](\cos\gamma_8(E,n))),$$

$$H_{162E,n}=([\Omega_{82}(E,n)(\sinh g_{12E,n})(\cos g_{22E,n})+\Omega_{72}(E,n)(\sin g_{12E,n})(\cosh g_{22E,n})](\sin\gamma_8(E,n)/2)),$$

$$\mathrm{H}_{192E,n}=[((2C_{182E,n}D_{182E,n})/L_0^2)]\ \text{and}\ D_{182E,n}=\sinh^{-1}(\omega_{182E,n})$$

The EEM can be written as

$$m^*(E_{F4}, n, \eta_g, B) = \frac{\hbar^2}{2} \; Real\ part\ of\ [G_{192E_{F4},n}]' \tag{7.18b}$$

where E_{F4} is the Fermi energy in this case.

From (7.18b) we observe that the EEM in this case is a function of Fermi energy, scattering potential, magnetic field and Landau quantum number.

The electron concentration is given by

$$n_0 = \frac{g_v eB}{\pi^2 \hbar} \sum_{n=0}^{n_{\max}} [\phi_{7C}(E_{F4}, n) + \phi_{8C}(E_{F4}, n)] \tag{7.19}$$

where

$$\phi_{7C}(E_{F4}, n) = \left[\left(G_{192E_{F4},n} + \sqrt{G^2_{192E_{F4},n} - H_{192E_{F4},n}}\right)/2\right]^{1/2}$$

and

$$\phi_{8C}(E_{F4}, n) = \sum_{r=1}^{s} \theta_{2r,4}[\phi_{7C}(E_{F4}, n)]$$

Thus using (7.5), (7.18a), (7.19) and the allied definitions we can study the ER in this case.

7.2.5 The ER in HD Stained Layer Superlattices with Graded Interfaces Under Magnetic Quantization

The dispersion relation of the conduction electrons of the constituent materials of the starined layer super lattices can be expressed as

$$[E - T_{1i}]k_x^2 + [E - T_{2i}]k_y^2 + [E - T_{3i}]k_z^2 = q_i E^3 - R_i E^2 + V_i E + \zeta_i \tag{7.20}$$

where

$$T_{1i} = \overline{\theta_i},\ \overline{\theta_i} = \left[E_{gi} - C_{1i}^{c}\varepsilon_i - (a_i + C_{1i}^{c})\varepsilon_i + \frac{3}{2}b_i\varepsilon_{xxi} - \frac{b_i\varepsilon_i}{2} + \frac{\sqrt{3}d_i\varepsilon_{xyi}}{2}\right],$$

$$T_{2i} = \omega_i,\ \omega_i = \left[E_{gi} - C_{1i}^{c}\varepsilon_i - (a_i + C_{1i}^{c})\varepsilon_i + \frac{3}{2}b_i\varepsilon_{xxi} - \frac{b_i\varepsilon_i}{2} - \frac{\sqrt{3}d_i\varepsilon_{xyi}}{2}\right],$$

$$T_{3i} = \delta_i,\ \delta_i = \left[E_{gi} - C_{1i}^{c}\varepsilon_i + (a_i + C_{1i}^{c})\varepsilon_i + \frac{3}{2}b_i\varepsilon_{zzi} - \frac{b_i\varepsilon_i}{2}\right]$$

$$R_i = q_i\left[2A_i + C_{1i}^{c}\varepsilon_i\right],\ q_i = \frac{3}{2B_{2i}^{2}},\ A_i = E_{gi} - C_{1i}^{c}\varepsilon_i,$$

$$V_i = q_i\left[A_i^2 - \frac{2C_{2i}^{2}\varepsilon_{xyi}}{3} + 2A_iC_{1i}^{c}\varepsilon_i\right],\ \zeta_i = q_i\left[\frac{2C_{2i}^{2}\varepsilon_{xyi}}{3} - C_{1i}^{c}\varepsilon_iA_i^2\right]$$

Therefore the electron energy spectrum in heavily doped stressed materials can be written as

$$\overline{P_i}(E,\eta_{gi})k_x^2 + \overline{Q_i}(E,\eta_{gi})k_y^2 + \overline{S_i}(E,\eta_{gi})k_z^2 = 1$$

where

$$\overline{P_i}(E,\eta_{gi}) = \frac{[\gamma_0(E,\eta_{gi}) - I_0T_{1i}]}{\overline{\Delta_i}(E,\eta_{gi})},$$

$$\overline{\Delta_i}(E,\eta_{gi}) = \left[\frac{-q_i\eta_{gi}^3}{2\sqrt{\pi}}\exp\left(\frac{-E^2}{\eta_{gi}^2}\right)\left[1+\frac{E^2}{\eta_{gi}^2}\right] - R_i\theta_0(E,\eta_{gi}) + V_i\gamma_0(E,\eta_{gi}) + \frac{\zeta_i}{2}\left[1 + Erf\left(\frac{E}{\eta_{gi}}\right)\right]\right],$$

$$I_0 = \frac{1}{2}\left[1 + Erf(E/\eta_{gi})\right],\ \overline{Q_i}(E,\eta_{gi}) = \frac{[\gamma_0(E,\eta_{gi}) - I_0T_{2i}]}{\overline{\Delta_i}(E,\eta_{gi})}$$

and

$$\overline{S_i}(E,\eta_{gi}) = \frac{[\gamma_0(E,\eta_{gi}) - I_0T_{3i}]}{\overline{\Delta_i}(E,\eta_{gi})}$$

The energy-wave vector dispersion relation of the conduction electrons in heavily doped strained layer SLs with graded interfaces can be expressed as

$$\cos(L_0k) = \frac{1}{2}\overline{\phi_6}(E,k_s) \tag{7.21}$$

where

$$\begin{aligned}\overline{\phi_6}(E,k_s) &= [2\cosh[T_4(E,\eta_{g2})]\cos[T_5(E,\eta_{g1})]] + [T_6(E,k_s)]\sinh[T_4(E,\eta_{g2})]\sin[T_5(E,\eta_{g1})]\\ &+\Delta_0\left[\left(\frac{k_0^2(E,\eta_{g2})}{k'(E,\eta_{g1})} - 3k'(E,\eta_{g1})\right)\cosh[T_4(E,\eta_{g2})]\sin[T_5(E,\eta_{g1})]\right.\\ &\left.+\left(3k_0(E,\eta_{g2}) - \frac{k'^2(E,\eta_{g1})}{k_0(E,\eta_{g2})}\right)\sinh[T_4(E,\eta_{g2})]\cos[T_5(E,\eta_{g1})]\right]\\ &+\Delta_0\Big[2\left(k_0^2(E,\eta_{g2}) - k'^2(E,\eta_{g1})\right)\cosh[T_4(E,\eta_{g2})]\cos[T_5(E,\eta_{g1})]\Big]\\ &+\frac{1}{12}\left(\frac{5k_0^3(E,\eta_{g2})}{k'(E,\eta_{g1})} + \frac{5k'^3(E,\eta_{g1})}{k_0(E,\eta_{g2})} - 34k_0(E,\eta_{g2})k'(E,\eta_{g1}))\sinh[T_4(E,\eta_{g2})]\sin[T_5(E,\eta_{g1})]\right)\Big]\end{aligned}$$

$$[T_4(E,\eta_{g2})] = k_0(E,\eta_{g2})[a_0 - \Delta_0],$$

$$k_0(E,\eta_{g2}) = [\overline{S_2}(E-V_0,\eta_{g2})]^{\frac{-1}{2}}\left[\overline{P_2}(E-V_0,\eta_{g2})k_x^2 + \overline{Q_2}(E-V_0,\eta_{g2})k_y^2 - 1\right]^{\frac{-1}{2}},$$

$$T_5(E,\eta_{g1}) = k'(E,\eta_{g1})[b_0 - \Delta_0],$$

$$k'(E,\eta_{g1}) = [\overline{S_1}(E,\eta_{g1})]^{\frac{-1}{2}}\left[1 - \overline{P_1}(E,\eta_{g1})k_x^2 - \overline{Q_1}(E,\eta_{g1})k_y^2\right]^{\frac{-1}{2}} \text{and}$$

$$T_6(E,k_s) = \left[\frac{k_0(E,\eta_{g2})}{k'(E,\eta_{g1})} - \frac{k'(E,\eta_{g1})}{k_0(E,\eta_{g2})}\right]$$

Therefore the dispersion relation of the conduction electrons in heavily doped strained layer QDSLs with graded interfaces can be expressed as

$$\cos(L_0k_0) = \frac{1}{2}\overline{\phi_6}(E,n) \tag{7.22a}$$

where

$$\begin{aligned}\overline{\phi_6}(E,n) &= [2\cosh[T_4(E,n,\eta_{g2})]\cos[T_5(E,n,\eta_{g1})]]\\ &+[T_6(E,n)]\sinh[T_4(E,n,\eta_{g2})]\sin[T_5(E,n,\eta_{g1})]\\ &+\Delta_0\left[\left(\frac{k_0^2(E,n,\eta_{g2})}{k_0'(E,n,\eta_{g1})} - 3k_0'(E,n,\eta_{g1})\right)\cosh[T_4(E,n,\eta_{g2})]\sin[T_5(E,n,\eta_{g1})]\right.\\ &\left.+\left(3k_0(E,n,\eta_{g2}) - \frac{k_0'^2(E,n,\eta_{g1})}{k_0(E,n,\eta_{g2})}\right)\sinh[T_4(E,n,\eta_{g2})]\cos[T_5(E,n,\eta_{g1})]\right]\\ &+\Delta_0[2(k_0^2(E,n,\eta_{g2}) - k_{0D}'^2(E,n,\eta_{g1}))\cosh[T_4(E,n,\eta_{g2})]\cos[T_5(E,n,\eta_{g1})]]\\ &+\frac{1}{12}\left(\frac{5k_0^3(E,n,\eta_{g2})}{k_0'(E,n,\eta_{g1})} + \frac{5k_0'3(E,n,\eta_{g1})}{k_0(E,n,\eta_{g2})} - 34k_0(E,n,\eta_{g2})k_0'(E,n,\eta_{g1})\right)\\ &\quad\sinh[T_4(E,n,\eta_{g2})]\sin[T_5(E,n,\eta_{g1})]\end{aligned}$$

$$[T_4(E,n,\eta_{g2})] = k_0(E,n,\eta_{g2})[a_0 - \Delta_0],$$
$$k_0(E,n,\eta_{g2}) = [\overline{S_2}(E - V_0,\eta_{g2})]^{\frac{-1}{2}}\left[\left[(n+1/2)\hbar eB/(\sqrt{\rho_1(E)\rho_2(E)})\right] - 1\right]^{\frac{-1}{2}},$$
$$\rho_1(E) = \hbar^2/(2\overline{P_2}(E - V_0,\eta_{g2})),$$
$$\rho_2(E) = \hbar^2/(2\overline{Q_2}(E - V_0,\eta_{g2})),\quad T_5(E,n,\eta_{g1}) = k_0'(E,n,\eta_{g1})[b_0 - \Delta_0],$$
$$k_0'(E,n,\eta_{g1}) = [\overline{S_1}(E,n,\eta_{g1})]^{\frac{-1}{2}}\left[1 - \left[(n+1/2)\hbar eB/(\sqrt{\rho_3(E)\rho_4(E)})\right]\right]^{\frac{-1}{2}},$$
$$\rho_3(E) = \hbar^2/(2\overline{P_1}(E,\eta_{g1})),\ \rho_4(E) = \hbar^2/(2\overline{Q_1}(E,\eta_{g1})),$$
$$T_6(E,n) = \left[\frac{k_0(E,n,\eta_{g2})}{k_0'(E,n,\eta_{g1})} - \frac{k_0'(E,n,\eta_{g1})}{k_0(E,n,\eta_{g2})}\right]$$

The EEM can be written as

$$m^*(E_{F6},n,\eta_g,B) = \hbar^2\ Real\ part\ of\ [\phi_{9C}(E_{F6},n)\phi'_{9C}(E_{F6},n)] \tag{7.22b}$$

where E_{F6} is the Fermi energy in this case and

$$\phi_{9C}(E_{F6},n) = \left[\frac{1}{L_0^2}\left[\cos^{-1}\left\{\frac{1}{2}\bar{\Phi}_6(E_{F6},n)\right\}\right]^2 - \frac{2eB}{\hbar}\left(n+\frac{1}{2}\right)\right]^{1/2}$$

From (7.22b) we note that the EEM is a function of Fermi energy, scattering potential, Landau quantum number and magnetic field.

The electron concentration is given by

$$n_0 = \frac{g_v eB}{\pi^2\hbar}\sum_{n=0}^{n_{\max}}[\phi_{9C}(E_{F6},n) + \phi_{10C}(E_{F6},n)] \tag{7.23}$$

where

$$\phi_{8C}(E_{F6},n) = \sum_{r=1}^{s}\theta_{2r,6}[\phi_{7C}(E_{F6},n)]$$

Thus using (7.5), (7.22a), (7.23) and the allied definitions we can study the ER in this case.

7.2.6 The ER in HD III-V Effective Mass Superlattices Under Magnetic Quantization

Following Sasaki [70], the electron dispersion law in III-V heavily doped effective mass superlattices (EMSLs) can be written as

$$k_x^2 = \left[\frac{1}{L_0^2}\left\{\cos^{-1}\left(f_{21}\left(E,k_y,k_z\right)\right)\right\}^2 - k_\perp^2\right] \tag{7.24}$$

In which

$$f_{21}\left(E,k_y,k_z\right) = a_1\cos\left[a_0C_{21}\left(E,k_\perp,\eta_{g1}\right)+b_0D_{21}\left(E,k_\perp,\eta_{g2}\right)\right] - a_2\cos\left[a_0C_{21}\left(E,k_\perp,\eta_{g1}\right)-b_0D_{21}\left(E,k_\perp,\eta_{g2}\right)\right],\; k_\perp^2 = k_y^2+k_z^2,$$

$$a_1 = \left[\sqrt{\frac{M_2(0,\eta_{g2})}{M_1(0,\eta_{g1})}}+1\right]^2\left[4\left(\frac{M_2(0,\eta_{g2})}{M_1(0,\eta_{g1})}\right)^{1/2}\right]^{-1},$$

$$a_2 = \left[\sqrt{\frac{M_2(0,\eta_{g2})}{M_1(0,\eta_{g1})}}-1\right]^2\left[4\left(\frac{M_2(0,\eta_{g2})}{M_1(0,\eta_{g1})}\right)^{1/2}\right]^{-1},$$

$$\begin{aligned} M_i(0,\eta_{gi}) = m_{ci}^*\Bigg[\frac{-2}{\sqrt{\pi}}T(0,\eta_{gi}) + 2\Bigg[\frac{\alpha_i b_i}{c_i}\frac{\eta_{gi}}{\sqrt{\pi}}+\frac{1}{2}\left(\frac{\alpha_i c_i + c_i b_i - \alpha_i b_i}{c_i^2}\right) \\ + \frac{1}{\sqrt{\pi}c_i}\left(1-\frac{\alpha_i}{c_i}\right)\left(1-\frac{b_i}{c_i}\right) - \frac{1}{c_i}\left(1-\frac{\alpha_i}{c_i}\right)\left(1-\frac{b_i}{c_i}\right) \\ \times\frac{2}{c_i\eta_{gi}\sqrt{\pi}}\Bigg\{\frac{-2}{c_i\eta_{gi}^2}\exp\left(\frac{-1}{c_i^2\eta_{gi}^2}\right)\left(\sum_{p=1}^{\alpha}\left(\exp\left(\frac{-p^2}{4}\right)\right)\frac{1}{p}\sinh\left(\frac{p}{c_i\eta_{gi}}\right)\right) \\ + \exp\left(\frac{-1}{c_i^2\eta_{gi}^2}\right)\left(\sum_{p=1}^{\alpha}\exp\left(\frac{-p^2}{4}\right)\frac{1}{\eta_{gi}}\cosh\left(\frac{p}{c_i\eta_{gi}}\right)\right)\Bigg\}\Bigg]\Bigg], \end{aligned}$$

$$\begin{aligned} \mathrm{T}(0,\eta_{gi}) = 2\Bigg[\frac{\alpha_i b_i}{c_i}\frac{\eta_{gi}^2}{4} + \left(\frac{\alpha_i c_i + b_i c_i - \alpha_i b_i}{c_i^2}\right)\frac{\eta_{gi}}{2\sqrt{\pi}} + \frac{1}{2c_i}\left(1-\frac{\alpha_i}{c_i}\right)\left(1-\frac{b_i}{c_i}\right) \\ -\frac{1}{c_i}\left(1-\frac{\alpha_i}{c_i}\right)\left(1-\frac{b_i}{c_i}\right)\frac{2}{c_i\eta_{gi}\sqrt{\pi}}\exp\left(\frac{-1}{c_i^2\eta_{gi}^2}\right)\sum_{p=1}^{\alpha}\frac{\exp(-p^2/4)}{p}\sinh\left(\frac{p}{c_i\eta_{gi}}\right)\Bigg], \end{aligned}$$

$$C_{21}(E,k_\perp,\eta_{g1}) = e_1 + ie_2,\ D_{21}(E,k_\perp,\eta_{g2}) = e_3 + ie_4,$$
$$e_1 = [((\sqrt{t_1^2+t_2^2}+t_1)/2)]^{\frac{1}{2}},\ e_2 = [((\sqrt{t_1^2+t_2^2}-t_1)/2)]^{\frac{1}{2}},$$
$$t_1 = \left[\frac{2m_{c1}^*}{\hbar^2}T_{11}(E,\Delta_1,\eta_{g1},E_{g1}) - k_\perp^2\right],\ t_2 = \frac{2m_{c1}^*}{\hbar^2}T_{21}(E,\Delta_1,\eta_{g1},E_{g1}),$$
$$e_3 = \left[\frac{\sqrt{t_3^2+t_4^2}+t_3}{2}\right]^{1/2},\ e_4 = \left[\frac{\sqrt{t_3^2+t_4^2}-t_3}{2}\right]^{1/2}$$
$$t_3 = \left[\frac{2m_{c2}^*}{\hbar^2}T_{12}(E,\Delta_2,\eta_{g2},E_{g2}) - k_\perp^2\right],\ t_4 = \frac{2m_{c2}^*}{\hbar^2}T_{22}(E,\Delta_2,\eta_{g2},E_{g2}),$$

Therefore (7.24) can be expressed as

$$k_x^2 = \delta_7 + i\delta_8 \tag{7.25}$$

where

$$\delta_7 = \left[\frac{1}{L_0^2}(\delta_5^2 - \delta_6^2) - k_\perp^2\right],\ \delta_5 = \cos^{-1}p_5,$$
$$p_5 = \left[\frac{1-\delta_3^2-\delta_4^2-\sqrt{(1-\delta_3^2-\delta_4^2)^2+4\delta_4^2}}{2}\right]^{1/2},$$
$$\delta_3 = (a_1\cos\Delta_1\cosh\Delta_2 - a_2\cos\Delta_3\cosh\Delta_4),$$
$$\delta_4 = (a_1\sin\Delta_1\sinh\Delta_2 - a_2\sin\Delta_3\sinh\Delta_4),$$
$$\Delta_1 = (a_0e_1 + b_0e_3),\ \Delta_2 = (a_0e_2 + b_0e_4),$$
$$\Delta_3 = (a_0e_1 - b_0e_3),\ \Delta_4 = (a_0e_2 - b_0e_4),$$
$$\delta_6 = \sinh^{-1}p_5 \text{ and } \delta_8 = \left[2\delta_5\delta_6/L_0^2\right]$$

Therefore the electron dispersion relation in heavily doped III-V QDSL assumes the form

$$(k_z)^2 = \delta_{7E,n} + i\delta_{8E,n} \tag{7.26a}$$

where

$$\delta_{7E,n} = \left[\frac{1}{L_0^2}\left(\delta_{5E,n}^2 - \delta_{6E,n}^2\right) - \left\{\frac{2eB}{\hbar}(n+\frac{1}{2})\right\}\right], \ \delta_{5E,n} = \cos^{-1} p_{5E,n},$$

$$p_{5E,n} = \left[\frac{1-\delta_{3E,n}^2-\delta_{4E,n}^2-\sqrt{\left(1-\delta_{3E,n}^2-\delta_{4E,n}^2\right)^2+4\delta_{4E,n}^2}}{2}\right]^{1/2},$$

$$\delta_{3E,n} = \left(a_1\cos\Delta_{1E,n}\cosh\Delta_{2E,n} - a_2\cos\Delta_{3E,n}\cosh\Delta_{4E,n}\right),$$

$$\delta_{4E,n} = \left(a_1\sin\Delta_{1E,n}\sinh\Delta_{2E,n} - a_2\sin\Delta_{3E,n}\sinh\Delta_{4E,n}\right),$$

$$\Delta_{1E,n} = \left(a_0 e_{1E,n} + b_0 e_{3E,n}\right),\ \Delta_{2E,n} = \left(a_0 e_{2E,n} + b_0 e_{4E,n}\right),$$

$$\Delta_{3E,n} = \left(a_0 e_{1E,n} - b_0 e_{3E,n}\right),\ \Delta_{4E,n} = \left(a_0 e_{2E,n} - b_0 e_{4E,n}\right),$$

$$\delta_{6E,n} = \sinh^{-1} \mathrm{p}_{5E,n} \text{ and } \delta_{8E,n} = \left[2\delta_{5E,n}\delta_{6E,n}/L_0^2\right],$$

$$e_{1E,n} = [((\sqrt{t_{1E,n}^2+t_2^2}+t_{1E,n})/2)]^{\frac{1}{2}},\ \mathrm{e}_{2E,n} = [((\sqrt{t_{1E,n}^2+t_2^2}-t_{1E,n})/2)]^{\frac{1}{2}},$$

$$e_{3E,n} = \left[\frac{\sqrt{t_{3E,n}^2+t_4^2}+t_{3E,n}}{2}\right]^{1/2},\ e_{4E,n} = \left[\frac{\sqrt{t_{3E,n}^2+t_4^2}-t_{3E,n}}{2}\right]^{1/2},$$

$$t_{1E,n} = \left[\frac{2m_{c1}^*}{\hbar^2}T_{11}\left(E,\Delta_1,\eta_{g1},E_{g1}\right) - \frac{2eB}{\hbar}(n+\frac{1}{2})\right],$$

$$t_{3E,n} = \left[\frac{2m_{c2}^*}{\hbar^2}T_{12}\left(E,\Delta_2,\eta_{g2},E_{g2}\right) - \frac{2eB}{\hbar}(n+\frac{1}{2})\right]$$

The EEM can be written as

$$m^*(E_{F7},n,\eta_g,B) = \frac{\hbar^2}{2}\,\textit{Real part of}\,\left[\delta_{7E_{F7},n}\right]' \tag{7.26b}$$

where E_{F7} is the Fermi energy in this case

Thus, from (7.26b) we can conclude that the EEM in this case is a function of the Fermi energy, scattering potential, magnetic field and the Landau quantum number.

The electron concentration is given by

$$n_0 = \frac{g_v eB}{\pi^2\hbar}\sum_{n=0}^{n_{\max}}\left[\phi_{11}(E_{F7},n) + \phi_{12}(E_{F7},n)\right] \tag{7.27}$$

where

$$\phi_{11}(E_{F7},n) = \left[\left(\delta_{7E_{F7},n} + \sqrt{\delta^2_{7E_{F7},n} - \delta_{8E_{F7},n}}\right)\Big/2\right]^{1/2}$$

$$\phi_{12}(E_{F7},n) = \sum_{r=1}^{s} \theta_{2r,7}[\phi_{11}(E_{F7},n)]$$

Thus using (7.5), (7.26a), (7.27) and the allied definitions, we can study the ER in this case.

7.2.7 The ER in HD II-VI Effective Mass Superlattices Under Magnetic Quantization

Following Sasaki [70], the electron dispersion law in heavily doped II-VI EMSLs can be written as

$$k_z^2 = \Delta_{13} + i\Delta_{14}, \tag{7.28}$$

where

$$\Delta_{13} = \left[\frac{1}{L_0^2}\left(\Delta_{11}^2 - \Delta_{12}^2\right) - k_s^2\right]$$

$$\Delta_{11} = \cos^{-1} p_6,\ p_6 = \left[\frac{1 - \Delta_9^2 - \Delta_{10}^2 - \sqrt{\left(1 - \Delta_9^2 - \Delta_{10}^2\right)^2 + 4\Delta_{10}^2}}{2}\right]^{1/2},$$

$$\Delta_9 = \left(\overline{a_1}\cos\Delta_6\cosh\Delta_7 - \overline{a_2}\cos\Delta_8\cosh\Delta_7\right),$$

$$\Delta_{10} = \left(\overline{a_1}\sin\Delta_6\sinh\Delta_7 + \overline{a_2}\sin\Delta_8\sinh\Delta_7\right),$$

$$\Delta_6 = \left[a_0 C_{22}(E,k_s,\eta_{g1}) + b_0 e_3\right],\ \Delta_7 = b_0 e_4,\ \Delta_8 = \left[a_0 C_{22}(E,k_s,\eta_{g1}) - b_0 e_3\right],$$

$$C_{22}(E,k_s,\eta_{g1}) = \left[\frac{2m^*_{\parallel,1}}{\hbar^2}\left\{\gamma_3(E,\eta_{g1}) - \frac{\hbar^2 k_s^2}{2m^*_{\perp,1}} \mp \bar{\lambda}_0 k_s\right\}\right]^{1/2},$$

$$\overline{a_1} = \left[\sqrt{\frac{M_2(0,\eta_{g2})}{\overline{M_1}(0,\eta_{g1})}} + 1\right]^2\left[4\left(\frac{M_2(0,\eta_{g2})}{\overline{M_1}(0,\eta_{g1})}\right)^{1/2}\right]^{-1},\ \overline{M_1}(0,\eta_{g1}) = m^*_{c1}\left(1 - \frac{2}{\pi}\right),$$

$$\overline{a_2} = \left[\sqrt{\frac{M_2(0,\eta_{g2})}{\overline{M_1}(0,\eta_{g1})}} - 1\right]^2\left[4\left(\frac{M_2(0,\eta_{g2})}{\overline{M_1}(0,\eta_{g1})}\right)^{1/2}\right]^{-1}$$

$$\Delta_{12} = \cos^{-1} p_6,\ \Delta_{14} = \frac{2\Delta_{11}\Delta_{12}}{L_0^2}$$

The electron dispersion law in heavily doped II-VI QDSL can be written as

$$(k_z)^2 = \Delta_{13E,n} + i\Delta_{14E,n}, \tag{7.29a}$$

where

$$\Delta_{13E,n} = \left[\frac{1}{L_0^2}\left(\Delta_{11E,n}^2 - \Delta_{12E,n}^2\right) - \left\{\frac{2eB}{\hbar}\left(n+\frac{1}{2}\right)\right\}\right]$$

$$\Delta_{11E,n} = \cos^{-1} p_{6E,n},\ p_{6E,n} = \left[\frac{1-\Delta_{9E,n}^2 - \Delta_{10E,n}^2 - \sqrt{\left(1-\Delta_{9E,n}^2-\Delta_{10E,n}^2\right)^2 + 4\Delta_{10E,n}^2}}{2}\right]^{1/2},$$

$$\Delta_{9E,n} = \left(\overline{a_1}\cos\Delta_{6E,n}\cosh\Delta_{7E,n} - \overline{a_2}\cos\Delta_{8E,n}\cosh\Delta_{7E,n}\right),$$
$$\Delta_{10E,n} = \left(\overline{a_1}\sin\Delta_{6E,n}\sinh\Delta_{7E,n} + \overline{a_2}\sin\Delta_{8E,n}\sinh\Delta_{7E,n}\right),$$
$$\Delta_{6E,n} = \left[a_0 C_{22E,n}(E_{E,n},\eta_{g1}) + b_0 e_{3E,n}\right],\ \Delta_{7E,n} = b_0 e_{4E,n},$$
$$\Delta_{8E,n} = \left[a_0 C_{22E,n}(E_{E,n},\eta_{g1}) - b_0 e_{3E,n}\right],$$

$$C_{22E,n}(E_{E,n},\eta_{g1}) = \left[\frac{2m_{\parallel,1}^*}{\hbar^2}\left\{\gamma_3(E_{E,n},\eta_{g1}) - \frac{\hbar^2}{2m_{\perp,1}^*}\left\{\frac{2eB}{\hbar}\left(n+\frac{1}{2}\right)\right\} \mp \bar{\lambda}_0\left[\left\{\frac{2eB}{\hbar}\left(n+\frac{1}{2}\right)\right\}\right]^{1/2}\right\}\right]^{1/2},$$

$$\Delta_{12E,n} = \cos^{-1} p_{6E,n},\ \Delta_{14E,n} = \frac{2\Delta_{11E,n}\Delta_{12E,n}}{L_0^2},$$

The EEM can be written as

$$m^*(E_{F8}, n, \eta_g, B) = \frac{\hbar^2}{2}\, Real\ part\ of\ [\Delta_{13E_{F8},n}]' \tag{7.29b}$$

where E_{F8} is the Fermi energy in this case

From (7.29b) we note that the EEM is the function of Fermi energy, scattering potential, magnetic field and the Landau quantum number.

The electron concentration is given by

$$n_0 = \frac{g_v eB}{\pi^2\hbar}\sum_{n=0}^{n_{\max}}\left[\phi_{13}(E_{F8},n) + \phi_{14}(E_{F8},n)\right] \tag{7.30}$$

where

$$\phi_{13}(E_{F8},n) = \left[\left(\Delta_{13E_{F8},n} + \sqrt{\Delta_{13E_{F8},n}^2 - \Delta_{14E_{F8},n}}\right)\Big/2\right]^{1/2},$$

and

$$\phi_{14}(E_{F8},n) = \sum_{r=1}^{s}\theta_{2r,7}[\phi_{13}(E_{F8},n)]$$

Thus using (7.5), (7.29a), (7.30) and the allied definitions, we can study the ER in this case.

7.2.8 The ER in HD IV-VI Effective Mass Superlattices Under Magnetic Quantization

Following Sasaki [70], the electron dispersion law in IV-VI, EMSLs can be written as

$$k_z^2 = \left[\frac{1}{L_0^2}\left\{\cos^{-1}\left(f_{23}(E, k_x, k_y)\right)\right\}^2 - k_s^2\right] \tag{7.31}$$

where

$$f_{23}(E, k_x, k_y) = a_3 \cos\left[a_0 C_{23}(E, k_x, k_y, \eta_{g1}) + b_0 D_{23}(E, k_x, k_y, \eta_{g1})\right] - a_4 \cos\left[a_0 C_{23}(E, k_x, k_y, \eta_{g2}) - b_0 D_{23}(E, k_x, k_y, \eta_{g2})\right],$$

$$a_3 = \left[\sqrt{\frac{M_3(0,\eta_{g2})}{M_3(0,\eta_{g1})}} + 1\right]^2 \left[4\left(\frac{M_3(0,\eta_{g2})}{M_3(0,\eta_{g1})}\right)^{1/2}\right]^{-1},$$

$$a_4 = \left[\sqrt{\frac{M_3(0,\eta_{g2})}{M_3(0,\eta_{g1})}} - 1\right]^2 \left[4\left(\frac{M_3(0,\eta_{g2})}{M_3(0,\eta_{g1})}\right)^{1/2}\right]^{-1}$$

$$M_3(0,\eta_{gi}) = (4\overline{p_{9,i}})^{-1}\left[\left\{\alpha_i\left(1-\frac{2}{\pi}\right)\left(\frac{1}{m_{l,i}^+} - \frac{1}{m_{l,i}^-}\right)\right\} + \left[[\overline{q_{9,i}}(0,\eta_{gi})]^2 + (4\overline{p_{9,i}})\overline{R_{9,i}}(0,\eta_{gi})\right]^{-1/2}\left[\alpha_i\left(1-\frac{2}{\pi}\right)\left(\frac{1}{m_{l,i}^+} - \frac{1}{m_{l,i}^-}\right)\overline{q_{9,i}}(0,\eta_{gi}) + 2\overline{p_{9,i}}\left(1-\frac{2}{\pi}+\frac{\alpha_i\eta_{gi}}{\sqrt{\pi}}\right)\right]\right],$$

$$\overline{p_{9,i}} = \frac{\alpha_i \hbar^4}{4m_{l,i}^+ m_{l,i}^-},\ \overline{q_{9,i}}(0,\eta_{gi}) = \left[\frac{\hbar^2}{2}\left(\frac{1}{m_{l,i}^+} + \frac{1}{m_{l,i}^-}\right) - \frac{\alpha_i \eta_{gi}}{\sqrt{\pi}}\left(\frac{1}{m_{l,i}^+} - \frac{1}{m_{l,i}^-}\right)\right],$$

$$\overline{R_{9,i}}(0,\eta_{gi}) = \left[\frac{\eta_{gi}}{\sqrt{\pi}} + \frac{\alpha_i \eta_{gi}^2}{2}\right],$$

$$C_{23}(E, k_x, k_y, \eta_{g1}) = \left[[2\overline{p_{9,1}}]^{-1}\left[-\overline{q_{9,1}}(E, k_x, k_y, \eta_{g1}) + \left[\{\overline{q_{9,1}}(E, k_x, k_y, \eta_{g1})\}^2 + (4\overline{p_{9,1}})\overline{R_{9,1}}(E, k_x, k_y, \eta_{g1})\right]^{1/2}\right]\right]^{1/2},$$

$$D_{23}(E, k_x, k_y, \eta_{g2}) = \left[[2\overline{p_{9,2}}]^{-1}\left[-\overline{q_{9,2}}(E, k_x, k_y, \eta_{g2}) + \left[\{\overline{q_{9,2}}(E, k_x, k_y, \eta_{g2})\}^2 + (4\overline{p_{9,2}})\overline{R_{9,2}}(E, k_x, k_y, \eta_{g2})\right]^{1/2}\right]\right]^{1/2},$$

$$\overline{q_{9,i}}(E, k_x, k_y, \eta_{gi}) = \left[\frac{\hbar^2}{2}\left(\frac{1}{m_{l,i}^*} + \frac{1}{m_{l,i}^-}\right) + \alpha_i \frac{\hbar^4}{4} k_s^2 \left(\frac{1}{m_{l,i}^+ m_{t,i}^-} + \frac{1}{m_{t,i}^+ m_{l,i}^-}\right) - \alpha_i \gamma_3(E,\eta_{gi})\left(\frac{1}{m_{l,i}^+} - \frac{1}{m_{l,i}^-}\right)\right],$$

$$\overline{R_{9,i}}(E, k_x, k_y, \eta_{gi}) = [\gamma_2(E,\eta_{gi}) + \gamma_3(E,\eta_{gi})\alpha_i \frac{\hbar^2}{2} k_s^2 \left(\frac{1}{m_{t,i}^+} - \frac{1}{m_{t,i}^-}\right) - \frac{\hbar^2}{2} k_s^2 \left(\frac{1}{m_{t,i}^*} - \frac{1}{m_{t,i}^-}\right) - \frac{\alpha \hbar^6}{4}\frac{k_s^4}{m_{t,i}^- m_{t,i}^+}],\ a_5 = \left[\sqrt{\frac{m_2^*}{m_1^*}} + 1\right]^2 \left[4\left(\frac{m_2^*}{m_1^*}\right)^{1/2}\right]^{-1}$$

Therefore the electron dispersion law in heavily doped IV-VI, EMSLs under magnetic quantization can be written as

$$(k_z)^2 = \left[[1/L_0^2]\{\cos^{-1}(f_{23}(E,n))\}^2 - \left(\frac{2eB}{\hbar}(n+\frac{1}{2})\right)\right] \tag{7.32a}$$

where

$$f_{23}(E,n) = a_3\cos\left[a_0 C_{23E,n}(E,n,\eta_{g1}) + b_0 D_{23E,n}(E,n,\eta_{g1})\right] - a_4\cos\left[a_0 C_{23E,n}(E,n,\eta_{g2}) - b_0 D_{23E,n}(E,n,\eta_{g2})\right],$$

$$C_{23}(E,n,\eta_{g1}) = \left[\left[2\overline{p_{9,1}}\right]^{-1}\left[-\overline{q_{9,1}}(E,n,\eta_{g1}) + \left[\{\overline{q_{9,1}}(E,n,\eta_{g1})\}^2 + (4\overline{p_{9,1}})\overline{R_{9,1}}(E,n,\eta_{g1})\right]^{1/2}\right]\right]^{1/2},$$

$$D_{23}(E,n,\eta_{g2}) = \left[\left[2\overline{p_{9,2}}\right]^{-1}\left[-\overline{q_{9,2}}(E,n,\eta_{g2}) + \left[\{\overline{q_{9,2}}(E,n,\eta_{g2})\}^2 + (4\overline{p_{9,2}})\overline{R_{9,2}}(E,n,\eta_{g2})\right]^{1/2}\right]\right]^{1/2},$$

$$\overline{q_{9,i}}(E,n,\eta_{gi}) = \left[\frac{\hbar^2}{2}\left(\frac{1}{m_{l,i}^*} + \frac{1}{m_{l,i}^-}\right) + \alpha_i\frac{\hbar^4}{4}\left(\frac{2eB}{\hbar}\left(n+\frac{1}{2}\right)\right)\left(\frac{1}{m_{l,i}^+ m_{t,i}^-} + \frac{1}{m_{t,i}^+ m_{l,i}^-}\right) - \alpha_i\gamma_3(E,\eta_{gi})\left(\frac{1}{m_{l,i}^+} - \frac{1}{m_{l,i}^-}\right)\right],$$

$$\overline{R_{9,i}}(E,n,\eta_{gi}) = \left[\gamma_2(E,\eta_{gi}) + \gamma_3(E,\eta_{gi})\alpha_i\frac{\hbar^2}{2}\left(\frac{2eB}{\hbar}\left(n+\frac{1}{2}\right)\right)\left(\frac{1}{m_{t,i}^+} - \frac{1}{m_{t,i}^-}\right) - \frac{\hbar^2}{2}\left(\frac{2eB}{\hbar}\left(n+\frac{1}{2}\right)\right)\left(\frac{1}{m_{t,i}^*} - \frac{1}{m_{t,i}^-}\right) - \frac{\alpha\hbar^6}{4}\frac{\left(\frac{2eB}{\hbar}\left(n+\frac{1}{2}\right)\right)^2}{m_{t,i}^- m_{t,i}^+}\right],$$

The EEM can be written as

$$m^*(E_{F9},n,\eta_g,B) = \hbar^2\ \mathit{Real\ part\ of}\ [\phi_{15}(E_{F9},n)\phi'_{15}(E_{F9},n)] \tag{7.32b}$$

where E_{F9} is the Fermi energy in this case and

$$\phi_{15}(E_{F9},n) = \left[\frac{1}{L_0^2}\left[\cos^{-1}\left\{\frac{1}{2}f_{23}(E_{F9},n)\right\}\right]^2 - \frac{2eB}{\hbar}\left(n+\frac{1}{2}\right)\right]^{1/2}$$

The (7.32b) tells us that the EEM in this case is a function of Fermi energy, scattering potential, magnetic field and the Landau quantum number.

The electron concentration is given by

$$n_0 = \frac{g_v eB}{\pi^2 \hbar} \sum_{n=0}^{n_{\max}} [\phi_{15}(E_{F9}, n) + \phi_{16}(E_{F9}, n)] \tag{7.33}$$

where

$$\phi_{16}(E_{F9}, n) = \sum_{r=1}^{s} \theta_{2r,6}[\phi_{15}(E_{F9}, n)]$$

Thus using (7.5), (7.32a) and (7.33) we can study the ER in this case.

7.2.9 The ER in HD HgTe/CdTe Effective Mass Superlattices Under Magnetic Quantization

Following Sasaki [70], the electron dispersion law in heavily doped HgTe/CdTe EMSLs can be written as

$$k_z^2 = \Delta_{13H} + i\Delta_{14H} \tag{7.34}$$

where

$$\Delta_{13H} = \left[\frac{1}{L_0^2}(\Delta_{11H}^2 - \Delta_{12H}^2) - k_s^2\right]$$

$$\Delta_{11H} = \cos^{-1} p_{6H},\ p_{6H} = \left[\frac{1 - \Delta_{9H}^2 - \Delta_{10H}^2 - \sqrt{(1 - \Delta_{9H}^2 - \Delta_{10H}^2)^2 + 4\Delta_{10H}^2}}{2}\right]^{1/2},$$

$$\Delta_{9H} = (\overline{a_{1H}} \cos\Delta_{5H} \cosh\Delta_{6H} - \overline{a_{2H}} \cos\Delta_{7H} \cosh\Delta_{6H}),$$
$$\Delta_{10H} = (\overline{a_{1H}} \sin\Delta_{5H} \sinh\Delta_{6H} + \overline{a_{2H}} \sin\Delta_{7H} \sinh\Delta_{6H}),$$
$$\Delta_{5H} = [a_0 C_{22H}(E, k_s, \eta_{g1}) + b_0 e_3],$$
$$\Delta_{6H} = b_0 e_4,\ \Delta_{7H} = [a_0 C_{22H}(E, k_s, \eta_{g1}) - b_0 e_3],$$

$$C_{22H}(E, k_s, \eta_{g1}) = \left[\frac{B_{01}^2 + 2A_1 E - B_{01}(B_{01}^2 + 4A_1 E)}{2A_1^2} - k_s^2\right]^{1/2},$$

$$\overline{a_{1H}} = \left[\sqrt{\frac{M_2(0, \eta_{g2})}{m_{c1}^*}} + 1\right]^2 \left[4\left(\frac{M_2(0, \eta_{g2})}{m_{c1}^*}\right)^{1/2}\right]^{-1},$$

$$\overline{a_{2H}} = \left[\sqrt{\frac{M_2(0, \eta_{g2})}{m_{c1}^*}} - 1\right]^2 \left[4\left(\frac{M_2(0, \eta_{g2})}{m_{c1}^*}\right)^{1/2}\right]^{-1}$$

$$\Delta_{12H} = \cos^{-1} p_{6H},\ \Delta_{14H} = \frac{2\Delta_{11H}\Delta_{12H}}{L_0^2}$$

The electron dispersion law in heavily doped HgTe/CdTe EMSLs under magnetic quantization can be written as

$$(k_z)^2 = \Delta_{13HE,n} + i\Delta_{14HE,n} \tag{7.35a}$$

where

$$\Delta_{13HE,n} = [(1/L_0^2)\left(\Delta_{11HE,n}^2 - \Delta_{12HE,n}^2\right) - \frac{2eB}{\hbar}\left(n+\frac{1}{2}\right)]$$

$$\Delta_{11HE,n} = \cos^{-1} p_{6HE,n},\ p_{6HE,n} = \left[\left(\left(1 - \Delta_{9HE,n}^2 - \Delta_{10HE,n}^2 - \sqrt{\left(1 - \Delta_{9HE,n}^2 - \Delta_{10HE,n}^2\right)^2 + 4\Delta_{10HE,n}^2}\right)/2\right)\right]^{\frac{1}{2}}$$

$$\Delta_{9HE,n} = \left(\overline{a_{1H}} \cos\Delta_{5HE,n} \cosh\Delta_{6HE,n} - \overline{a_{2H}} \cos\Delta_{7HE,n} \cosh\Delta_{6HE,n}\right),$$

$$\Delta_{10HE,n} = \left(\overline{a_{1H}} \sin\Delta_{5HE,n} \sinh\Delta_{6HE,n} + \overline{a_{2H}} \sin\Delta_{7HE,n} \sinh\Delta_{6HE,n}\right),$$

$$\Delta_{5HE,n} = \left[a_0 C_{22HE,n}\left(E_{E,n}, \eta_{g1}\right) + b_0 e_3\right],\ \Delta_{6HE,n} = b_0 e_4,\ \Delta_{7HE,n} = \left[a_0 C_{22HE,n}\left(E_{E,n}, \eta_{g1}\right) - b_0 e_3\right],$$

$$C_{22HE,n}\left(E_{E,n}, \eta_{g1}\right) = \left[\frac{B_{01}^2 + 2A_1 E_{E,n} - B_{01}(B_{01}^2 + 4A_1 E_{E,n})}{2A_1^2} - \left[\frac{2eB}{\hbar}\left(n+\frac{1}{2}\right)\right]\right]^{1/2},$$

$$\Delta_{12HE,n} = \cos^{-1} p_{6HE,n},\ \Delta_{14HE,n} = \frac{2\Delta_{11HE,n}\Delta_{12HE,n}}{L_0^2}$$

The EEM can be written as

$$m^*(E_{F10}, n, \eta_g, B) = \frac{\hbar^2}{2}\ \mathit{Real\ part\ of}\ [\Delta_{13HE_{F10},n}]' \tag{7.35b}$$

where $E_{F10}$is the Fermi energy in this case.

The electron concentration is given by

$$n_0 = \frac{g_v eB}{\pi^2\hbar}\sum_{n=0}^{n_{\max}}[\phi_{17}(E_{F10}, n) + \phi_{18}(E_{F10}, n)] \tag{7.36}$$

where

$$\phi_{17}(E_{F10}, n) = \left[\left(\Delta_{13HE_{F10},n} + \sqrt{\Delta_{13HE_{F10},n}^2 - \Delta_{14HE_{F10},n}}\right)\Big/2\right]^{1/2}$$

and

$$\phi_{18}(E_{F10}, n) = \sum_{r=1}^{s}\theta_{2r,7}[\phi_{17}(E_{F10}, n)]$$

Thus using (7.5), (7.35a) and (7.36) we can study the ER in this case.

7.2.10 The ER in HD Stained Layer Effective Mass Superlattices Under Magnetic Quantization

The dispersion relation of the constituent materials of heavily doped III-V super lattices can be written as

$$\bar{P}_i(E,\eta_{gi})k_x^2 + \bar{Q}_i(E,\eta_{gi})k_y^2 + \bar{S}_i(E,\eta_{gi})k_z^2 = 1 \tag{7.37}$$

where

$$\bar{P}_i(E,\eta_{gi}) = (\gamma_0(E,\eta_{gi}) - I_0T_{1i})(\overline{\Delta}_i(E,\eta_{gi}))^{-1}, I_0 = (1/2)[1+Erf(E/\eta_{gi})],$$
$$T_{1i} = [E_{gi} - C_{1i}^c\varepsilon_i - (a_i + C_{1i}^c)\varepsilon_i + (3/2)b_i\varepsilon_{xxi} - (b_i\varepsilon_i/2) + (\sqrt{3}d_i\varepsilon_{xyi}/2)],$$
$$\overline{\Delta}_i(E,\eta_{gi}) = [(-q_i\eta_{gi}^3/2\sqrt{\pi})\exp(-(E^2/\eta_{gi}^2))[1+(E^2/\eta_{gi}^2)] - R_i\theta_0(E,\eta_{gi}) + V_i\gamma_0(E,\eta_{gi})$$
$$+ (\zeta_i/2)[1+Erf(E/\eta_{gi})]], q_i = (3/2B_{2i}^2), R_i = q_i[2A_i + C_{1i}^c\varepsilon_i], A_i = E_{gi} - C_{1i}^c\varepsilon_i,$$
$$V_i = q_i[A_i^2 - (2C_{2i}^2\varepsilon_{xyi}/3) + 2A_iC_{1i}^c\varepsilon_i], \zeta_i = q_i[(2C_{2i}^2\varepsilon_{xyi}/3) - C_{1i}^c\varepsilon_iA_i^2],$$
$$\bar{Q}_i(E,\eta_{gi}) = (\gamma_0(E,\eta_{gi}) - I_0T_{2i})(\overline{\Delta}_i(E,\eta_{gi}))^{-1}, T_{2i} = [E_{gi} - C_{1i}^c\varepsilon_i - (a_i + C_{1i}^c)\varepsilon_i + (3/2)b_i\varepsilon_{xxi} - (b_i\varepsilon_i/2)$$
$$- (\sqrt{3}d_i\varepsilon_{xyi}/2)], \bar{S}_i(E,\eta_{gi}) = (\gamma_0(E,\eta_{gi}) - I_0T_{3i})(\overline{\Delta}_i(E,\eta_{gi}))^{-1},$$
$$T_{3i} = [E_{gi} - C_{1i}^c\varepsilon_i + (a_i + C_{1i}^c)\varepsilon_i + (3/2)b_i\varepsilon_{zzi} - (b_i\varepsilon_i/2)],$$

The electron energy spectrum in heavily doped strained layer effective mass superlattices can be written as

$$k_z^2 = \left[\frac{1}{L_0^2}\left\{\cos^{-1}(f_{40}(E,k_x,k_y))\right\}^2 - k_s^2\right] \tag{7.38}$$

where

$$f_{40}(E,k_x,k_y) = a_{20}\cos\left[a_0C_{40}(E,k_x,k_y,\eta_{g1}) + b_0D_{40}(E,k_x,k_y,\eta_{g1})\right]$$
$$- a_{21}\cos\left[a_0C_{40}(E,k_x,k_y,\eta_{g2}) - b_0D_{40}(E,k_x,k_y,\eta_{g2})\right],$$
$$a_{20} = \left[\sqrt{\frac{M_{s2}(0,\eta_{g2})}{M_{s1}(0,\eta_{g1})}} + 1\right]^2\left[4\left(\frac{M_{s2}(0,\eta_{g2})}{M_{s1}(0,\eta_{g1})}\right)^{1/2}\right]^{-1},$$
$$M_{si}(0,\eta_{gi}) = (\hbar/2)\rho_i(\eta_{gi})$$
$$\rho_i(\eta_{gi}) = [(\eta_{gi}/2\sqrt{\pi}) - (T_{3i}/2)]^{-2} \times [\{(\eta_{gi}/2\sqrt{\pi})$$
$$- (T_{3i}/2)\}\{(V_i/2) - (R_i\eta_{gi}/\sqrt{\pi}) + (\zeta_i/\eta_{gi}\sqrt{\pi})\}$$
$$- ((1/2) - (T_{3i}/\eta_{gi}\sqrt{\pi}))\{(\zeta_i/2) + (V_i\eta_{gi}/2\sqrt{\pi})$$
$$- (R_i\eta_{gi}^2/4) - (q_i\eta_{gi}^3/2\sqrt{\pi})\}]$$

$$a_{20} = \left[\sqrt{\frac{M_{s2}(0,\eta_{g2})}{M_{s1}(0,\eta_{g1})}} + 1\right]^2 \left[4\left(\frac{M_{s2}(0,\eta_{g2})}{M_{s1}(0,\eta_{g1})}\right)^{1/2}\right]^{-1},$$

$$a_{21} = \left[\sqrt{\frac{M_{s2}(0,\eta_{g2})}{M_{s1}(0,\eta_{g1})}} - 1\right]^2 \left[4\left(\frac{M_{s2}(0,\eta_{g2})}{M_{s1}(0,\eta_{g1})}\right)^{1/2}\right]^{-1}$$

$$C_{40}(E,k_x,k_y,\eta_{g1}) = [1 - \bar{P}_1(E,\eta_{g1})k_x^2 - \bar{Q}_1(E,\eta_{g1})k_y^2]^{1/2}[\bar{S}_1(E,\eta_{g1})]^{-1/2}$$

$$D_{40}(E,k_x,k_y,\eta_{g2}) = [1 - \bar{P}_2(E,\eta_{g2})k_x^2 - \bar{Q}_2(E,\eta_{g2})k_y^2]^{1/2}[\bar{S}_2(E,\eta_{g2})]^{-1/2}$$

Therefore, the electron dispersion law in heavily doped strained layer effective mass superlattices under magnetic quantization can be expressed as

$$(k_z)^2 = \left[\frac{1}{L_0^2}\{\cos^{-1}(f_{40}(E,n))\}^2 - \left(\frac{2eB}{\hbar}\left(n+\frac{1}{2}\right)\right)\right] \quad (7.39a)$$

where

$$f_{40}(E,n) = a_{20}\cos\left[a_0C_{40}(E,n,\eta_{g1}) + b_0D_{40}(E,n,\eta_{g1})\right] - a_{21}\cos\left[a_0C_{40}(E,n,\eta_{g2}) - b_0D_{40}(E,n,\eta_{g2})\right],$$

$$C_{40}(E,n,\eta_{g1}) = [1 - \frac{\hbar eB}{\phi_{50}(E,\eta_{g1})}(n+\frac{1}{2})]^{1/2}[\bar{S}_1(E,\eta_{g1})]^{-1/2},$$

$$\phi_{50}(E,\eta_{g1}) = \sqrt{\psi_{50}(E,\eta_{g1})\psi_{51}(E,\eta_{g1})},$$

$$\psi_{50}(E,\eta_{g1}) = \frac{\hbar^2}{2\bar{P}_1(E,\eta_{g1})},\ \psi_{51}(E,\eta_{g1}) = \frac{\hbar^2}{2\bar{Q}_1(E,\eta_{g1})}$$

$$D_{40}(E,n,\eta_{g2}) = [1 - \frac{\hbar eB}{\phi_{501}(E,\eta_{g2})}(n+\frac{1}{2})]^{1/2}[\bar{S}_2(E,\eta_{g2})]^{-1/2}$$

$$\phi_{501}(E,\eta_{g2}) = \sqrt{\psi_{501}(E,\eta_{g2})\psi_{511}(E,\eta_{g2})}$$

$$\psi_{501}(E,\eta_{g2}) = \frac{\hbar^2}{2\bar{P}_2(E,\eta_{g2})},\ \psi_{511}(E,\eta_{g2}) = \frac{\hbar^2}{2\bar{Q}_2(E,\eta_{g2})}$$

The EEM can be written as

$$m^*(E_{F11},n,\eta_g,B) = \hbar^2\,Real\ part\ of\ [\phi_{19}(E_{F11},n)\phi'_{19}(E_{F11},n)] \quad (7.39b)$$

where E_{F11} is the Fermi energy in this case and

$$\phi_{19}(E_{F11},n) = [\frac{1}{L_0^2}\left[\cos^{-1}\left\{\frac{1}{2}f_{40}(E_{F11},n)\right\}\right]^2 - \frac{2eB}{\hbar}\left(n+\frac{1}{2}\right)]^{1/2}$$

The electron concentration is given by

$$n_0 = \frac{g_v eB}{\pi^2 \hbar} \sum_{n=0}^{n_{\max}} [\phi_{19}(E_{F11}, n) + \phi_{20}(E_{F11}, n)] \tag{7.40}$$

where

$$\phi_{20}(E_{F11}, n) = \sum_{r=1}^{s} \theta_{2r,6}[\phi_{19}(E_{F11}, n)]$$

Thus using (7.5), (7.39a) and (7.40) we can study the ER in this case.

7.3 Open Research Problem

(R.B.1) Investigate the ER for all types of HD super-lattices under alternating magnetic field.

References

1. S. Mukherjee, S.N. Mitra, P.K. Bose, A.R. Ghatak, A. Neoigi, J.P. Banerjee, A. Sinha, M. Pal, S. Bhattacharya, K.P. Ghatak, J. Compu, Theor. Nanosc. **4**, 550 (2007)
2. N.G. Anderson, W.D. Laidig, R.M. Kolbas, Y.C. Lo, J. Appl. Phys. **60**(2361), (1986)
3. N. Paitya, K.P. Ghatak, J. Adv. Phys. **1**(161), (2012)
4. N. Paitya, S.Bhattacharya, D. De, K.P. Ghatak, Adv. Sci. Eng. Med. **4**(96), (2012)
5. S. Bhattacharya, D. De, S. M. Adhikari, K. P. Ghatak, Superlatt. Microst. **51**(203), (2012)
6. D. De, S. Bhattacharya, S.M. Adhikari, A. Kumar, P.K. Bose, K.P. Ghatak, Beilstein, J. Nanotech. **2**(339), (2012)
7. D. De, A. Kumar, S.M. Adhikari, S. Pahari, N. Islam, P. Banerjee, S.K. Biswas, S. Bhattacharya, K.P. Ghatak, Superlatt. Microstruct. 47(377), (2010)
8. S. Pahari, S. Bhattacharya, S. Roy, A. Saha, D. De, K.P. Ghatak, Superlatt. Microstruct. **46**(760), (2009)
9. S. Pahari, S. Bhattacharya, K.P. Ghatak, J. Comput. Theor. Nanosci. **6**(2088), (2009)
10. S.K. Biswas, A.R. Ghatak, A. Neogi, A. Sharma, S. Bhattacharya, K.P. Ghatak, Phys. E: Low-dimen. Syst. Nanostruct. **36**(163), (2007)
11. L.J. Singh, S. Choudhury, D. Baruah, S.K. Biswas, S. Pahari, K.P. Ghatak, Phys. B: Condens. Matter, **368**(188), (2005)
12. S. Chowdhary, L.J. Singh, K.P. Ghatak, Phys. B: Condens. Matter, **365**(5), (2005)
13. L.J. Singh, S. Choudhary, A. Mallik, K.P. Ghatak, J. Comput. Theor. Nanosci. **2**(287), (2005)
14. K.P. Ghatak, J Mukhopadhyay, J.P. Banerjee, SPIE Proc. **4746**(1292), (2002)
15. K.P. Ghatak, S. Dutta, D.K. Basu, B. Nag, Il Nuovo Cimento. **D20**(227), (1998)
16. K.P. Ghatak, D.K. Basu, B. Nag, J. Phys. Chem. Solids **58**(133), (1997)
17. K.P. Ghatak, B. De, Mat. Resc. Soc. Proc. **300**(513), (1993)
18. K.P. Ghatak, B. Mitra, Il Nuovo Cimento. **D15**(97), (1993)
19. K.P. Ghatak, Inter. Soc. Opt. Photon. Proc. Soc. Photo Opt. Instrum. Eng. **1626**(115), (1992)
20. K.P. Ghatak, A. Ghoshal, Phys. Stat. Sol. **170**(K27), (1992)

21. K.P. Ghatak, S. Bhattacharya, S.N. Biswas, Proc. Soc. Photo opt. instru. Engg. **836**, (72) (1988)
22. K.P. Ghatak, A Ghoshal, S.N. Biswas, M. Mondal, Proc. Soc. Photo Opt. Instru. Eng. **1308**(356), (1990)
23. K.P. Ghatak, B. De, Proc. Wide Bandgap Semin. Symp., Matt. Res. Soc. (377), (1992)
24. K.P. Ghatak, B. De, Defect Eng. Semin. Growth, Process. Device Tech. Symp. Mat. Res. Soc. **262**(911), (1992)
25. S.N. Biswas, K.P. Ghatak, Int. J. Electron. Theor. Exp. **70**(125), (1991)
26. B. Mitra, K.P. Ghatak, Phys. Lett. A. **146**(357), (1990)
27. B. Mitra, K.P. Ghatak, Phys. Lett. A. **142**(401), (1989)
28. K.P. Ghatak, B. Mitra, A. Ghoshal, Phys. Status Solidi **154**(K121), (1989)
29. B. Mitra, K.P. Ghatak, Phys. Status Solidi **149**(K117), (1988)
30. K.P. Ghatak, S.N. Biswas, Proc. Soc. Photo Opt. Instrum. Eng. **792**(239), (1987)
31. S. Bhattacharyya, K.P. Ghatak, S. Biswas, OE/Fibers' 87, Int. Soc. Opt. Photon. (73), (1987)
32. M. Mondal, K.P. Ghatak, Czech. J. Phys. B. **36**(1389), (1986)
33. K.P. Ghatak, A.N. Chakravarti, Phys. Status Solidi **117**(707), (1983)
34. L.V. Keldysh, Sov. Phys. Solid State **4**, 1658 (1962)
35. L. Esaki, R. Tsu, IBM J. Res. Dev. **14**, 61 (1970)
36. G. Bastard, *Wave Mechanics Applied to Heterostructures* (Editions de Physique, Les Ulis, 1990)
37. E.L. Ivchenko, G. Pikus, *Superlattices and Other Heterostructures*, (Springer, Berlin, 1995)
38. R. Tsu, *Superlattices to Nanoelectronics* (Elsevier, The Netherlands, 2005)
39. P. Fürjes, C.S. Dücs, M. Ádám, J. Zettner, I. Bársony, Superlattices Microstruct. **35**, 455 (2004)
40. T. Borca-Tasciuc, D. Achimov, W.L. Liu, G. Chen, H.W. Ren, C.H. Lin, S.S. Pei, Microscale Thermophys. Eng. **5**, 225 (2001)
41. B.S. Williams, Nat. Photonics **1**, 517 (2007)
42. A. Kosterev, G. Wysocki, Y. Bakhirkin, S. So, R. Lewicki, F. Tittel, R.F. Curl, Appl. Phys. B **90**, 165 (2008)
43. M.A. Belkin, F. Capasso, F. Xie, A. Belyanin, M. Fischer, A. Wittmann, J. Faist, Appl. Phys. Lett. **92**, 201101 (2008)
44. G.J. Brown, F. Szmulowicz, R. Linville, A. Saxler, K. Mahalingam, C.-H. Lin, C.H. Kuo, W.Y. Hwang, IEEE Photonics Technol. Lett. **12**, 684 (2000)
45. H.J. Haugan, G.J. Brown, L. Grazulis, K. Mahalingam, D.H. Tomich, Physica E **20**, 527 (2004)
46. S.A. Nikishin, V.V. Kuryatkov, A. Chandolu, B.A. Borisov, G.D. Kipshidze, I. Ahmad, M. Holtz, H. Temkin, Jpn. J. Appl. Phys. **42**, L1362 (2003)
47. Y.K. Su, H.C. Wang, C.L. Lin, W.B. Chen, S.M. Chen, Jpn. J. Appl. Phys. **42**, L751 (2003)
48. C.H. Liu, Y.K. Su, L.W. Wu, S.J. Chang, R.W. Chuang, Semicond. Sci. Technol. **18**, 545 (2003)
49. S.B. Che, I. Nomura, A. Kikuchi, K. Shimomura, K. Kishino, Phys. Status Solidi **229**, 1001 (2002)
50. C.P. Endres, F. Lewen, T.F. Giesen, S. SchlEEMer, D.G. Paveliev, Y.I. Koschurinov, V.M. Ustinov, A.E. Zhucov, Rev. Sci. Instrum. **78**, 043106 (2007)
51. F. Klappenberger, K.F. Renk, P. Renk, B. Rieder, Y.I. Koshurinov, D.G. Pavelev, V. Ustinov, A. Zhukov, N. Maleev, A. Vasilyev, Appl. Phys. Lett. **84**, 3924 (2004)
52. X. Jin, Y. Maeda, T. Saka, M. Tanioku, S. Fuchi, T. Ujihara, Y. Takeda, N. Yamamoto, Y. Nakagawa, A. Mano, S. Okumi, M. Yamamoto, T. Nakanishi, H. Horinaka, T. Kato, T. Yasue, T. Koshikawa, J. Crystal Growth **310**, 5039 (2008)
53. X. Jin, N. Yamamoto, Y. Nakagawa, A. Mano, T. Kato, M. Tanioku, T. Ujihara, Y. Takeda, S. Okumi, M. Yamamoto, T. Nakanishi, T. Saka, H. Horinaka, T. Kato, T. Yasue, T. Koshikawa, Appl. Phys. Express **1**, 045002 (2008)
54. B.H. Lee, K.H. Lee, S. Im, M.M. Sung, Org. Electron. **9**, 1146 (2008)

55. P.H. Wu, Y.K. Su, I.L. Chen, C.H. Chiou, J.T. Hsu, W.R. Chen, Jpn. J. Appl. Phys. **45**, L647 (2006)
56. A.C. Varonides, Renew. Energy **33**, 273 (2008)
57. M. Walther, G. Weimann, Phys. Status Solidi **203**, 3545 (2006)
58. R. Rehm, M. Walther, J. Schmitz, J. Fleiβner, F. Fuchs, J. Ziegler, W. Cabanski, Opto-Electron. Rev. **14**, 19 (2006)
59. R. Rehm, M. Walther, J. Scmitz, J. Fleissner, J. Ziegler, W. Cabanski, R. Breiter, Electron. Lett. **42**, 577 (2006)
60. G.J. Brown, F. Szmulowicz, H. Haugan, K. Mahalingam, S. Houston, Microelectron. J. **36**, 256 (2005)
61. K.V. Vaidyanathan, R.A. Jullens, C.L. Anderson, H.L. Dunlap, Solid State Electron. **26**, 717 (1983)
62. B.A. Wilson, IEEE J. Quantum Electron. **24**, 1763 (1988)
63. M. Krichbaum, P. Kocevar, H. Pascher, G. Bauer, IEEE. J. Quantum Electron. **24**, 717 (1988)
64. J.N. Schulman, T.C. McGill, Appl. Phys. Lett. **34**, 663 (1979)
65. H. Kinoshita, T. Sakashita, H. Fajiyasu, J. Appl. Phys. **52**, 2869 (1981)
66. L. Ghenin, R.G. Mani, J.R. Anderson, J.T. Cheung, Phys. Rev. B **39**, 1419 (1989)
67. C.A. Hoffman, J.R. Mayer, F.J. Bartoli, J.W. Han, J.W. Cook, J.F. Schetzina, J.M. Schubman, Phys. Rev. B. **39**, 5208 (1989)
68. V.A. Yakovlev, Sov. Phys. Semicon. **13**, 692 (1979)
69. E.O. Kane, J. Phys. Chem. Solids **1**, 249 (1957)
70. H.X. Jiang, J.Y. Lin, J. Appl. Phys. **61**, 624 (1987)
71. H. Sasaki, Phys. Rev. B **30**, 7016 (1984)

Chapter 8
Appendix C: The ER in HDS and Their Nano-Structures Under Cross-Fields Configuration

8.1 Introduction

The influence of crossed electric and quantizing magnetic fields on the transport properties of semiconductors having various band structures are relatively less investigated as compared with the corresponding magnetic quantization, although, the cross-fields are fundamental with respect to the addition of new physics and the related experimental findings. In 1966, Zawadzki and Lax [1] formulated the electron dispersion law for III-V semiconductors in accordance with the two band model of Kane under cross fields configuration which generates the interest to study this particular topic of semiconductor science in general [2–38].

In Sect. 8.2.1 of theoretical background, the ER in HD nonlinear optical materials in the presence of crossed electric and quantizing magnetic fields has been investigated by formulating the electron dispersion relation. The Sect. 8.2.2 reflects the study of the ER in HD III-V, ternary and quaternary compounds as a special case of Sect. 8.2.1. The Sect. 8.2.3 contains the study of the ER for the HD II-VI semiconductors in the present case. In Sect. 8.2.4, the ER under cross field configuration in HD IV-VI semiconductors has been investigated in accordance with the models of the Cohen, the Lax nonparabolic ellipsoidal and the parabolic ellipsoidal respectively. In the Sect. 8.2.5, the ER for the HD stressed Kane type semiconductors has been investigated. The Sects. 8.2.6–8.2.10 discuss the ERs' in QWs of the above HD semiconductors in the presence of cross-fields configuration respectively. The last Sect. 8.3 contains three open research problems.

K. P. Ghatak and S. Bhattacharya, *Heavily-Doped 2D-Quantized Structures and the Einstein Relation*, Springer Tracts in Modern Physics 260,
DOI 10.1007/978-3-319-08380-3_8, © Springer International Publishing Switzerland 2015

8.2 Theoretical Background

8.2.1 The ER in HD Nonlinear Optical Semiconductors Under Cross-Fields Configuration

The (1.2) of Chap. 1 can be expressed as

$$T_{22}(E,\eta_g) = \frac{p_s^2}{2m_\perp^*} + \frac{p_z^2}{2M_{||}} T_{22}(E,\eta_g)[T_{21}(E,\eta_g)]^{-1} \tag{8.1}$$

where, $p_s = \hbar k_s$ and $p_z = \hbar k_z$.

We know that from electromagnetic theory that,

$$\vec{B} = \nabla \times \vec{A} \tag{8.2}$$

where, $\vec{A}$ is the vector potential. In the presence of quantizing magnetic field B along z direction, the (8.2) assumes the form

$$0\hat{i} + 0\hat{j} + B\hat{k} = \begin{vmatrix} \hat{i} & \hat{j} & \hat{k} \\ \frac{\partial}{\partial x} & \frac{\partial}{\partial y} & \frac{\partial}{\partial z} \\ A_x & A_y & A_z \end{vmatrix} \tag{8.3}$$

where $\hat{i}$, $\hat{j}$ and $\hat{k}$ are orthogonal triads. Thus, we can write

$$\begin{aligned} \frac{\partial A_z}{\partial y} - \frac{\partial A_y}{\partial z} &= 0 \\ \frac{\partial A_x}{\partial z} - \frac{\partial A_z}{\partial x} &= 0. \\ \frac{\partial A_y}{\partial x} - \frac{\partial A_x}{\partial y} &= B \end{aligned} \tag{8.4}$$

This particular set of equations is being satisfied for $A_x = 0$, $A_y = Bx$ and $A_z = 0$.

Therefore in the presence of the electric field E_o along x axis and the quantizing magnetic field B along z axis for the present case following (8.1) one can approximately write,

$$T_{22}(E,\eta_g) + |e|E_o\hat{x}\rho(E,\eta_g) = \frac{\hat{p}_x^2}{2m_\perp^*} + \frac{(\hat{p}_y - |e|B\hat{x})^2}{2m_\perp^*} + \frac{\hat{p}_z^2}{2a(E,\eta_g)} \tag{8.5}$$

where

$$\rho(E) \equiv \frac{\partial}{\partial E}[\mathrm{T}_{22}(E,\eta_g)] \text{ and } a(E,\eta_g) \equiv m_{||}^*[\mathrm{T}_{22}(E,\eta_g)]^{-1}[\mathrm{T}_{21}(E,\eta_g)]$$

Let us define the operator $\hat{\theta}$ as

$$\hat{\theta} = -\hat{p}_y + |e|B\hat{x} - \frac{m_\perp^* E_o \rho(E,\eta_g)}{B} \tag{8.6}$$

Eliminating the operator $\hat{x}$, between (8.5) and (8.6) the dispersion relation of the conduction electron in tetragonal semiconductors in the presence of cross fields configuration is given by

$$\mathrm{T}_{22}(E,\eta_g) = \left[\left((n+\frac{1}{2})\hbar\omega_{01}\right) + \left(\frac{[\hbar k_z(E)]^2}{2a(E,\eta_g)}\right) - \left(\frac{E_o \hbar k_y \rho(E,\eta_g)}{B}\right) - \left(\frac{M_\perp \rho^2(E,\eta_g)E_o^2}{2B^2}\right)\right] \tag{8.7}$$

where,

$$\omega_{01} = \frac{|e|B}{m_\perp^*} \tag{8.8a}$$

The EEMs along Z and Y directions can, respectively be expressed from (8.7) as

$$\begin{aligned} m_z^*(\bar{E}_{FBHD},\eta_g,n,E_0) = {} & \textit{Real part of } \left[a'(\bar{E}_{FBHD},\eta_g)\left[T_{22}(\bar{E}_{FBHD},\eta_g)\right.\right. \\ & \left. - \left(n+\frac{1}{2}\right)\hbar\omega_{01} + \frac{M_\perp \rho^2(\bar{E}_{FBHD},\eta_g)E_0^2}{2B^2}\right] \\ & \left. + a(\bar{E}_{FBHD},\eta_g)\left[T'_{22}(\bar{E}_{FBHD},\eta_g) + \frac{M_\perp \rho(\bar{E}_{FBHD},\eta_g)\rho'(\bar{E}_{FBHD},\eta_g)E_0^2}{B^2}\right]\right] \end{aligned} \tag{8.8b}$$

and

$$\begin{aligned} m_y^*(\bar{E}_{FBHD},\eta_g,n,E_0) = {} & \left(\frac{B}{E_0}\right)^2 \textit{Real part of } \left[\rho(\bar{E}_{FBHD},\eta_g)\right]^{-3}\left[T_{22}(\bar{E}_{FBHD},\eta_g)\right. . \\ & \left. - \left(n+\frac{1}{2}\right)\hbar\omega_{01} + \frac{M_\perp \rho^2(\bar{E}_{FBHD},\eta_g)E_0^2}{2B^2}\right] \\ & \left[\left[\rho(\bar{E}_{FBHD},\eta_g)\left[T'_{22}(\bar{E}_{FBHD},\eta_g)\right.\right.\right. \\ & \left. + \frac{M_\perp \rho(\bar{E}_{FBHD},\eta_g)\rho'(\bar{E}_{FBHD},\eta_g)E_0^2}{B^2}\right] - \left[T_{22}(\bar{E}_{FBHD},\eta_g)\right. \\ & \left.\left.\left. - \left(n+\frac{1}{2}\right)\hbar\omega_{01} + \frac{M_\perp \rho^2(\bar{E}_{FBHD},\eta_g)E_0^2}{2B^2}\right]\rho'(\bar{E}_{FBHD},\eta_g)\right]\right] \end{aligned} \tag{8.8c}$$

where $\bar{E}_{FBHD}$ is the Fermi energy in the presence of cross-fields configuration and heavy doping as measured from the edge of the conduction band in the vertically upward direction in the absence of any quantization.

When $E_0 \to 0, m_z^*(\bar{E}_{FBHD}, \eta_g, n, E_0) \to \infty$, which is a physically justified result. The dependence of the EEM along y direction on the Fermi energy, electric field, magnetic field and the magnetic quantum number is an intrinsic property of cross fields together with the fact in the present case of heavy doping, the EEM exists in the band gap. Another characteristic feature of cross field is that various transport coefficients will be sampled dimension dependent. These conclusions are valid for even isotropic parabolic energy bands and cross fields introduce the index dependent anisotropy in the effective mass.

The formulation of ER requires the expression of the electron concentration which can, in general, be written excluding the electron spin as

$$n_o = \frac{-g_v}{L_x \pi^2} \sum_{n=0}^{n_{\max}} \int_{\bar{E}_0}^{\infty} I(E, \eta_g) \frac{\delta f_o}{\delta E} dE \tag{8.9}$$

where L_x is the sample length along x direction, $\bar{E}_0$ is determined by the equation

$$I(\bar{E}_0, \eta_g) = 0$$

where

$$I(E, \eta_g) = \int_{x_l(E,\eta_g)}^{x_h(E,\eta_g)} k_z(E) dk_y \tag{8.10}$$

in which, $x_l(E, \eta_g) \equiv \frac{-E_0 M_\perp \rho(E,\eta_g)}{\hbar B}$ and $x_h(E, \eta_g) \equiv \frac{|e| B L_x}{\hbar} + x_l(E, \eta_g)$.

Thus we get

$$\begin{aligned} I(E, \eta_g) = \frac{2}{3}\left[\frac{B\sqrt{2a(E, \eta_g)}}{\hbar^2 E_0 \rho(E, \eta_g)}\right] \Bigg[& \left[T_{22}(E, \eta_g) - \left(n + \frac{1}{2}\right)\frac{\hbar|e|B}{m_\perp^*} + |e| E_0 L_x \rho(E, \eta_g) - \frac{m_\perp^* E_0^2 [\rho(E, \eta_g)]^2}{2B^2}\right]^{\frac{3}{2}} \\ & - \left[T_{22}(E, \eta_g) - \left(n + \frac{1}{2}\right)\frac{\hbar|e|B}{m_\perp^*} - \frac{m_\perp^* E_0^2 [\rho(E, \eta_g)]^2}{2B^2}\right]^{\frac{3}{2}} \Bigg] \end{aligned} \tag{8.11}$$

Therefore the electron concentration is given by

$$n_0 = \left(\frac{2g_v B\sqrt{2}}{3 L_x \pi^2 \hbar^2 E_o}\right) \text{Real part of} \sum_{n=0}^{n_{\max}} \left[T_{41HD}(n, \bar{E}_{FBHD}, \eta_g) + T_{42HD}(n, \bar{E}_{FBHD}, \eta_g)\right] \tag{8.12}$$

where

$$T_{41}\left(n,\bar{E}_{FBHD},\eta_g\right)\equiv\frac{\sqrt{a\left(\bar{E}_{FBHD},\eta_g\right)}}{\rho\left(\bar{E}_{FBHD},\eta_g\right)}\left[\left[T_{22}\left(\bar{E}_{FBHD},\eta_g\right)-\left(n+\frac{1}{2}\right)\frac{\hbar|e|B}{M_\perp}\right.\right.$$
$$\left.+|e|E_0L_x\rho\left(\bar{E}_{FBHD},\eta_g\right)-\frac{m_\perp^*E_0^2\left[\rho\left(\bar{E}_{FBHD},\eta_g\right)\right]^2}{2B^2}\right]^{3/2}$$
$$\left.-\left[T_{22}\left(\bar{E}_{FBHD},\eta_g\right)-\left(n+\frac{1}{2}\right)\frac{\hbar|e|B}{m_\perp^*}-\frac{m_\perp^*E_0^2\left[\rho\left(\bar{E}_{FBHD},\eta_g\right)\right]^2}{2B^2}\right]^{3/2}\right]$$

where $\overline{E}_{FBHD}$ is the Fermi energy in this case and

$$T_{42HD}\left(n,\bar{E}_{FBHD},\eta_g\right)\equiv\sum_{r=1}^{s}\left[L(r)T_{41HD}\left(n,\bar{E}_{FBHD},\eta_g\right)\right]$$

The ER in this case can be written as

$$\frac{D}{\mu}=Real\,part\,of\left[\frac{n_0}{|e|}\left[\frac{\partial n_0}{\partial(\overline{E}_{FBHD}-e'')}\right]^{-1}\right] \tag{8.13}$$

where e'' is the Landau sub-band energy under cross-fields configuration.

Thus using (8.12), (8.13) and the allied definitions we can study the ER in this case.

8.2.2 The ER in HD Kane Type III-V Semiconductors Under Cross-Fields Configuration

(a) Under the conditions $\delta=0$, $\Delta_{||}=\Delta_\perp=\Delta$ and $m_{||}^*=m_\perp^*=m_c$, (8.7) assumes the form

$$T_{33}\left(E,\eta_g\right)=\left(n+\frac{1}{2}\right)\hbar\omega_0+\frac{[\hbar k_z(E)]^2}{2m_c}$$
$$-\frac{E_0}{B}\hbar k_y\left\{T_{33}\left(E,\eta_g\right)\right\}'-\frac{m_cE_0^2\left[\left\{T_{33}\left(E,\eta_g\right)\right\}'\right]^2}{2B^2} \tag{8.14a}$$

where

$$\mathrm{T}_{33}(\mathrm{E},\eta_g)=\mathrm{T}_{31}(\mathrm{E},\eta_g)+\mathrm{i}\mathrm{T}_{32}(\mathrm{E},\eta_g)$$

The use of (8.14a) leads to the expressions of the EEMs' along z and y directions as

$$m_z^*(\bar{E}_{FBHD},\eta_g,n,E_0)=m_c\,Real\,part\,of\,[\{T_{33}(\bar{E}_{FBHD},\eta_g)\}'' + \frac{m_cE_0^2\{T_{33}(\bar{E}_{FBHD},\eta_g)\}'\{T_{33}(\bar{E}_{FBHD},\eta_g)\}''}{B^2}] \quad (8.14b)$$

$$m_y^*(\bar{E}_{FBHD},\eta_g,n,E_0)=\left(\frac{B}{E_0}\right)^2 Real\ part\ of\ \left[\left\{T_{33}(\bar{E}_{FBHD},\eta_g)\right\}'\right]^{-1} \left[T_{33}(\bar{E}_{FBHD},\eta_g)-\left(n+\frac{1}{2}\right)\hbar\omega_0+\frac{m_cE_0^2\left[\{T_{33}(\bar{E}_{FBHD},\eta_g)\}'\right]^2}{2B^2}\right] \left[\frac{-\{T_{33}(\bar{E}_{FBHD},\eta_g)\}''}{[\{T_{33}(\bar{E}_{FBHD},\eta_g)\}']^2}\left[T_{33}(\bar{E}_{FBHD},\eta_g)-\left(n+\frac{1}{2}\right)\hbar\omega_0 +\frac{m_cE_0^2\left[\{T_{33}(\bar{E}_{FBHD},\eta_g)\}'\right]^2}{2B^2}\right]+1+\frac{m_cE_0^2\{T_{33}(\bar{E}_{FBHD},\eta_g)\}''}{B^2}\right] \quad (8.14c)$$

The Landau energy $(\bar{E}_{n_1})$ can be written as

$$T_{33}(\bar{E}_{n_1},\eta_g)=\left(n+\frac{1}{2}\right)\hbar\omega_0-\frac{m_cE_0^2\left[\{T_{33}(\bar{E}_{n_1},\eta_g)\}'\right]^2}{2B^2} \quad (8.15)$$

The electron concentration in this case assumes the form

$$n_0=\frac{2g_vB\sqrt{2m_c}}{3L_x\pi^2\hbar^2E_0}\,Real\,part\,of\sum_{n=0}^{n_{\max}}\left[T_{43HD}(n,\bar{E}_{FB},\eta_g)+T_{44HD}(n,\bar{E}_{FB},\eta_g)\right] \quad (8.16)$$

where,

$$T_{43HD}(n,\bar{E}_{FBHD},\eta_g)\equiv\left[\left[T_{33}(\bar{E}_{FB},\eta_g)-\left(n+\frac{1}{2}\right)\hbar\omega_0-\frac{m_cE_0^2}{2B^2}\left[\{T_{33}(\bar{E}_{FB},\eta_g)\}'\right]^2 +|e|E_0L_x\left[\{T_{33}(\bar{E}_{FB},\eta_g)\}'\right]\right]^{3/2}-\left[T_{33}(\bar{E}_{FB},\eta_g)-\left(n+\frac{1}{2}\right)\hbar\omega_0 -\frac{m_cE_0^2}{2B^2}\left[\{T_{33}(\bar{E}_{FB},\eta_g)\}'\right]^2\right]^{3/2}\right]\frac{1}{\left[\{T_{33}(\bar{E}_{FB},\eta_g)\}'\right]}$$

and

$$T_{44HD}(n,\bar{E}_{FBHD},\eta_g)\equiv\sum_{r=1}^{s}\left[L(r)T_{43HD}(n,\bar{E}_{FBHD},\eta_g)\right].$$

Thus using (8.13), (8.16) and the allied definitions we can study the ER in this case.

(b) Under the condition $\Delta \gg E_g$, (8.14a–8.14c) assumes the form

$$\gamma_2(E,\eta_g) = \left(n+\frac{1}{2}\right)\hbar\omega_0 - \frac{E_0}{B}\hbar k_y \gamma_2'(E,\eta_g) - \frac{m_c E_0^2}{2B^2}\left(\gamma_2'(E,\eta_g)\right)^2 + \frac{[\hbar k_z(E)]^2}{2m_c} \tag{8.17a}$$

The use of (8.17a) leads to the expressions of the EEMs' along z and y directions as

$$m_z^*(\bar{E}_{FBHD},\eta_g,n,E_0) = m_c\left[\{\gamma_2(\bar{E}_{FBHD},\eta_g)\}'' + \frac{m_c E_0^2\{\gamma_2(\bar{E}_{FBHD},\eta_g)\}'\{\gamma_2(\bar{E}_{FBHD},\eta_g)\}''}{B^2}\right] \tag{8.17b}$$

$$\begin{aligned} m_y^*(\bar{E}_{FBHD},\eta_g,n,E_0) = {} & \left(\frac{B}{E_0}\right)^2 \frac{1}{[\{\gamma_2(\bar{E}_{FBHD},\eta_g)\}']}\left[\gamma_2(\bar{E}_{FBHD},\eta_g) - \left(n+\frac{1}{2}\right)\hbar\omega_0\right. \\ & \left. + \frac{m_c E_0^2[\{\gamma_2(\bar{E}_{FBHD},\eta_g)\}']^2}{2B^2}\right]\left[\frac{-\{\gamma_2(\bar{E}_{FBHD},\eta_g)\}''}{\left[\{\gamma_2(\bar{E}_{FBHD},\eta_g)\}'\right]^2}\left[\gamma_2(\bar{E}_{FBHD},\eta_g).\right.\right. \\ & \left. - \left(n+\frac{1}{2}\right)\hbar\omega_0 + \frac{m_c E_0^2[\{\gamma_2(\bar{E}_{FBHD},\eta_g)\}']^2}{2B^2}\right] \\ & \left. + 1 + \frac{m_c E_0^2\{\gamma_2(\bar{E}_{FBHD},\eta_g)\}''}{B^2}\right] \end{aligned} \tag{8.17c}$$

The Landau energy $(\bar{E}_{n_2})$ can be written as

$$\gamma_2(\bar{E}_{n_2},\eta_g) = \left(n+\frac{1}{2}\right)\hbar\omega_0 - \frac{m_c E_0^2}{2B^2}\left(\gamma_2'(\bar{E}_{n_2},\eta_g)\right)^2 \tag{8.18}$$

The expressions for n_0 in this case assume the forms

$$n_0 = \frac{2g_v B\sqrt{2m_c}}{3L_x \pi^2\hbar^2 E_0}\sum_{n=0}^{n_{max}}\left[T_{47HD}\left(n,\bar{E}_{FBHD},\eta_g\right) + T_{48HD}\left(n,\bar{E}_{FBHD},\eta_g\right)\right] \tag{8.19}$$

where

$$\begin{aligned} T_{47HD}(n,\bar{E}_{FBHD},\eta_g) \equiv {} & \left[\left[\gamma_2(\bar{E}_{FBHD},\eta_g) - \left(n+\frac{1}{2}\right)\hbar\omega_0 + |e|E_0 L_x\left(\gamma_2'(\bar{E}_{FBHD},\eta_g)\right)\right.\right. \\ & \left. - \frac{m_c E_0^2}{2B^2}\left(\gamma_2'(\bar{E}_{FBHD},\eta_g)\right)^2\right]^{3/2} - \left[\left(\gamma_2(\bar{E}_{FBHD},\eta_g)\right) - \left(n+\frac{1}{2}\right)\hbar\omega_0\right. \\ & \left.\left. - \frac{m_c E_0^2}{2B^2}\left(\gamma_2'(\bar{E}_{FBHD},\eta_g)\right)^2\right]^{3/2}\right]\left[\gamma_2'(\bar{E}_{FBHD},\eta_g)\right]^{-1} \end{aligned}$$

and

$$T_{48HD}\left(n,\bar{E}_{FBHD},\eta_g\right)\equiv\sum_{r=0}^{s}L(r)\left[T_{47HD}\left(n,\bar{E}_{FBHD},\eta_g\right)\right].$$

Thus using (8.13), (8.19) and the allied definitions we can study the ER in this case.

(c) For $\alpha\to 0$ and we can write,

$$\gamma_3(E,\eta_g)=\left(n+\frac{1}{2}\right)\hbar\omega_0-\frac{E_0}{B}\hbar k_y\gamma_3'(E,\eta_g)-\frac{m_cE_0^2}{2B^2}\left(\gamma_3'(E,\eta_g)\right)^2+\frac{\left[\hbar k_z(E)\right]^2}{2m_c} \tag{8.20a}$$

The use of (8.20a) leads to the expressions of the EEMs' along z and y directions as

$$m_z^*(\bar{E}_{FBHD},\eta_g,n,E_0)=m_c\left[\left\{\gamma_3(\bar{E}_{FBHD},\eta_g)\right\}''+\frac{m_cE_0^2\left\{\gamma_3(\bar{E}_{FBHD},\eta_g)\right\}'\left\{\gamma_3(\bar{E}_{FBHD},\eta_g)\right\}''}{B^2}\right] \tag{8.20b}$$

$$\begin{aligned}m_y^*(\bar{E}_{FBHD},\eta_g,n,E_0)=&\left(\frac{B}{E_0}\right)^2\frac{1}{\left[\left\{\gamma_3(\bar{E}_{FBHD},\eta_g)\right\}'\right]}\left[\gamma_3(\bar{E}_{FBHD},\eta_g)-\left(n+\frac{1}{2}\right)\hbar\omega_0.\right.\\&+\frac{m_cE_0^2\left[\left\{\gamma_3(\bar{E}_{FBHD},\eta_g)\right\}'\right]^2}{2B^2}\Bigg]\left[\frac{-\left\{\gamma_3(\bar{E}_{FBHD},\eta_g)\right\}''}{\left[\left\{\gamma_3(\bar{E}_{FBHD},\eta_g)\right\}'\right]^2}\left[\gamma_3(\bar{E}_{FBHD},\eta_g)-\left(n+\frac{1}{2}\right)\hbar\omega_0.\right.\right.\\&\left.+\frac{m_cE_0^2\left[\left\{\gamma_3(\bar{E}_{FBHD},\eta_g)\right\}'\right]^2}{2B^2}\right]+1+\frac{m_cE_0^2\left\{\gamma_3(\bar{E}_{FBHD},\eta_g)\right\}''}{B^2}\Bigg]\end{aligned} \tag{8.20c}$$

The Landau energy $(\bar{E}_{n_3})$ can be written as

$$\gamma_3(\bar{E}_{n_3},\eta_g)=\left(n+\frac{1}{2}\right)\hbar\omega_0-\frac{m_cE_0^2}{2B^2}\left(\gamma_3'(\bar{E}_{n_3},\eta_g)\right)^2 \tag{8.21}$$

The expressions for n_0 in this case assume the forms

$$n_0=\frac{2g_vB\sqrt{2m_c}}{3L_x\pi^2\hbar^2E_0}\sum_{n=0}^{n_{\max}}\left[T_{49HD}\left(n,\bar{E}_{FBHD},\eta_g\right)+T_{50HD}\left(n,\bar{E}_{FBHD},\eta_g\right)\right] \tag{8.22}$$

where

$$T_{49HD}(n, \bar{E}_{FBHD}, \eta_g) \equiv \left[\left[\gamma_3(\bar{E}_{FBHD}, \eta_g) - \left(n+\frac{1}{2}\right)\hbar\omega_0 + |e|E_0 L_x(\gamma_3'(\bar{E}_{FBHD}, \eta_g)) - \frac{m_c E_0^2}{2B^2}(\gamma_3'(\bar{E}_{FBHD}, \eta_g))^2\right]^{3/2} - \left[(\gamma_3(\bar{E}_{FBHD}, \eta_g)) - \left(n+\frac{1}{2}\right)\hbar\omega_0 - \frac{m_c E_0^2}{2B^2}(\gamma_3'(\bar{E}_{FBHD}, \eta_g))^2\right]^{3/2}\right]\left[\gamma_3'(\bar{E}_{FBHD}, \eta_g)\right]^{-1}$$

and

$$T_{50HD}(n, \bar{E}_{FBHD}, \eta_g) \equiv \sum_{r=0}^{s} L(r)\left[T_{49HD}(n, \bar{E}_{FBHD}, \eta_g)\right].$$

Thus using (8.13), (8.22) and the allied definitions we can study the ER in this case.

8.2.3 The ER in HD II-VI Semiconductors Under Cross-Fields Configuration

The electron energy spectrum in HD II–VI semiconductors in the presence of electric field E_0 along x direction and quantizing magnetic field B along z direction can approximately be written as

$$\gamma_3(E, \eta_g) = \beta_1(n, E_0) - \frac{E_0}{B}\hbar k_y \gamma_3'(E, \eta_g) - \frac{m_\parallel^* E_0^2}{2B^2}(\gamma_3'(E, \eta_g))^2 + \frac{[\hbar k_z(E)]^2}{2m_\parallel^*} \quad (8.23a)$$

where

$$\beta_1(n, E_0) \equiv \left[\left(n+\frac{1}{2}\right)\hbar\omega_{02} - \left(\frac{E_0^2 m_\perp^*}{2B^2}\right) + D\left\{\left(n+\frac{1}{2}\right)\hbar\omega_{02} + \left(\frac{E_0^2 m_\perp^*}{2B^2}\right)\right\}^{1/2}\right], \omega_{02} \equiv \frac{|e|B}{m_\perp^*}.$$

and

$$D \equiv \pm\frac{\bar{\lambda}_0\sqrt{2m_\perp^*}}{\hbar}$$

The use of (8.23a) leads to the expressions of the EEMs' along z and y directions as

$$m_z^*(\bar{E}_{FBHD}, \eta_g, n, E_0) = m_\parallel^*\left[\{\gamma_3(\bar{E}_{FBHD}, \eta_g)\}'' + \frac{m_\parallel^* E_0^2\{\gamma_3(\bar{E}_{FBHD}, \eta_g)\}'\{\gamma_3(\bar{E}_{FBHD}, \eta_g)\}''}{B^2}\right] \quad (8.23b)$$

$$m_y^*(\bar{E}_{FBHD},\eta_g,n,E_0)=\left(\frac{B}{E_0}\right)^2\frac{1}{[\{\gamma_3(\bar{E}_{FBHD},\eta_g)\}']}\left[\gamma_3(\bar{E}_{FBHD},\eta_g)-\beta_1(n,E_0)\right.$$
$$\left.+\frac{m_\parallel^*E_0^2[\{\gamma_3(\bar{E}_{FBHD},\eta_g)\}']^2}{2B^2}\right]\left[\frac{-\{\gamma_3(\bar{E}_{FBHD},\eta_g)\}''}{[\{\gamma_3(\bar{E}_{FBHD},\eta_g)\}']^2}\left[\gamma_3(\bar{E}_{FBHD},\eta_g)\right.\right.$$
$$\left.\left.-\beta_1(n,E_0)+\frac{m_\parallel^*E_0^2[\{\gamma_3(\bar{E}_{FBHD},\eta_g)\}']^2}{2B^2}\right]+1+\frac{m_\parallel^*E_0^2\{\gamma_3(\bar{E}_{FBHD},\eta_g)\}''}{B^2}\right] \tag{8.23c}$$

The Landau energy $(\bar{E}_{n_4})$ can be written as

$$\gamma_3(\bar{E}_{n_4},\eta_g)=\beta_1(n,E_0)-\frac{m_\parallel^*E_0^2}{2B^2}\left(\gamma_3'(\bar{E}_{n_4},\eta_g)\right)^2 \tag{8.24}$$

The expression for n_0 in this case assumes the form

$$n_0=\frac{2g_vB\sqrt{2m_\parallel^*}}{3L_x\pi^2\hbar^2E_0}\sum_{n=0}^{n_{\max}}\left[T_{53HD}(n,\bar{E}_{FBHD},\eta_g)+T_{54HD}(n,\bar{E}_{FBHD},\eta_g)\right] \tag{8.25}$$

where

$$T_{53HD}(n,\bar{E}_{FBHD},\eta_g)\equiv\left[\left[\gamma_3(\bar{E}_{FBHD},\eta_g)-\beta_1(n,E_0)+|e|E_0L_x\left(\gamma_3'(\bar{E}_{FBHD},\eta_g)\right)\right.\right.$$
$$\left.-\frac{m_\parallel^*E_0^2}{2B^2}\left(\gamma_3'(\bar{E}_{FBHD},\eta_g)\right)^2\right]^{3/2}-\left[\left(\gamma_3(\bar{E}_{FBHD},\eta_g)\right)-\beta_1(n,E_0)\right.$$
$$\left.\left.-\frac{m_\parallel^*E_0^2}{2B^2}\left(\gamma_3'(\bar{E}_{FBHD},\eta_g)\right)^2\right]^{3/2}\right]\left[\gamma_3'(\bar{E}_{FBHD},\eta_g)\right]^{-1}$$

and

$$T_{54HD}(n,\bar{E}_{FBHD},\eta_g)\equiv\sum_{r=0}^{s}L(r)\left[T_{53HD}(n,\bar{E}_{FBHD},\eta_g)\right].$$

Thus using (8.13), (8.25) and the allied definitions we can study the ER in this case.

8.2.4 The ER in HD IV-VI Semiconductors Under Cross-Fields Configuration

The (3.68) can be written as

$$\frac{p_s^2}{2M_1^*(E,\eta_g)} + \frac{p_z^2}{2M_3^*(E,\eta_g)} = g(E,\eta_g) \tag{8.26a}$$

where

$$M_1^*(E,\eta_g) = \left[\frac{(\bar{R})^2}{E_g}\{c_1(\alpha_1,E,E_g) - iD_1(\alpha_1,E,E_g)\} + \frac{(\bar{S})^2}{\Delta_c'}\{c_2(\alpha_2,E,E_g) - iD_2(\alpha_2,E,E_g)\} + \frac{(\bar{Q})^2}{\Delta_c''}\{c_3(\alpha_3,E,E_g) - iD_3(\alpha_3,E,E_g)\}\right]^{-1}$$

$$M_3^*(E,\eta_g) = \left[\frac{2(\bar{A})^2}{E_g}\{c_1(\alpha_1,E,E_g) - iD_1(\alpha_1,E,E_g)\} + \frac{(\bar{S}+\bar{Q})^2}{\Delta_c''}\{c_3(\alpha_3,E,E_g) - iD_3(\alpha_3,E,E_g)\}\right]^{-1}$$

and

$$g^*(E,\eta_g) = 2\hbar^2\gamma_0(E,\eta_g)$$

In the presence of quantizing magnetic field B along z direction and the electric field along x-axis, from above equation one obtains

$$\frac{\hat{p}_x^2}{2M_1^*(E,\eta_g)} + \frac{(\hat{p}_y - |e|B\hat{x})^2}{2M_1^*(E,\eta_g)} + \frac{\hat{p}_z^2}{2M_3^*(E,\eta_g)} = g^*(E,\eta_g) + |e|E_0\hat{x}\rho_1^*(E,\eta_g) \tag{8.26b}$$

where

$$\rho_1^*(E,\eta_g) = \frac{\partial}{\partial E}\left[g^*(E,\eta_g)\right]$$

Let us define the operator $\hat{\theta}$ as

$$\hat{\theta} = -\hat{p}_y + |e|B\hat{x} - \frac{\rho_1^*(E,\eta_g)E_0\left[M_1^*(E,\eta_g)\right]}{B} \tag{8.27}$$

Eliminating $\hat{x}$, between the above two equations, the dispersion relation of the conduction electrons in HD stressed Kane type semiconductors in the presence of cross fields configuration can be expressed as

$$g^*(E,\eta_g) = \left(n+\frac{1}{2}\right)\hbar\overline{\omega_{i1}}(E,\eta_g) + \frac{\hbar^2 k_z^2}{2M_3^*(E,\eta_g)} - \frac{E_0}{B}\rho_1^*(E,\eta_g)\hbar k_y - \frac{E_0^2}{2B^2}\left[\rho_1^*(E,\eta_g)\right]^2 M_1^*(E,\eta_g) \tag{8.28a}$$

where

$$\overline{\omega_{i1}}(E,\eta_g) = eB[M_1^*(E,\eta_g)]^{-1}$$

The use of (8.28a) leads to the expressions of the EEMs' along z and y directions as

$$\begin{aligned} m_z^*(\bar{E}_{FBHD},\eta_g,n,E_0) = {} & Real\,part\,of\Big[\left[M_3^*(\bar{E}_{FBHD},\eta_g)\right]'\Big[g^*(\bar{E}_{FBHD},\eta_g) - \left(n+\frac{1}{2}\right)\hbar\overline{\omega_{i1}}(\bar{E}_{FBHD},\eta_g) \\ & + \frac{E_0^2}{2B^2}\left[\rho_1^*(\bar{E}_{FBHD},\eta_g)\right]^2 M_1^*(\bar{E}_{FBHD},\eta_g)\Big] + \left[M_3^*(\bar{E}_{FBHD},\eta_g)\right]\Big[\left[g^*(\bar{E}_{FBHD},\eta_g)\right]' \\ & - (n+\frac{1}{2})\hbar\left[\overline{\omega_{i1}}(\bar{E}_{FBHD},\eta_g)\right]' + \frac{E_0^2}{2B^2}\Big[\left[\rho_1^*(\bar{E}_{FBHD},\eta_g)\right]^2\left[M_1^*(\bar{E}_{FBHD},\eta_g)\right]'\Big] \\ & + 2\left[M_3^*(\bar{E}_{FBHD},\eta_g)\right]\left[\rho_1^*(\bar{E}_{FBHD},\eta_g)\right]\left[\rho_1^*(\bar{E}_{FBHD},\eta_g)\right]'\Big]\Big] \end{aligned} \tag{8.28b}$$

and

$$\begin{aligned} m_y^*(\bar{E}_{FBHD},\eta_g,n,E_0) = {} & (B/E_0)^2\ Real\ part\ of\ [[\rho_1^*(\bar{E}_{FBHD},\eta_g)]^{-3}[g^*(\bar{E}_{FBHD},\eta_g) - \left(n+\frac{1}{2}\right)\hbar\overline{\omega_{i1}}(\bar{E}_{FBHD},\eta_g) \\ & + \frac{E_0^2}{2B^2}[\rho_1^*(\bar{E}_{FBHD},\eta_g)]^2 M_1^*(\bar{E}_{FBHD},\eta_g)][[\rho_1^*(\bar{E}_{FBHD},\eta_g)][[g^*(\bar{E}_{FBHD},\eta_g)]' \\ & - \left(n+\frac{1}{2}\right)\hbar[\overline{\omega_{i1}}(\bar{E}_{FBHD},\eta_g)]' + \frac{E_0^2}{2B^2}[[\rho_1^*(\bar{E}_{FBHD},\eta_g)]^2[M_1^*(\bar{E}_{FBHD},\eta_g)]'] \\ & - [\rho_1^*(\bar{E}_{FBHD},\eta_g)]'[g^*(\bar{E}_{FBHD},\eta_g) - \left(n+\frac{1}{2}\right)\hbar\overline{\omega_{i1}}(\bar{E}_{FBHD},\eta_g) \\ & + \frac{E_0^2}{2B^2}[\rho_1^*(\bar{E}_{FBHD},\eta_g)]^2 M_1^*(\bar{E}_{FBHD},\eta_g)]] \end{aligned} \tag{8.28c}$$

The Landau level energy (E_{n9}) in this case can be expressed through the equation

$$g^*(E_{n_9},\eta_g) = (n+\frac{1}{2})\hbar\overline{\omega_{i1}}(E_{n_9},\eta_g) - \frac{E_0^2}{2B^2}[\rho_1^*(E_{n9},\eta_g)]^2 M_1^*(E_{n9},\eta_g) \tag{8.28d}$$

The electron concentration can be written as

$$n_0 = \frac{2B}{3L_x\pi^2\hbar^2 E_0}\ Real\,part\,of \sum_{n=0}^{n_{\max}}\left[T_{4131HD}(n,\overline{E}_{FBHD},\eta_g) + T_{4141HD}(n,\overline{E}_{FBHD},\eta_g)\right] \tag{8.28e}$$

where

$$T_{4131HD}(n,\overline{E}_{FBHD},\eta_g)=\left[\frac{\sqrt{2M_3^*(\overline{E}_{FBHD},\eta_g)}}{\rho_1^*(\overline{E}_{FBHD},\eta_g)}\right]\left[[T_{51}(n,\overline{E}_{FBHD},\eta_g)\right.$$
$$\left.+\frac{E_0}{B}\rho_1^*(\overline{E}_{FBHD},\eta_g)\hbar x_{hHD1}(\overline{E}_{FBHD},\eta_g)\rho_1^*(\overline{E}_{FBHD},\eta_g)\right]^{\frac{3}{2}}$$
$$-\left[T_{51}(n,\overline{E}_{FBHD},\eta_g)+\frac{E_0}{B}\rho_1^*(\overline{E}_{FBHD},\eta_g)\right.$$
$$\left.\hbar x_{lHD1}(\overline{E}_{FBHD},\eta_g)\rho_1^*(\overline{E}_{FBHD},\eta_g)]^{\frac{3}{2}}\right],$$

$$T_{51}(n,\overline{E}_{FBHD},\eta_g)=\left[g^*(\overline{E}_{FBHD},\eta_g)-\left(n+\frac{1}{2}\right)\hbar\overline{\omega_{i1}}(\overline{E}_{FBHD},\eta_g)\right.$$
$$\left.+\frac{M_1^*(\overline{E}_{FBHD},\eta_g)E_0^2}{2B^2}[\rho_1^*(\overline{E}_{FBHD},\eta_g)]^2\right]$$

$$x_{lHD1}(\overline{E}_{FBHD},\eta_g)=\frac{-M_1^*(\overline{E}_{FBHD},\eta_g)E_0[\rho_1^*(\overline{E}_{FBHD},\eta_g)]}{B},\quad x_{hHD1}(\overline{E}_{FBHD},\eta_g)$$
$$=\frac{|e|BL_x}{\hbar}+x_{lHD1}(\overline{E}_{FBHD},\eta_g)$$

and

$$T_{4141HD}(n,\bar{E}_{FBHD},\eta_g)\equiv\sum_{r=1}^{s}L(r)T_{4131HD}(n,\bar{E}_{FBHD},\eta_g)$$

Thus using (8.13), (8.28c) and the allied definitions we can study the ER in this case.

8.2.5 The ER in HD Stressed Semiconductors Under Cross-Fields Configuration

The use of (2.48) can be written as

$$\frac{p_x^2}{2m_1^*(E,\eta_g)}+\frac{p_y^2}{2m_2^*(E,\eta_g)}+\frac{p_z^2}{2m_3^*(E,\eta_g)}=G^*(E,\eta_g)\tag{8.28f}$$

where

$$m_1^*(E,\eta_g)=[2\hbar^2[\gamma_0(E,\eta_g)-I(1)T_{17}]]^{-1},\quad T_{17}\equiv\left[E_g-C_1\varepsilon-(\bar{a}_0+C_1)\varepsilon+\frac{3}{2}\bar{b}_0\varepsilon_{xx}-\frac{\bar{b}_0}{2}\varepsilon+\left(\sqrt{3}/2\right)\varepsilon_{xy}\bar{d}_0\right],$$
$$m_2^*(E,\eta_g)=[2\hbar^2[\gamma_0(E,\eta_g)-I(1)T_{27}]]^{-1},\quad T_{27}\equiv\left[E_g-C_1\varepsilon-(\bar{a}_0+C_1)\varepsilon+\frac{3}{2}\bar{b}_0\varepsilon_{xx}-\frac{\bar{b}_0}{2}\varepsilon-\left(\sqrt{3}/2\right)\varepsilon_{xy}\bar{d}_0\right],$$
$$m_3^*(E,\eta_g)=[2\hbar^2[\gamma_0(E,\eta_g)-I(1)T_{37}]]^{-1},\quad T_{37}\equiv\left[E_g-C_1\varepsilon-(\bar{a}_0+C_1)\varepsilon+\frac{3}{2}\bar{b}_0\varepsilon_{zz}-\frac{\bar{b}_0}{2}\varepsilon\right]$$

and the other symbols are written in (1.196a) of Chap. 1.

In the presence of quantizing magnetic field B along z direction and the electric field along x-axis, from (8.28d) one obtains

$$\frac{\hat{p}_x^2}{2m_1^*(E,\eta_g)} + \frac{(\hat{p}_y - |e|B\hat{x})^2}{2m_2^*(E,\eta_g)} + \frac{\hat{p}_z^2}{2m_3^*(E,\eta_g)} = G^*(E,\eta_g) + |e|E_0\hat{x}\left[\frac{m_1^*(E,\eta_g)}{m_2^*(E,\eta_g)}\right]^{\frac{1}{2}}\rho^*(E,\eta_g) \tag{8.29}$$

where

$$\rho^*(E,\eta_g) = \frac{\partial}{\partial E}[G^*(E,\eta_g)]$$

Let us define the operator $\hat{\theta}$ as

$$\hat{\theta} = -\hat{p}_y + |e|B\hat{x} - \frac{\rho^*(E,\eta_g)E_0\left[m_1^*(E,\eta_g)m_2^*(E,\eta_g)\right]^{\frac{1}{2}}}{B} \tag{8.30}$$

Eliminating $\hat{x}$, between the above two equations, the dispersion relation of the conduction electrons in HD stressed Kane type semiconductors in the presence of cross fields configuration can be expressed as

$$G^*(E,\eta_g) = \left(n+\frac{1}{2}\right)\hbar\overline{\omega_i}(E,\eta_g) + \frac{\hbar^2k_z^2}{2m_3^*(E,\eta_g)} - \frac{E_0}{B}\rho^*(E,\eta_g)\left[\frac{m_1^*(E,\eta_g)}{m_2^*(E,\eta_g)}\right]^{\frac{1}{2}}\hbar k_y - \frac{E_0^2}{2B^2}\left[\rho^*(E,\eta_g)\right]^2 m_1^*(E,\eta_g) \tag{8.31a}$$

where

$$\overline{\omega_i}(E,\eta_g) = eB\left[m_1^*(E,\eta_g)m_2^*(E,\eta_g)\right]^{-\frac{1}{2}}$$

The use of (8.31a) leads to the expressions of the EEMs' along z and y directions as

$$m_z^*(\bar{E}_{FBHD},\eta_g,n,E_0) = [[m_3^*(\bar{E}_{FBHD},\eta_g)]'[G^*(\bar{E}_{FBHD},\eta_g) - \left(n+\frac{1}{2}\right)\hbar\overline{\omega_i}(\bar{E}_{FBHD},\eta_g)$$
$$+\frac{E_0^2}{2B^2}[\rho^*(\bar{E}_{FBHD},\eta_g)]^2 m_1^*(\bar{E}_{FBHD},\eta_g)]$$
$$+[m_3^*(\bar{E}_{FBHD},\eta_g)][[G^*(\bar{E}_{FBHD},\eta_g)]' - (n+\frac{1}{2})\hbar[\overline{\omega_i}(\bar{E}_{FBHD},\eta_g)]'$$
$$+\frac{E_0^2}{2B^2}[2[\rho^*(\bar{E}_{FBHD},\eta_g)][\rho^*(\bar{E}_{FBHD},\eta_g)]'[m_1^*(\bar{E}_{FBHD},\eta_g)]$$
$$+[m_1^*(\bar{E}_{FBHD},\eta_g)]'[\rho^*(\bar{E}_{FBHD},\eta_g)]^2]]] \tag{8.31b}$$

$$m_y^*(\bar{E}_{FBHD},\eta_g,n,E_0) = (B/E_0)^2\,[m_4^*(\bar{E}_{FBHD},\eta_g)]^{-3}[G^*(\bar{E}_{FBHD},\eta_g)$$
$$-\left(n+\frac{1}{2}\right)\hbar\overline{\omega_i}(\bar{E}_{FBHD},\eta_g)$$
$$+\frac{E_0^2}{2B^2}[\rho^*(\bar{E}_{FBHD},\eta_g)]^2 m_1^*(\bar{E}_{FBHD},\eta_g)][[m_4^*(\bar{E}_{FBHD},\eta_g)]$$
$$[[G^*(\bar{E}_{FBHD},\eta_g)]' - \left(n+\frac{1}{2}\right)\hbar[\overline{\omega_i}(\bar{E}_{FBHD},\eta_g)]'$$
$$+\frac{E_0^2}{2B^2}[[\rho^*(\bar{E}_{FBHD},\eta_g)]^2[m_1^*(\bar{E}_{FBHD},\eta_g)]']$$
$$-[m_4^*(\bar{E}_{FBHD},\eta_g)]'[G^*(\bar{E}_{FBHD},\eta_g) - (n+\frac{1}{2})\hbar\overline{\omega_i}(\bar{E}_{FBHD},\eta_g)$$
$$+\frac{E_0^2}{2B^2}[\rho^*(\bar{E}_{FBHD},\eta_g)]^2 m_1^*(\bar{E}_{FBHD},\eta_g)]] \tag{8.31c}$$

where

$$m_4^*(\bar{E}_{FBHD},\eta_g) = \left[[\rho^*(\bar{E}_{FBHD},\eta_g)]\left[\frac{m_1^*(\bar{E}_{FBHD},\eta_g)}{m_2^*(\bar{E}_{FBHD},\eta_g)}\right]^{\frac{1}{2}}\right]$$

The Landau level energy (E_{n_8})in this case can be expressed through the equation

$$G^*(E_{n_8},\eta_g) = \left(n+\frac{1}{2}\right)\hbar\overline{\omega_i}(E_{n_8},\eta_g) - \frac{E_0^2}{2B^2}[\rho^*(E_{n_8},\eta_g)]^2 m_1^*(E_{n_8},\eta_g) \tag{8.31d}$$

The electron concentration can be written as

$$n_0 = \frac{2B}{3L_x\pi^2\hbar^2 E_0}\sum_{n=0}^{n_{\max}}\left[T_{413HD}(n,\overline{E}_{FBHD},\eta_g) + T_{414HD}(n,\overline{E}_{FBHD},\eta_g)\right] \tag{8.31e}$$

where

$$\begin{aligned}
T_{413HD}(n,\overline{E}_{FBHD},\eta_g) = {} & [\frac{\sqrt{2m_3^*(\overline{E}_{FBHD},\eta_g)}}{\rho^*(\overline{E}_{FBHD},\eta_g)}][[T_5(n,\overline{E}_{FBHD},\eta_g) \\
& + \frac{E_0}{B}\rho^*(\overline{E}_{FBHD},\eta_g)\hbar x_{hHD}(\overline{E}_{FBHD},\eta_g)\rho^*(\overline{E}_{FBHD},\eta_g)]^{\frac{3}{2}} \\
& - [T_5(n,\overline{E}_{FBHD},\eta_g) + \frac{E_0}{B}\rho^*(\overline{E}_{FBHD},\eta_g)\hbar x_{lHD}(\overline{E}_{FBHD},\eta_g) \\
& \rho^*(\overline{E}_{FBHD},\eta_g)]^{\frac{3}{2}}],
\end{aligned}$$

$$\begin{aligned}
T_5(n,\overline{E}_{FBHD},\eta_g) = {} & [G^*(\overline{E}_{FBHD},\eta_g) - \left(n+\frac{1}{2}\right)\hbar\overline{\omega}_l(\overline{E}_{FBHD},\eta_g) \\
& + \frac{m_1^*(\overline{E}_{FBHD},\eta_g)E_0^2}{2B^2}[\rho^*(\overline{E}_{FBHD},\eta_g)]^2]
\end{aligned}$$

$$x_{lHD}(\overline{E}_{FBHD},\eta_g) = \frac{-m_1^*(\overline{E}_{FBHD},\eta_g)E_0[\rho^*(\overline{E}_{FBHD},\eta_g)]}{B},$$

$$x_{hHD}(\overline{E}_{FBHD},\eta_g) = \frac{|e|BL_x}{\hbar} + x_{lHD}(\overline{E}_{FBHD},\eta_g)$$

and

$$T_{414HD}(n,\bar{E}_{FBHD},\eta_g) \equiv \sum_{r=1}^{s} L(r)T_{413HD}(n,\bar{E}_{FBHD},\eta_g)$$

Thus using (8.13), (8.31e) and the allied definitions we can study the ER in this case.

8.2.6 The ER in Ultrathin Films of HD Nonlinear Optical Semiconductors Under Cross-Fields Configuration

The dispersion relation of the conduction electrons in HD ultrathin films of nonlinear optical material in the presence of cross-fields configuration can be written as

$$T_{22}(E,\eta_g) = \left[\left(\left(n+\frac{1}{2}\right)\hbar\omega_{01}\right) + \left(\frac{[\hbar]^2}{2a(E,\eta_g)}\right)\left(\frac{\pi n_z}{d_z}\right)^2 - \left(\frac{E_o\hbar k_y\rho(E,\eta_g)}{B}\right) - \left(\frac{M_\perp\rho^2(E,\eta_g)E_o^2}{2B^2}\right)\right] \tag{8.31f}$$

The use of (8.31f) leads to the expression of EEM along y direction as

$$m_y^*(e_{fA1},\eta_g,n,E_0,n_z) = Real\ part\ of\ (B/E_0)^2 T_{49}(e_{fA1},\eta_g,n_z)[T_{49}(e_{fA1},\eta_g,n_z)]' \tag{8.31g}$$

where e_{fA1} is the Fermi energy in this case and

$$T_{49}(e_{fA1},\eta_g,n_z) = [T_{22}(e_{fA1},\eta_g)] - \left[\left(\left(n+\frac{1}{2}\right)\hbar\omega_{01}\right) + \left(\frac{[\hbar]^2}{2a(e_{fA1},\eta_g)}\right)\left(\frac{\pi n_z}{d_z}\right)^2 - \left(\frac{M_\perp\rho^2(e_{fA1},\eta_g)E_o^2}{2B^2}\right)\right][\rho(e_{fA1},\eta_g)]^{-1}$$

The investigation of the ER in this case requires an expression of electron statistics which, in turn, can be written as

$$n_0 = \frac{g_v eB}{\pi\hbar}\ Real\, part\ of \left[\sum_{n=0}^{n_{\max}}\sum_{n_z=1}^{n_{z\max}} F_{-1}(\eta_{A1})\right] \tag{8.32a}$$

where $\eta_{A1} = \frac{e_{fA1}-e_{A1}}{k_B T}$ and e_{A1} is defined by the following equation

$$T_{22}(e_{A1},\eta_g) = \left[\left((n+\frac{1}{2})\hbar\omega_{01}\right) + \left(\frac{[\hbar]^2}{2a(e_{A1},\eta_g)}\right)\left(\frac{\pi n_z}{d_z}\right)^2 - \left(\frac{E_o\hbar k_y\rho(e_{A1},\eta_g)}{B}\right) - \left(\frac{M_\perp\rho^2(e_{A1},\eta_g)E_o^2}{2B^2}\right)\right] \tag{8.32b}$$

The ER in this case can be written as

$$\left(\frac{D}{\mu}\right) = Real\, part\ of \left(\left(\frac{n_0}{e}\right)\left[\frac{\partial n_0}{\partial(e_{fAi}-e_{Ai})}\right]^{-1}\right) \tag{8.32c}$$

where e_{fAi} is the Fermi energy level e_{Ai} is the subband energy.

Thus using (8.32a), (8.32c) and the allied definitions we can study the ER in this case.

8.2.7 The ER in Ultrathin Films of HD Kane Type III-V, Ternary and Quaternary Semiconductors Under Cross-Fields Configuration

(a) Under the conditions $\delta = 0$, $\Delta_{\|} = \Delta_{\perp} = \Delta$ and $m_{\|}^{*} = m_{\perp}^{*} = m_c$, (8.31f) assumes the form

$$T_{33}(E,\eta_g) = \left(n+\frac{1}{2}\right)\hbar\omega_0 + \frac{\left[\hbar\frac{\pi n_z}{d_z}\right]^2}{2m_c} - \frac{E_0}{B}\hbar k_y\{T_{33}(E,\eta_g)\}' - \frac{m_c E_0^2\left[\{T_{33}(E,\eta_g)\}'\right]^2}{2B^2} \tag{8.33a}$$

The use of (8.33a) leads to the expression of EEM along y direction as

$$m_y^{*}(e_{fA2},\eta_g,n,E_0,n_z) = Real\,part\,of\,(B/E_0)^2 T_{50}(e_{fA2},\eta_g,n_z)[T_{50}(e_{fA2},\eta_g,n_z)]' \tag{8.33b}$$

where e_{fA2} is the Fermi energy in this case and

$$T_{50}(e_{fA2},\eta_g,n_z) = \left[T_{33}(e_{fA2},\eta_g) - \left(n+\frac{1}{2}\right)\hbar\omega_0 - \frac{\left[\hbar\frac{\pi n_z}{d_z}\right]^2}{2m_c} + \frac{m_c E_0^2\left[\{T_{33}(e_{fA2},\eta_g)\}'\right]^2}{2B^2}\right]\left[\{T_{33}(e_{fA2},\eta_g)\}'\right]^{-1}$$

The electron concentration is given by

$$n_0 = \frac{g_v eB}{\pi\hbar}\,Real\,part\,of\left[\sum_{n=0}^{n_{\max}}\sum_{n_z=1}^{n_{z\max}} F_{-1}(\eta_{A2})\right] \tag{8.34a}$$

where $\eta_{A2} = \frac{e_{fA2}-e_{A2}}{k_B T}$ and e_{A2} is defined by the following equation

$$T_{33}(e_{A2},\eta_g) = \left(n+\frac{1}{2}\right)\hbar\omega_0 + \frac{\left[\hbar\frac{\pi n_z}{d_z}\right]^2}{2m_c} - \frac{m_c E_0^2\left[\{T_{33}(e_{A2},\eta_g)\}'\right]^2}{2B^2} \tag{8.34b}$$

Thus using (8.32b), (8.34a, 8.34b) and the allied definitions we can study the ER in this case.

(b) HD two band model of Kane

Under the condition $\Delta \gg E_g$, (8.33a) assumes the form

$$\gamma_2(E,\eta_g) = \left(n+\frac{1}{2}\right)\hbar\omega_0 - \frac{E_0}{B}\hbar k_y \gamma_2'(E,\eta_g) - \frac{m_c E_0^2}{2B^2}\left(\gamma_2'(E,\eta_g)\right)^2 + \frac{[\hbar k_z(E)]^2}{2m_c} \tag{8.35a}$$

The use of (8.35a) leads to the expression of EEM along y direction as

$$m_y^*(\mathrm{e}_{fA3},\eta_g,n,E_0,n_z) = (B/E_0)^2 T_{51}(\mathrm{e}_{fA3},\eta_g,n_z)[T_{51}(\mathrm{e}_{fA3},\eta_g,n_z)]' \tag{8.35b}$$

where e_{fA3} is the fermi energy in this case and

$$T_{51}(\mathrm{e}_{fA3},\eta_g,n_z) = \left[\gamma_2(\mathrm{e}_{fA3},\eta_g) - \left(n+\frac{1}{2}\right)\hbar\omega_0 - \frac{\left[\hbar\frac{\pi n_z}{d_z}\right]^2}{2m_c} + \frac{m_c E_0^2\left[\{\gamma_2(\mathrm{e}_{fA3},\eta_g)\}'\right]^2}{2B^2}\right]\left[\{\gamma_2(\mathrm{e}_{fA3},\eta_g)\}'\right]^{-1}$$

The electron concentration is given by

$$n_0 = \frac{g_v eB}{\pi\hbar}\sum_{n=0}^{n_{\max}}\sum_{n_z=1}^{n_{z\max}} F_{-1}(\eta_{A3}) \tag{8.36a}$$

where $\eta_{A3} = \frac{e_{fA3}-e_{A3}}{k_B T}$ and e_{A3} is the lowest positive root of the following equation

$$\gamma_2(e_{A3},\eta_g) = \left(n+\frac{1}{2}\right)\hbar\omega_0 - \frac{m_c E_0^2}{2B^2}\left(\gamma_2'(e_{A3},\eta_g)\right)^2 \tag{8.36b}$$

Thus using (8.32c), (8.36b) and the allied definitions we can study the ER in this case.

(c) HD Parabolic energy bands

The dispersion relation, the EEM and the electron statistics for this model under this condition
$\alpha \to 0$ can be written as

$$\gamma_1(E,\eta_g) = \left(n+\frac{1}{2}\right)\hbar\omega_0 - \frac{E_0}{B}\hbar k_y \gamma_1'(E,\eta_g) - \frac{m_c E_0^2}{2B^2}\left(\gamma_1'(E,\eta_g)\right)^2 + \frac{[\hbar k_z(E)]^2}{2m_c} \tag{8.37a}$$

The use of (8.37a) leads to the expression of EEM along y direction as

$$m_y^*(\mathrm{e}_{fA4}, \eta_g, n, E_0, n_z) = (B/E_0)^2 T_{52}(\mathrm{e}_{fA4}, \eta_g, n_z)[T_{52}(\mathrm{e}_{fA4}, \eta_g, n_z)]' \quad (8.37b)$$

where e_{fA4} is the Fermi energy in this case and

$$T_{52}(\mathrm{e}_{fA4}, \eta_g, n_z) = \left[\gamma_3(\mathrm{e}_{fA4}, \eta_g) - \left(n + \frac{1}{2}\right)\hbar\omega_0 - \frac{\left[\hbar \frac{\pi n_z}{d_z}\right]^2}{2m_c} + \frac{m_c E_0^2 \left[\{\gamma_3(\mathrm{e}_{fA4}, \eta_g)\}'\right]^2}{2B^2}\right] \left[\{\gamma_3(\mathrm{e}_{fA4}, \eta_g)\}'\right]^{-1}$$

The electron concentration is given by

$$n_0 = \frac{g_v eB}{\pi\hbar} \sum_{n=0}^{n_{\max}} \sum_{n_z=1}^{n_{z\max}} F_{-1}(\eta_{A4}) \quad (8.38a)$$

where $\eta_{A4} = \frac{e_{fA4} - e_{A4}}{k_B T}$ and e_{A4} is the lowest positive root of the equation

$$\gamma_1(e_{A4}, \eta_g) = \left(n + \frac{1}{2}\right)\hbar\omega_0 - \frac{m_c E_0^2}{2B^2}\left(\gamma_1'(e_{A4}, \eta_g)\right)^2 \quad (8.38b)$$

Thus using (8.32c), (8.38b) and the allied definitions we can study the ER in this case.

8.2.8 The ER in Ultrathin Films of HD II-VI Semiconductors Under Cross-Fields Configuration

The dispersion relation in this case in ultrathin films of HD II-VI semiconductors can be written as

$$\gamma_3(E, \eta_g) = \beta_1(n, E_0) - \frac{E_0}{B}\hbar k_y \gamma_3'(E, \eta_g) - \frac{m_\parallel^* E_0^2}{2B^2}\left(\gamma_3'(E, \eta_g)\right)^2 + \frac{\left[\hbar \frac{n_z \pi}{d_z}\right]^2}{2m_\parallel^*} \quad (8.39a)$$

The use of (8.39a) leads to the expression of EEM along y direction as

$$m_y^*(\mathrm{e}_{fA5}, \eta_g, n, E_0, n_z) = (B/E_0)^2 T_{53}(\mathrm{e}_{fA5}, \eta_g, n_z)[T_{53}(\mathrm{e}_{fA5}, \eta_g, n_z)]' \quad (8.39b)$$

where e_{fA5} is the Fermi energy in this case and

$$T_{53}(e_{fA5}, \eta_g, n_z) = \left[\gamma_3(e_{fA5}, \eta_g) - \beta_1(n, E_0) - \frac{\left[\hbar \frac{\pi n_z}{d_z}\right]^2}{2m_{\|}^*} + \frac{m_{\|}^* E_0^2 \left[\{\gamma_3(e_{fA5}, \eta_g)\}'\right]^2}{2B^2}\right] \left[\{\gamma_3(e_{fA5}, \eta_g)\}'\right]^{-1}$$

The electron concentration per unit area in this case assumes the form

$$n_0 = \frac{g_v eB}{\pi\hbar} \sum_{n=0}^{n_{\max}} \sum_{n_z=1}^{n_{z\max}} F_{-1}(\eta_{A5}) \tag{8.39c}$$

where $\eta_{A5} = \frac{e_{fA5} - e_{A5}}{k_B T}$ and e_{A5} is determined from the equation

$$\gamma_3(e_{A5}, \eta_g) = \beta_1(n, E_0) - \frac{m_{\|}^* E_0^2}{2B^2} \left(\gamma_3'(e_{A5}, \eta_g)\right)^2 \tag{8.39d}$$

Thus using (8.32c), (8.39c) and the allied definitions we can study the ER in this case.

8.2.9 The ER in Ultrathin Films of HD IV-VI Semiconductors Under Cross-Fields Configuration

The dispersion relation in this case is given by

$$g^*(E, \eta_g) = \left(n + \frac{1}{2}\right)\hbar\overline{\omega_{i1}}(E, \eta_g) + \frac{\hbar^2 \left(\frac{n_z \pi}{d_z}\right)^2}{2M_3^*(E, \eta_g)} - \frac{E_0}{B}\rho_1^*(E, \eta_g)\hbar k_y - \frac{E_0^2}{2B^2}\left[\rho_1^*(E, \eta_g)\right]^2 M_1^*(E, \eta_g) \tag{8.40a}$$

The use of (8.40a) leads to the expression of EEM along y direction as

$$m_y^*(e_{fA6}, \eta_g, n, E_0, n_z) = Real\,part\ of\ (B/E_0)^2 T_{54}(e_{fA6}, \eta_g, n_z)[T_{54}(e_{fA6}, \eta_g, n_z)]' \tag{8.40b}$$

where e_{fA6} is the Fermi energy in this case and

$$T_{49}(e_{fA6},\eta_g,n_z) = [g^*(e_{fA6},\eta_g) - (n+\frac{1}{2})\hbar\overline{\omega_{i1}}(e_{fA6},\eta_g) - \frac{\hbar^2(\frac{n_z\pi}{d_z})^2}{2M_3^*(e_{fA6},\eta_g)} + \frac{E_0^2}{2B^2}[\rho_1^*(e_{fA6},\eta_g)]^2 M_1^*(e_{fA6},\eta_g)][\rho_1^*(e_{fA6},\eta_g)]^{-1}$$

The surface electron concentration in this case assumes the form

$$n_0 = \frac{g_v eB}{\pi\hbar}\, Real\,part\,of\,[\sum_{n=0}^{n_{\max}}\sum_{n_z=1}^{n_{z\max}} F_{-1}(\eta_{A6})] \tag{8.40c}$$

where $\eta_{A6} = \frac{e_{fA6}-e_{A6}}{k_B T}$ and e_{A6} is defined by the following equation

$$g^*(e_{A6},\eta_g) = \left(n+\frac{1}{2}\right)\hbar\overline{\omega_{i1}}(e_{A6},\eta_g) + \frac{\hbar^2\left(\frac{n_z\pi}{d_z}\right)^2}{2M_3^*(e_{A6},\eta_g)} - \frac{E_0^2}{2B^2}\left[\rho_1^*(e_{A6},\eta_g)\right]^2 M_1^*(e_{A6},\eta_g) \tag{8.41}$$

Thus using (8.32c), (8.40c) and the allied definitions we can study the ER in this case.

8.2.10 The ER in Ultrathin Films of HD Stressed Semiconductors Under Cross-Fields Configuration

The dispersion relation in this case assumes the form

$$G^*(E,\eta_g) = \left(n+\frac{1}{2}\right)\hbar\overline{\omega_i}(E,\eta_g) + \frac{\hbar^2\left(\frac{n_z\pi}{d_z}\right)^2}{2m_3^*(E,\eta_g)} - \frac{E_0}{B}\rho^*(E,\eta_g)\left[\frac{m_1^*(E,\eta_g)}{m_2^*(E,\eta_g)}\right]^{\frac{1}{2}}\hbar k_y - \frac{E_0^2}{2B^2}\left[\rho^*(E,\eta_g)\right]^2 m_1^*(E,\eta_g) \tag{8.42a}$$

The use of (8.42a) leads to the expression of EEM along y direction as

$$m_y^*(e_{fA7},\eta_g,n,E_0,n_z) = Real\,part\,of\;(B/E_0)^2 T_{55}(e_{fA7},\eta_g,n_z)\left[T_{55}(e_{fA7},\eta_g,n_z)\right]' \tag{8.42b}$$

where e_{fA7} is the Fermi energy in this case and

$$T_{55}(e_{fA7},\eta_g,n_z) = \left[G^*(e_{fA7},\eta_g) - \left(n+\frac{1}{2}\right)\hbar\overline{\omega_i}(e_{fA7},\eta_g). - \frac{\hbar^2\left(\frac{n_z\pi}{d_z}\right)^2}{2m_3^*(e_{fA7},\eta_g)} + \frac{E_0^2}{2B^2}\left[\rho^*(e_{fA7},\eta_g)\right]^2 m_1^*(e_{fA7},\eta_g)\right]\left[m_4^*(e_{fA7},\eta_g)\right]^{-1}$$

The electron concentration can be written as

$$n_0 = \frac{g_v eB}{\pi\hbar}\sum_{n=0}^{n_{\max}}\sum_{n_z=1}^{n_{z\max}} F_{-1}(\eta_{A7}) \tag{8.42c}$$

where $\eta_{A7} = \frac{e_{fA7}-e_{A7}}{k_B T}$ and e_{A7} is defined by the following equation

$$G^*(e_{A7},\eta_g) = (n+\frac{1}{2})\hbar\overline{\omega_i}(e_{A7},\eta_g) + \frac{\hbar^2\left(\frac{n_z\pi}{d_z}\right)^2}{2m_3^*(e_{A7},\eta_g)} - \frac{E_0^2}{2B^2}\left[\rho^*(e_{A7},\eta_g)\right]^2 m_1^*(e_{A7},\eta_g) \tag{8.43}$$

Thus using (8.32c), (8.42c) and the allied definitions we can study the ER in this case.

8.3 Open Research Problems

R.C.1 Investigate the ER in the presence of an arbitrarily oriented quantizing magnetic and crossed electric fields in HD tetragonal semiconductors by including broadening and the electron spin. Study all the special cases for HD III–V, ternary and quaternary materials in this context.

R.C.2 Investigate the ERs for all models of HD IV–VI, II–VI and stressed Kane type compounds in the presence of an arbitrarily oriented quantizing magnetic and crossed electric fields by including broadening and electron spin.

R.C.3 Investigate the ER for all the materials as stated in R.2.1 of Chap. 2 in the presence of an arbitrarily oriented quantizing magnetic and crossed electric fields by including broadening and electron spin.

References

1. W. Zawadzki, B. Lax, Phys. Rev. Lett. **16**, 1001 (1966)
2. M.J. Harrison, Phys. Rev. A **29**, 2272 (1984)
3. J. Zak, W. Zawadzki, Phys. Rev. **145**, 536 (1966)
4. W. Zawadzki, Q.H. Vrehen, B. Lax, Phys. Rev. **148**, 849 (1966)
5. Q.H. Vrehen, W. Zawadzki, M. Reine, Phys. Rev. **158**, 702 (1967)
6. M.H. Weiler, W. Zawadzki, B. Lax, Phys. Rev. **163**(733), 333 (1967)
7. W. Zawadzki, J. Kowalski, Phys. Rev. Lett. **27**, 1713 (1971)
8. C. Chu, M.S. Chu, T. Ohkawa, Phys. Rev. Lett. **41**, 653 (1978)
9. P. Hu, C.S. Ting, Phys. Rev. B **36**, 9671 (1987)
10. E.I. Butikov, A.S. Kondratev, A.E. Kuchma, Sov. Phys. Sol. State **13**, 2594 (1972)
11. K.P. Ghatak, J.P. Banerjee, B. Goswami, B. Nag, Non. Optics and Quantum Optics **16**, 241 (1996)
12. M. Mondal, K.P. Ghatak, Phys. Stat. Sol. (b) **133**, K67 (1986)
13. M. Mondal, N. Chattopadhyay, K.P. Ghatak, Jour. Low Temp. Phys. **66**, 131 (1987)
14. K.P. Ghatak, M. Mondal, Zeitschrift fur Physik B **69**, 471 (1988)
15. M. Mondal, K.P. Ghatak, Phys. Lett. A **131A**, 529 (1988)
16. M. Mondal , K.P. Ghatak, in Phys. Stat. Sol. (b) Germany, **147**, K179 (1988)
17. B. Mitra, K.P. Ghatak, Phys. Lett. **137A**, 413 (1989)
18. B. Mitra, A. Ghoshal, K.P. Ghatak, Phys. Stat. Soli. (b) **154**, K147 (1989)
19. B. Mitra, K.P. Ghatak, Phys. Stat. Sol. (b) **164**, K13 (1991)
20. K.P. Ghatak, B. Mitra, Int. J. Electron. **70**, 345 (1991)
21. S.M. Adhikari, D. De, J.K. Baruah, S. Chowdhury, KP Ghatak Adv. Sci. Focus **1**, 57 (2013)
22. S Pahari, S Bhattacharya, D. De, S. M. Adhikari, A. Niyogi, A. Dey, N. Paitya, S. C. Saha, K. P. Ghatak, P. K. Bose, Phys. B Condens. Matter **405**, 4064 (2010)
23. S. Bhattaacharya, S. Choudhary, S. Ghoshal, S.K. Bishwas, D. De, K.P. Ghatak, J. Comp. Theo. Nanoscience. **3**, 423 (2006)
24. M. Mondal, K.P. Ghatak, Ann. der Physik. **46**, 502 (1989)
25. K.P. Ghatak, B. De MRS Proceedings **242**, 377 (1992)
26. K.P. Ghatak, B. Mitra, Internat. Jour. Electron. Theo. Exp. **70**, 343 (1991)
27. B. Mitra, K.P. Ghatak, Phys. Lett. A **141**, 81 (1989)
28. B. Mitra, K.P. Ghatak, Phys. Lett. A **137**, 413 (1989)
29. M. Mondal, N. Chattopadhyay, K.P. Ghatak, J. low Temp. Phys. **73**, 321 (1988)
30. M. Mondal, K.P. Ghatak, Phys. Lett. A **131**, 529 (1988)
31. K. P. Ghatak, N. Chattopadhyay, S. N. Biswas, International society for optics and photonics. Proc. Soc. Photo-Optical Instru. Engg. USA, 203 (1987)
32. K.P. Ghatak, M. Mitra, B. Goswami, B. Nag, Nonlinear Optics **16**, 167 (1996)
33. K.P. Ghatak, D.K. Basu, B. Nag, J. Phys. Chem. Sol. **58**, 133 (1997)
34. S. Biswas, N. Chattopadhyay, K. P. Ghatak Internat. Soc. Optics and Photonics, Proc. Soc. Photo-Optical Instru. Engg. USA **836**, 175 (1987)
35. K.P. Ghatak, M. Mondal, S. Bhattacharyya, SPIE **1284**, 113 (1990)
36. K.P. Ghatak, Photonic materials and optical bistability. SPIE. **1280,** 53 (1990)
37. K.P. Ghatak, S.N. Biswas, Growth and characterization of materials for infrared detectors and nonlinear optical switches. SPIE. **1484**, 149 (1991)
38. K.P. Ghatak, Fiber optic and laser sensors IX. SPIE. **1584**, 435 (1992)

Chapter 9
Appendix D: The ER in HD III-V, Ternary and Quaternary Semiconductors Under Strong Electric Field

9.1 Introduction

In the investigation of transport properties of nano-devices under electric field, we assumed that the electron energy spectrum becomes an invariant quantity, which is not true specially in the presence of strong electric field. In nano-devices the in-built electric field is so large that the electron dispersion relation changes fundamentally and in this appendix we shall investigate the influence of intense electric field on the ER under various physical conditions in III-V, ternary and quaternary materials. In Sect. 9.2.1 of theoretical background Sect. 9.2, we shall study the ER under strong electric field in HD said semiconductors. The Sect. 9.2.2, explores the ER in the presence of quantizing magnetic field under strong electric field in HD said materials. In Sect. 9.2.3, we study the ER in ultrathin films of HD III-V, ternary and quaternary materials under strong electric field. In Sect. 9.2.4, the ER has been investigated in accumulation layers of HD III-V, ternary and quaternary materials. In Sect. 9.2.5, the ER in doping superlattices of HD III-V, ternary and quaternary materials under strong electric field has been studied. In Sect. 9.2.6, the magneto ER in effective mass superlattices of HD said materials under strong electric field has been investigated. In Sect. 9.2.7, the electro ER in ultrathin films of HD aforementioned compounds under cross fields configuration has been investigated. The last Sect. 9.3 contains 43 open research problems.

9.2 Theoretical Background

9.2.1 The ER Under Strong Electric Field in HD III-V, Ternary and Quaternary Materials

In the presence of strong electric field F_s along x direction, the electron energy spectrum in Kane type III-V semiconductors whose unperturbed conduction electrons obey the three band models of Kane can be expressed following [1] as

K. P. Ghatak and S. Bhattacharya, *Heavily-Doped 2D-Quantized Structures and the Einstein Relation*, Springer Tracts in Modern Physics 260,
DOI 10.1007/978-3-319-08380-3_9, © Springer International Publishing Switzerland 2015

$$\frac{\hbar^2k^2}{2m_c} = \left[e_1E^4 + e_2E^3 + e_3E^2 + e_4E + e_5 - \frac{e_6}{1+CE} + e_7(1+CE)^{-2}\right] \tag{9.1}$$

where

$$e_1 = Q_f\omega_1, \quad Q_f = \frac{m_c}{m_r}E_g^{-4}\left[5e_fE_g^{-2} - 6G_f + 7h_fE_g^{-4}\right], \quad m_r\left(= \left(\frac{1}{m_c}+\frac{1}{m_v}\right)^{-1}\right)$$

is the reduced mass, m_v is the effective heavy hole mass at the edge of the valance band, $e_f = A_fP_f$,

$$A_f = \left[F\hbar E_g(E_g - \delta')\right]^2 m_c\left(6m_\gamma^2(\delta')^4\right)^{-1},$$

$F = eF_s$, F_s is the electric field along x direction.

$$\delta' = \frac{\Delta E_g^2}{\chi}, \ \chi = 6E_g^2 + 9\Delta E_g + 4\Delta^2, \quad \frac{1}{m_r} = \left(\frac{1}{m_c}+\frac{1}{m_v}\right), \quad G_f = e_f(4\delta' + C_f),$$

$$C_f = (2E_gQ^2 + PQ(E_g - E_g') - 2p^2E_g), \quad P = \frac{r_0^2}{2}\left(\frac{E_g - \delta'}{E_g + \delta'}\right),$$

$$r_0 = \left[\frac{6}{\chi}(E_g + \Delta)\left(E_g + \frac{2}{3}\Delta\right)\right]^{\frac{1}{2}}$$

$$Q = \frac{t^2}{2}, \quad t = \left[\frac{6}{\chi}\left(E_g + \frac{2}{3}\Delta\right)\right]^{\frac{1}{2}}, \quad h_f = (4\delta' e_fC_f)(B_f)^{-1}, \quad B_f = (P+Q)^2$$

$$P_f = E_g^{-3}(e_fE_g^{-2} - G_f + h_fE_g^{-4}), \quad \omega_1 = a_1^2, \quad a_1 = \frac{ab}{c}, \quad a = \frac{1}{E_g},$$

$$b = \frac{1}{E_g + \Delta}, \quad c = \left(E_g + \frac{2}{3}\Delta\right)^{-1},$$

$$e_2 = Q_f\omega_2, \quad \omega_2 = 2a_1b_1, \quad b_1 = (c)^{-2}(ac + bc - ab),$$

$$e_3 = (1 - P_f)a_1 + Q_f\omega_3, \quad \omega_3 = (b_1^2 + 2a_1c_1),$$

$$c_1 = \left[\frac{1}{c}\left(1 - \frac{a}{c}\right)\left(1 - \frac{b}{c}\right)\right], \quad e_4 = [(1 - P_f)b_1 + Q_f\omega_4],$$

$$\omega_6 = \frac{2c_1b_1}{c}\left(1 - \frac{cc_1}{b_1}\right), \quad \omega_4 = 2b_1c_1,$$

$$e_5 = [(1 - P_f)c_1 + Q_f\omega_5], \quad \omega_5 = (c_1^2 - 2c_1b_1),$$

$$e_7 = Q_f\omega_7, \quad \omega_7 = c_1^2, \quad e_6 = [(1 - P_f)c_1 - Q_f\omega_6],$$

Using (9.1) and (1.4) we get

$$\frac{\hbar^2 k^2}{2m_c}\int\limits_{-\infty}^{E} F(v)dv = e_1\int\limits_{-\infty}^{E}(E-v)^4F(v)dv + e_2 \int\limits_{-\infty}^{E}(E-v)^3F(v)dv + e_3\int\limits_{-\infty}^{E}(E-v)^2F(v)dv + e_4\int\limits_{-\infty}^{E}(E-v)F(v)dv + e_5\int\limits_{-\infty}^{E}F(v)dv - e_6 \int\limits_{-\infty}^{E}\frac{F(v)dv}{1+c(E-v)} + e_7\int\limits_{-\infty}^{E}F(v)dv[1+c(E-v)]^2 \quad (9.2)$$

Let us put $I(11) = \int\limits_{-\infty}^{E}(E-v)^4F(v)dv$

$$= E^4\int\limits_{-\infty}^{E}F(v)dv + \int\limits_{-\infty}^{E}v^4F(v)dv + 6E^2\int\limits_{-\infty}^{E}v^2F(v)dv - 2E^3\int\limits_{-\infty}^{E}vF(v)dv - 4E\int\limits_{-\infty}^{E}v^3F(v)dv \quad (9.3)$$

Now

$$\int\limits_{-\infty}^{E}v^4(F(v))dv = \left[\frac{3\eta_g^4}{8\pi}\left[1+Erf\left(\frac{E}{\eta_g}\right)\right] - \frac{\eta_g^5}{4E\pi}\exp\left(\frac{-E^2}{\eta_g^2}\right)\right]$$
$$= \frac{3\eta_g^4}{8\pi}\left[1+Erf\left(\frac{E}{\eta_g}\right) - \frac{2\eta_g}{3E}\exp\left(\frac{-E^2}{\eta_g^2}\right)\right] = \delta_0'(E,\eta_g) \quad (9.4)$$

$$\int\limits_{-\infty}^{E}v^2F(v)dv = \frac{\eta_g^2}{4}\left[1+Erf\left(\frac{E}{\eta_g}\right)\right] \quad (9.5)$$

$$\int\limits_{-\infty}^{E}v(F(v))dv = \frac{-\eta_g}{2\sqrt{\pi}}\exp\left(\frac{-E^2}{\eta_g^2}\right) \quad (9.6)$$

$$\int_{-\infty}^{E} v^3 F(v)dv = \frac{-\eta_g^3}{2\sqrt{\pi}} \exp\left(\frac{-E^2}{\eta_g^2}\right)\left[1 + \frac{E^2}{\eta_g^2}\right] \tag{9.7}$$

Thus

$$\begin{aligned} I(11) = {} & \frac{E^4}{2}\left[1 + Erf\left(\frac{E}{\eta_g}\right)\right] + \frac{3\eta_g^4}{8\pi}\left[1 + Erf\left(\frac{E}{\eta_g}\right) - \frac{2\eta_g}{3E}\exp\left(\frac{-E^2}{\eta_g^2}\right)\right] \\ & + \frac{3}{2}(E\eta_g)^2\left[1 + Erf\left(\frac{E}{\eta_g}\right)\right] \\ & + \frac{E^3\eta_g}{\sqrt{\pi}}\exp\left(\frac{-E^2}{\eta_g^2}\right) + \frac{2E\eta_g^3}{\sqrt{\pi}}\exp\left(\frac{-E^2}{\eta_g^2}\right)\left[1 + \frac{E^2}{\eta_g^2}\right] = \varphi_0(E, \eta_g) \end{aligned} \tag{9.8}$$

In Chap. 1 we have proved that

$$I(\alpha, E, \eta_g) = \int_{-\infty}^{E} \frac{F(v)dv}{1 + e(E - v)} = \frac{2}{c\eta_g\sqrt{\pi}} e^{-u^2}\left[\sum_{p=1}^{\infty} \frac{e^{\frac{-p^2}{4}}}{p}\sinh(pu)\right] - i\frac{\sqrt{\pi}}{c\eta_g} e^{-u^2} \tag{9.9}$$

where $u = \frac{1+cE}{c\eta_g}$.

The theorem of differentiation under the sign of integration tells us

$$\frac{\partial}{\partial x}\int_{A(x)}^{B(x)} F(x, y)dy = \int_{A(x)}^{B(x)} \frac{\partial}{\partial x}[F(x, y)]dy + F\left(x, B(x)\frac{\partial B(x)}{\partial x}\right) - F\left(x, A(x)\frac{\partial A(x)}{\partial x}\right) \tag{9.10}$$

where the notations have their usual meaning and the integrals are convergent.

Using (9.9) and (9.10) and differentiating (9.9) with respect to E we get

$$\begin{aligned} -c\int_{-\infty}^{E} \frac{F(v)dv}{[1 + c(E - v)]^2} + \frac{1}{\sqrt{\pi\eta_g^2}}\exp\left(\frac{-E^2}{\eta_g^2}\right) = {} & \frac{2}{c\eta_g\sqrt{\pi}} e^{-u^2} 2u \frac{1}{\eta_g}\left[\sum_{p=1}^{\infty} \frac{e^{\frac{-p^2}{4}}}{p}\sinh(pu)\right] \\ & + \frac{2}{c\eta_g\sqrt{\pi}} e^{-u^2} \frac{1}{\eta_g}\sum_{p=1}^{\infty} e^{\frac{-p^2}{4}}\cosh(pu) - i\frac{\sqrt{\pi}}{c\eta_g} e^{-u^2} 2u\frac{1}{\eta_g} \end{aligned}$$

$$\therefore \quad \int_{-\infty}^{E} \frac{F(v)dv}{[1 + c(E - v)]^2} = c_3(E, \eta_g, c) + iD_3(E, \eta_g, c) \tag{9.11}$$

where

$$c_3(E,\eta_g,c) = [\frac{1}{\pi c\eta_g}\exp\left(\frac{-E^2}{\eta_g^2}\right) - \frac{4ue^{-u^2}}{c^2\eta_g^2\sqrt{\pi}}[\sum_{p=1}^{\infty}\exp(\frac{-p^2}{4})p^{-1}\sinh(pu)]$$
$$- \frac{2}{c^2\eta_g^2\sqrt{\pi}}e^{-u^2}\sum_{p=1}^{\infty}e^{\frac{-p^2}{4}}\cosh(pu)]$$
$$D_3(E,\eta_g,c) = \frac{2u}{c^2\eta_g^2}\exp(-u^2), \quad u = \frac{1+cE}{c\eta_g}$$

Again (9.9) can be written as

$$\int_{-\infty}^{E}\frac{F(v)dv}{1+c(E-v)} = c_1(c,E,\eta_g) - ic_2(c,E,\eta_g) \tag{9.12}$$

where $c_1(c,E,\eta_g) = \frac{2}{c\eta_g\sqrt{\pi}}e^{-u^2}\left[\sum_{p=1}^{\infty}\frac{e^{\frac{-p^2}{4}}}{p}\sinh(pu)\right]$ and $c_2(c,E,\eta_g) = \frac{\sqrt{\pi}}{c\eta_g}e^{-u^2}$

We know that

$$\int_{-\infty}^{E}(E-v)^3F(v)dv = \left[\frac{E}{2}\left[1+Erf\left(\frac{E}{\eta_g}\right)\right]\left[E^2+\frac{3}{2}\eta_g^2\right]+\frac{\eta_g}{2\sqrt{\pi}}e^{-\frac{E^2}{\eta_g^2}}(4E^2+\eta_g^2)\right]$$
$$= \varphi_1(E,\eta_g) \tag{9.13}$$

Therefore the dispersion relation in heavily doped III-IV semiconductors whose unperturbed conduction electrons obey the three band models of Kane in the presence of an electric field along x axis can be expressed as

$$\frac{\hbar^2k^2}{2m_c} = J_4(E,c,\eta_g) \tag{9.14}$$

where $J_4(E,c,\eta_g) = J_1(E,c,\eta_g) + iJ_2(E,c,\eta_g)$,

$$J_1(E,c,\eta_g) = 2\left[1+Erf\left(\frac{E}{\eta_g}\right)\right]^{-1}[e_1\varphi_0(E,\eta_g)+e_2\varphi_1(E,\eta_g)+e_3\theta_0(E,\eta_g)+e_4\gamma_0(E,\eta_g)$$
$$+ g_{(\frac{e_5}{2})}\left[1+Erf\left(\frac{E}{\eta_g}\right)\right] - e_6c_1(E,c,\eta_g)+e_7c_3(E,c,\eta_g)\Bigg]$$

and $J_2(E,c,\eta_g) = 2\left[1+Erf\left(\frac{E}{\eta_g}\right)\right]^{-1}[e_6c_2(E,c,\eta_g)+e_7D_3(E,c,\eta_g)]$.

For two band model of Kane, the dispersion relation in the presence of electric field F_s along x direction is given by

$$\frac{\hbar^2 k^2}{2m_c} = P_{11f} E(1 + \alpha E) - Q_{11f} \tag{9.15}$$

where $P_{11f} = \left[1 + (Q_{11f})\left(\frac{5m_c}{m_r E_g}\right)\right]$, $Q_{11f} = \frac{(\hbar F)^2}{12 m_r E_g^2}$.

Therefore under the condition of heavy doping (9.15) assumes the form

$$\frac{\hbar^2 k^2}{2m_c} = J_5(E, \eta_g) \tag{9.16a}$$

where $J_5(E, \eta_g) = P_{11f}\gamma_2(E, \eta_g) - Q_{11f}$.

Thus (9.14) and (9.16a) are key equations for investigating the electronic properties in III-V Kane type heavily doped semiconductors in the presence of a strong electric field.

The EEM in III-V Kane type heavily doped semiconductors in the presence of a strong electric field whose energy band structures in the absence of any perturbation are defined by three and two band models of kane can be written from (9.14) and (9.16a) as

$$m^*(E_{FE}, F) = m_c \, Real\, part\, of\, [J_4(E_{FE}, c, \eta_g)]' \tag{9.16b}$$

and

$$m^*(E_{FE}, F) = m_c [J_5(E_{FE}, \eta_g)]' \tag{9.16c}$$

where E_{FE} is the Fermi energy in the present case for this chapter.

Thus following (9.14) electron concentration is given by

$$n_0 = \frac{g_v}{3\pi^2}\left(\frac{2m_c}{\hbar^2}\right)^{\frac{3}{2}} Real\ part\ of\ [J_6(E_{FE}, c, \eta_g) + J_7(E_{FE}, c, \eta_g)] \tag{9.17}$$

where

$J_6(E_{FE}, c, \eta_g) = [J_4(E_{FE}, c, \eta_g)]^{\frac{3}{2}}$ and $J_7(E_{FE}, c, \eta_g) = \sum_{r=1}^{s} L(r) J_6(E_{FE}, c, \eta_g)$.

The ER in this case is given by

$$\frac{D}{\mu} = Real\ part\ of \left[\left(\frac{n_0}{e}\right)\left(\frac{\partial n_0}{\partial(E_{FE} - e'_{EF})}\right)^{-1}\right] \tag{9.18}$$

where e'_{EF} is the energy when $k = 0$ in the dispersion relation.

Thus using (9.17), (9.18) and the allied definitions, we can study the ER in this case.

For the dispersion relation (9.16a) the corresponding electron concentration can be written as

$$n_0 = \frac{g_v}{3\pi^2}\left(\frac{2m_c}{\hbar^2}\right)^{\frac{3}{2}}\left[J_8(E_{FE},\eta_g) + J_9(E_{FE},\eta_g)\right] \tag{9.19}$$

where $J_8(E_{FE},\eta_g) = [J_5(E_{FE},\eta_g)]^{\frac{3}{2}}$ and $J_9(E_{FE},\eta_g) = \sum_{r=1}^{s} L(r)J_8(E_{FE},\eta_g)$.

Thus using (9.18), (9.19) and the allied definitions, we can study the ER in this case.

9.2.2 The ER in the Presence of Quantizing Magnetic Field Under Strong Electric Field in HD III-V, Ternary and Quaternary Materials

The electron energy spectrum under magnetic quantization can be written as

$$\frac{\hbar^2 k_z^2}{2m_c} + \left(n + \frac{1}{2}\right)\hbar\omega_0 = J_4(E, c, \eta_g) \tag{9.20}$$

$$\frac{\hbar^2 k_z^2}{2m_c} + \left(n + \frac{1}{2}\right)\hbar\omega_0 = J_5(E, \eta_g) \tag{9.21a}$$

The EEM in this case can be written using (9.20) and (9.21a) as

$$m^*(E_{FEB}, F) = m_c\, Real\, part\, of\, [J_4(E_{FEB}, c, \eta_g)]' \tag{9.21b}$$

and

$$m^*(E_{FEB}, F) = m_c [J_5(E_{FEB}, \eta_g)]' \tag{9.21c}$$

where E_{FEB} is the Fermi energy in the present case

The electron concentration for the dispersion relation (9.20) is given by

$$n_0 = \frac{eBg_v\sqrt{2m_c}}{\pi^2\hbar^2}\, Real\ part\ of \sum_{n=0}^{n_{max}}\left[J_{10}(E_{FEB}, c, \eta_g, n) + J_{11}(E_{FEB}, c, \eta_g, n)\right] \tag{9.22}$$

where $J_{10}(E_{FEB}, c, \eta_g, n) = \sqrt{J_4(E_{FEB}, c, \eta_g, n) - \left(n + \frac{1}{2}\right)\hbar\omega_0}$ and $J_{11}(E_{FEB}, c, \eta_g, n) = \sum_{r=1}^{s} L(r) J_{10}(E_{FEB}, c, \eta_g, n)$

The ER in this case can be written as

$$\frac{D}{\mu} = Real\,part\,of\left[\left(\frac{n_0}{e}\right)\left(\frac{\partial n_0}{\partial(E_{FEB} - e'_{EFB})}\right)^{-1}\right] \tag{9.23}$$

where e'_{EFB} is the Landau sub band energy in this case.

Thus, using (9.22), (9.23) and the allied definitions we can study the ER in this case.

The electron concentration for the dispersion relation (9.21) is given by

$$n_0 = \frac{eBg_v\sqrt{2m_c}}{\pi^2\hbar^2}\sum_{n=0}^{n_{\max}}[J_{12}(E_{FEB},\eta_g,n) + J_{13}(E_{FEB},\eta_g,n)] \tag{9.24}$$

where $J_{12}(E_{FEB},\eta_g,n) = \sqrt{J_5(E_{FEB},\eta_g,n) - (n+\frac{1}{2})\hbar\omega_0}$ and $J_{13}(E_{FEB},\eta_g,n) = \sum_{r=1}^{s} L(r)J_{12}(E_{FEB},\eta_g,n)$.

Thus using (9.22), (9.24) and the allied definitions, we can study the ER in this case.

9.2.3 The ER in Ultrathin Films of HD III-V, Ternary and Quaternary Materials Under Strong Electric Field

For ultrathin films in the presence of size quantization, the 2D dispersion laws for (9.14) and (9.16a) assume the forms

$$\frac{\hbar^2 k_s^2}{2m_c} + \frac{\hbar^2}{2m_c}\left(\frac{n_z\pi}{d_z}\right)^2 = J_4(E,c,\eta_g) \tag{9.25}$$

and

$$\frac{\hbar^2 k_s^2}{2m_c} + \frac{\hbar^2}{2m_c}\left(\frac{n_z\pi}{d_z}\right)^2 = J_5(E,\eta_g) \tag{9.26a}$$

The EEM in this case can be written using (9.25) and (9.26a) as

$$m^*(E_{FESQ},F) = m_c\,Real\,part\,of\,[J_4(E_{FESQ},c,\eta_g)]' \tag{9.26b}$$

and

$$m^*(E_{FESQ},F) = m_c[J_5(E_{FESQ},\eta_g)]' \tag{9.26c}$$

where E_{FESQ} is the Fermi energy in this case.

The 2D electron concentration for (9.25) can be written as

$$n_0 = \frac{m_c g_v}{\pi\hbar^2} \, Real\,part\,of \sum_{n_z=1}^{n_{z\max}} [J_{17}(E_{FESQ}, c, \eta_g, n_z) + J_{18}(E_{FESQ}, c, \eta_g, n_z)] \quad (9.27)$$

where $J_{17}(E_{FESQ}, c, \eta_g, n_z) = \left[J_4(E_{FESQ}, c, \eta_g) - \frac{\hbar^2}{2m_c}\left(\frac{n_z\pi}{d_z}\right)^2\right]$ and $J_{18}(E_{FESQ}, c, \eta_g, n_z) = \sum_{r=1}^{s} L(r)[J_{17}(E_{FESQ}, c, \eta_g, n_z)]$

The ER in this case is given by

$$\frac{D}{\mu} = Real\,part\,of \left[\left(\frac{n_0}{e}\right)\left(\frac{\partial n_0}{\partial(E_{FESQ} - e'_{EFSQ})}\right)^{-1}\right] \quad (9.28)$$

where e'_{EFSQ} is the subband energy in this case.

Thus, by using (9.27), (9.28) and the allied definitions we can study the ER in this case.

The 2D electron concentration for (9.26a) can be written as

$$n_0 = \frac{m_c g_v}{\pi\hbar^2} \sum_{n_z=1}^{n_{z\max}} [J_{19}(E_{FESQ}, c, \eta_g, n_z) + J_{20}(E_{FESQ}, c, \eta_g, n_z)] \quad (9.29)$$

where $J_{19}(E_{FESQ}, c, \eta_g, n_z) = [J_5(E_{FESQ}, c, \eta_g) - \frac{\hbar^2}{2m_c}(\frac{n_z\pi}{d_z})^2]$ and $J_{20}(E_{FESQ}, c, \eta_g, n_z) = \sum_{r=1}^{s} L(r)[J_{19}(E_{FESQ}, c, \eta_g, n_z)]$

Thus by using (9.28), (9.29) and the allied definitions, we can study the ER in this case.

9.2.4 The ER in Accumulation Layers of HD III-V, Ternary and Quaternary Materials

The 2D dispersion relations corresponding to (9.14) and (9.16a) are given by

$$J_4(E, c, \eta_g) = \frac{\hbar^2 k_s^2}{2m_c^*} + S_i\left(\frac{\hbar e \bar{F}_s}{\sqrt{2m_c}} J'_4(E, c, \eta_g)\right)^{\frac{2}{3}} \quad (9.30)$$

and

$$J_5(E, \eta_g) = \frac{\hbar^2 k_s^2}{2m_c} + S_i\left(\frac{\hbar e \bar{F}_s}{\sqrt{2m_c}} J'_5(E, \eta_g)\right)^{\frac{2}{3}} \quad (9.31a)$$

where $\bar{F}_s$ is the surface electric field along Z-direction.

Using (9.30) and (9.31a), the EEMs can be expressed as

$$m^*(E_{FEIL}, c, \eta_g, i) = m_c \, Real\, part\, of \, [J_{21}(E_{FEIL}, c, \eta_g, i)]' \tag{9.31b}$$

$$m^*(E_{FEIL}, \eta_g, i) = m_c[J_{23}(E_{FEIL}, \eta_g, i)]' \tag{9.31c}$$

where E_{FEIL} is the Fermi energy in this case, $J_{21}(E_{FEIL}, c, \eta_g, i) = \left[J_4(E_{FEIL}, c, \eta_g) - S_i\left(\frac{\hbar e \bar{F}_s}{\sqrt{2m_c}} J_4'(E_{FEIL}, c, \eta_g)\right)^{\frac{2}{3}}\right]$ and $J_{23}(E_{FEIL}, \eta_g, i) = \left[J_5(E_{FEIL}, \eta_g) - S_i\left(\frac{\hbar e \bar{F}_s}{\sqrt{2m_c}} J_5'(E_{FEIL}, \eta_g)\right)^{\frac{2}{3}}\right]$.

The 2D electron concentration for (9.30) can be written as

$$\begin{aligned} n_0 = Real\, part\, of &\left[\frac{m_c g_v}{\pi\hbar^2}\sum_{i=0}^{i_{\max}}\left[[J_{21}(E_{FEIL}, c, \eta_g, i) + J_{22}(E_{FEIL}, c, \eta_g, i)]\right.\right. \\ &\left.\left. + \frac{g_v}{3\pi^2}\left(\frac{2m_c}{\hbar^2}\right)^{\frac{3}{2}} t_i J_4(E_{FE}, c, \eta_g)\right]\right] \end{aligned} \tag{9.32}$$

where $J_{22}(E_{FEIL}, c, \eta_g, i) = \sum_{r=1}^{s} L(r)[J_{21}(E_{FEIL}, c, \eta_g, i)]$and E_{FE} is determined from (9.17).

The ER in this case is given by

$$\frac{D}{\mu} = Real\ part\ of\ \left[\left(\frac{n_0}{e}\right)\left(\frac{\partial n_0}{\partial(E_{FEIL} - e'_{EFIL})}\right)^{-1}\right] \tag{9.33}$$

where e'_{EFIL} is the subband energy in this case.

Thus, by using (9.32), (9.33) and the allied definitions we can study the ER in this case.

The 2D electron concentration for (9.31a) can be written as

$$\begin{aligned} n_0 = Real\, part\, of &\left[\frac{m_c g_v}{\pi\hbar^2}\sum_{i=0}^{i_{\max}}\left[[J_{23}(E_{FEIL}, \eta_g, i) + J_{25}(E_{FEIL}, \eta_g, i)]\right.\right. \\ &\left.\left. + \frac{g_v}{3\pi^2}\left(\frac{2m_c}{\hbar^2}\right)^{\frac{3}{2}} t_i J_5(E_{FE}, \eta_g)\right]\right] \end{aligned} \tag{9.34}$$

where $J_{24}(E_{FEIL}, \eta_g, i) = \sum_{r=1}^{s} L(r)[J_{23}(E_{FEIL}, \eta_g, i)]$and E_{FE} is determined from (9.19).

Thus by using (9.33), (9.34) and the allied definitions we can study the ER in this case.

9.2.5 The ER in Doping Super lattices of HD III-V, Ternary and Quaternary Materials Under Strong Electric Field

For doping super lattices the 2D dispersion relations corresponding to (9.14) and (9.16a) assume the forms

$$J_4(E,c,\eta_g) = \frac{\hbar^2 k_s^2}{2m_c} + \left(n_i + \frac{1}{2}\right)\hbar\omega_1(E,c,\eta_g) \tag{9.35}$$

$$J_5(E,\eta_g) = \frac{\hbar^2 k_s^2}{2m_c} + \left(n_i + \frac{1}{2}\right)\hbar\omega_2(E,\eta_g) \tag{9.36a}$$

where $n_i = (0,1,2,\ldots)$ is the mini-band index, d_0 is the superlattice period, $\omega_1(E,c,\eta_g) = \left(\frac{n_s e^2}{\varepsilon_{sc} d_0 J_4'(E,c,\eta_g)}\right)^{\frac{1}{2}}$ and $\omega_2(E,\eta_g) = \left(\frac{n_s e^2}{\varepsilon_{sc} d_0 J_5'(E,\eta_g)}\right)^{\frac{1}{2}}$.

Using (9.35) and (9.36a), the EEM can be written as

$$m^*(E_{FEDSL},c,\eta_g,n_i) = m_c \, Real\, part\, of\, [J_{25}(E_{FEDSL},c,\eta_g,n_i)]' \tag{9.36b}$$

$$m^*(E_{FEDSL},\eta_g,n_i) = m_c[J_{27}(E_{FEDSL},\eta_g,n_i)]' \tag{9.36c}$$

where E_{FEDSL} is the Fermi energy in this case,

$$J_{25}(E_{FEDSL},c,\eta_g,n_i) = [J_4(E_{FEDSL},c,\eta_g) - (n_i + \frac{1}{2})\hbar\omega_1(E_{FEDSL},c,\eta_g)]$$

and

$$J_{27}(E_{FEDSL},\eta_g,n_i) = [J_5(E_{FEDSL},\eta_g) - (n_i + \frac{1}{2})\hbar\omega_2(E_{FEDSL},\eta_g)]$$

The electron statistics for (9.35) can be written as

$$n_0 = \frac{m_c g_v}{\pi\hbar^2} Real\, part\, of \sum_{n_i=0}^{n_i \max} [J_{25}(E_{FEDSL},c,\eta_g,n_i) + J_{26}(E_{FEDSL},c,\eta_g,n_i)] \tag{9.37}$$

where

$[J_{26}(E_{FEDSL},c,\eta_g,n_i) = \sum_{r=1}^{s} L(r)[J_{25}(E_{FEDSL},c,\eta_g,n_i)]$.

The ER in this case is given by

$$\frac{D}{\mu} = Real\, part\, of\, [(\frac{n_0}{e})(\frac{\partial n_0}{\partial(E_{FEDSL} - e'_{EFDSL})})^{-1}] \tag{9.38}$$

where e'_{EFDSL} is the sub-band energy.

Using (9.37), (9.38) and the allied definitions, we can study the ER in this case.

For (9.36a) the electron statistics can be written as

$$n_0 = \frac{m_c g_v}{\pi\hbar^2} \sum_{n_i=0}^{n_i \max} \left[J_{27}(E_{FEDSL}, \eta_g, n_i) + J_{28}(E_{FEDSL}, \eta_g, n_i)\right] \tag{9.39}$$

where $J_{28}(E_{FEDSL}, \eta_g, n_i) = \sum_{r=1}^{s} L(r)[J_{27}(E_{FEDSL}, \eta_g, n_i)]$

Using (9.38), (9.39) and the allied definitions, we can study the ER in this case.

9.2.6 The Magneto ER in Effective Mass Superlattices of HD III-V, Ternary and Quaternary Materials Under Strong Electric Field

The electron dispersion law in III-V effective mass super lattices can be written as

$$k_x^2 = \left[\frac{1}{L_0^2}\left[Cos^{-1}\{f_{HD}(E, c, \eta_g, k_y, k_z)\}\right]^2 - k_\perp^2\right] \tag{9.40}$$

where $f_{HD}(E, c, \eta_g, k_y, k_z) = \left[a_{1HD} Cos[a_0 C_{1HD}(E, \eta_{g1}, c_1, k_\perp) + b_0 D_{1HD}(E, \eta_{g2}, c_2, k_\perp)]\right] - \left[a_{2HD} Cos[a_0 C_{1HD}(E, \eta_{g1}, c_1, k_\perp) - b_0 D_{1HD}(E, \eta_{g2}, c_2, k_\perp)]\right], a_{1HD} = \left[\sqrt{\frac{m_{c2} J_4'(0, c_2, \eta_{g2})}{m_{c1} J_4'(0, c_1, \eta_{g1})}} + 1\right]^2 \cdot \left[4\sqrt{\frac{m_{c2} J_4'(0, c_2, \eta_{g2})}{m_{c1} J_4'(0, c_1, \eta_{g1})}}\right]^{-1}, a_{2HD} = \left[\sqrt{\frac{m_{c2} J_4'(0, c_2, \eta_{g2})}{m_{c1} J_4'(0, c_1, \eta_{g1})}} - 1\right]^2 \cdot \left[4\sqrt{\frac{m_{c2} J_4'(0, c_2, \eta_{g2})}{m_{c1} J_4'(0, c_1, \eta_{g1})}}\right]^{-1}, C_{1HD}(E, \eta_{g1}, c_1, k_\perp) = \left[\frac{2m_{c1}}{\hbar^2} J_4(E, \eta_{g1}, c_1) - k_\perp^2\right]$ and $D_{1HD}(E, \eta_{g1}, c_1, k_\perp) = \left[\frac{2m_{c2}}{\hbar^2} J_4\ (E, \eta_{g2}, c_2) - k_\perp^2\right]$.

In the presence of a quantizing magnetic field B along k_x direction, the magneto electron energy spectrum can be written as

$$k_x^2 = \omega_{HD}(E, \eta_g, c, n) \tag{9.41a}$$

where $\omega_{HD}(E, \eta_g, c, n) = \left[\frac{1}{L_0^2}\left[Cos^{-1}\{f_{HD}(E, c, \eta_g, n)\}\right]^2 - \frac{2eB}{\hbar}\left(n + \frac{1}{2}\right)\right]$,

$$f_{HD}(E, c, \eta_g, n) = [[a_{1HD} Cos[a_0 C_{1HD}(E, \eta_{g1}, c_1, n) + b_0 D_{1HD}(E, \eta_{g2}, c_2, n)]] - [a_{2HD} Cos[a_0 C_{1HD}(E, \eta_{g1}, c_1, n) - b_0 D_{1HD}(E, \eta_{g2}, c_2, n)]]],$$

$$C_{1HD}(E, \eta_{g1}, c_1, n) = \left[\frac{2m_{c1}}{\hbar^2} J_4(E, \eta_{g1}, c_1) - \frac{2eB}{\hbar}\left(n + \frac{1}{2}\right)\right]$$

and

$$D_{1HD}(E, \eta_{g1}, c_1, n) = [\frac{2m_{c2}}{\hbar^2} J_4(E, \eta_{g2}, c_2) - \frac{2eB}{\hbar}(n + \frac{1}{2})]$$

The EEM in this case can be written from (9.41a) as

$$m^*(E_{SL},\eta_g,c,n)=\frac{\hbar^2}{2}\,Real\,part\,of\,[\omega_{HD}(E_{SL},\eta_g,c,n)]' \tag{9.41b}$$

where E_{SL} is the Fermi energy in this case.

The electron concentration is given by

$$n_0=\frac{g_v eB}{\pi^2\hbar}Real\,part\,of\sum_{n=0}^{n_{\max}}[J_{40}(E_{SL},\eta_g,n)+J_{41}(E_{SL},\eta_g,n)] \tag{9.42}$$

where

$J_{40}(E_{SL},\eta_g,n)=[\omega_{HD}(E_{SL},\eta_g,c,n)]^{\frac{1}{2}}$ and $J_{41}(E_{SL},\eta_g,n)=\sum_{r=1}^{s}L(r)[J_{40}(E_{SL},\eta_g,n)]$

The ER in this case is given by

$$\frac{D}{\mu}=\frac{n_0}{e}Real\,part\,of\left[\frac{\partial n_0}{\partial(E_{SL}-\omega'_{SL})}\right]^{-1} \tag{9.43}$$

where ω'_{SL} is the sub-band energy in this case.

Using (9.42), (9.43) and the allied definitions, we can study the ER in this case.

The electron dispersion law in III-V effective mass super lattices whose constituent materials obey (9.16a) can be expressed as

$$k_x^2=\left[\frac{1}{L_0^2}[Cos^{-1}\{f_{HD5}(E,\eta_g,k_y,k_z)\}]^2-k_\perp^2\right] \tag{9.44}$$

where $f_{HD5}(E,\eta_g,k_y,k_z)=[[a_{1HD5}Cos[a_0C_{1HD5}(E,\eta_{g1},k_\perp)+b_0D_{1HD5}(E,\eta_{g2},k_\perp)]]-[a_{2HD5}$ $\%Cos[a_0C_{1HD5}(E,\eta_{g1},k_\perp)-b_0D_{1HD5}(E,\eta_{g2},k_\perp)]]],a_{1HD5}=\left[\sqrt{\frac{m_{c2}J_5'(0,\eta_{g2})}{m_{c1}J_5'(0,\eta_{g1})}}+1\right]^2\cdot$ $\left[4\sqrt{\frac{m_{c2}J_5'(0,\eta_{g2})}{m_{c1}J_5'(0,\eta_{g1})}}\right]^{-1},a_{2HD5}=\left[\sqrt{\frac{m_{c2}J_5'(0,\eta_{g2})}{m_{c1}J_5'(0,\eta_{g1})}}-1\right]^2\cdot\left[4\sqrt{\frac{m_{c2}J_5'(0,\eta_{g2})}{m_{c1}J_5'(0,\eta_{g1})}}\right]^{-1},C_{1HD5}(E,\eta_{g1},k_\perp)$ $=\left[\frac{2m_{c1}}{\hbar^2}J_5(E,\eta_{g1})-k_\perp^2\right]$ and $D_{1HD5}\ (E,\eta_{g1},k_\perp)=\left[\frac{2m_{c2}}{\hbar^2}J_5(E,\eta_{g2},)-k_\perp^2\right]$

In the presence of a quantizing magnetic field B along k_x direction, the magneto-electron energy spectrum can be written as

$$k_x^2=\omega_{HD5}(E,\eta_g,n) \tag{9.45a}$$

where

$$\omega_{HD5}(E,\eta_g,n) = \left[\frac{1}{L_0^2}[Cos^{-1}\{f_{HD5}(E,\eta_g,n)\}]^2 - \frac{2eB}{\hbar}\left(n+\frac{1}{2}\right)\right]$$

$$f_{HD5}(E,\eta_g,n) = [[a_{1HD5}Cos[a_0C_{1HD5}(E,\eta_{g1},n) + b_0D_{1HD5}(E,\eta_{g2},n)]] - [a_{2HD5}Cos[a_0C_{1HD5}(E,\eta_{g1},n) - b_0D_{1HD5}(E,\eta_{g2},n)]]],$$

$$C_{1HD5}(E,\eta_{g1},n) = \left[\frac{2m_{c1}}{\hbar^2}J_5(E,\eta_{g1}) - \frac{2eB}{\hbar}\left(n+\frac{1}{2}\right)\right]$$

and

$$D_{1HD5}(E,\eta_{g1},n) = \left[\frac{2m_{c2}}{\hbar^2}J_5(E,\eta_{g2}) - \frac{2eB}{\hbar}\left(n+\frac{1}{2}\right)\right]$$

The EEM in this case can be written from (9.45a) as

$$m^*(E_{SL},\eta_g,n) = \frac{\hbar^2}{2}[\omega_{HD5}(E_{SL},\eta_g,n)]' \tag{9.45b}$$

The electron concentration is given by

$$n_0 = \frac{g_v eB}{\pi^2\hbar}\sum_{n=0}^{n_{\max}}[J_{50}(E_{SL},\eta_g,n) + J_{51}(E_{SL},\eta_g,n)] \tag{9.46}$$

where

$$J_{50}(E_{SL},\eta_g,n) = [\omega_{HD5}(E_{SL},\eta_g,n)]^{\frac{1}{2}} \text{ and } J_{51}(E_{SL},\eta_g,n) = \sum_{r=1}^{s}L(r)[J_{50}(E_{SL},\eta_g,n)]$$

Using (9.43), (9.46) and the allied definitions, we can study the ER in this case.

9.2.7 The Electro ER in Ultrathin Films of HD III-V, Ternary and Quaternary Materials Under Cross Fields Configuration

In the presence of an crossed electric field E_0 along x axis and quantizing magnetic field B along z axis, (9.14) and (9.16a) assume the forms

$$J_4(E,c,\eta_g) = (n+\frac{1}{2})\hbar\omega_0 + \frac{\hbar^2k_z^2(E)}{2m_c} - \frac{E_0}{B}\hbar k_y[J_4'(E,c,\eta_g)] - \frac{m_cE_0^2[J_4'(E,c,\eta_g)]^2}{2B^2} \tag{9.47}$$

and

$$J_5(E,c,\eta_g) = (n+\frac{1}{2})\hbar\omega_0 + \frac{\hbar^2 k_z^2(E)}{2m_c} - \frac{E_0}{B}\hbar k_y[J_5'(E,c,\eta_g)] - \frac{m_c E_0^2 [J_5'(E,c,\eta_g)]^2}{2B^2} \quad (9.48)$$

For ultra thin film under crossed field configuration we get

$$J_4(E,c,\eta_g) = \left(n+\frac{1}{2}\right)\hbar\omega_0 + \left(\frac{\hbar^2}{2m_c}\right)\left(\frac{n_z}{d_z}\right)^2 - \frac{E_0}{B}\hbar k_y[J_4'(E,c,\eta_g)] - \frac{m_c E_0^2}{2B^2}[J_4'(E,c,\eta_g)]^2 \quad (9.49)$$

and

$$J_5(E,\eta_g) = \left(n+\frac{1}{2}\right)\hbar\omega_0 + \left(\frac{\hbar^2}{2m_c}\right)\left(\frac{n_z}{d_z}\right)^2 - \frac{E_0}{B}\hbar k_y[J_5'(E,\eta_g)] - \frac{m_c E_0^2}{2B^2}[J_5'(E,\eta_g)]^2 \quad (9.50a)$$

Using (9.49) and (9.50a), the respective EEMs can be written as

$$m^*(E_{FHDUL},c,\eta_g,n,n_z) = Real\,part\,of \left(\frac{B}{E_0}\right)^2 T_{99}(E_{FHDUL},c,\eta_g,n,n_z)[T_{99}(E_{FHDUL},c,\eta_g,n,n_z)]' \quad (9.50b)$$

$$m^*(E_{FHDUL},\eta_g,n,n_z) = \left(\frac{B}{E_0}\right)^2 T_{100}(E_{FHDUL},\eta_g,n,n_z)[T_{100}(E_{FHDUL},\eta_g,n,n_z)]' \quad (9.50c)$$

where E_{FHDUL} is the Fermi energy in the present case,

$$T_{99}(E_{FHDUL},c,\eta_g,n,n_z) = \left[J_4(E_{FHDUL},c,\eta_g) - \left(n+\frac{1}{2}\right)\hbar\omega_0 - \left(\frac{\hbar^2}{2m_c}\right)\left(\frac{n_z}{d_z}\right)^2 + \frac{m_c E_0^2}{2B^2}[J_4'(E_{FHDUL},c,\eta_g)]^2\right][J_4'(E_{FHDUL},c,\eta_g)]^{-1}]$$

and

$$T_{100}(E_{FHDUL},\eta_g,n,n_z) = \left[J_5(E_{FHDUL},\eta_g) - \left(n+\frac{1}{2}\right)\hbar\omega_0 - \left(\frac{\hbar^2}{2m_c}\right)\left(\frac{n_z}{d_z}\right)^2 + \frac{m_c E_0^2}{2B^2}[J_5'(E_{FHDUL},\eta_g)]^2\right][J_5'(E_{FHDUL},\eta_g)]^{-1}]$$

The electron statistics for (9.49) and (9.50a) assume the forms

$$n_0 = \frac{g_v eB}{\pi\hbar} \, Real\, part\, of \sum_{n=0}^{n_{\max}} \sum_{n_z=1}^{n_{z\max}} F_{-1}(\eta_{1HD}) \tag{9.51}$$

and

$$n_0 = \frac{g_v eB}{\pi\hbar} \sum_{n=0}^{n_{\max}} \sum_{n_z=1}^{n_{z\max}} F_{-1}(\eta_{2HD}) \tag{9.52}$$

where $\eta_{1HD} = \frac{E_{FHDUL} - E'_{FHDUL}}{k_B T}$, E_{FHDUL} is the Fermi energy in the present case, E'_{FHDUL} is the subband energy obtained by substituting $k_y = 0$ and $E = E'_{FHDUL}$ in (9.49), $\eta_{2HD} = \frac{E_{FHDUL} - E''_{FHDUL}}{k_B T}$ and E''_{FHDUL} is the subband energy obtained by substituting $k_y = 0$ and $E = E''_{FHDUL}$ in (9.50a).

The ER in this case is given by

$$\frac{D}{\mu} = \frac{n_0}{e} \, Real\, part\, of \left[\frac{\partial n_0}{\partial(E_{FHDUL} - e_{FHDUL})}\right]^{-1} \tag{9.53}$$

where e_{FHDUL} is the Fermi energy in the present case.

Thus by using (9.51),(9.52), (9.53) and the allied definitions, we can study the ER in this case.

9.3 Open Research Problems

(R.D.1) Investigate the ER for the HD bulk materials whose respective dispersion relations of the carriers in the absence of any field is given in Chap. 1 in the presence of intense electric field which change the original band structure and consider its effect in the subsequent study in each case.

(R.D.2) Investigate the ER as defined in (R.D.1) in the presence of an arbitrarily oriented non-uniform light waves for all the HD materials as considered R.D.1.

(R.D.3) Investigate the ER as defined in (R.D.1) in the presence of an arbitrarily oriented non-quantizing alternating non-uniform electric field for all the cases of R.D.1.

(R.D.4) Investigate the ER as defined in (R.D.1) for all the materials in the presence of arbitrarily oriented non-quantizing non-uniform electric field for all the appropriate cases.

(R.D.5) Investigate the ER as defined in (R.D.1) for all the materials in the presence of arbitrarily oriented non-quantizing alternating electric field for all the appropriate cases of problem R.D.4.

(R.D.6) Investigate the ER as defined in (R.D.1) for the negative refractive index, organic, magnetic and other advanced optical materials in the presence of arbitrarily oriented electric field.

(R.D.7) Investigate the ER as defined in (R.D.1) in the presence of alternating non-quantizing electric field for all the problems of R.D.6.

(R.D.8) Investigate the ER as defined in (R.D.1) for all the quantum confined HD materials (i.e., multiple quantum wells and wires) whose unperturbed carrier energy spectra are defined in R.D.1 in the presence of arbitrary oriented quantizing magnetic field by including the effects of spin and broadening respectively.

(R.D.9) Investigate the ER as defined in (R.D.1) in the presence of an additional arbitrarily oriented alternating quantizing magnetic field respectively for all the problems of R.D.8.

(R.D.10) Investigate the ER as defined in (R.D.1) in the presence of arbitrarily oriented alternating quantizing magnetic field and arbitrary oriented non-quantizing non-uniform electric field respectively for all the problems of R.D.8.

(R.D.11) Investigate the ER as defined in (R.D.1) in the presence of arbitrary oriented alternating non- uniform quantizing magnetic field and additional arbitrary oriented non-quantizing alternating electric field respectively for all the problems of R.D.8.

(R.D.12) Investigate the ER as defined in (R.D.1) in the presence of arbitrary oriented and crossed quantizing magnetic and electric fields respectively for all the problems of R.D.8.

(R.D.13) Investigate the ER as defined in (R.D.1) for all the appropriate HD low-dimensional systems of this chapter in the presence of finite potential wells.

(R.D.14) Investigate the ER as defined in (R.D.1) for all the appropriate HD low-dimensional systems of this chapter in the presence of parabolic potential wells.

(R.D.15) Investigate the ER as defined in (R.D.1) for all the appropriate HD systems of this chapter forming quantum rings.

(R.D.16) Investigate the ER as defined in (R.D.1) for all the above appropriate problems in the presence of elliptical Hill and quantum square rings respectively.

(R.D.17) Investigate the ER as defined in (R.D.1) for multiple HD carbon nano-tubes.

(R.D.18) Investigate the ER as defined in (R.D.1) for multiple HD carbon nano-tubes in the presence of non-quantizing non-uniform alternating light waves.

(R.D.19) Investigate the ER as defined in (R.D.1) for multiple HD carbon nano-tubes in the presence of non-quantizing non-uniform alternating magnetic field.

(R.D.20) Investigate the ER as defined in (R.D.1) for multiple HD carbon nano-tubes in the presence of crossed electric and quantizing magnetic fields.

(R.D.21) Investigate the ER as defined in (R.D.1) for HD semiconductor nanotubes for all the materials whose unperturbed carrier dispersion laws are defined in Chap. 1.

(R.D.22) Investigate the ER as defined in (R.D.1) for HD semiconductor nanotubes in the presence of non-quantizing alternating light waves for all the materials whose unperturbed carrier dispersion laws is defined in Chap. 1.

(R.D.23) Investigate the ER as defined in (R.D.1) for HD semiconductor nanotubes in the presence of non-quantizing alternating magnetic field for all the materials whose unperturbed carrier dispersion laws are defined in Chap. 1.

(R.D.24) Investigate the ER as defined in (R.D.1) for HD semiconductor nanotubes in the presence of non-uniform light waves for all the materials whose unperturbed carrier dispersion laws are defined in Chap. 1.

(R.D.25) Investigate the ER as defined in (R.D.1) for HD semiconductor nanotubes in the presence of alternating quantizing magnetic fields for all the materials whose unperturbed carrier dispersion laws are defined in Chap. 1.

(R.D.26) Investigate the ER as defined in (R.D.1) for HD semiconductor nanotubes in the presence of crossed electric and quantizing magnetic fields for all the materials whose unperturbed carrier dispersion laws are defined in Chap. 1.

(R.D.27) Investigate the ER as defined in (R7.1) for all the appropriate nipi structures of the HD materials whose unperturbed carrier energy spectra are defined in Chap. 1.

(R.D.28) Investigate the ER as defined in (R.D.1) for all the appropriate nipi structures of the HD materials whose unperturbed carrier energy spectra are defined in Chap. 1, in the presence of an arbitrarily oriented non-quantizing non-uniform additional electric field.

(R.D.29) Investigate the ER as defined in (R.D.1) for all the appropriate nipi structures of the HD materials whose unperturbed carrier energy spectra are defined in Chap. 1 in the presence of non-quantizing alternating additional magnetic field.

(R.D.30) Investigate the ER as defined in (R.D.1) for all the appropriate nipi structures of the HD materials whose unperturbed carrier energy spectra are defined in Chap. 1 in the presence of quantizing alternating additional magnetic field.

(R.D.31) Investigate the ER as defined in (R.D.1) for all the appropriate nipi structures of the HD materials whose unperturbed carrier energy spectra are defined in Chap. 1 in the presence of crossed electric and quantizing magnetic fields.

(R.D.32) Investigate the ER as defined in (R.D.1) for HD nipi structures for all the appropriate cases of all the above problems.

(R.D.33) Investigate the ER as defined in (R.D.1) for the appropriate accumulation layers of all the materials whose unperturbed carrier energy spectra are defined in Chap. 1 in the presence of crossed electric and quantizing magnetic fields by considering electron spin and broadening of Landau levels.

(R.D.34) Investigate the ER as defined in (R.D.1) for quantum confined HD III-V, II-VI, IV-VI, HgTe/CdTe effective mass super-lattices together with short period, strained layer, random, Fibonacci, poly-type and saw-tooth super-lattices

(R.D.35) Investigate the ER as defined in (R.D.1) in the presence of quantizing magnetic field respectively for all the cases of R.D.35.

(R.D.36) Investigate the ER as defined in (R.D.1) in the presence of non-quantizing non-uniform additional electric field respectively for all the cases of R.D.35.

(R.D.37) Investigate the ER as defined in (R.D.1) in the presence of non-quantizing alternating electric field respectively for all the cases of R.D.35

(R.D.38) Investigate the ER as defined in (R.D.1) in the presence of crossed electric and quantizing magnetic fields respectively for all the cases of R.D.35.

(R.D.39) Investigate the ER as defined in (R.D.1) for heavily doped quantum confined super-lattices for all the problems of R.D.35.

(R.D.40) Investigate the ER as defined in (R.D.1) in the presence of quantizing non-uniform magnetic field respectively for all the cases of R.D.40.

(R.D.41) Investigate the ER as defined in (R.D.1) in the presence of crossed electric and quantizing magnetic fields respectively for all the cases of R.D.40.

(R.D.42) Investigate the ER as defined in (R.D.1) for all the systems in the presence of strain.

(R.D.43) Investigate all the problems of this chapter by removing all the mathematical approximations and establishing the respective appropriate uniqueness conditions.

Reference

1. K.P. Ghatak, S. Bhattacharya, *Debye Screening Length: Effects of Nanostructured Materials, Springer Tracts in Modern Physics*, vol. 255 (Springer, Heidelberg, 2014)

Chapter 10
Appendix E: The ER for HD III-V, Ternary and Quaternary Semiconductors Under External Photo-Excitation

10.1 Introduction

With the advent of nano-photonics, there has been a considerable interest in studying the optical processes in semiconductors and their nanostructures [1]. It appears from the literature, that the investigations have been carried out on the assumption that the carrier energy spectra are invariant quantities in the presence of intense light waves, which is not fundamentally true. The physical properties of semiconductors in the presence of light waves which change the basic dispersion relation have relatively less investigated in the literature [2–12]. In this chapter we shall study the ER in HD III-V, ternary and quaternary semiconductors on the basis of newly formulated electron dispersion law under external photo excitation.

In Sect. 10.2.1 of the theoretical background Sect. 10.2, we have formulated the dispersion relation of the conduction electrons of HD III-V, ternary and quaternary materials in the presence of light waves whose unperturbed electron energy spectrum is described by the three-band model of Kane in the absence of band tailing. In the same section, we have studied the dispersion relations for the said HD materials in the presence of external photo-excitation when the unperturbed energy spectra are defined by the two band model of Kane and that of parabolic energy bands in the absence of band tails respectively for the purpose of relative comparison. In Sect. 10.2.2, we have studied the ER for all the aforementioned cases. In Sect. 10.2.3, we have studied the opto ER in the said HD materials under magnetic quantization. In Sect. 10.2.4, we have studied the opto ER in the presence of crossed electric and quantizing magnetic fields. In Sect. 10.2.5, we have studied the opto ER in the presence of size quantization. In Sect. 10.2.6, we have studied the opto ER in HD doping superlattices. In Sect. 10.2.7, we have studied the opto ER in the accumulation layers. The Sect. 10.3 contains the open research problems.

K. P. Ghatak and S. Bhattacharya, *Heavily-Doped 2D-Quantized Structures and the Einstein Relation*, Springer Tracts in Modern Physics 260,
DOI 10.1007/978-3-319-08380-3_10, © Springer International Publishing Switzerland 2015

10.2 Theoretical Background

10.2.1 The Formulation of the Electron Dispersion Law in the Presence of Light Waves in III-V, Ternary and Quaternary Semiconductors

The Hamiltonian ($\hat{H}$) of an electron in the presence of light wave characterized by the vector potential $\vec{A}$ can be written following [11–13] as

$$\hat{H} = \left[\left| \left(\hat{p} + |e|\vec{A} \right) \right|^2 \Big/ 2m \right] + V(\vec{r}) \tag{10.1}$$

in which, $\hat{p}$ is the momentum operator, $V(\vec{r})$ is the crystal potential and m is the free electron mass. Equation (10.1) can be expressed as

$$\hat{H} = \hat{H}_0 + \hat{H}' \tag{10.2}$$

where,

$$\hat{H}_0 = \frac{\hat{p}^2}{2m} + V(\vec{r}) \quad \text{and} \quad \hat{H}' = \frac{|e|}{2m}\vec{A} \cdot \hat{p} \tag{10.3}$$

The perturbed Hamiltonian $\hat{H}'$ can be written as

$$\hat{H}' = \left(\frac{-i\hbar|e|}{2m} \right) \left(\vec{A} \cdot \nabla \right) \tag{10.4}$$

where, $i = \sqrt{-1}$ and $\hat{p} = -i\hbar\nabla$

The vector potential ($\vec{A}$) of the monochromatic light of plane wave can be expressed as

$$\vec{A} = A_0 \vec{\varepsilon}_s \cos(\vec{s}_0 \cdot \vec{r} - \omega t) \tag{10.5}$$

where A_0 is the amplitude of the light wave, $\vec{\varepsilon}_s$ is the polarization vector, $\vec{s}_0$ is the momentum vector of the incident photon, $\vec{r}$ is the position vector, ω is the angular frequency of light wave and t is the time scale. The matrix element of $\hat{H}'_{nl}$ between initial state, $\psi_l(\vec{q}, \vec{r})$ and final state $\psi_n\left(\vec{k}, \vec{r}\right)$ in different bands can be written as

$$\hat{H}'_{nl} = \frac{|e|}{2m} \left\langle n\vec{k} | \vec{A} \cdot \hat{p} | l\vec{q} \right\rangle \tag{10.6}$$

Using (10.4) and (10.5), we can re-write (10.6) as

$$\hat{H}'_{nl} = \left(\frac{-i\hbar|e|A_0}{4m}\right)\vec{\varepsilon}_s \cdot \left[\left\{\left\langle n\vec{k}\middle|e^{(i\vec{s}_0\cdot\vec{r})}\nabla\middle|l\vec{q}\right\rangle e^{-i\omega t}\right\} + \left\{\left\langle n\vec{k}\middle|e^{(i\vec{s}_0\cdot\vec{r})}\nabla\middle|l\vec{q}\right\rangle e^{i\omega t}\right\}\right] \tag{10.7}$$

The first matrix element of (10.7) can be written as

$$\begin{aligned}\left\langle n\vec{k}\middle|e^{(i\vec{s}_0\cdot\vec{r})}\nabla\middle|l\vec{q}\right\rangle &= \int e^{\left(i\left[\vec{q}+\vec{s}_0-\vec{k}\right]\cdot\vec{r}\right)} i\vec{q}u_n^*\left(\vec{k},\vec{r}\right)u_l(\vec{q},\vec{r})d^3r \\ &+ \int e^{\left(i\left[\vec{q}+\vec{s}_0-\vec{k}\right]\cdot\vec{r}\right)} u_n^*\left(\vec{k},\vec{r}\right)\nabla u_l(\vec{q},\vec{r})d^3r\end{aligned} \tag{10.8}$$

The functions $u_n^* u_1$ and $u_n^* \nabla u_1$ are periodic. The integral over all space can be separated into a sum over unit cells times an integral over a single unit cell. It is assumed that the wave length of the electromagnetic wave is sufficiently large so that if $\vec{k}$ and $\vec{q}$ are within the Brillouin zone, $\left(\vec{q}+\vec{s}_0-\vec{k}\right)$ is not a reciprocal lattice vector.

Therefore, we can write (10.8) as

$$\begin{aligned}\left\langle n\vec{k}\middle|e^{(i\vec{s}_0\cdot\vec{r})}\nabla\middle|l\vec{q}\right\rangle &= \left[\frac{(2\pi)^3}{\Omega}\right]\left\{i\vec{q}\delta\left(\vec{q}+\vec{s}_0-\vec{k}\right)\delta_{nl} + \delta\left(\vec{q}+\vec{s}_0-\vec{k}\right)\int_{cell} u_n^*\left(\vec{k},\vec{r}\right)\nabla u_l(\vec{q},\vec{r})d^3r\right\} \\ &= \left[\frac{(2\pi)^3}{\Omega}\right]\left\{\delta\left(\vec{q}+\vec{s}_0-\vec{k}\right)\int_{cell} u_n^*\left(\vec{k},\vec{r}\right)\nabla u_l(\vec{q},\vec{r})d^3r\right\}\end{aligned} \tag{10.9}$$

where, Ω is the volume of the unit cell and $\int u_n^*(\vec{k},\vec{r})u_l(\vec{q},\vec{r})d^3r = \delta(\vec{q}-\vec{k})\delta_{nl} = 0$, since $n \neq l$.

The delta function expresses the conservation of wave vector in the absorption of light wave and $\vec{s}_0$ is small compared to the dimension of a typical Brillouin zone and we set $\vec{q} = \vec{k}$.

From (10.8) and (10.9), we can write,

$$\hat{H}'_{nl} = \frac{|e|A_0}{2m}\vec{\varepsilon}_s \cdot \hat{p}_{nl}(\vec{k})\delta(\vec{q}-\vec{k})\cos(\omega t) \tag{10.10}$$

where, $\hat{p}_{nl}(\vec{k}) = -i\hbar\int u_n^*\nabla u_l d^3r = \int u_n^*(\vec{k},\vec{r})\hat{p}u_l(\vec{k},\vec{r})d^3r$.

Therefore, we can write

$$\hat{H}'_{nl} = \frac{|e|A_0}{2m}\vec{\varepsilon} \cdot \hat{p}_{nl}(\vec{k}) \tag{10.11}$$

where, $\vec{\varepsilon} = \vec{\varepsilon}_s \cos\omega t$.

When a photon interacts with a semiconductor, the carriers (i.e., electrons) are generated in the bands which are followed by the inter-band transitions. For example, when the carriers are generated in the valence band, the carriers then make inter-band transition to the conduction band. The transition of the electrons within the same band i.e., $\hat{H}'_{nn} = \left\langle n\vec{k}|\hat{H}'|n\vec{k}\right\rangle$ is neglected. Because, in such a case, i.e., when the carriers are generated within the same bands by photons, are lost by recombination within the aforementioned band resulting zero carriers.

Therefore,

$$\left\langle n\vec{k}|\hat{H}'|n\vec{k}\right\rangle = 0 \tag{10.12}$$

With $n = c$ stands for conduction band and $l = v$ stand for valance band, the energy equation for the conduction electron can approximately be written as

$$I_{11}(E) = \left(\frac{\hbar^2 k^2}{2m_c}\right) + \frac{\left(\frac{|e|A_0}{2m}\right)^2 \left\langle \left|\vec{\varepsilon}\cdot\hat{p}_{cv}(\vec{k})\right|^2\right\rangle_{av}}{E_c(\vec{k}) - E_v(\vec{k})} \tag{10.13}$$

where, $I_{11}(E) \equiv E(aE+1)(bE+1)/(cE+1)$, $a \equiv 1/E_{g_0}$, E_{g_0} is the un-perturbed band-gap, $b \equiv 1/(E_{g_0}+\Delta)$, $c \equiv 1/(E_{g_0}+2\Delta/3)$, and $\left\langle \left|\vec{\varepsilon}\cdot\hat{p}_{cv}(\vec{k})\right|^2\right\rangle_{av}$ represents the average of the square of the optical matrix element (OME).

For the three-band model of Kane, we can write,

$$\xi_{1k} = E_c(\vec{k}) - E_v(\vec{k}) = (E_{g_0}^2 + E_{g_0}\hbar^2 k^2/m_r)^{1/2} \tag{10.14}$$

where, m_r is the reduced mass and is given by $m_r^{-1} = (m_c)^{-1} + m_v^{-1}$, and m_v is the effective mass of the heavy hole at the top of the valance band in the absence of any field.

The doubly degenerate wave functions $u_1(\vec{k},\vec{r})$ and $u_2(\vec{k},\vec{r})$ can be expressed as [12]

$$u_1(\vec{k},\vec{r}) = a_{k+}[(is)\downarrow'] + b_{k+}\left[\frac{X'-iY'}{\sqrt{2}}\uparrow'\right] + c_{k+}[Z'\downarrow_{\prime}] \tag{10.15}$$

and

$$u_2(\vec{k},\vec{r}) = a_{k-}[(is)\uparrow'] - b_{k-}\left[\frac{X'+iY'}{\sqrt{2}}\downarrow'\right] + c_{k-}[Z'\uparrow'] \tag{10.16}$$

s is the s-type atomic orbital in both unprimed and primed coordinates, $\downarrow'$ indicates the spin down function in the primed coordinates,

$$
\begin{aligned}
a_{k\pm} &\equiv \beta\Big[E_{g_0} - (\gamma_{0k\pm})^2 (E_{g_0} - \delta')\big]^{1/2} (E_{g_0} + \delta')^{-1/2}\Big], \\
\beta &\equiv \left[(6(E_{g_0} + 2\Delta/3)(E_{g_0} + \Delta))/\chi\right]^{1/2}, \\
\chi &\equiv \left(6E_{g_0}^2 + 9E_{g_0}\Delta + 4\Delta^2\right), \\
\gamma_{0k\pm} &\equiv \left[\frac{(\xi_{1k} \mp E_{g_0})}{2(\xi_{1k} + \delta')}\right], \\
\xi_{1k} &\equiv E_c(\vec{k}) - E_v(\vec{k}) = E_{g_0}\left[1 + 2\left(1 + \frac{m_c}{m_v}\right)\frac{I_{11}(E)}{E_{g_0}}\right]^{1/2}, \\
\delta' &\equiv \left(E_{g_0}^2 \Delta\right)(\chi)^{-1},\ X',\ Y',\ \mathit{and}\ Z'
\end{aligned}
$$

are the p-type atomic orbitals in the primed coordinates, $\uparrow'$ indicates the spin-up function in the primed coordinates, $b_{k\pm} \equiv \rho\gamma_{0k\pm}$, $\rho \equiv \left(4\Delta^2/3\chi\right)^{1/2}$, $c_{k\pm} \equiv t\gamma_{0k\pm}$ and $t \equiv \left[6(E_{g_0} + 2\Delta/3)^2/\chi\right]^{1/2}$.

We can, therefore, write the expression for the optical matrix element (OME) as

$$
\text{OME} = \hat{p}_{cv}\left(\vec{k}\right) = \left\langle u_1(\vec{k}, \vec{r})|\hat{p}|u_2(\vec{k}, \vec{r})\right\rangle \tag{10.17}
$$

Since the photon vector has no interaction in the same band for the study of inter-band optical transition, we can therefore write

$$
\langle S|\hat{p}|S\rangle = \langle X|\hat{p}|X\rangle = \langle Y|\hat{p}|Y\rangle = \langle Z|\hat{p}|Z\rangle = 0
$$

and

$$
\langle X|\hat{p}|Y\rangle = \langle Y|\hat{p}|Z\rangle = \langle Z|\hat{p}|X\rangle = 0
$$

There are finite interactions between the conduction band (CB) and the valance band (VB) and we can obtain

$$
\begin{aligned}
\langle S|\hat{p}|X\rangle &= \hat{i}\cdot\hat{P} = \hat{i}\cdot\hat{P}_x \\
\langle S|\hat{p}|Y\rangle &= \hat{j}\cdot\hat{P} = \hat{j}\cdot\hat{P}_y \\
\langle S|\hat{p}|Z\rangle &= \hat{k}\cdot\hat{P} = \hat{k}\cdot\hat{P}_z
\end{aligned}
$$

where, $\hat{i}$, $\hat{j}$ and $\hat{k}$ are the unit vectors along x, y and z axes respectively.

It is well known [12] that

$$
\begin{bmatrix} \uparrow' \\ \downarrow' \end{bmatrix} = \begin{bmatrix} e^{-i\phi/2}\cos(\theta/2) & e^{i\phi/2}\sin(\theta/2) \\ e^{-i\phi/2}\sin(\theta/2) & e^{i\phi/2}\cos(\theta/2) \end{bmatrix} \begin{bmatrix} \uparrow \\ \downarrow \end{bmatrix}
$$

and

$$\begin{bmatrix} X' \\ Y' \\ Z' \end{bmatrix} = \begin{bmatrix} \cos\theta\cos\phi & \cos\theta\sin\phi & -\sin\phi \\ -\sin\phi & \cos\phi & 0 \\ \sin\theta\cos\phi & \sin\theta\sin\phi & \cos\theta \end{bmatrix} \begin{bmatrix} X \\ Y \\ Z \end{bmatrix}$$

Besides, the spin vector can be written as

$$\vec{S} = \frac{\hbar}{2}\vec{\sigma}, \quad \text{where, } \sigma_x = \begin{bmatrix} 0 & 1 \\ 1 & 0 \end{bmatrix}, \ \sigma_y = \begin{bmatrix} 0 & -i \\ i & 0 \end{bmatrix} \text{ and } \sigma_Z = \begin{bmatrix} 1 & 0 \\ 0 & -1 \end{bmatrix}.$$

From above, we can write

$$\begin{aligned} \hat{p}_{CV}(\vec{k}) &= \left\langle u_1\left(\vec{k},\vec{r}\right)|\hat{P}|u_2\left(\vec{k},\vec{r}\right)\right\rangle \\ &= \left\langle\left\{a_{k_+}[(iS)\downarrow'] + b_{k_+}\left[\left(\frac{X'-iY'}{\sqrt{2}}\right)\uparrow'\right] + c_{k_+}[Z'\downarrow']\right\}|\hat{P}|a_{k_-}[(is)\uparrow'] \right. \\ &\quad \left. -b_{k_-}\left(\frac{X'+iY'}{\sqrt{2}}\right)\downarrow' + c_{k_-}[Z'\uparrow']\right\rangle \end{aligned}$$

Using above relations, we get

$$\begin{aligned} \hat{p}_{CV}\left(\vec{k}\right) &= \left\langle u_1\left(\vec{k},\vec{r}\right)|\hat{P}|u_2\left(\vec{k},\vec{r}\right)\right\rangle \\ &= \frac{b_{k_+}a_{k_-}}{\sqrt{2}}\{\langle(X'-iY')|\hat{P}|iS\rangle\langle\uparrow' \mid \uparrow'\rangle\} + c_{k_+}a_{k_-}\{\langle Z'|\hat{P}|iS\rangle\langle\downarrow' \mid \uparrow'\rangle\} \\ &\quad - \frac{a_{k_+}b_{k_-}}{\sqrt{2}}\{\langle iS|\hat{P}|(X'+iY')\rangle\langle\downarrow' \mid \downarrow'\rangle\} + a_{k_+}c_{k_-}\{\langle iS|\hat{P}|Z'\rangle\langle\downarrow' \mid \uparrow'\rangle\} \end{aligned} \tag{10.18}$$

From (10.18), we can write

$$\begin{aligned} \langle(X'-iY')|\hat{P}|iS\rangle &= \langle(X')|\hat{P}|iS\rangle - \langle(iY')|\hat{P}|iS\rangle \\ &= i\int u_{X'}^*\hat{P}S - \int -iu_{Y'}^*\hat{P}iu_X = i\langle X'|\hat{P}|S\rangle - \langle Y'|\hat{P}|S\rangle \end{aligned}$$

From the above relations, for X', Y' and Z', we get

$$|X'\rangle = \cos\theta\cos\phi|X\rangle + \cos\theta\sin\phi|Y\rangle - \sin\theta|Z\rangle$$

Thus,

$$\langle X'|\hat{P}|S\rangle = \cos\theta\cos\phi\langle X|\hat{P}|S\rangle + \cos\theta\sin\phi\langle Y|\hat{P}|S\rangle - \sin\theta\langle Z|\hat{P}|S\rangle = \hat{P}\hat{r}_1$$

where, $\hat{r}_1 = \hat{i}\cos\theta\cos\phi + \hat{j}\cos\theta\sin\phi - \hat{k}\sin\theta$

$$|Y'\rangle = -\sin\phi|X\rangle + \cos\phi|Y\rangle + 0|Z\rangle$$

Thus,

$$\langle Y'|\hat{P}|S\rangle = -\sin\phi\langle X|\hat{P}|S\rangle + \cos\phi\langle Y|\hat{P}|S\rangle + 0\langle Z|\hat{P}|S\rangle = \hat{P}\hat{r}_2$$

where $\hat{r}_2 = -\hat{i}\sin\phi + \hat{j}\cos\phi$
so that $\langle (X' - iY')|\hat{P}|S\rangle = \hat{P}(i\hat{r}_1 - \hat{r}_2)$
Thus,

$$\frac{a_{k_-}b_{k_+}}{\sqrt{2}}\langle (X' - iY')|\hat{P}|S\rangle\left\langle \uparrow'|\uparrow'\right\rangle = \frac{a_{k_-}b_{k_+}}{\sqrt{2}}\hat{P}(i\hat{r}_1 - i\hat{r}_2)\left\langle \uparrow'|\uparrow'\right\rangle \tag{10.19}$$

Now since,

$$\langle iS|\hat{P}|(X' + iY')\rangle = i\langle S|\hat{P}|X'\rangle - \langle S|\hat{P}|Y'\rangle = \hat{P}(i\hat{r}_1 - \hat{r}_2)$$

We can write,

$$-\left[\frac{a_{k_+}a_{k_-}}{\sqrt{2}}\{\langle iS|\hat{P}|(X' + iY')\rangle\langle\downarrow' \mid \downarrow'\rangle\}\right] = -\left[\frac{a_{k_+}b_{k_-}}{\sqrt{2}}\hat{P}(i\hat{r}_1 - \hat{r}_2)\langle\downarrow' \mid \downarrow'\rangle\right] \tag{10.20}$$

Similarly, we get

$$|Z'\rangle = \sin\theta\cos\phi\,|X\rangle + \sin\theta\sin\phi\,|Y\rangle + \cos\theta\,|Z\rangle$$

So that,

$$\langle Z'|\hat{P}|iS\rangle = i\langle Z'|\hat{P}|S\rangle = i\hat{P}\{\sin\theta\cos\phi\,\hat{i} + \sin\theta\sin\phi\,\hat{j} + \cos\theta\,\hat{k}\} = i\hat{P}\hat{r}_3$$

where, $\hat{r}_3 = \hat{i}\sin\theta\cos\phi + \hat{j}\sin\theta\sin\phi + \hat{k}\cos\theta$
Thus,

$$c_{k_+}a_{k_-}\langle Z'|\hat{P}|iS\rangle\langle\downarrow' \mid \uparrow'\rangle = c_{k_+}a_{k_-}i\hat{P}\hat{r}_3\langle\downarrow' \mid \uparrow'\rangle \tag{10.21}$$

Similarly, we can write,

$$c_{k_-}a_{k_+}\langle iS|\hat{P}|Z'\rangle\langle\downarrow'\mid\uparrow'\rangle = c_{k_-}a_{k_+}i\hat{P}\hat{r}_3\langle\downarrow'\mid\uparrow'\rangle \tag{10.22}$$

Therefore, we obtain

$$\begin{aligned}&\frac{a_{k_-}b_{k_+}}{\sqrt{2}}\{\langle(X'-iY')|\hat{P}|S\rangle\langle\uparrow'\mid\uparrow'\rangle\} - \frac{a_{k_+}b_{k_-}}{\sqrt{2}}\{\langle iS|\hat{P}|(X'+iY')\rangle\langle\downarrow'\mid\downarrow'\rangle\}\\ &= \frac{\hat{P}}{\sqrt{2}}\left(-a_{k_+}b_{k_-}\langle\downarrow'\mid\downarrow'\rangle + a_{k_-}b_{k_+}\langle\uparrow'\mid\uparrow'\rangle\right)(i\hat{r}_1-\hat{r}_2)\end{aligned} \tag{10.23}$$

Also, we can write,

$$\begin{aligned}&c_{k_+}a_{k_-}\langle Z'|\hat{P}|iS\rangle\langle\uparrow'\mid\uparrow'\rangle + c_{k_-}a_{k_+}\langle iS|\hat{P}|Z'\rangle\langle\downarrow'|\uparrow'\rangle\\ &= i\hat{p}\left(c_{k_+}a_{k_-}+c_{k_-}a_{k_+}\right)\hat{r}_3[\langle\downarrow'\mid\downarrow'\rangle]\end{aligned} \tag{10.24}$$

Combining (10.23) and (10.24), we find

$$\begin{aligned}\hat{p}_{CV}(\vec{k}) = &\frac{\hat{P}}{\sqrt{2}}(i\hat{r}_1-\hat{r}_2)\left\{(b_{k_+}a_{k_-})\langle\uparrow'|\uparrow'\rangle - (b_{k_-}a_{k_+})\langle\downarrow'\mid\downarrow'\rangle\right\}\\ &+ i\hat{P}\hat{r}_3\left(c_{k_+}a_{k_-}-c_{k_-}a_{k_+}\right)\langle\downarrow'\mid\uparrow'\rangle\end{aligned} \tag{10.25}$$

From the above relations, we obtain,

$$\left.\begin{aligned}\uparrow' &= e^{-i\phi/2}\cos(\theta/2)\uparrow + e^{i\phi/2}\sin(\theta/2)\downarrow\\ \downarrow' &= -e^{-i\phi/2}\sin(\theta/2)\uparrow + e^{i\phi/2}\cos(\theta/2)\downarrow\end{aligned}\right\} \tag{10.26}$$

Therefore,

$$\begin{aligned}\langle\downarrow'\mid\uparrow'\rangle_x = &-\sin(\theta/2)\cos(\theta/2)\langle\uparrow\mid\uparrow\rangle_x + e^{-i\phi}\cos^2(\theta/2)\langle\downarrow\mid\uparrow\rangle_x\\ &- e^{-i\phi}\sin^2(\theta/2)\langle\uparrow\mid\downarrow\rangle_x + \sin(\theta/2)\cos(\theta/2)\langle\downarrow\mid\downarrow\rangle_x\end{aligned} \tag{10.27}$$

But we know from above that

$$\langle\uparrow\mid\uparrow\rangle_x = 0, \langle\downarrow\mid\uparrow\rangle = \frac{1}{2}, \langle\downarrow\mid\uparrow\rangle_x = \frac{1}{2} \text{ and } \langle\downarrow\mid\downarrow\rangle_x = 0$$

Thus, from (10.27), we get

$$\begin{aligned}\left\langle \downarrow' | \uparrow' \right\rangle_x &= \frac{1}{2}\left[e^{-i\phi}\cos^2(\theta/2) - e^{i\phi}\sin^2(\theta/2)\right] \\ &= \frac{1}{2}\left[(\cos\phi - i\sin\phi)\cos^2(\theta/2) - (\cos\phi + i\sin\phi)\sin^2(\theta/2)\right] \\ &= \frac{1}{2}[\cos\phi\cos\theta - i\sin\phi] \end{aligned} \tag{10.28}$$

Similarly, we obtain

$$\langle\downarrow' \mid \uparrow'\rangle_y = \frac{1}{2}[i\cos\phi + \sin\phi\cos\theta] \text{ and } \langle\downarrow' \mid \uparrow'\rangle_z = \frac{1}{2}[-\sin\theta]$$

Therefore,

$$\begin{aligned}\langle\downarrow' \mid \uparrow'\rangle &= \hat{i}\langle\downarrow' \mid \uparrow'\rangle_x + \hat{j}\langle\downarrow' \mid \uparrow'\rangle_y + \hat{k}\langle\downarrow' \mid \uparrow'\rangle_z \\ &= \frac{1}{2}\left\{(\cos\theta\cos\phi - i\sin\phi)\hat{i} + (i\cos\phi + \sin\phi\cos\theta)\hat{j} - \sin\theta\hat{k}\right\} \\ &= \frac{1}{2}\left[\left\{(\cos\theta\cos\phi)\hat{i} + (\sin\phi\cos\theta)\hat{j} - \sin\theta\hat{k}\right\} + i\left\{-\hat{i}\sin\phi + \hat{j}\cos\phi\right\}\right] \\ &= \frac{1}{2}[\hat{r}_1 + i\hat{r}_2] = -\frac{1}{2}i[i\hat{r}_1 - \hat{r}_2]\end{aligned}$$

Similarly, we can write

$$\left\langle \uparrow' | \uparrow' \right\rangle = \frac{1}{2}\left[\hat{i}\sin\theta\cos\phi + \hat{j}\sin\theta\sin\phi + \hat{k}\cos\theta\right] = \frac{1}{2}\hat{r}_3 \quad \text{and} \quad \langle\downarrow' \mid \downarrow'\rangle = -\frac{1}{2}\hat{r}_3$$

Using the above results and following (10.25) we can write

$$\begin{aligned}\hat{p}_{CV}\left(\vec{k}\right) &= \frac{\hat{P}}{\sqrt{2}}(i\hat{r}_1 - \hat{r}_2)\left\{\left(a_{k_-}b_{k_+}\right)\langle\uparrow' \mid \uparrow'\rangle - \left(b_{k_-}a_{k_+}\right)\langle\downarrow' \mid \downarrow'\rangle\right\} \\ &\quad + i\hat{P}\hat{r}_3\left\{\left(c_{k_+}a_{k_-} - c_{k_-}a_{k_+}\right)\langle\downarrow' \mid \uparrow'\rangle\right\} \\ &= \frac{\hat{P}}{2}\hat{r}_3(i\hat{r}_1 - \hat{r}_2)\left\{\left(\frac{a_{k_-}b_{k_+}}{\sqrt{2}} + \frac{b_{k_-}a_{k_+}}{\sqrt{2}}\right)\right\} \\ &\quad + \frac{\hat{P}}{2}\hat{r}_3(i\hat{r}_1 - \hat{r}_2)\left\{\left(c_{k_+}a_{k_-} + c_{k_-}a_{k_+}\right)\right\}\end{aligned}$$

Thus,

$$\hat{p}_{CV}\left(\vec{k}\right) = \frac{\hat{P}}{2}\hat{r}_3(i\hat{r}_1 - \hat{r}_2)\left\{a_{k_+}\left(\frac{b_{k_-}}{\sqrt{2}} + c_{k_-}\right) + a_{k_-}\left(\frac{b_{k_+}}{\sqrt{2}} + c_{k_+}\right)\right\} \tag{10.29}$$

We can write that,

$$|\hat{r}_1| = |\hat{r}_2| = |\hat{r}_3| = 1, \text{also}, \hat{P}\hat{r}_3 = \hat{P}_x \sin\theta\cos\phi\hat{i} + \hat{P}_y \sin\theta\sin\phi\hat{j} + \hat{P}_z\cos\theta\hat{k}$$

where, $\hat{P} = \langle S|\hat{P}|X\rangle = \langle S|\hat{P}|Y\rangle = \langle S|\hat{P}|Z\rangle$,

$$\langle S|\hat{P}|X\rangle = \int u_C^*(0,\vec{r})\hat{P}u_{VX}(0,\vec{r})d^3r = \hat{P}_{CVX}(0) \quad \text{and} \quad \langle S|\hat{P}|Z\rangle = \hat{P}_{CVZ}(0)$$

Thus,

$$\hat{P} = \hat{P}_{CVX}(0) = \hat{P}_{CVY}(0) = \hat{P}_{CVZ}(0) = \hat{P}_{CV}(0)$$

where, $\hat{P}_{CV}(0) \equiv \int u_c^*(0,\vec{r})\hat{P}u_V(0,\vec{r})d^3r \equiv \hat{P}$

For a plane polarized light wave, we have the polarization vector $\vec{\varepsilon}_s = \hat{k}$, when the light wave vector is traveling along the z-axis. Therefore, for a plane polarized light-wave, we have considered $\vec{\varepsilon}_s = \hat{k}$.

Then, from (10.29) we get

$$\left(\vec{\varepsilon}\cdot\hat{p}_{CV}\left(\vec{k}\right)\right) = \vec{k}\cdot\frac{\hat{P}}{2}\hat{r}_3(i\hat{r}_1 - \hat{r}_2)\left[A\left(\vec{k}\right) + B\left(\vec{k}\right)\right]\cos\omega t \tag{10.30}$$

and

$$\left.\begin{aligned} A(\vec{k}) &= a_{k_-}\left(\tfrac{b_{k_+}}{\sqrt{2}} + c_{k_+}\right) \\ B(\vec{k}) &= a_{k_+}\left(\tfrac{b_{k_-}}{\sqrt{2}} + c_{k_-}\right) \end{aligned}\right\} \tag{10.31}$$

Thus,

$$\begin{aligned} \left|\vec{\varepsilon}\cdot\hat{p}_{cv}\left(\vec{k}\right)\right|^2 &= \left|\hat{k}\cdot\frac{\hat{P}}{2}\hat{r}_3\right|^2 |i\hat{r}_1 - \hat{r}_2|^2\left[A(\vec{k}) + B(\vec{k})\right]^2\cos^2\omega t \\ &= \frac{1}{4}\left|\hat{P}_z\cos\theta\right|^2\left[A(\vec{k}) + B(\vec{k})\right]^2\cos^2\omega t \end{aligned} \tag{10.32}$$

So, the average value of $\left|\vec{\varepsilon}\cdot\hat{p}_{cv}\left(\vec{k}\right)\right|^2$ for a plane polarized light-wave is given by

$$\begin{aligned}\left\langle\left|\vec{\varepsilon}\cdot\hat{P}_{cv}\left(\vec{k}\right)\right|^2\right\rangle_{av} &= \frac{2}{4}\left|\hat{P}_z\right|^2\left[A(\vec{k})+B(\vec{k})\right]^2\left(\int_0^{2\pi}d\phi\int_0^{\pi}\cos^2\theta\sin\theta d\theta\right)\left(\frac{1}{2}\right)\\ &= \frac{2\pi}{3}\left|\hat{P}_z\right|^2\left[A(\vec{k})+B(\vec{k})\right]^2\end{aligned} \tag{10.33}$$

where $\left|\hat{P}_z\right|^2=\left(\frac{1}{2}\right)\left|\vec{k}\cdot\hat{p}_{cv}(0)\right|^2$ and

$$\left|\vec{k}\cdot\hat{p}_{cv}(0)\right|^2=\frac{m^2}{4m_r}\frac{E_{g_0}\left(E_{g_0}+\Delta\right)}{\left(E_{g_0}+\frac{2}{3}\Delta\right)} \tag{10.34}$$

We shall express $A(\vec{k})$ and $B(\vec{k})$ in terms of constants of the energy spectra in the following way:

Substituting $a_{k\pm}$, $b_{k\pm}$, $c_{k\pm}$ and $\gamma_{0k\pm}$ in $A(\vec{k})$ and $B(\vec{k})$ in (10.31) we get

$$A(\vec{k})=\beta\left(t+\frac{\rho}{\sqrt{2}}\right)\left\{\left(\frac{E_{g0}}{E_{g_0}+\delta'}\right)\gamma_{0k_+}^2-\gamma_{0k_+}^2\gamma_{0k_-}^2\left(\frac{E_{g_0}-\delta'}{E_{g_0}+\delta'}\right)\right\}^{1/2} \tag{10.35}$$

$$B(\vec{k})=\beta\left(t+\frac{\rho}{\sqrt{2}}\right)\left\{\left(\frac{E_{g0}}{E_{g_0}+\delta'}\right)\gamma_{0k_-}^2-\gamma_{0k_+}^2\gamma_{0k_-}^2\left(\frac{E_{g_0}-\delta'}{E_{g_0}+\delta'}\right)\right\}^{1/2} \tag{10.36}$$

in which, $\gamma_{0k_+}^2\equiv\frac{\xi_{1k}-E_{g0}}{2(\xi_{1k}+\delta')}\equiv\frac{1}{2}\left[1-\left(\frac{E_{g0}+\delta'}{\xi_{1k}+\delta'}\right)\right]$ and $\gamma_{0k_-}^2\equiv\frac{\xi_{1k}+E_{g0}}{2(\xi_{1k}+\delta')}\equiv\frac{1}{2}\left[1+\left(\frac{E_{g0}-\delta'}{\xi_{1k}+\delta'}\right)\right]$
Substituting $x\equiv\xi_{1k}+\delta'$ in $\gamma_{0k\pm}^2$, we can write,

$$A(\vec{k})=\beta\left(t+\frac{\rho}{\sqrt{2}}\right)\left\{\left(\frac{E_{g0}}{E_{g_0}+\delta'}\right)\frac{1}{2}\left(1-\frac{E_{g_0}+\delta'}{x}\right)-\frac{1}{4}\left(\frac{E_{g0}-\delta'}{E_{g_0}+\delta'}\right)\left(1-\frac{E_{g_0}+\delta'}{x}\right)\left(1+\frac{E_{g0}-\delta'}{x}\right)\right\}^{1/2}$$

Thus,

$$A(\vec{k})=\frac{\beta}{2}\left(t+\frac{\rho}{\sqrt{2}}\right)\left\{1-\frac{2a_0}{x}+\frac{a_1}{x^2}\right\}^{1/2}$$

where, $a_0\equiv\left(E_{g_0}^2+\delta'^2\right)\left(E_{g_0}+\delta'\right)^{-1}$ and $a_1\equiv\left(E_{g_0}-\delta'\right)^2$.

After tedious algebra, one can show that

$$A(\vec{k}) = \frac{\beta}{2}\left(t + \frac{\rho}{\sqrt{2}}\right)(E_{g_0} - \delta')\left[\frac{1}{\xi_{1k} + \delta'} - \frac{1}{\xi_{g_0} + \delta'}\right]^{1/2}\left[\frac{1}{\xi_{1k} + \delta'} - \frac{(E_{g_0} + \delta')}{(E_{g_0} - \delta')}\right]^{1/2} \tag{10.37}$$

Similarly, from (10.36), we can write,

$$B(\vec{k}) = \beta\left(t + \frac{\rho}{\sqrt{2}}\right)\left\{\left(\frac{E_{g0}}{E_{g_0} + \delta'}\right)\frac{1}{2}\left[1 + \frac{E_{g_0} - \delta'}{x}\right] - \frac{1}{4}\left(\frac{E_{g0} - \delta'}{E_{g_0} + \delta'}\right)\left(1 - \frac{E_{g0} + \delta'}{x}\right)\left(1 + \frac{E_{g0} - \delta'}{x}\right)\right\}^{1/2}$$

So that, finally we get,

$$B(\vec{k}) = \frac{\beta}{2}\left(t + \frac{\rho}{\sqrt{2}}\right)\left(1 + \frac{(E_{g_0} - \delta')}{\xi_{1k} + \delta'}\right) \tag{10.38}$$

Using (10.33), (10.34), (10.37) and (10.38), we can write

$$\left(\frac{|e|A_0}{2m}\right)^2 \frac{\left\langle\left|\vec{\varepsilon}\cdot\hat{p}_{cv}(\vec{k})\right|^2\right\rangle_{av}}{E_c(\vec{k}) - E_v(\vec{k})} = \left(\frac{|e|A_0}{2m}\right)^2 \frac{2\pi}{3}\left|\vec{k}\cdot\hat{p}_{cv}(0)\right|^2 \frac{\beta^2}{4}\left(t + \frac{\rho}{\sqrt{2}}\right)^2$$
$$\frac{1}{\xi_{1k}}\left\{\left(1 + \frac{E_{g_0} - \delta'}{\xi_{1k} + \delta'}\right) + (E_{g_0} - \delta')\left[\frac{1}{\xi_{1k} + \delta'} - \frac{1}{E_{g_0} + \delta'}\right]^{1/2}\left[\frac{1}{\xi_{1k} + \delta'} - \frac{E_{g_0} + \delta'}{(E_{g_0} - \delta')^2}\right]^{1/2}\right\}^2 \tag{10.39}$$

Following Nag [12], it can be shown that

$$A_0^2 = \frac{I\lambda^2}{2\pi^2 c^3\sqrt{\varepsilon_{sc}\varepsilon_0}} \tag{10.40}$$

where, I is the light intensity of wavelength λ, ε_0 is the permittivity of free space and c is the velocity of light. Thus, the simplified electron energy spectrum in III-V, ternary and quaternary materials in the presence of light waves can approximately be written as

$$\frac{\hbar^2 k^2}{2m_c} = \beta_0(E, \lambda) \tag{10.41}$$

where, $\beta_0(E, \lambda) \equiv [I_{11}(E) - \theta_0(E, \lambda)]$,

$$\theta_0(E,\lambda) \equiv \frac{|e|^2}{96 m_r \pi c^3} \frac{I\lambda^2}{\sqrt{\varepsilon_{sc}\varepsilon_0}} \frac{E_{g_0}(E_{g_0}+\Delta)}{(E_{g_0}+\frac{2}{3}\Delta)} \frac{\beta^2}{4}\left(t+\frac{\rho}{\sqrt{2}}\right)^2 \frac{1}{\phi_0(E)}$$

$$\left\{\left(1+\frac{E_{g_0}-\delta'}{\phi_0(E)+\delta'}\right)+(E_{g_0}-\delta')\left[\frac{1}{\phi_0(E)+\delta'}-\frac{1}{E_{g_0}+\delta'}\right]^{1/2}\left[\frac{1}{\phi_0(E)+\delta'}-\frac{E_{g_0}+\delta'}{(E_{g_0}-\delta')^2}\right]^{1/2}\right\}^2$$

and $\phi_0(E) \equiv E_{g_0}\left(1+2\left(1+\frac{m_c}{m_v}\right)\frac{I_{11}(E)}{E_{g_0}}\right)^{1/2}$

Thus, under the limiting condition $\vec{k} \to 0$, from (10.41), we observe that $E \neq 0$ and is positive. Therefore, in the presence of external light waves, the energy of the electron does not tend to zero when $\vec{k} \to 0$, where as for the un-perturbed three band model of Kane, $I_{11}(E) = \left[\hbar^2 k^2/(2m_c)\right]$ in which $E \to 0$ for $\vec{k} \to 0$. As the conduction band is taken as the reference level of energy, therefore the lowest positive value of E for $\vec{k} \to 0$ provides the increased band gap(ΔE_g) of the semiconductor due to photon excitation. The values of the increased band gap can be obtained by computer iteration processes for various values of I and λ respectively.

Special Cases:

1. For the two-band model of Kane, we have $\Delta \to 0$. Under this condition, $I_{11}(E) \to E(1+aE) = \frac{\hbar^2 k^2}{2m_c}$. Since, $\beta \to 1$, $t \to 1$, $\rho \to 0$, $\delta' \to 0$ for $\Delta \to 0$, from (10.41), we can write the energy spectrum of III-V, ternary and quaternary materials in the presence of external photo-excitation whose unperturbed conduction electrons obey the two band model of Kane as

$$\frac{\hbar^2 k^2}{2m_c} = \tau_0(E,\lambda) \tag{10.42}$$

where, $\tau_0(E,\lambda) \equiv E(1+\alpha E) - B_0(E,\lambda)$,

$$B_0(E,\lambda) \equiv \frac{|e|^2 I\lambda^2 E_{g0}}{384\pi c^3 m_r \sqrt{\varepsilon_{sc}\varepsilon_0}} \frac{1}{\phi_1(E)}\left\{\left(1+\frac{E_{g0}}{\phi_1(E)}\right)+E_{g0}\left[\frac{1}{\phi(E)}-\frac{1}{E_{g0}}\right]\right\}^2,\ \phi_1(E)$$
$$\equiv E_{g0}\left\{1+\frac{2m_c}{m_r}aE(1+aE)\right\}^{1/2}.$$

2. For relatively wide band gap semiconductors, one can write, $a \to 0$, $b \to 0$, $c \to 0$ and $I_{11}(E) \to E$.

Thus, from (10.42), we get,

$$\frac{\hbar^2 k^2}{2m_c} = \rho_0(E, \lambda) \tag{10.43}$$

where, $\rho_0(E, \lambda) \equiv E - \frac{|e|^2 I \lambda^2}{96\pi c^3 m_r \sqrt{\varepsilon_{sc}\varepsilon_0}} \left[1 + \left(\frac{2m_c}{m_r}\right) aE\right]^{-3/2}$.

10.2.2 The ER in the Presence of Light Waves in HD III-V, Ternary and Quaternary Semiconductors

The (10.40), (10.41) and (10.42) can approximately be written as

$$\frac{\hbar^2 k^2}{2m_c^*} = U_\lambda I_{11}(\mathrm{E}) - \mathrm{P}_\lambda \tag{10.44}$$

$$\frac{\hbar^2 k^2}{2m_c^*} = t_{1\lambda} E + t_{2\lambda} E^2 - \delta_\lambda \tag{10.45}$$

and

$$\frac{\hbar^2 k^2}{2m_c^*} = t_{1\lambda} E - \delta_\lambda \tag{10.46a}$$

where

$$U_\lambda = (1 + \theta_\lambda), \theta_\lambda = \frac{C_0}{A}(t_\lambda + \frac{BJ_\lambda}{A}), C_0 = [\frac{|e|^2 I \lambda^2 E_{g0} + \Delta}{96 m_r \pi c^3 \sqrt{\varepsilon_{sc}\varepsilon_0}(E_{g0} + \frac{2}{3}\Delta)} \frac{\beta^2}{4}(1 + \frac{\rho}{\sqrt{2}})^2]$$

$$B = [1 + \frac{m^c}{m_V}], \mathrm{G}_\lambda = [\frac{2B}{(A + \delta')^3} - \frac{BC_\lambda}{(A + \delta')}]$$

$$C_\lambda = [(\mathrm{E}_{g0} + \delta')^{-1} + (\mathrm{E}_{g0} + \delta')(\mathrm{E}_{g0} - \delta')^{-2}](A + \delta')^{-1}$$

$$P_\lambda = \frac{C_0}{A} J_\lambda, J_\lambda = (\mathrm{D}_\lambda + 2(\mathrm{E}_{g0} - \delta')\sqrt{f_\lambda}),$$

$$D_\lambda = (1 + \frac{2(\mathrm{E}_{g0} - \delta')}{(A + \delta')}), \mathrm{f}_\lambda = [\frac{1}{(A + \delta')^2} + \frac{1}{(\mathrm{E}_{g0} - \delta')^2} - C_\lambda],$$

$$\mathrm{t}_{1\lambda} = (1 + \frac{3m_c}{m_r}\alpha\delta_\lambda), \alpha = \frac{1}{E_{g0}}, \delta_\lambda = \frac{|e|^2 I \lambda^2}{96 m_r \pi c^3 \sqrt{\varepsilon_{sc}\varepsilon_0}} \text{ and } t_{2\lambda} = \alpha t_{1\lambda}$$

Under the condition of heavy doping, following the methods as developed in Chap. 1, the HD dispersion relations in this case in the presence of light waves can be written as

$$\frac{\hbar^2 k^2}{2m_c} = T_1(\mathrm{E}, \eta_g, \lambda) \tag{10.46b}$$

$$\frac{\hbar^2 k^2}{2m_c} = T_2(\mathrm{E}, \eta_g, \lambda) \tag{10.47}$$

$$\frac{\hbar^2 k^2}{2m_c} = T_3(\mathrm{E}, \eta_g, \lambda) \tag{10.48}$$

where

$$T_1(\mathrm{E}, \eta_g, \lambda) = [U_\lambda[\mathrm{T}_{31}(\mathrm{E}, \eta_g) + i\mathrm{T}_{32}(\mathrm{E}, \eta_g)] - \mathrm{P}_\lambda],$$
$$\mathrm{T}_2(\mathrm{E}, \eta_g, \lambda) = [t_{1\lambda}\gamma_3(\mathrm{E}, \eta_g) + (t_{2\lambda})2\theta_0(\mathrm{E}, \eta_g)[1 + \mathrm{Erf}(E/\eta_g)]^{-1} - \delta_\lambda]$$

and

$$\mathrm{T}_3(\mathrm{E}, \eta_g, \lambda) = [t_{1\lambda}\gamma_3(\mathrm{E}, \eta_g) - \delta_\lambda]$$

The EEM can be expressed in this case by using (10.46a, 10.46b), (10.47) and (10.48).

$$m^*(\mathrm{E}_{FHDL}, \eta_g, \lambda) = \mathrm{m}_c \; \textit{Real part of} \; [\mathrm{T}_1(\mathrm{E}_{FHDL}, \eta_g, \lambda)]' \tag{10.49}$$

$$m^*(\mathrm{E}_{FHDL}, \eta_g, \lambda) = \mathrm{m}_c[\mathrm{T}_2(\mathrm{E}_{FHDL}, \eta_g, \lambda)]' \tag{10.50}$$

$$m^*(\mathrm{E}_{FHDL}, \eta_g, \lambda) = \mathrm{m}_c[\mathrm{T}_3(\mathrm{E}_{FHDL}, \eta_g, \lambda)]' \tag{10.51}$$

The electron concentration at low temperatures is given by

$$n_0 = \frac{g_v}{3\pi^2}\left(\frac{2m_c}{\hbar^2}\right)^{\frac{3}{2}} \textit{Real part of} \; [\mathrm{T}_1(\mathrm{E}_{FHDL}, \eta_g, \lambda)]^{\frac{3}{2}} \tag{10.52}$$

$$n_0 = \frac{g_v}{3\pi^2}\left(\frac{2m_c}{\hbar^2}\right)^{\frac{3}{2}}[\mathrm{T}_2(\mathrm{E}_{FHDL}, \eta_g, \lambda)]^{\frac{3}{2}} \tag{10.53}$$

$$n_0 = \frac{g_v}{3\pi^2}\left(\frac{2m_c}{\hbar^2}\right)^{\frac{3}{2}}[\mathrm{T}_3(\mathrm{E}_{FHDL}, \eta_g, \lambda)]^{\frac{3}{2}} \tag{10.54}$$

where E_{FHDL} is the Fermi energy in HD III-V semiconductors in the presence of light waves as measured from the age of the unperturbed conduction band in the vertically upward direction. The ER in this case is given by

$$\left(\frac{D}{\mu}\right)_L = Real\ part\ of \left[\frac{n_0}{e}\left[\frac{\partial n_0}{\partial(E_{FHDL} - E_{0HDL})}\right]^{-1}\right] \tag{10.55}$$

where E_{0HDL} can be obtained by substituting $k = 0$ and $E = E_{0HDL}$ in (10.46b), (10.47) and (10.48) respectively.

Using (10.52)–(10.55) and the allied definitions, we can study the ER in the presence of light waves in HD III-V semiconductors whose unperturbed conduction bands are described by three and two band models of Kane together with parabolic energy bands respectively.

10.2.3 The Opto ER Under Magnetic Quantization in HD III-V, Ternary and Quaternary Semiconductors

(i) Using (10.46b), the magneto-dispersion law, in the absence of spin, for HD III-V, ternary and quaternary semiconductors, in the presence of photo-excitation, whose unperturbed conduction electrons obey the three band model of Kane, is given by

$$T_1(E, \eta_g) = \left(n + \frac{1}{2}\right)\hbar\omega_0 + \frac{\hbar^2 k_z^2}{2m_c} \tag{10.56}$$

Using (10.52), the DOS function in the present case can be expressed as

$$D_B(E, \eta_g, \lambda) = \frac{g_v|e|\sqrt{2m_c}}{2\pi^2\hbar^2}\sum_{n=0}^{n_{\max}}\left[\{T_1(E, \eta_g, \lambda)\}'\left\{T_1(E, \eta_g, \lambda) - \left(n + \frac{1}{2}\right)\hbar\omega_0\right\}^{-1/2} H(E - E_{n_{l1}})\right] \tag{10.57}$$

where, $E_{n_{l1}}$ is the Landau sub-band energies in this case and is given as

$$T_1(E_{n_{l1}}, \eta_g, \lambda) = \left(n + \frac{1}{2}\right)\hbar\omega_0 \tag{10.58}$$

The EEM in this case assumes the form

$$m^*(E_{FHDLB}, \eta_g, \lambda) = m_c\ Real\ part\ of\ \{T_1(E_{FHDLB}, \eta_g, \lambda)\}' \tag{10.59}$$

where, E_{FHDLB} is the Fermi energy under quantizing magnetic field in the presence of light waves as measured from the edge of the conduction band in the vertically upward direction in the absence of any quantization. Combining (10.57) with the Fermi-Dirac occupation probability factor and using the generalized Sommerfeld's lemma [13], the electron concentration can be written as

$$n_0 = \frac{g_v|e|B\sqrt{2m_c}}{\pi^2\hbar^2}\sum_{n=0}^{n_{\max}}\left[M_{13}(E_{FHDLB},\eta_g,\lambda) + N_{13}(E_{FHDLB},\eta_g,\lambda)\right] \quad (10.60)$$

where,
$M_{13}(E_{FHDLB},\eta_g,\lambda) \equiv$ *Real part of* $\left[T_1(E_{FHDLB},\eta_g,\lambda) - \left(n+\frac{1}{2}\right)\hbar\omega_0\right]^{1/2}$
and

$$N_{13}(E_{FHDLB},\eta_g,\lambda) \equiv \sum_{r=1}^{s} L(r)M_{13}(E_{FHDLB},\eta_g,\lambda).$$

The ER in this case is given by

$$\left(\frac{D}{\mu}\right)_{LB} = \text{Real part of}\left[\frac{n_0}{e}\left[\frac{\partial n_0}{\partial(E_{FHDLB} - E_{0HDLB})}\right]^{-1}\right] \quad (10.61)$$

where, E_{0HDLB} can be obtained from the magneto dispersion law under the conditions $k = 0$ and $E = E_{0HDLB}$.
Thus by using (10.56), (10.60), (10.61) and the allied equations, we can study the ER in this case.

(ii) Using (10.47), the magneto-dispersion law, in the absence of spin, for HD III-V, ternary and quaternary semiconductors, in the presence of photo-excitation, whose unperturbed conduction electrons obey the two band model of Kane, is given by

$$T_2(E,\eta_g) = \left(n+\frac{1}{2}\right)\hbar\omega_0 + \frac{\hbar^2k_z^2}{2m_c} \quad (10.62)$$

Using (10.62), the DOS function in the present case can be expressed as

$$D_B(E,\eta_g,\lambda) = \frac{g_v|e|\sqrt{2m_c}}{2\pi^2\hbar^2}\sum_{n=0}^{n_{\max}}\left[\{T_2(E,\eta_g,\lambda)\}'\{T_2(E,\eta_g,\lambda). - \left(n+\frac{1}{2}\right)\hbar\omega_0\right\}^{-1/2} H(E - E_{n_{l2}})\Bigg] \quad (10.63)$$

where, $E_{n_{l2}}$ is the Landau sub-band energies in this case and is given as

$$T_2(E_{n_{l2}},\eta_g,\lambda) = \left(n+\frac{1}{2}\right)\hbar\omega_0 \quad (10.64)$$

The EEM in this case assumes the form

$$m^*(E_{FHDLB}, \eta_g, \lambda) = m_c\{T_2(E_{FHDLB}, \eta_g, \lambda)\}' \tag{10.65}$$

where, E_{FHDLB} is the Fermi energy under quantizing magnetic field in the presence of light waves as measured from the edge of the conduction band in the vertically upward direction in the absence of any quantization. Combining (10.65) with the Fermi-Dirac occupation probability factor and using the generalized Sommerfeld's lemma, the electron concentration can be written as

$$n_0 = \frac{g_v|e|B\sqrt{2m_c}}{\pi^2\hbar^2}\sum_{n=0}^{n_{\max}}\left[M_{23}(E_{FHDLB}, \eta_g, \lambda) + N_{23}(E_{FHDLB}, \eta_g, \lambda)\right] \tag{10.66}$$

where, $M_{23}(E_{FHDLB}, \eta_g, \lambda) \equiv \left[T_2(E_{FHDLB}, \eta_g, \lambda) - \left(n + \frac{1}{2}\right)\hbar\omega_0\right]^{1/2}$ and $N_{23}(E_{FHDLB}, \eta_g, \lambda) \equiv \sum_{r=1}^{s} L(r)M_{23}(E_{FHDLB}, \eta_g, \lambda)$.

Using (10.61), (10.66) and the allied definitions, we can study the ER in this case, where E_{0HDLB} can be obtained from the magneto dispersion law under the conditions $k = 0$ and $\mathrm{E} = E_{0HDLB}$ in (10.47).

(iii) Using (10.48), the magneto-dispersion law, in the absence of spin, for HD III-V, ternary and quaternary semiconductors, in the presence of photo-excitation, whose unperturbed conduction electrons obey the parabolic energy bands, is given by

$$T_2(E, \eta_g) = \left(n + \frac{1}{2}\right)\hbar\omega_0 + \frac{\hbar^2 k_z^2}{2m_c} \tag{10.67}$$

Using (10.62), the DOS function in the present case can be expressed as

$$D_B(E, \eta_g, \lambda) = \frac{g_v|e|\sqrt{2m_c}}{2\pi^2\hbar^2}\sum_{n=0}^{n_{\max}}\left[\{T_3(E, \eta_g, \lambda)\}'\left\{T_3(E, \eta_g, \lambda) - \left(n + \frac{1}{2}\right)\hbar\omega_0\right\}^{-1/2} H(E - E_{n_{12}})\right] \tag{10.68}$$

where, $E_{n_{12}}$ is the Landau sub-band energies in this case and is given as

$$T_3(E_{n_{12}}, \eta_g, \lambda) = \left(n + \frac{1}{2}\right)\hbar\omega_0 \tag{10.69}$$

The EEM in this case assumes the form

$$m^*(E_{FHDLB}, \eta_g, \lambda) = m_c\{T_3(E_{FHDLB}, \eta_g, \lambda)\}' \tag{10.70}$$

where, E_{FHDLB} is the Fermi energy under quantizing magnetic field in the presence of light waves as measured from the edge of the conduction band in the vertically upward direction in the absence of any quantization. Combining

(10.69) with the Fermi-Dirac occupation probability factor and using the generalized Sommerfeld's lemma, the electron concentration can be written as

$$n_0 = \frac{g_v|e|B\sqrt{2m_c}}{\pi^2\hbar^2}\sum_{n=0}^{n_{\max}}\left[M_{33}(E_{FHDLB},\eta_g,\lambda)+N_{33}(E_{FHDLB},\eta_g,\lambda)\right] \quad (10.71)$$

where,

$$M_{33}(E_{FHDLB},\eta_g,\lambda) \equiv \left[T_3(E_{FHDLB},\eta_g,\lambda)-\left(n+\frac{1}{2}\right)\hbar\omega_0\right]^{1/2}$$

and

$$N_{33}(E_{FHDLB},\eta_g,\lambda) \equiv \sum_{r=1}^{s}L(r)M_{33}(E_{FHDLB},\eta_g,\lambda).$$

Using (10.61), (10.71) and the allied definitions, we can study the ER in this case, where, E_{0HDLB} can be obtained from the magneto dispersion law under the conditions $k = 0$ and $E = E_{0HDLB}$ in (10.48).

10.2.4 The ER in the Presence of Cross-Field Configuration Under External Photo-Excitation in HD III-V, Ternary and Quaternary Semiconductors

(i) The electron dispersion law in the present case is given by

$$T_1(E,\eta_g,\lambda) = \left(n+\frac{1}{2}\right)\hbar\omega_0 + \frac{[\hbar k_z(E)]^2}{2m_c} - \frac{E_0}{B}\hbar k_y\{T_1(E,\eta_g,\lambda)\}' - \left\{\frac{m_c E_0^2\left[\{T_1(E,\eta_g,\lambda)\}\right]'}{2B^2}\right\} \quad (10.72)$$

The use of (10.72) leads to the expressions of the EEMs' along z and y directions as

$$m_z^*(E_{F_{BLHDC}},n,E_0,B,\lambda) = Real\,part\,of\,m_c\left[\{T_1(E_{F_{BLHDC}},\eta_g,\lambda)\}'' + \frac{m_c E_0^2\{T_1(E_{F_{BLHDC}},\eta_g,\lambda)\}'\{T_1(E_{F_{BLHDC}},\eta_g,\lambda)\}''}{B^2}\right] \quad (10.73)$$

$$m_y^*(E_{F_{BLHDC}}, n, E_0B, \lambda) = Real\ part\ of(\frac{B}{E_0})^2[\{T_1(E_{F_{BLHDC}}, \eta_g, \lambda)\}']^{-1}[T_1(E_{F_{BLHDC}}, \eta_g, \lambda) - (n+\frac{1}{2})\hbar\omega_0 + \frac{m_c E_0^2[\{T_1(E_{F_{BLHDC}}, \eta_g, \lambda)\}']^2}{2B^2}][\frac{\{T_1(E_{F_{BLHDC}}, \eta_g, \lambda)\}''}{[\{T_1(E_{F_{BLHDC}}, \eta_g, \lambda\}']^2}[T_1(E_{F_{BLHDC}}, \eta_g, \lambda) - (n+\frac{1}{2})\hbar\omega_0 + \frac{m_c E_0^2[\{T_1(E_{F_{BLHDC}}, \eta_g, \lambda)\}']^2}{2B^2}] + 1 + \frac{m_c E_0^2\{T_1(E_{F_{BLHDC}}, \eta_g, \lambda)\}''}{B^2}] \quad (10.74)$$

where, $E_{F_{BLHDC}}$ is the Fermi energy in this case.
The Landau energy $(E_{n_{l41}})$ can be written as

$$T_1(E_{n_{l41}}, \eta_g, \lambda) = (n+\frac{1}{2})\hbar\omega_0 - \left\{\frac{m_c E_0^2[\{T_1(E_{n_{l41}}, \eta_g, \lambda)\}']^2}{2B^2}\right\} \quad (10.75)$$

The electron concentration in this case assume the forms

$$n_0 = \frac{2g_v B\sqrt{2m_c}}{3L_x\pi^2\hbar^2 E_0}\sum_{n=0}^{n_{\max}}\left[M_{16HDC}(E_{F_{BLHDC}}, \eta_g, \lambda) + N_{16HDC}(E_{F_{BLHDC}}, \eta_g, \lambda)\right] \quad (10.76)$$

where,

$$M_{16HDC}(E_{F_{BLHDC}}, \eta_g, \lambda) \equiv \text{Real part of}[[T_1(E_{F_{BLHDC}}, \eta_g, \lambda) - (n+\frac{1}{2})\hbar\omega_0 - \frac{m_c E_0^2}{2B^2}[\{T_1(E_{F_{BLHDC}}, \eta_g, \lambda)\}']^2 + |e|E_0L_x[\{T_1(E_{F_{BLHDC}}, \eta_g, \lambda\}']]^{3/2} - [T_1(E_{F_{BLHDC}}, \eta_g, \lambda) - (n+\frac{1}{2})\hbar\omega_0 - \frac{m_c E_0^2}{2B^2}[\{T_1(E_{F_{BLHDC}}, \eta_g, \lambda)\}']^2]^{3/2}]\frac{1}{[\{T_1(E_{F_{BLHDC}}, \eta_g, \lambda)\}']}$$

and

$$N_{16HDC}(E_{F_{BLHDC}}, \eta_g, \lambda) \equiv \text{Real part of}\sum_{r=1}^{s}\left[L(r)M_{16HDC}(E_{F_{BLHDC}}, \eta_g, \lambda)\right].$$

The ER in this case can be written as

$$(\frac{D}{\mu})_{LB} = Real\ part\ of\left[\frac{n_0}{e}\left[\frac{\partial n_0}{\partial(E_{F_{BLHDC}} - E_{0HDLBC})}\right]^{-1}\right] \quad (10.77)$$

where, E_{0HDLBC} can be obtained from 10.72 under the conditions $k = 0$ and $E = E_{0HDLBC}$.
Thus by using (10.77), (10.76), (10.72) and the allied definitions, we can study the ER in this case.

(ii) Similarly, the electron dispersion law in this case is given by

$$T_2(E,\eta_g,\lambda) = \left(n+\frac{1}{2}\right)\hbar\omega_0 - \frac{E_0}{B}\hbar k_y\{T_2(E,\eta_g,\lambda)\}' - \frac{m_c E_0^2}{2B^2}\left[\{T_2(E,\eta_g,\lambda)\}'\right] + \frac{[\hbar k_z(E)]^2}{2m_c} \tag{10.78}$$

The use of (10.78) leads to the expressions of the EEMs' along z and y directions as

$$m_z^*(E_{F_{BLHDC}},n,E_0,B,\lambda) = m_c\left[\{T_2(E_{F_{BLHDC}},\eta_g,\lambda)\}'' + \frac{m_c E_0^2\{T_2(E_{F_{BLHDC}},\eta_g,\lambda)\}'\{T_2(E_{F_{BLHDC}},\eta_g,\lambda)\}''}{B^2}\right] \tag{10.79}$$

$$m_y^*(E_{F_{BLHDC}},n,E_0B,\lambda) = (\frac{B}{E_0})^2[\{T_2(E_{F_{BLHDC}},\eta_g,\lambda)\}']^{-1}[T_2(E_{F_{BLHDC}},\eta_g,\lambda) - (n+\frac{1}{2})\hbar\omega_0 + \frac{m_c E_0^2[\{T_2(E_{F_{BLHDC}},\eta_g,\lambda)\}']^2}{2B^2}][\frac{\{T_2(E_{F_{BLHDC}},\eta_g,\lambda)\}''}{[\{T_2(E_{F_{BLHDC}},\eta_g,\lambda\}']^2}[T_2(E_{F_{BLHDC}},\eta_g,\lambda) - (n+\frac{1}{2})\hbar\omega_0 + \frac{m_c E_0^2[\{T_2(E_{F_{BLHDC}},\eta_g,\lambda)\}']^2}{2B^2}] + 1 + \frac{m_c E_0^2\{T_2(E_{F_{BLHDC}},\eta_g,\lambda)\}''}{B^2}] \tag{10.80}$$

where, $E_{F_{BLHDC}}$ is the Fermi energy in this case.
The Landau energy $(E_{n_{l42}})$ can be written as

$$T_2(E_{n_{l42}},\eta_g,\lambda) = (n+\frac{1}{2})\hbar\omega_0 - \{\frac{m_c E_0^2[\{T_2(E_{n_{l42}},\eta_g,\lambda)\}']^2}{2B^2}\} \tag{10.81}$$

The electron concentration in this case assume the forms

$$n_0 = \frac{2g_v B\sqrt{2m_c}}{3L_x\pi^2\hbar^2 E_0}\sum_{n=0}^{n_{\max}}\left[M_{26HDC}(E_{F_{BLHDC}},\eta_g,\lambda) + N_{26HDC}(E_{F_{BLHDC}},\eta_g,\lambda)\right] \tag{10.82}$$

where,

$$
\begin{aligned}
& M_{26HDC}(E_{F_{BLHDC}}, \eta_g, \lambda) \\
& \equiv [[T_2(E_{F_{BLHDC}}, \eta_g, \lambda) - (n+\frac{1}{2})\hbar\omega_0 - \frac{m_c E_0^2}{2B^2}[\{T_2(E_{F_{BLHDC}}, \eta_g, \lambda)\}']^2 \\
& + |e|E_0 L_x[\{T_2(E_{F_{BLHDC}}, \eta_g, \lambda\}']]^{3/2} - [T_2(E_{F_{BLHDC}}, \eta_g, \lambda) \\
& - (n+\frac{1}{2})\hbar\omega_0 - \frac{m_c E_0^2}{2B^2}[\{T_2(E_{F_{BLHDC}}, \eta_g, \lambda)\}']^2]^{3/2}]\frac{1}{[\{T_2(E_{F_{BLHDC}}, \eta_g, \lambda)\}']}
\end{aligned}
$$

Thus by using (10.77), (10.78), (10.82) and the allied definitions, we can study the ER in this case.

(iii) Similarly, the electron dispersion law in this case is given by

$$
\begin{aligned}
T_2(E, \eta_g, \lambda) = & \left(n+\frac{1}{2}\right)\hbar\omega_0 - \frac{E_0}{B}\hbar k_y\{T_3(E, \eta_g, \lambda)\}' \\
& - \frac{m_c E_0^{2'}}{2B^2}[\{T_3(E, \eta_g, \lambda)\}'] + \frac{[\hbar k_z(E)]^2}{2m_c}
\end{aligned} \tag{10.83}
$$

The use of (10.78) leads to the expressions of the EEMs' along z and y directions as

$$
\begin{aligned}
m_z^*(E_{F_{BLHDC}}, n, E_0, B, \lambda) = m_c \Bigg[& \{T_3(E_{F_{BLHDC}}, \eta_g, \lambda)\}'' \\
& + \frac{m_c E_0^2\{T_3(E_{F_{BLHDC}}, \eta_g, \lambda)\}'\{T_3(E_{F_{BLHDC}}, \eta_g, \lambda)\}''}{B^2}\Bigg]
\end{aligned} \tag{10.84}
$$

$$
\begin{aligned}
& m_y^*(E_{F_{BLHDC}}, n, E_0 B, \lambda) = (\frac{B}{E_0})^2[\{T_3(E_{F_{BLHDC}}, \eta_g, \lambda)\}']^{-1}[T_3(E_{F_{BLHDC}}, \eta_g, \lambda) - (n+\frac{1}{2})\hbar\omega_0 \\
& + \frac{m_c E_0^2[\{T_3(E_{F_{BLHDC}}, \eta_g, \lambda)\}']^2}{2B^2}][\frac{\{T_3(E_{F_{BLHDC}}, \eta_g, \lambda)\}'}{[\{T_3(E_{F_{BLHDC}}, \eta_g, \lambda\}']^2}[T_3(E_{F_{BLHDC}}, \eta_g, \lambda) \\
& - (n+\frac{1}{2})\hbar\omega_0 + \frac{m_c E_0^2[\{T_3(E_{F_{BLHDC}}, \eta_g, \lambda)\}']^2}{2B^2}] + 1 \\
& + \frac{m_c E_0^2\{T_3(E_{F_{BLHDC}}, \eta_g, \lambda)\}'}{B^2}]
\end{aligned} \tag{10.85}
$$

where, $E_{F_{BLHDC}}$ is the Fermi energy in this case.

The Landau energy $(E_{n_{l43}})$ can be written as

$$T_3(E_{n_{l42}}, \eta_g, \lambda) = (n + \frac{1}{2})\hbar\omega_0 - \{\frac{m_c E_0^2[\{T_3(E_{n_{l42}}, \eta_g, \lambda)\}']^2}{2B^2}\} \quad (10.86)$$

The electron concentration in this case assume the forms

$$n_0 = \frac{2g_v B\sqrt{2m_c}}{3L_x \pi^2 \hbar^2 E_0} \sum_{n=0}^{n_{\max}} \left[M_{36HDC}(E_{F_{BLHDC}}, \eta_g, \lambda) + N_{36HDC}(E_{F_{BLHDC}}, \eta_g, \lambda)\right] \quad (10.87)$$

where,

$$M_{36HDC}(E_{F_{BLHDC}}, \eta_g, \lambda) \equiv [[T_3(E_{F_{BLHDC}}, \eta_g, \lambda) - (n + \frac{1}{2})\hbar\omega_0 - \frac{m_c E_0^2}{2B^2}[\{T_3(E_{F_{BLHDC}}, \eta_g, \lambda)\}']^2 + |e|E_0 L_x[\{T_3(E_{F_{BLHDC}}, \eta_g, \lambda\}']]^{3/2} - [T_3(E_{F_{BLHDC}}, \eta_g, \lambda) - (n + \frac{1}{2})\hbar\omega_0 - \frac{m_c E_0^2}{2B^2}[\{T_3(E_{F_{BLHDC}}, \eta_g, \lambda)\}']^2]^{3/2}] \frac{1}{[\{T_3(E_{F_{BLHDC}}, \eta_g, \lambda)\}']}$$

Thus by using (10.77), (10.83), (10.87) and the allied definitions, we can study the ER in this case.

10.2.5 The ER in Ultrathin Films of HD III-V, Ternary and Quaternary Semiconductors Under External Photo-Excitation

(i) The 2D electron energy spectrum in ultra-thin films of HD III-V, ternary and quaternary materials, whose unperturbed band structure is defined by the three band model of Kane, in the presence of light waves can be expressed following (10.46b) as

$$\frac{\hbar^2 k_s^2}{2m_c} + \frac{\hbar^2}{2m_c}\left(\frac{n_z \pi}{d_z}\right)^2 = T_1(E, \eta_g, \lambda) \quad (10.88)$$

The sub band energies ($E_{n_{l7HD}}$) can be written as

$$T_1(E_{n_{l7HD}}, \eta_g, \lambda) = \frac{\hbar^2}{2m_c}(n_z \pi/d_z)^2 \quad (10.89)$$

The expression of the EEM in this case is given by

$$m^*(E_{F2DLHD}, \eta_z, \lambda) = m_c \; Real\ part\ of \{T_1(E_{F2DLHD}, \eta_g, \lambda)\}' \quad (10.90)$$

where, E_{F2DLHD} is the Fermi energy in the present case as measured from the edge of the conduction band in the vertically upward direction in absence of any quantization.
The DOS function can be written as

$$N_{2D}(E, \eta_g, \lambda) = \left(\frac{m_c g_v}{\pi \hbar^2}\right) \sum_{n_z=1}^{n_{z\max}} [T_1(E, \eta_g, \lambda)]' H(E - E_{nl7HD}) \quad (10.91)$$

Combining (10.91), with the Fermi Dirac occupation probability factor, the two dimensional electron concentration can be expressed as

$$n_{2D} = \frac{m_c g_v}{\pi \hbar^2} \sum_{n_z=1}^{n_{z\max}} [M_{18HD}(n_z, E_{F2DLHD}), \eta_g, \lambda) + N_{18HD}(n_z, E_{F2DLHD}, \eta_g, \lambda)] \quad (10.92a)$$

where,

$$M_{18HD}(n_z, E_{F2DLHD}, \eta_g, \lambda) \equiv Real\ part\ of \left[T_1(E_{F2DLHD}, \eta_g, \lambda) - \frac{\hbar^2}{2m_c}\left(\frac{n_z \pi}{d_z}\right)^2\right],$$

$$N_{18HD}(n_z, E_{F2DLHD}, \eta_g, \lambda) \equiv \sum_{r=1}^{s} L(r) M_{18HD}(n_z, E_{F2DLHD}, \eta_g, \lambda),$$

The ER in this case can be written as

$$\left(\frac{D}{\mu}\right)_{LS} = Real\ part\ of \left[\frac{n_0}{e}\left[\frac{\partial n_0}{\partial(E_{F2DLHD} - E_{0HDLS})}\right]^{-1}\right] \quad (10.92b)$$

where, E_{0HDLS} is obtained from (10.88) by substituting $k_s = 0$ and $E = E_{0HDLS}$.
Thus using (10.92a), (10.92b), (10.88) and the allied definitions, we can study the ER in this case.

(ii) The 2D electron energy spectrum in ultra-thin films of HD III-V, ternary and quaternary materials, whose unperturbed band structure is defined by the two band model of Kane, in the presence of light waves can be expressed following (10.47) as

$$\frac{\hbar^2 k_s^2}{2m_c} + \frac{\hbar^2}{2m_c}\left(\frac{n_z \pi}{d_z}\right)^2 = T_2(E, \eta_g, \lambda) \quad (10.93)$$

The sub band energies ($E_{n_{l8HD}}$) can be written as

$$T_2(E_{n_{l8HD}}, \eta_g, \lambda) = \frac{\hbar^2}{2m_c}(n_z\pi/d_z)^2 \tag{10.94}$$

The expression of the EEM in this case is given by

$$m^*(E_{F2DLHD}, \eta_z, \lambda) = m_c\{T_2(E_{F2DLHD}, \eta_g, \lambda)\}' \tag{10.95}$$

The DOS function can be written as

$$N_{2D}(E, \eta_g, \lambda) = \left(\frac{m_c g_v}{\pi\hbar^2}\right)\sum_{n_z=1}^{n_{z\max}}\left[T_2(E, \eta_g, \lambda)\right]' H(E - E_{n_{l8HD}}) \tag{10.96}$$

Combining (10.96), with the Fermi Dirac occupation probability factor, the two dimensional electron concentration can be expressed as

$$n_{2D} = \frac{m_c g_v}{\pi\hbar^2}\sum_{n_z=1}^{n_{z\max}}\left[M_{19HD}(n_z, E_{F2DLHD}), \eta_g, \lambda) + N_{19HD}(n_z, E_{F2DLHD}, \eta_g, \lambda)\right] \tag{10.97}$$

where

$$M_{19HD}(n_z, E_{F2DLHD}, \eta_g, \lambda) \equiv \left[T_2(E_{F2DLHD}, \eta_g, \lambda) - \frac{\hbar^2}{2m_c}\left(\frac{n_z\pi}{d_z}\right)^2\right],$$

$$N_{19HD}(n_z, E_{F2DLHD}, \eta_g, \lambda) \equiv \sum_{r=1}^{s} L(r) M_{19HD}(n_z, E_{F2DLHD}, \eta_g, \lambda),$$

Thus using (10.97), (10.93) and the allied definitions, we can study the ER in this case.

(iii) The 2D electron energy spectrum in ultra-thin films of HD III-V, ternary and quaternary materials, whose unperturbed band structure is defined by the parabolic energy bands, in the presence of light waves can be expressed following (10.48) as

$$\frac{\hbar^2 k_s^2}{2m_c} + \frac{\hbar^2}{2m_c}\left(\frac{n_z\pi}{d_z}\right)^2 = T_3(E, \eta_g, \lambda) \tag{10.98}$$

The sub band energies ($E_{n_{l9HD}}$) can be written as

$$T_3(E_{n_{l9HD}}, \eta_g, \lambda) = \frac{\hbar^2}{2m_c}(n_z\pi/d_z)^2 \tag{10.99}$$

The expression of the EEM in this case is given by

$$m^*(E_{F2DLHD}, n_z, \lambda) = m_c\{T_3(E_{F2DLHD}, \eta_g, \lambda)\}' \tag{10.100}$$

The DOS function can be written as

$$N_{2D}(E, \eta, \lambda) = \left(\frac{m_c g_v}{\pi\hbar^2}\right) \sum_{n_z=1}^{n_{z\max}} \left[T_3(E, \eta_g, \lambda)\right]' H(E - E_{n_{19HD}}) \tag{10.101}$$

Combining (10.101), with the Fermi Dirac occupation probability factor, the two dimensional electron concentration can be expressed as

$$n_{2D} = \frac{m_c g_v}{\pi\hbar^2} \sum_{n_z=1}^{n_{z\max}} \left[M_{20HD}(n_z, E_{F2DLHD}, \eta_g, \lambda) + N_{20HD}(n_z, E_{F2DLHD}, \eta_g, \lambda)\right] \tag{10.102}$$

where

$$M_{20HD}(n_z, E_{F2DLHD}, \eta_g, \lambda) \equiv \left[T_3(E_{F2DLHD}, \eta_g, \lambda) - \frac{\hbar^2}{2m_c}\left(\frac{n_z\pi}{d_z}\right)^2\right],$$

$$N_{20HD}(n_z, E_{F2DLHD}, \eta_g, \lambda) \equiv \sum_{r=1}^{s} L(r) M_{20HD}(n_z, E_{F2DLHD}, \eta_g, \lambda),$$

Thus using (10.98), (10.102) and the allied definitions, we can study the ER in this case.

10.2.6 The ER in Doping Superlattices of HD III-V, Ternary and Quaternary Semiconductors Under External Photo-Excitation

(i) The electron energy spectrum in doping superlattices of HD III-V, ternary and quaternary materials in the presence of external photo-excitation whose unperturbed electrons are defined by the three band model of Kane can be expressed following (10.46b) as

$$T_1(E, \eta_g, \lambda) = \left(n_i + \frac{1}{2}\right)\hbar\omega_{91HD}(E, \eta_g, \lambda) + \frac{\hbar^2 k_s^2}{2m_c} \tag{10.103}$$

where $\omega_{91HD}(E, \eta_g, \lambda) \equiv \left(\frac{n_0|e|^2}{d_0\varepsilon_{sc}T_1'(E, \eta_g, \lambda)m_c}\right)^{1/2}$.

The sub band energies $(E_{n_{i10HD}})$ can be written as

$$T_1(E_{n_{i10HD}}, \eta_g, \lambda) = \left(n_i + \frac{1}{2}\right)\hbar\omega_{91HD}(E_{n_{i10HD}}, \eta_g, \lambda) \tag{10.104}$$

The expression of the EEM in this case is given by

$$m^*(E_{F2DLHDD}, \eta_g, \lambda, n_i) = m_c\{\mathrm{M}_{40HD}(E_{F_{2DLHDD}}, \eta_g, \lambda, n_i)\}' \tag{10.105}$$

where,

$$\mathrm{M}_{40HD}(E_{F2DLHDD}, \eta_g, \lambda, n_i) = Real\ part\ of\{T_1(E_{F2DLHDD}, \eta_g, \lambda) - \left(n_i + \frac{1}{2}\right)\hbar\omega_{91HD}(E_{F2DLHDD}, \eta_g, \lambda)\}$$

and $E_{F2DLHDD}$ is the Fermi energy in the present case as measured from the edge of the conduction band in the vertically upward direction in absence of any quantization.
The DOS function can be written as

$$N_{2D}(E, \eta_g, \lambda) = \left(\frac{m_c g_v}{\pi\hbar^2}\right)\sum_{n_z=1}^{n_{z\max}}[\mathrm{M}_{40HD}(E, \eta_g, \lambda)]' H(E - E_{n_{i10HD}}) \tag{10.106}$$

Combining (10.106), with the Fermi Dirac occupation probability factor, the two dimensional electron concentration can be expressed as

$$n_{2D} = \left(\frac{m_c g_v}{\pi\hbar^2}\right)\sum_{n_i=1}^{n_{i\max}}[\mathrm{M}_{40HD}(E_{F2DLHDD}, \eta_g, \lambda, n_i) + N_{40HD}(E_{F2DLHDD}, \eta_g, \lambda, n_i)] \tag{10.107}$$

where $N_{40HD}(E_{F2DLHDD}, \eta_g, \lambda, n_i) \equiv \sum_{r=1}^{S} L(r)\mathrm{M}_{40HD}(E_{F2DLHDD}, \eta_g, \lambda, n_i)$,
The ER in this case can be written as

$$\left(\frac{D}{\mu}\right)_{LS} = Real\ part\ of\left[\frac{n_0}{e}\left[\frac{\partial n_0}{\partial(E_{F2DLHDD} - E_{0HDLSDS})}\right]^{-1}\right] \tag{10.108}$$

where $E_{0HDLSDS}$ is obtained from (10.103) by substituting $k_s = 0$ and $E = E_{0HDLSDS}$.
Thus using (10.103), (10.107), (10.108) and the allied definitions, we can study the ER in this case.

(ii) The electron energy spectrum in doping superlattices of HD III-V, ternary and quaternary materials in the presence of external photo-excitation whose unperturbed electrons are defined by the two band model of Kane can be expressed following (10.47) as

$$T_2(E,\eta_g,\lambda) = \left(n_i + \frac{1}{2}\right)\hbar\omega_{92HD}(E,\eta_g,\lambda) + \frac{\hbar^2 k_s^2}{2m_c} \quad (10.109)$$

where $\omega_{92HD}(E,\eta_g,\lambda) \equiv \left(\frac{n_0|e|^2}{d_0 \varepsilon_{sc} T_2'(E,\eta_g,\lambda)m_c}\right)^{1/2}$.
The sub band energies $(E_{n_{i11HD}})$ can be written as

$$T_2(E_{n_{i11HD}},\eta_g,\lambda) = \left(n_i + \frac{1}{2}\right)\hbar\omega_{92HD}(E_{n_{i11HD}},\eta_g,\lambda) \quad (10.110)$$

The expression of the EEM in this case is given by

$$m^*(E_{F2DLHHD},\eta_g,\lambda,n_i) = m_c\{\mathrm{M}_{41HD}(E_{F2DLHDD},\eta_g,\lambda,n_i)\}' \quad (10.111)$$

where

$$\mathrm{M}_{41HD}(E_{F2HLDD},\eta_g,\lambda,n_i) = \{T_2(E_{F2DLHDD},\eta_g,\lambda) - \left(n_i + \frac{1}{2}\right)\hbar\omega_{92HD}(E_{F2DLHDD},\eta_g,\lambda)\}$$

and $E_{F2DLHDD}$ is the Fermi energy in the present case as measured from the edge of the conduction band in the vertically upward direction in absence of any quantization.The DOS function can be written as

$$N_{2D}(E,\eta_g,\lambda) = \left(\frac{m_c g_v}{\pi\hbar^2}\right)\sum_{n_z=1}^{n_{z\max}}[\mathrm{M}_{41HD}(E,\eta_g,\lambda)]' H(E - E_{n_{i11HD}}) \quad (10.112)$$

Combining (10.112), with the Fermi Dirac occupation probability factor, the two dimensional electron concentration can be expressed as

$$n_{2D} = \left(\frac{m_c g_v}{\pi\hbar^2}\right)\sum_{n_i=1}^{n_{i\max}}[\mathrm{M}_{41HD}(E_{F2DLHDD},\eta_g,\lambda,n_i) + N_{41HD}(E_{F2DLHDD},\eta_g,\lambda,n_i)] \quad (10.113)$$

where, $N_{41HD}(E_{F2DLHDD},\eta_g,\lambda,n_i) \equiv \sum_{r=1}^{S} L(r)\mathrm{M}_{41HD}(E_{F2DLHDD},\eta_g,\lambda,n_i)$,
Thus using (10.108), (10.109) and (10.113), we can study the ER in this case.

(iii) The electron energy spectrum in doping superlattices of HD III-V, ternary and quaternary materials in the presence of external photo-excitation whose unperturbed electrons are defined by the two band model of Kane can be expressed following (10.48) as

$$T_3(E,\eta_g,\lambda) = \left(n_i + \frac{1}{2}\right)\hbar\omega_{93HD}(E,\eta_g,\lambda) + \frac{\hbar^2 k_s^2}{2m_c} \quad (10.114)$$

where $\omega_{93HD}(E,\eta_g,\lambda) \equiv \left(\frac{n_0|e|^2}{d_0\varepsilon_{sc}T_3'(E,\eta_g,\lambda)m_c}\right)^{1/2}$.
The sub band energies $(E_{n_{i12HD}})$ can be written as

$$T_3(E_{n_{i12HD}},\eta_g,\lambda) = \left(n_i + \frac{1}{2}\right)\hbar\omega_{93HD}(E_{n_{i12HD}},\eta_g,\lambda) \quad (10.115)$$

The expression of the EEM in this case is given by

$$m^*(E_{F2DLHHD},\eta_g,\lambda,n_i) = m_c\{\mathrm{M}_{42HD}(E_{F2DLHDD},\eta_g,\lambda,n_i)\}' \quad (10.116)$$

where $\mathrm{M}_{42HD}(E_{F2HLDD},\eta_g,\lambda,n_i) = \{T_3(E_{F2DLHDD},\eta_g,\lambda) - (n_i + \frac{1}{2})\hbar\omega_{93HD}(E_{F2DLHDD},\eta_g,\lambda)\}$ and $E_{F2DLHDD}$ is the Fermi energy in the present case as measured from the edge of the conduction band in the vertically upward direction in absence of any quantization.
The DOS function can be written as

$$N_{2D}(E,\eta_g,\lambda) = \left(\frac{m_c g_v}{\pi\hbar^2}\right)\sum_{n_z=1}^{n_{z\max}}\left[\mathrm{M}_{42HD}(E,\eta_g,\lambda)\right]' H(E - E_{n_{i12HD}}) \quad (10.117)$$

Combining (10.117), with the Fermi Dirac occupation probability factor, the two dimensional electron concentration can be expressed as

$$n_{2D} = \left(\frac{m_c g_v}{\pi\hbar^2}\right)\sum_{n_i=1}^{n_{i\max}}\left[\mathrm{M}_{42HD}(E_{F2DLHDD},\eta_g,\lambda,n_i) + N_{42HD}(E_{F2DLHDD},\eta_g,\lambda,n_i)\right] \quad (10.118)$$

where, $N_{42HD}(E_{F2DLHDD},\eta_g,\lambda,n_i) \equiv \sum_{r=1}^{S} L(r)\mathrm{M}_{42HD}(E_{F2DLHDD},\eta_g,\lambda,n_i)$,
Thus using (10.108), (10.114), (10.118) and the allied definitions, we can study the ER in this case.

10.2.7 The Magneto ER in Effective Mass Superlattices of HD III-V, Ternary and Quaternary Semiconductors Under External Photo-Excitation

(a) Following Sasaki [9.15], the electron dispersion law in HD III-V effective mass superlattices (EMSLs) in the presence of light waves, the dispersion relations of whose constituent materials in the absence of any perturbation are defined by the three band model of Kane can be written as.

$$k_x^2 = \left[\frac{1}{L_0^2}\{\cos^{-1}(f_{HD1}(E, k_y, k_z, \lambda))\}^2 - k_\perp^2\right] \tag{10.119}$$

in which, $f_{HD1}(E, k_y, k_z, \lambda) = a_{1HD1}\cos[a_0 C_{1HD1}(E, k_\perp) + b_0 D_{1HD1}(E, k_\perp)]$ $-a_{2HD1}\cos[a_0 C_{1HD1}(E, k_\perp) - b_0 D_{1HD1}(E, k_\perp)]$, $k_\perp^2 = k_y^2 + k_z^2, L_0 = a_0 + b_0$,

$$a_{1HD1} = \left[\sqrt{\frac{m_{c2}\ \textit{Real part of}\ [T_1(0, \eta_{g2}, \lambda)]}{m_{c1}\ \textit{Real part of}\ [T_1(0, \eta_{g1}, \lambda)]}} + 1\right]^2 \left[4\left(\frac{m_{c2}\ \textit{Real part of}\ [T_1(0, \eta_{g2}, \lambda)]}{m_{c1}\ \textit{Real part of}\ [T_1(0, \eta_{g1}, \lambda)]}\right)^{1/2}\right]^{-1},$$

$$a_{2HD1} = \left[-1 + \sqrt{\frac{m_{c2}\ \textit{Real part of}\ [T_1(0, \eta_{g2}, \lambda)]}{m_{c1}\ \textit{Real part of}\ [T_1(0, \eta_{g1}, \lambda)]}}\right]^2 \left[4\left(\frac{m_{c2}\ \textit{Real part of}\ [T_1(0, \eta_{g2}, \lambda)]}{m_{c1}\ \textit{Real part of}\ [T_1(0, \eta_{g1}, \lambda)]}\right)^{1/2}\right]^{-1},$$

$$C_{1HD1}(E, k_\perp\lambda) \equiv \left[\left(\frac{2m_{c1}}{\hbar^2}\right)T_1(E, \eta_{g1}, \lambda) - k_\perp^2\right]^{1/2}$$

and

$$D_{1HD1}(E, k_\perp\lambda) \equiv \left[\left(\frac{2m_{c2}}{\hbar^2}\right)T_1(E, \eta_{g2}, \lambda) - k_\perp^2\right]^{1/2}.$$

In the presence of an external magnetic field along x-direction, the simplified magneto dispersion law in this case can be written as

$$k_x^2 = [\rho_{4HD1}(n, E, \lambda)] \tag{10.120}$$

in which, $\rho_{4HD1}(n, E, \lambda) = \frac{1}{L_0^2}[\cos^{-1}(f_{HD1}(n, E, \lambda))]^2 - \left\{\frac{2|e|B}{\hbar}\left(n + \frac{1}{2}\right)\right\}$,

$$f_{HD1}(E, n, \lambda) = a_{1HD1}\cos[a_0 C_{1HD1}(E, n, \lambda) + b_0 D_{1HD1}(E, n, \lambda)] - a_{2HD1}\cos[a_0 C_{1HD1}(E, n, \lambda) + b_0 D_{1HD1}(E, n, \lambda)]$$

$$C_{1HD1}(E, n, \lambda) \equiv \left[\left(\frac{2m_{c1}}{\hbar^2}\right)T_1(E, \eta_{g_1}, \lambda) - \left\{\frac{2|e|B}{\hbar}\left(n + \frac{1}{2}\right)\right\}\right]^{1/2}$$

and

$$D_{1HD1}(E,n,\lambda) \equiv \left[\left(\frac{2m_{c2}}{\hbar^2}\right)T_1(E,\eta_{g_2},\lambda) - \left\{\frac{2|e|B}{\hbar}\left(n+\frac{1}{2}\right)\right\}\right]^{1/2}.$$

The EEM in this case assumes the form

$$m^*(n, E_{fSLHDB}, \lambda) = Real\,part\,of\,\frac{\hbar^2}{2}[\rho_{4HD}(n, E_{fSLHDB}, \lambda)]' \tag{10.121}$$

where E_{fSLHDB} is the Fermi energy in this case.

The EEM in III-V EMSLs under magnetic quantization depends on both the Fermi energy, magnetic quantum number and wavelength which is the intrinsic property of such SLs.

The Landau Subband energies $(E_{n_{SL5HD}})$ can be written as

$$\rho_{4HD1}(n, E_{n_{SL5HD}}, \lambda) = 0 \tag{10.122}$$

The electron concentration in this case can be expressed as

$$n_0 = (\frac{|e|Bg_v}{\pi^2\hbar})\;Real\,part\,of\sum_{n=0}^{n_{\max}}[T_{99HD1}(n, E_{fSLHDB}, \lambda) + T_{910HD1}(n, E_{fSLHDB}, \lambda)] \tag{10.123}$$

where, $T_{99HD1}(n, E_{fSLHDB}, \lambda) = \sqrt{\rho_{4HD1}(n, E_{fSLDHB}, \lambda)}$
and $T_{910HD1}(n,\ E_{fSLHDB}, \lambda) \equiv \sum_{r=1}^{s} L(r)\left[T_{99HD1}(n, E_{fSLHDB}, \lambda)\right]$.

The ER in this case is given by

$$(\frac{D}{\mu})_{LSLHD1} = Real\ part\ of\ [\frac{n_0}{e}[\frac{\partial n_0}{\partial(E_{fSLHDB} - \Delta_{99})}]^{-1}] \tag{10.124}$$

where Δ_{99} is obtained by substituting $k_x = 0$ and $E = \Delta_{99}$ in (10.120)
Thus using (10.120), (10.123), (10.124) and the allied definitions, we can study the ER in this case.

(b) Following Sasaki [9.15], the electron dispersion law in HD III-V effective mass superlattices (EMSLs) in the presence of light waves, the dispersion relations of whose constituent materials in the absence of any perturbation are defined by the two band model of Kane can be written as.

$$k_x^2 = \left[\frac{1}{L_0^2}\{\cos^{-1}(f_{HD2}(E, k_y, k_z, \lambda))\}^2 - k_\perp^2\right] \tag{10.125}$$

In which

$$f_{HD2}(E,k_y,k_z\lambda) = a_{1HD2}\cos[a_0C_{1HD2}(E,k_\perp) + b_0D_{1HD2}(E,k_\perp)] - a_{2HD2}\cos[a_0C_{1HD2}(E,k_\perp) - b_0D_{1HD2}(E,k_\perp)],$$

$$a_{1HD2} = \left[\sqrt{\frac{m_{c2}[\mathrm{T}_2(0,\eta_{g2},\lambda)]}{m_{c1}[\mathrm{T}_2(0,\eta_{g1},\lambda)]}}+1\right]^2\left[4\left(\frac{m_{c2}[\mathrm{T}_2(0,\eta_{g2},\lambda)]}{m_{c1}[\mathrm{T}_2(0,\eta_{g1},\lambda)]}\right)^{1/2}\right]^{-1},$$

$$a_{2HD2} = \left[-1+\sqrt{\frac{m_{c2}[\mathrm{T}_2(0,\eta_{g2},\lambda)]}{m_{c1}[\mathrm{T}_2(0,\eta_{g1},\lambda)]}}\right]^2\left[4\left(\frac{m_{c2}[\mathrm{T}_2(0,\eta_{g2},\lambda)]}{m_{c1}[\mathrm{T}_2(0,\eta_{g1},\lambda)]}\right)^{1/2}\right]^{-1},$$

$$C_{1HD2}(E,k_\perp\lambda) \equiv \left[\left(\frac{2m_{c1}}{\hbar^2}\right)T_2(E,\eta_{g1},\lambda) - k_\perp^2\right]^{1/2}$$

and

$$D_{1HD2}(E,k_\perp\lambda) \equiv \left[\left(\frac{2m_{c2}}{\hbar^2}\right)T_2(E,\eta_{g2},\lambda) - k_\perp^2\right]^{1/2}.$$

In the presence of an external magnetic field along x-direction, the simplified magneto dispersion law in this case can be written as

$$k_x^2 = [\rho_{4HD2}(n,E,\lambda)] \tag{10.126}$$

in which, $\rho_{4HD2}(n,E,\lambda) = \frac{1}{L_0^2}[\cos^{-1}(f_{HD2}(n,E,\lambda))]^2 - \left\{\frac{2|e|B}{\hbar}\left(n+\frac{1}{2}\right)\right\}$,

$$f_{HD2}(E,n,\lambda) = a_{1HD2}\cos[a_0C_{1HD2}(E,n,\lambda) + b_0D_{1HD2}(E,n,\lambda)] - a_{2HD2}\cos[a_0C_{1HD2}(E,n,\lambda) - b_0D_{1HD2}(E,n,\lambda)]$$

$$C_{1HD2}(E,n,\lambda) \equiv \left[\left(\frac{2m_{c1}}{\hbar^2}\right)T_2(E,\eta_{g_1},\lambda) - \left\{\frac{2|e|B}{\hbar}\left(n+\frac{1}{2}\right)\right\}\right]^{1/2}$$

and

$$D_{1HD2}(E,n,\lambda) \equiv \left[\left(\frac{2m_{c2}}{\hbar^2}\right)T_2(E,\eta_{g_2},\lambda) - \left\{\frac{2|e|B}{\hbar}\left(n+\frac{1}{2}\right)\right\}\right]^{1/2}.$$

The EEM in this case assumes the form

$$m^*(n, E_{fSLHDB}, \lambda) = \frac{\hbar^2}{2}[\rho_{4HD2}(n, E_{fSLHDB}, \lambda)]' \tag{10.127}$$

where E_{fSLHDB} is the Fermi energy in this case.

The EEM in III-V EMSLs under magnetic quantization depends on both the Fermi energy, magnetic quantum number and wavelength which is the intrinsic property of such SLs.

The Landau Subband energies $(E_{n_{SL5HD}})$ can be written as

$$\rho_{4HD2}(n, E_{n_{SL5HD2}}, \lambda) = 0 \tag{10.128}$$

The electron concentration in this case can be expressed as

$$n_0 = \left(\frac{|e|Bg_v}{\pi^2\hbar}\right)\sum_{n=0}^{n_{\max}}[T_{99HD2}(n, E_{fSLHDB}, \lambda) + T_{910HD2}(n, E_{fSLHDB}, \lambda)] \tag{10.129}$$

where, $T_{99HD2}(n, E_{fSLHDB}, \lambda) = \sqrt{\rho_{4HD2}(n, E_{fSLDB}, \lambda)}$ and $T_{910HD2}(n, E_{fSLHDB}, \lambda) \equiv \sum_{r=1}^{s} L(r)\left[T_{99HD2}(n, E_{fSLHDB}, \lambda)\right]$.

Thus using (10.124), (10.128), (10.129) and the alied definitions, we can study the ER in this case.

(c) Following Sasaki [9.15], the electron dispersion law in HD III-V effective mass superlattices (EMSLs) in the presence of light waves, the dispersion relations of whose constituent materials in the absence of any perturbation are defined by the parabolic energy bands can be written as.

$$k_x^2 = \left[\frac{1}{L_0^2}\{\cos^{-1}(f_{HD3}(E, k_y, k_z, \lambda))\}^2 - k_\perp^2\right] \tag{10.130}$$

In which

$$f_{HD3}(E, k_y, k_z\lambda) = a_{1HD3}\cos[a_0 C_{1HD3}(E, k_\perp) + b_0 D_{1HD3}(E, k_\perp)] - a_{2HD3}\cos[a_0 C_{1HD3}(E, k_\perp) - b_0 D_{1HD3}(E, k_\perp)],$$

$$a_{1HD3} = \left[\sqrt{\frac{m_{c2}[T_3(0, \eta_{g2}, \lambda)]}{m_{c1}[T_3(0, \eta_{g1}, \lambda)]}} + 1\right]^2 \left[4\left(\frac{m_{c2}[T_3(0, \eta_{g2}, \lambda)]}{m_{c1}[T_3(0, \eta_{g1}, \lambda)]}\right)^{1/2}\right]^{-1},$$

$$a_{2HD3} = \left[-1 + \sqrt{\frac{m_{c2}[T_3(0, \eta_{g2}, \lambda)]}{m_{c1}[T_3(0, \eta_{g1}, \lambda)]}}\right]^2 \left[4\left(\frac{m_{c2}[T_3(0, \eta_{g2}, \lambda)]}{m_{c1}[T_3(0, \eta_{g1}, \lambda)]}\right)^{1/2}\right]^{-1},$$

$$C_{1HD3}(E, k_\perp\lambda) \equiv \left[\left(\frac{2m_{c1}}{\hbar^2}\right)T_3(E, \eta_{g1}, \lambda) - k_\perp^2\right]^{1/2}$$

and

$$D_{1HD3}(E, k_{\perp}\lambda) \equiv \left[\left(\frac{2m_{c2}}{\hbar^2}\right)T_3(E, \eta_{g2}, \lambda) - k_{\perp}^2\right]^{1/2}.$$

In the presence of an external magnetic field along x-direction, the simplified magneto dispersion law in this case can be written as

$$k_x^2 = [\rho_{4HD3}(n, E, \lambda)] \tag{10.131}$$

in which,

$$\rho_{4HD3}(n, E, \lambda) = \frac{1}{L_0^2}\left[\cos^{-1}(f_{HD3}(n, E, \lambda))\right]^2 - \left\{\frac{2|e|B}{\hbar}\left(n + \frac{1}{2}\right)\right\},$$

$$f_{HD3}(E, n, \lambda) = a_{1HD3}\cos[a_0 C_{1HD3}(E, n, \lambda) + b_0 D_{1HD3}(E, n, \lambda)] - a_{2HD3}\cos[a_0 C_{1HD3}(E, n, \lambda) - b_0 D_{1HD3}(E, n, \lambda)]$$

$$C_{1HD3}(E, n, \lambda) = \left[\left(\frac{2m_{c1}}{\hbar^2}\right)T_3(E, \eta_{g_1}, \lambda) - \left\{\frac{2|e|B}{\hbar}\left(n + \frac{1}{2}\right)\right\}\right]^{1/2}$$

and

$$D_{1HD3}(E, n, \lambda) = \left[\left(\frac{2m_{c2}}{\hbar^2}\right)T_3(E, \eta_{g_2}, \lambda) - \left\{\frac{2|e|B}{\hbar}\left(n + \frac{1}{2}\right)\right\}\right]^{1/2}.$$

The EEM in this case assumes the form

$$m^*(n, E_{fSLHDB}, \lambda) = \frac{\hbar^2}{2}[\rho_{4HD3}(n, E_{fSLHDB}, \lambda)]' \tag{10.132}$$

where E_{fSLHDB} is the Fermi energy in this case.
The EEM in III-V EMSLs under magnetic quantization depends on both the Fermi energy, magnetic quantum number and wavelength which is the intrinsic property of such SLs.
The Landau Subband energies $(E_{n_{SL5\,HD3}})$ can be written as

$$\rho_{4HD3}(n, E_{n_{SL5\,HD3}}, \lambda) = 0 \tag{10.133}$$

The electron concentration in this case can be expressed as

$$n_0 = \left(\frac{|e|Bg_v}{\pi^2\hbar}\right)\sum_{n=0}^{n_{\max}}[T_{99HD3}(n, E_{fSLHDB}, \lambda) + T_{910HD3}(n, E_{fSLHDB}, \lambda)] \tag{10.134}$$

where, $T_{99HD3}(n, E_{fSLHDB}, \lambda) = \sqrt{\rho_{4HD3}(n, E_{fSLHDB}, \lambda)}$ and $T_{910HD3}(n, E_{fSLHDB}, \lambda) \equiv \sum_{r=1}^{s} L(r)\left[T_{99HD3}(n, E_{fSLHDB}, \lambda)\right]$.
Thus using (10.124), (10.133), (10.134) and the allied definitions, we can study the ER in this case.

10.3 Open Research Problems

(R.E.1) Investigate the ER in the presence of intense external light waves for all the HD materials whose respective dispersion relations of the carriers in the absence of any field are given in R 1.1.

(R.E.2) Investigate the ER for the heavily–doped semiconductors in the presences of Gaussian, exponential, Kane, Halperian, Lax and Bonch-Burevich types of band tails [16] for all systems whose unperturbed carrier energy spectra are defined in (R 1.1) in the presence of external light waves.

(R.E.3) Investigate the ER in the presence of external light waves for bulk specimens of the negative refractive index, organic, magnetic and other advanced optical materials in the presence of an arbitrarily oriented alternating electric field.

(R.E.4) Investigate all the appropriate problems of this chapter for a Dirac electron.

(R.E.5) Investigate all the appropriate problems of this chapter by including the many body, broadening and hot carrier effects respectively.

(R.E.6) Investigate all the appropriate problems of this chapter by removing all the mathematical approximations and establishing the respective appropriate uniqueness conditions.

References

1. P.K. Basu, *Theory of Optical Process in Semiconductors, Bulk and Microstructures* (Oxford University Press, Oxford, 1997)
2. K.P. Ghatak, S. Bhattacharya, S. Bhowmik, R. Benedictus, S. Chowdhury, J. Appl. Phys. **103**, 094314 (2008)
3. K.P. Ghatak, S. Bhattacharya, J. Appl. Phys. **102**, 073704 (2007)
4. K.P. Ghatak, S. Bhattacharya, S.K. Biswas, A. De, A.K. Dasgupta, Phys. Scr. **75**, 820 (2007)
5. P.K. Bose, N. Paitya, S. Bhattacharya, D. De, S. Saha, K.M. Chatterjee, S. Pahari, K.P. Ghatak, Quantum Matter **1**, 89 (2012)
6. K.P. Ghatak, S. Bhattacharya, A. Mondal, S. Debbarma, P. Ghorai, A. Bhattacharjee, Quantum Matter **2**, 25 (2013)
7. S. Bhattacharya, D. De, S. Ghosh, J.P. Bannerje, M. Mitra, B. Nag, S. Saha, S.K. Bishwas, M. Paul, J. Comput. Theor. Nanosci. **7**, 1066 (2010)
8. K.P. Ghatak, S. Bhattacharya, S. Pahari, S.N. Mitra, P.K. Bose, J. Phys. Chem. Solids **70**, 122 (2009)
9. S. Bhattacharya, D. De, R. Sarkar, S. Pahari, A. De, A.K. Dasgupta, S.N. Biswas, K.P. Ghatak, J. Comput. Theor. Nanosci. **5**, 1345 (2008)
10. S. Mukherjee, D. De, D. Mukherjee, S. Bhattacharya, A. Sinha, K.P. Ghatak, Phys. B **393**, 347 (2007)
11. K. Seeger, *Semiconductor Physics*, 7th edn. (Springer, Berlin, 2006)
12. B.R. Nag, *Physics of Quantum Well Devices* (Kluwer Academic Publishers, The Netherlands, 2000)
13. R.K. Pathria, *Statistical Mechanics*, 2nd edn. (Butterworth-Heinemann, Oxford, 1996)

Material Index

K. P. Ghatak and S. Bhattacharya, *Heavily-Doped 2D-Quantized Structures and the Einstein Relation*, Springer Tracts in Modern Physics 260,
DOI 10.1007/978-3-319-08380-3, © Springer International Publishing Switzerland 2015

Subject Index

K. P. Ghatak and S. Bhattacharya, *Heavily-Doped 2D-Quantized Structures and the Einstein Relation*, Springer Tracts in Modern Physics 260,
DOI 10.1007/978-3-319-08380-3, © Springer International Publishing Switzerland 2015

MIX
Papier aus verantwortungsvollen Quellen
Paper from responsible sources
FSC® C105338

If you have any concerns about our products,
you can contact us on
ProductSafety@springernature.com

In case Publisher is established outside the EU,
the EU authorized representative is:
Springer Nature Customer Service Center GmbH
Europaplatz 3, 69115 Heidelberg, Germany

Printed by Libri Plureos GmbH
in Hamburg, Germany